T0133876

Instrumentation

Operation, Measurement, Scope and Application of Instruments

Instrumentation

Operation, Measurement, Scope and Application of Instruments

N.V.S. Raju

B.E. (Mech.), PGDIRPM, M.Tech., MBA., Ph.D.

Vice Principal
Professor of Mechanical Engineering,
JNTUH College of Engineering, Jagitial,
Karimnagar (Dist.), Telangana State, India.

BSP BS Publications

CRC Press is an imprint of the
Taylor & Francis Group, an **informa** business

Instrumentation-Operation, Measurement, Scope and Application of Instruments ***by N.V.S. Raju***

© 2016, *by Publisher, Hyderabad, India.*

Published by

BS Publications

A unit of **BSP Books Pvt. Ltd.**
4-4-309/316, Giriraj Lane, Sultan Bazar,
Hyderabad - 500 095, India.

For

CRC Press

Taylor & Francis Group, an **informa** business

6000 Broken Sound Parkway, NW

Suite 300, Boca Raton, FL 33487

711 Third Avenue

New York, NY 10017

2 Park Square, Milton Park

Abigdon, Oxon OX14 4RN, UK

www.taylorandfrancisgroup.com

For distribution in rest of the world other than the India, Pakistan, Nepal, Myanmar (Burma), Bhutan, Bangladesh and Sri Lanka.

ISBN: 978-1-138-62655-3

All rights reserved. No part of this publication may be reproduced, store a retrieval system, or transmitted in any form or by means, electronic, mechanical, photocopying, recording and/or otherwise, without the prior written permission of the publishers. This book may not be lent, resold, hired out or otherwise disposed of by way of trade in any form, binding or cover other than that in which it is published, without the prior consent of the publishers.

British Library Cataloguing in Publication Data

A Catalogue record for this book is available from the British Library

Printed at: Sanat Printers, Kundli, Haryana, India.

Gururbrahma Gururvishnu Gururdevo Maheswarah
Gurursakshaath Parabrahma Thasmai Sree Gurave Namah

Dedicated to

The holy feet of

Satchidananda Sadguru Sainath

The ultimate Guru of All those Gurus who taught me

And My first gurus who brought me up

My Mother (Late.Smt. Ammaayi) &

My Father (Sri N.V.Murali Manohar Raju)

Art work by: SSRK, M. Pradeep Kumar and B. Madhukar

Preface

I am glad to present this book titled "Instrumentation-Operation, Measurement, Scope, and Application of Instruments (IOMSAI)" for the students of Mechanical Engineering (ME), Production Engineering (PE), Metallurgical Engineering (Met.E), Mechatronics and so forth. Though the syllabus is extracted from Jawaharlal Nehru Technological Universities of Telangana (JNTUH) and A.P. (JNTUK & JNTUA), high care is taken to cover maximum topics of syllabi of almost all universities of India and abroad.

During my long experience and association with my students, I observed many students feeling difficult in understanding many topics of the subject from existing books. Perhaps, this is due to the fact that the subject is blend of several & various applications of the technological and engineering concepts that have been learnt in the lower classes. Most of the students do not realize this fact and/or might have forgotten or do not remember/recall/recap such concepts before studying subject. No book of this field is providing the prerequisites and fundamentals nor appropriately directs the readers. Having understood this shortfall in the present titles, at the very beginning of each chapter, the feature "**STARTERS**" is given, so that the students may reckon, recall and revise before studying this subject. Further, the topics are dealt from grass-root level by providing prerequisites wherever required.

Particularly, some students brought to my notice that they are unable to remember the contents of the subject. In fact this is the basic motive for writing this book. In view of this, extra care is taken to explain with easy language and used certain techniques wherever required, but without compromising with standard and quality. One of such techniques used in this book for memorizing the contents of each chapter is that the entire chapter is summarily sculpted on the first page of the chapter in the form of "**INFOGRAPHICS**" and "**Instrumentation-Operation, Measurement, Scope & Application of Instruments (I-O-M-S-A-I)**" of the entire chapter is accommodated in a single page as a **infographics**. These special features of this book will help the students not only to memorize at faster rate but also becomes easy for the quick, last minute preparation for the exam. Further, these pages at beginning of each chapter will even help the Professors also to one minute preparation for their class.

The questions and problems appeared in previous examinations of JNTUs and other Universities are also incorporated with solutions at appropriate places right at the content (**Self Assessment Questionnaire-SAQ**) to ease the students to searching for their answer, as well they can assess themselves and can also understand how the topic may be questioned in their examinations.

The basic objective of bringing out this book, as mentioned earlier is that there is no textbook to cater the exact needs of the students in this course. Though there are numerous books available in instrumentation they are not able to satisfy the student needs completely. Moreover, sufficient literature is not available on many topics. The interesting information relevant to the topics, are provided under the '**MILE**' (**More Instruments for Learning Engineers**), which would become the tonic to relieve the monotonic fatigue & stresses on the students while studying it. These will help the students answer well in their viva-voce of lab/practical exams and interviews of competitive exams. These unique and value addition features are not be available in any of the present books but are given to the students 'for free' i.e., more instrumentation for no extra cost.

The chapter end '**Review Questions**' are given in different varieties such as short answer type and long answer type questions. Further, the **objective type** questions are also included at the end of each chapter to prepare for the competitive exams and viva-voce.

This book doesn't claim for its originality but for the presentation. It is a sincere and humble attempt to collect the requisite information and present it in a form suitable and useful to students and teachers. I am greatly indebted to the authors of numerous books and websites that are referred to.

In spite of utmost care taken to make this book error free, still some mistakes might have crept in for which I seek your approval for the omission and commissions. Any suggestions in improvement of this book will be gratefully acknowledged.

Dr. N.V.S. Raju
enviousraju@rediffmail.com

Acknowledgement

A bit of inspiration from a Guru leads to terra byte of achievement for all.

At the outset, I express my deep sense of gratitude for encouragement and support extended by my gurus by whose efforts I am today.

- Dr. K. Narayana Rao, *Former Member Secretary, AICTE, New Delhi.*
- Smt. Shailaja Ramayyar, *Vice Chancellor I/C, JNTU Hyderabad.*
- Dr. C.B. Krishna Murthy, Prof. *(Retd.) in Mechanical Engg., JNTU.*
- Dr. *(Col. Retd.)* K. Prabhakar Rao, *Principal, Global Inst. of Engg. & Tech.*
- Dr. N. Yadaiah, *Registrar, JNTU Hyderabad.*
- Dr. A.Vinaya Babu, *Prof. of CSE, Former Director(Adm), Former Principal, JNTUH CEH.*
- Dr. M.Thirumalachary, *Former Principal, JNTUH CE Jagitial.*
- Dr. N.V. Ramana, *Principal, JNTUH CE Jagitial.*
- Pandit. N.V. Murali Manohar Raju, *(Head Master, Retd.), My father & My Guru.*

For the timely suggestions and help in various forms to make this book a success my warmest thanks are due to my colleagues Mrs. M. Shailaja, AP & Head ME, Dr. K. Vasanth Kumar, APME, Dr. Ch. Sridhar Reddy, Vice Principal, JNTUH CEM, Dr. Suresh Arjula, APME, Mr. B. Kranthi Kiran, AP CSE.

Thanks are due to my former colleagues and friends Dr. A. Prasada Raju, Principal (SCIT) Mr. N. Venkateshwarlu Assoc.Prof., MED-SOET, IGNOU, Mr. Somasundaram Kalaga, Mr. Ch. N. Srinivasa Rao, Dr. D. V. Ravi Shankar, Principal, TKREC, Dr. A. Suresh Rao, PCSE (TKREC), Mr. N.L. Narayana, PMED, SVIT and my beloved students.

I wish to offer my special thanks to the help extended by Mr. S. Shiva Rama Krishna, Cartoonist & APME-SCIT Karimnagar, JNTUH CEJ, Mr. M. Pradeep Kumar, APME-KITS Huzurabad, Mr. B. Madhukar, Mr. A.Vinod Kumar, Peshi to Principal-JNTUH CEJ, Mr. Shivraj Chawaria, APME-KEC-Armoor, Mr. D. Sandeep Kumar, AP-SVIT, Hyderabad, my students Mr. K. Vikas (ME) and Mr. S. Nikhil (CSE) for adding highest value to this book in the forms of MILEs, Bird's Eye views, Infographics at JNTUH CEJ and questionnaire, and also Mr. M. Anjaiah, Mr. L. Naresh for energizing as and when required. I am greatly indebted to them.

I am grateful to many of my fellow colleagues (apology for brevity, as the list is big) who inspired and had given suggestions by which the book has taken a good shape and also the students of JNTUH CEJ who contributed by the way of constructive criticism, discussions and help in need.

Acknowledgement

While authoring this book, I had to go through numerous books and met many professors/ practitioners. I thank all those authors and authorities and who directly or indirectly helped me while authoring this book. I feel sorry if I missed any inadvertently.

I further acknowledge the entire team of the Publishers (BSP) in general, Mr. Nikhil Shah, Mr. Vasudeva Rao, Mr. Hariprasad, Mr. Naresh Davergave, and Mrs. K. Sandhya in particular for their untiring efforts without which this book would not have seen any light.

Last but not least, I thank my wife Mrs. Prasanna and daughters Kum. Srija Nataraj, Kum. Himaja Padmapriya and Kum. Sanja Karunamayi, and all Kasu & Chennamadhavuni family members for their co-operation and patience.

Dr. N.V.S. Raju
enviousraju@gmail.com

Contents

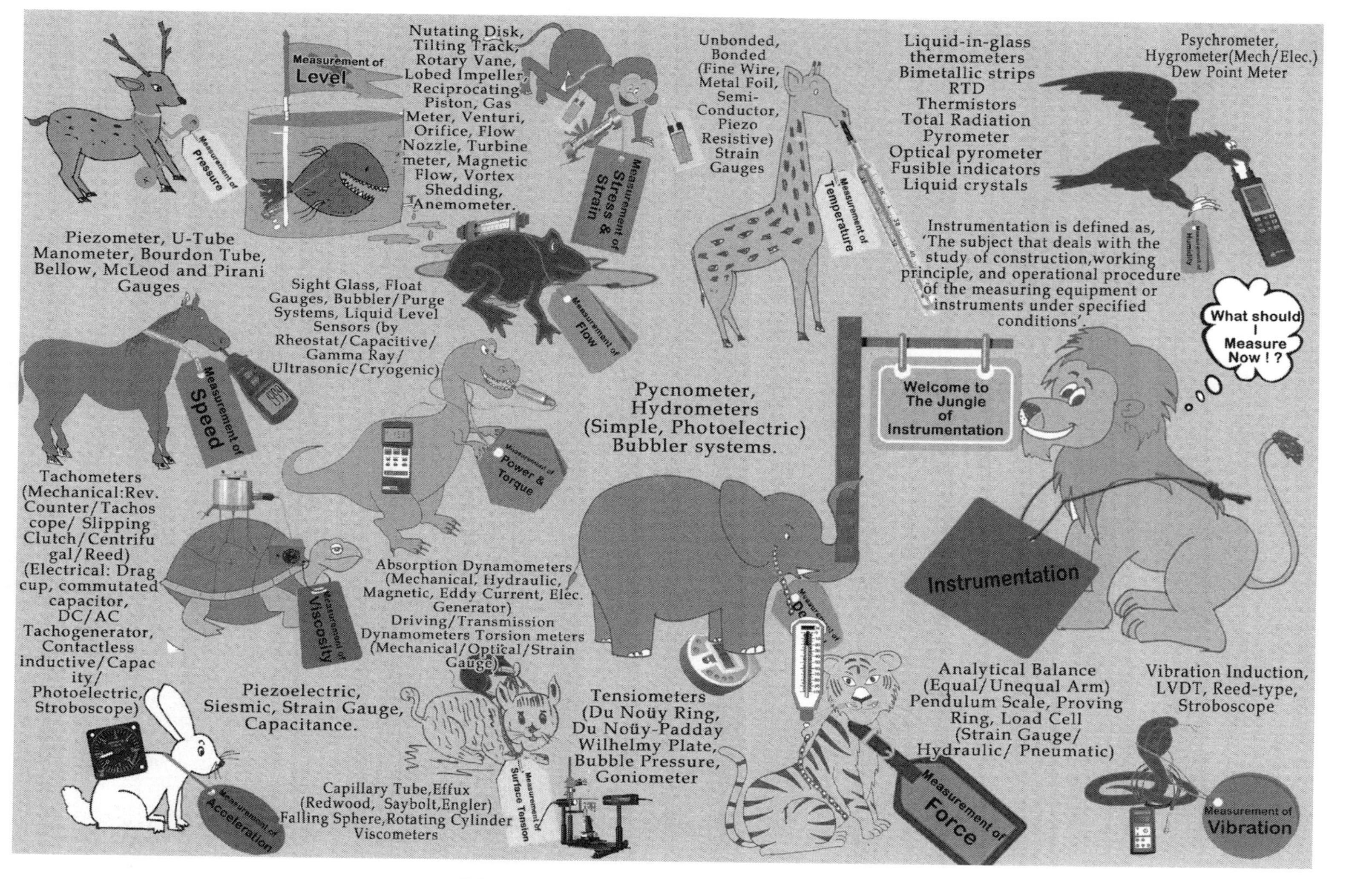

CHAPTER INFOGRAPHIC

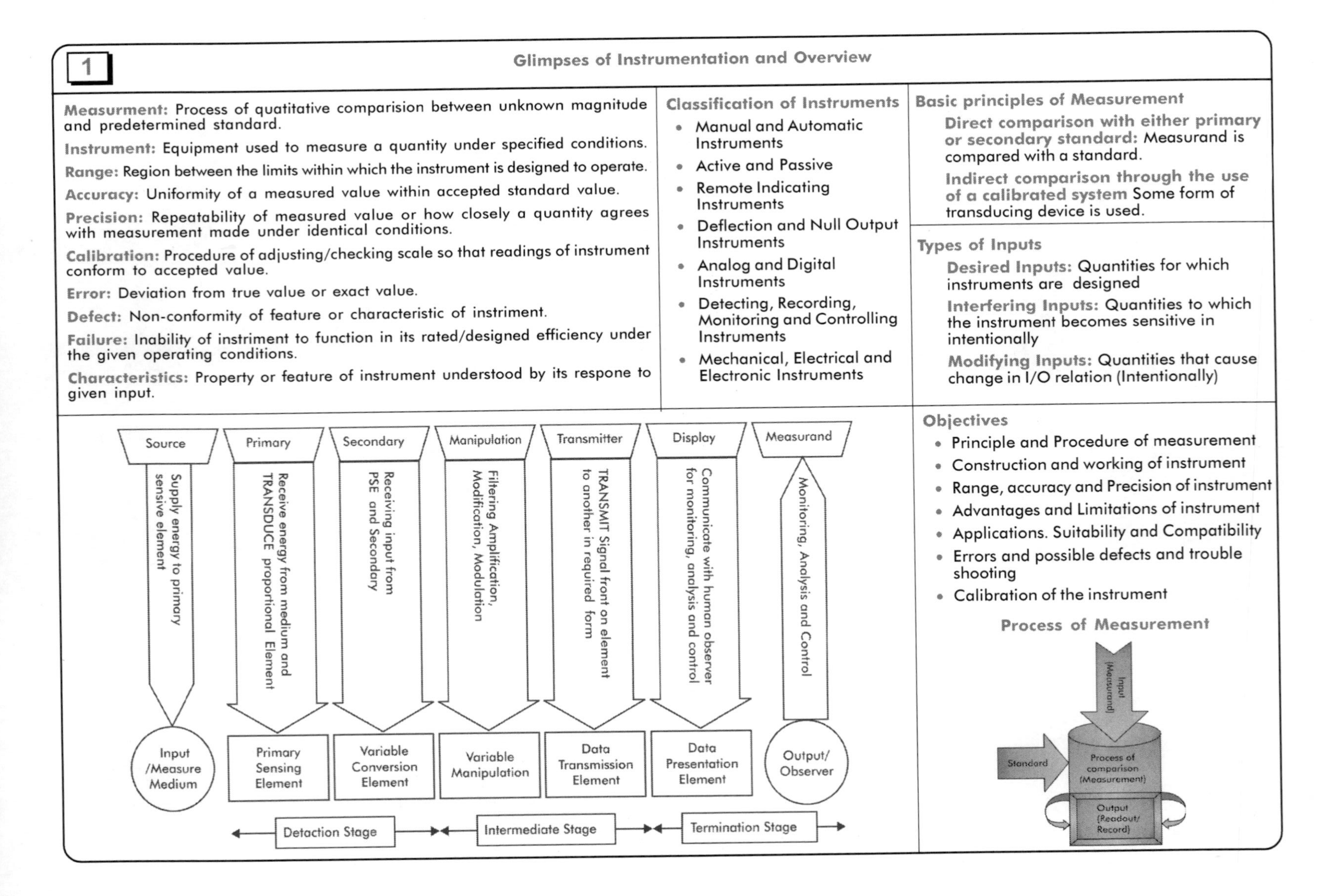

1 Glimpses of Instrumentation and Overview

Measurment: Process of quatitative comparision between unknown magnitude and predetermined standard.

Instrument: Equipment used to measure a quantity under specified conditions.

Range: Region between the limits within which the instrument is designed to operate.

Accuracy: Uniformity of a measured value within accepted standard value.

Precision: Repeatability of measured value or how closely a quantity agrees with measurement made under identical conditions.

Calibration: Procedure of adjusting/checking scale so that readings of instrument conform to accepted value.

Error: Deviation from true value or exact value.

Defect: Non-conformity of feature or characteristic of instriment.

Failure: Inability of instriment to function in its rated/designed efficiency under the given operating conditions.

Characteristics: Property or feature of instrument understood by its respone to given input.

Classification of Instruments

- Manual and Automatic Instruments
- Active and Passive
- Remote Indicating Instruments
- Deflection and Null Output Instruments
- Analog and Digital Instruments
- Detecting, Recording, Monitoring and Controlling Instruments
- Mechanical, Electrical and Electronic Instruments

Basic principles of Measurement

Direct comparison with either primary or secondary standard: Measurand is compared with a standard.

Indirect comparison through the use of a calibrated system Some form of transducing device is used.

Types of Inputs

Desired Inputs: Quantities for which instruments are designed

Interfering Inputs: Quantities to which the instrument becomes sensitive in intentionally

Modifying Inputs: Quantities that cause change in I/O relation (Intentionally)

Objectives

- Principle and Procedure of measurement
- Construction and working of instrument
- Range, accuracy and Precision of instrument
- Advantages and Limitations of instrument
- Applications. Suitability and Compatibility
- Errors and possible defects and trouble shooting
- Calibration of the instrument

Process of Measurement

Chapter 1

MEASUREMENT SYSTEMS – AN OVERVIEW AND BASIC CONCEPTS

STARTERS

To study this chapter, you should have awareness on the following concepts. For a better understanding, it is always a good idea to revise these prerequisites.

- Various fundamental and derived quantities and their denotations, and units.
- Standards of measurement and methods of standardizing.
- Notations, and units of various quantities.
- International standards, National standards and Local standards.
- Definitions and basics of various physical quantities.
- Conversion factors of one system of units to the other for various quantities.
- Basic principles of solid mechanics, fluid mechanics, elasticity, heat, light, sound, magnetism, electricity and semi-conductors.
- Awareness on various (generally used) materials such as metals, alloys, semi-conductors, rubbers, plastics, adhesives, abrasives, ceramics, cermets etc., and their general properties.
- Awareness on atomic physics, electromagnetic waves, X-rays, radioactive (gamma) rays, ultrasonic and infrared rays.
- A brief idea on structure and working of the electrical and electronic components such as resistor, capacitor, transformer, transistor etc.
- A thorough idea on fundamental laws of physics, such as Newton's laws, Boyle's law, gas laws, Bernoulli's principle etc.
- An understanding on system approach i.e., input-output models.

LEARNING OBJECTIVES

After studying this chapter you should be able to

- Describe what a measurement system is.
- Understand and define the terms – instrument and instrumentation and the related terminology.
- Scope and applications of measuring systems.
- Describe Generalized I/O model of measuring system and different types of inputs.
- Understand the classification of the instruments.

1.1 INTRODUCTION

Scientific era started with invention of wheel and the thought of measuring its consequential outputs has become pivotal during nineteenth century. Many instruments and *equipment* have been discovered during this period, and the list is being added every day in the light of sophistication, precision and accuracy.

Perhaps, Archimedes is the first scientist to give a scope of thinking to measure physical quantity (such as density) without destructing (Non Destructing Testing Measurement). Since then there is a revolutionary development in discoveries to add to the development in discoveries to list of measuring instruments.

Though this subject is concerned with construction and working of various instruments, the readers of this subject need a wide range of fundamental concepts and various relationships among the physical quantities. For example, the temperature can be measured by using the change in resistance. In this case, it is necessary for us to know the fundamental concepts of temperature, resistance, the relationship between them, the extent to which this relation holds good and so on. Similarly, the temperature can also be measured by using the concept of expansion of liquids and in this case one should know these concepts thoroughly to construct or calibrate such equipment using this principle.

Therefore, one who wants to measure the temperature needs to know about how this physical quantity can be influenced by manipulating various physical quantities either directly or indirectly. Thus it is essential for an instrumentation engineer to know widespread conceptual ideas for constructing the equipment. This text book makes an attempt to provide this conceptual framework to maximum extent possible.

1.2 THE PROCESS OF MEASUREMENT

The process of measurement is concerned with the input compared with a standard resulting in some output as shown in the Figure 1.1. Thus the salient terms involved in the process of measurement are:

1. The Measurement

More Instruments for Learning Engineers

A measuring cup/jug is the simplest instrument often seen as kitchen ware made up of plastic, glass, or metal that measures quantity (volume) of liquid or bulk solid cooking ingredients such as flour, rice and sugar etc. (also for alcoholic beverages, milk, food grains, washing powder, liquid detergents, bleach, syrups, chemicals in labs, hospitals etc.) The cup/jar consists a scale marked in terms of weights/volumes. Transparent/ translucent cups can be read from outside. For smaller measurements scoops/ spoons are used which do not have any scale but are filled and leveled to maximum capacity to use as a unit.

Measuring Cup or Jug

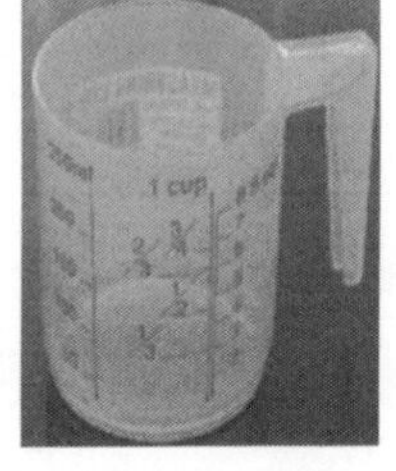

2. The Measurand
3. The Standard
4. The Output
5. The Process of Measurement
6. The Instrument and
7. The Instrumentation

Before studying the subject instrumentation, one should be familiar with the above terms. Let us discuss about the above terms now.

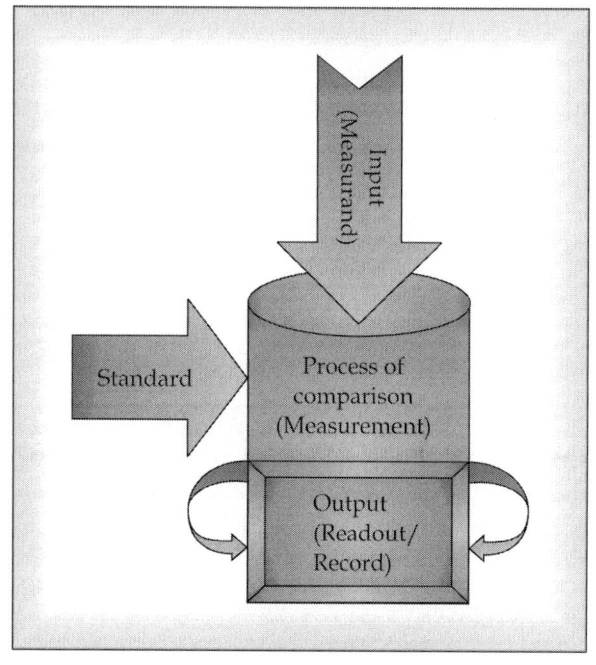

Fig. 1.1 The Process of Measurement

The Measurement

The term *'measurement'* is used to describe the quantity in some numeric value or units of physical entities such as length, weight, temperature etc., or a change in one of these physical entities of a material.

The Measurand

The physical quantity or the characteristic condition that is to be measured in instrumentation system is variously named as *measurand, measurement variable, instrumentation variable* and *process variable.* The measurand may be a fundamental quantity such as length, mass and time or a derived quantity like speed, velocity, acceleration, power, etc., or a qualitative condition like color, brightness, roughness etc.

Types of Instrumentation Variables

With reference to the spatial dependence and the points of measurements, the instrumentation variables can be described in three categories as follows.

1. ***Independent variables:*** This type of variable has a spatial independence. Example: Time.
2. ***Pervariables or through variables:*** These variables can be specified and measured at one point in space. Examples: force, momentum, current and charge.
3. ***Transvariables or across variables:*** These kind of variable require two points (usually one point is the reference) to specify or measure them. Examples: displacement, velocity, temperature and voltage.

The Standard

This result of measurement is expressed as a number representing the ratio of the unknown quantity to the adopted standard. This number indicates the value of the measured quantity. For example, 5 m length of an object means that the object is 10 times as large as 1 m.

Of course, for consistent quantitative comparison of physical parameters, certain standards of mass, length, time, temperature and electrical quantities have already been established that are internationally accepted and well-preserved under controlled environmental conditions.

The Output

The *output* is readout by the observer after comparing the object with a quantity of same kind, called *standard.*

The Process of Measurement

The process of measurement is concerned with comparison of input with a standard. The most important thing to be noted here is that there is high responsibility on the shoulders of the observer in choosing the standard for the measurement or comparison which must be accurately known and universally accepted. Further, the procedure should be verifiable and apparatus employed for obtaining the comparison must be demonstrable, i.e., accuracy can be reproduced anywhere on the globe. In other words, the measurements obtained must be accepted with confidence.

The Instrument

The human senses sometimes mislead or cannot provide accurate quantitative information about the knowledge of events. Further, it may neither be recorded permanently nor be reproduced exactly. Thence, the strict requirements of precise and accurate measurements in the technological fields led to the development of mechanical supports called instruments. Instruments can be described as the essential extensions of human sensing and perception. The man-made instruments are accurate and sensitive in their response. Moreover, they retain their characteristics for extended periods of time, unlike human senses.

An instrument is one which would sense a physical parameter (such as length, pressure, temperature, velocity etc.), process and translate it into a format and range that can be interpreted by the observer. The instrument also consist the controls, by which the operator can obtain, respond to, and manipulate the information.

Instruments may be very simple, such as clinical thermometer or extremely complex such as the device to sense the physiological reactions of a human in a spaceship. A good instrument should have the following compulsory features or characteristics.

1. It should be able to provide required range of measurement
2. It should be controllable, able to monitor
3. The outputs should be visible, legible and clear
4. It should be sufficiently accurate and should show the results with required precision
5. It should be comfortable to humans to use
6. It must be safe to humans as well as to the environment.

The Instrumentation

We understood that when the input is given to the instrument, it measures and compares with the standard and results in an output as shown in the Fig. 1.1.

Summarily, the process of measurement is composed of three basic quantitative issues namely, the input (often named as measurand), the predefined standards (with which the input is compared) and the output (often called readout or record). The instrumentation is the subject that deals with the study of such processes of measurements.

Instrument and Instrumentation

The term instrument is used in this context as the measuring equipment and is defined as: *The equipment used to measure a physical quantity under the specified conditions and defined procedure is called an 'Instrument'.*

Thus Instrumentation is defined as, *'The subject that deals with the study of construction, working principle, and operational procedure of the measuring equipment or instruments under specified conditions'.*

SELF ASSESSMENT QUESTIONS-1.1

1. What is instrumentation? Briefly describe the salient terms used while measuring with an instrument?
2. Describe the process of measurement by an instrument.
3. Define and explain the following terms.
   (a) Measurand (b) Measurement
   (c) Standard (d) Instrument
4. What are different types of variables used in the process of measurement through an instrument?
5. What are the desirable features that a good instrument must contain?
6. Describe the essential requirements and pre-requisites that a designer of instrument should have. What precautions should be taken while designing and operating an instrument?

1.3 SCOPE OF INSTRUMENTATION

Some time back when I was teaching this subject, I had an interesting and witty experience with one of my smart students. Let me share that here.

While I was teaching about measurement of pressures, barometer, manometer etc., on that day, I observed a student sleepy in the last bench. I called him to stand up and asked what I was teaching about. He murmured 'b...a...r...o...m...e...t...e...r'

I was not happy with his answer and I wanted to know how much attentive he was. So, I asked a question to him and the conversation was very interesting which went on as below.

I : How do you measure the height of the building opposite to us using a barometer?

Student : I drop this barometer from top of the building and note the time of flash of sound. Having known the velocity of sound, acceleration due to gravity, we can calculate the height of the building.

I : No! You must use barometer...

Student : Then, I take this barometer to the top of the building and tie it to a thread. Then I release it slowly till it just touches the ground. Now the length of the thread measures the height of the building.

I was bit surprised to his cleverness...

I : No! Is there any other method?

Student : Yes sir! I have. I take this barometer under sun and I measure the shadow of the barometer and the shadow of the building. Now the ratio of shadow of barometer to that of building will be equal to ratio of height of barometer to that of building.

That was again surprise for me, but I wanted a better answer from him.

I : But, you have to use barometer in a better way. Think any other method.

Student : Yes! I go to the steps of the building and count how many barometers are fit per step and the height of the building is product of number of steps with the number of barometers per step.

I : You naughty boy! You got to use barometer and some physics or engineering principles.

Student : Then I have one more method. From a known distance from the foot of the building, I look at the tip of barometer and tip of the building to match on a straight line to obtain the angle of elevation. The tangent of this angle multiplied by the distance gives height.

I wanted to extract the answer using the principle of barometric pressure i.e., the product of height of mercury column, density and the acceleration due to gravity. The difference of pressures divided by density and gravity is to be calculated. So, I asked him again.

I : No! My dear pretty smart student! Use your barometer effectively and get the best way to find the height.

Student : Sir! The only best and effective way of using this barometer is that I present this to the owner or builder of the building and ask the height of the building.

Of course, the whole class was immersed in laughter for a minute. But the element of truth I too have learnt in the class was that there is no one hard and fast single method to measure a physical quantity and the innovative application of engineering principles widen the boundaries of instrumentation.

In the early days, various instruments and their measurement principles were usually studied along with the individual machine or the field of relevance. Now, the list of measuring equipment or instruments is so large that there is a need to study these as a separate subject. Moreover, the need for more accuracy in reading the measurement is another dimension to necessitate to studying this subject under separate head, raising curtains to a wide scope.

The subject 'Instrumentation' is concerned with working principles and operation or procedure of experimentation of various instruments. The physical quantities measured by these instruments include temperature, distance, speed, acceleration, force, pressure, humidity, torque, torsion, vibration, stress, strain, flow rate etc. Again the characters such as range of measurement, accuracy, precision, response etc., categorize the above measuring instruments into variety and classes of their kind. So, it is definitely good idea to understand and familiarize the terminology of instrumentation before going into depth of the subject.

More Instruments for Learning Engineers

sextâns,-antis (in Latin means $^1/_6$ of a turn or 60°); octant ($^1/_8$ turn or 45°), quintant ($^1/_5$ turn or 72°) & quadrant ($^1/_4$ turn or 90°) measure the angle between a celestial body & horizon (*altitude*), by *sighting* or *shooting*. It uses the angle and the time to find a position line on aeronautical chart, to sight the sun at solar noon, Polaris at night, and the lunar distance between the moon and a celestial body to find Greenwich time and hence the longitude. Sighting the height of landmark is measure of *distance*. Newton's principle of the doubly reflecting navigation instrument is basis for reflecting quadrant. John Hadley and Thomas Godfrey invented octant, John Bird made the 1st sextant (still found on US Naval Warships). First, the octant, later sextant, replaced the Davis quadrant.

Quadrant-Quitant-Sextant-Octant

1.4 JARGON OF INSTRUMENTATION

To understand a subject easily, one should first become familiar with the terms used in the subject. The most general terminology used in instrumentation is enlisted below. However it is not an exhaustive. The detailed list of terms and their explanations are dealt at appropriate place in this text.

1. **Measurand:** The physical variable such as length, temperature, pressure etc., which is the object of measurement of an instrument, is called measurand or the measured variable.
2. **Measurement:** It is defined as the process of obtaining a quantitative comparison between a predetermined standard and an unknown magnitude of the same parameter.
3. **Instrument:** The equipment used to measure a physical quantity under the specified conditions and defined procedures is called an instrument.
4. **Instrumentation:** Instrumentation is defined as the subject that deals with the study of construction, working principle, and operational procedures of the measuring equipment or instruments under specified conditions.
5. **Range:** The region between the limits within which an instrument is designed to operate for measuring a physical quantity is called the range.
6. **Accuracy:** It is defined as the uniformity of a measured value with an accepted standard value.
7. **Precision:** It refers to how closely the individual measured values of a quantity agree with each other when the measurements are carried out under identical conditions at a short interval of time. In simple words it refers to repeatability.
8. **Calibration:** The procedure laid down for making adjusting, or checking a scale, so that readings of an instrument confirm to an accepted standard is called the calibration.
9. **Error:** It is defined as the deviation from the true or exact value.
10. **Characteristics:** These are properties or features or behaviour of an instrument which are understood by its response to a given input.

1.5 BASIC PRINCIPLES OF MEASUREMENT

Measurement can be made in two ways, these are:

1. Direct comparison with either a primary or a secondary standard and
2. Indirect comparison through the use of a calibrated system.

These methods are elaborated in the following sections.

1.5.1 Direct Comparison

In this method the measured quantity is compared (directly) with a standard. The result is usually expressed in a number of a certain unit. The direct comparison method is often used to measure the fundamental physical quantities such as length, mass, time etc.

For example, to measure the length of a steel bar or a ribbon, we compare the length of the bar with a standard, and find that the bar is so many millimeters long because the standard is also marked in *mm*. This is what we mean by direct comparison. The standard used for measurement is called secondary standard. The direct comparison method is not always adequate. The human senses are not sensitive enough to make direct comparisons of all quantities with equal facility. For instance, we can measure small distances with 1 mm accuracy using a steel rule, but we require greater accuracy in several occasions such as measurement of a pitch or length between of gear teeth. In such cases we must have to take assistance from some more complex form of measuring system.

1.5.2 Indirect Comparison

In many applications, the comparison is made indirectly by using certain device or a chain of devices called 'transducer' which converts one form of physical quantity to another. The chain of devices converts the basic input into an analogous form, which processes and presents in a measurable format at the output as a known function of the original output. Such conversion is necessary so that the information can be interpreted with a great ease and clarity.

For example to detect strain in a machine member; assistance is required from a device or system that senses, converts, and finally presents an analogous output in the form of a displacement on a scale or chart. To find the level of water in a water tank, it is converted into length by a float that makes the measurement easy, simple and clear.

1.6 COMPLEXITY OF MEASUREMENT METHODS

As discussed earlier in this chapter, (in section 1.3) the design of instrument may vary from the simplest to complex measurement method. Some measurements can be easily made just by human senses while some others need complicated conversions and calculations by using various physical relationships. The complexity of an instrument mostly depends upon

- The measurement being made
- The characteristics of the measurand and
- The accuracy level required.

Based on the above three factors, the measurements can be categorized into three types namely the primary, secondary and tertiary measurements.

1.6.1 Primary Measurement

In the primary measurement, a physical parameter is determined by simply comparing it directly with reference standards. The required information can be obtained through human senses such as sense of sight or touch.

Examples:

1. Judging a liquid is acidic or basic by color on litmus paper
2. Assessment of the temperature of red hot iron by matching the color
3. Compare masses by simply supported beam at center (as fulcrum)
4. Knowing coldness or hotness of a body by sense of touch
5. Estimating the length of a wire using a stick
6. Time measurement by counting the number of bells (or strokes) of a clock

In all the above cases, we can notice that the observer can indicate only that a given liquid is acidic or basic; the iron rod is hotter or not; one object contains more or less mass than the other and so on. Thus the primary measurements provide subjective information only.

1.6.2 Secondary Measurement

In many technological activities, it is often not easy to directly measure or observe the exact quantity. The human senses do not suffice to make direct comparison of all the quantities. Sometimes, it may not even be possible to measure directly. Further, such measurements may be time consuming, unsafe and tedious also. Thence, a suitable indirect method may be employed in which the measurand is converted into some effect which can be directly measurable. The indirect methods make comparison with a standard through use of a calibrated system through an empirical relation between the actual measurement made and the desired result. Such indirect measurements involving one translation are called *secondary measurements.*

Examples:

1. A spring balance measures the weight of an object by converting the weight into equivalent displacement of pointer on scale due to elongation in the spring.
2. For liquid level measurement, the volume of liquid is converted to displacement due to the raise of float valve.
3. The conversion of pressure into displacement by means of bellows.
4. In a photo voltaic cell, the light beam is converted to voltage.

Electrical methods are usually preferred in the indirect methods owing to their high speed of operation and simple processing of the measured variable.

1.6.3 Tertiary Measurement

Sometimes the measurement is not easy at the secondary stage. In such cases one more translation is required. Such indirect measurements involving two conversions are called *tertiary measurements.*

1. For measurement of pressure in bourdon tube pressure gauge, the pressure is first converted to linear displacement (secondary) through bourdon tube pushing the mechanical linkage and then to angular displacement (tertiary) by rack and pinion arrangement.
2. Measurement of speed of rotating shaft by means of an electrical tachometer. Here, the shaft speed is converted to voltage and then voltage is converted to length.
3. A system developing an electrical voltage proportional to a physical variable say, temperature and then converting the measured voltage to the corresponding value of the displacement.

Self Assessment Questions-1.2

1. Differentiate between direct measurement and indirect measurement.
2. Define the following terms and briefly describe with examples.
   (a) Range (b) Accuracy (c) Precision (d) Calibration
3. Define the following terms and briefly describe with examples.
   (a) Measurand (b) Measurement (c) Error (d) Calibration
4. Define the following terms and briefly describe with examples.
   (a) Instrument (b) Error (c) Accuracy (d) Precision
5. Distinguish between
   (a) Precision and accuracy (b) Measurand and measurement
6. What are the various methods of measurements? Explain them.
7. State the factors on which the complexity of an instrument depends. Also explain the types of measurements, which can be categorized based on those factors.
8. What are primary, secondary and territory measurements? Explain with suitable examples.

1.7 APPLICATIONS OF MEASURING INSTRUMENTS

Measuring instruments are generally applied in three ways. These are

(i) Monitoring of a process or operation

(ii) Controlling a process or an operation

(iii) Experimental analysis and engineering

1.7.1 Instruments for Monitoring

Some of the measuring instruments performing monitoring function are

- the instruments which simply indicate the condition or status of the situation or environment: e.g., thermometers, pressure gauges
- the instruments that indicate the quantity of commodity used: e.g., gas meter, water meter, electricity meter
- the instruments which monitor the usage or output such as speed: e.g., odometer, tachometer, speedometer etc.

1.7.2 Instruments for Controlling

Some of the measuring instruments performing controlling function are

- the instruments used as a component or part of an automatic system: e.g., a home heating system, an automatic control unit in a refrigerator, a fuse/ stabilizer etc. (These consist a thermostatic control such as bimetallic element or a thermocouple which give the required information for proper functioning of the controlling system.

1.7.3 Instruments for Experiments and Analysis

Some of the measuring instruments performing function of experiments and analysis are

- the instruments which theoretically interpret the results based on the analysis or logical conclusion: e.g., a program given in a computer or a mobile phone that indicates cost of a call etc.
- the instruments which show the output based on the experimental results e.g., the white blood cells (WBC) or red blood cells (RBC) and platelets count in a blood test.

More Instruments for Learning Engineers

Pocket watch was the time measuring device from 16th century until wristwatches became popular after World War-I during which a transitional design, trench watches were used by the military. Pocket watches are connected with chain to secure it to waistcoat, lapel, or belt loop, and thus prevent from being dropped. Women's watch chains were more decorative than protective and were frequently decorated with silver or enamel pendant, often representing the association with some club or society. Further, gadgets like watch winding key, vesta case or a cigar cutter also appeared on watch chains. Also some fasteners were designed to put through a buttonhole of jacket or waistcoat.

Pocket Watch

1.8 GENERALIZED INPUT-OUTPUT MODEL OF MEASUREMENT SYSTEM

All instruments are neither identical nor alike. The measurement principle also may be different in different instruments. However, when it is described with a system approach, we can assume as model with certain processing between the two ends i.e., input and output. Now, if the process is also generalized, we can understand the measurement system very easily.

How do we generalize?

Suppose, you have seen 'a cat killed a rat'.

This is one case you found. But now, if you question 'does any cat kill any rat or only this cat kills that rat?' Obviously, you get a general answer that, any cat kills any rat. But more you think on it, you can further generalize by putting a question why and how a cat kills a rat i.e., what is fundamental principle behind it? Then the more generalized answer is that 'a stronger conquers a weaker'.

Similarly, a clinical thermometer measures temperature due to expansion of mercury in a graduated column with temperature. This can be extended by applying the principle of expansion of liquids and thus extended to other liquids such as alcohol, thence can be considered to describe all liquid-in-glass thermometers.

Further, considering this principle of the conversion of one form to the other as 'transduction', we can further generalize and describe any instrument. Thus we shall now understand a generalized input-output model of measurement system.

As we all know, measurement is obtained using a measuring instrument which is an assembly of physical facilities in an order. Further, an instrument designed to carry out certain task, is described in terms of its physical elements. But, in this approach, we have to give separate description for instrument instead of each physical element that it constitutes as instrument can be identified with functional elements. It is possible that a physical element may do many functions. Now, considering the basic action of the functional elements, in generalized approach, the description of an instrument is carried out in terms of the basic functional elements.

On close observation, we can notice the basic functions between the input and output in an instrument as sensing, conversion, manipulation, transmission and presentation. For our ease of understanding, the sensing and conversion are considered under detection stage while manipulation, transmission and presentation are considered to be in the intermediate stage and the presented data along with final output make the termination stage.

A block diagram of generalized measurement instrumentation is shown Fig. 1.2.

We shall now identify and define the basic actions of the functional elements in a measurement system.

1. **Medium:** It is the input to a measurement system. It supplies energy to the primary sensing element.
2. **Measured Quantity:** This is also called measurand. This is the physical variable whose measurement is under consideration.
3. **Primary Sensing Element:** Primary sensing element receives energy/input from the medium to be measured and produces a proportional output. The output from the primary sensing

element is usually a physical variable such as displacement or voltage. Thus primary sensing element is a primary transducer which converts one physical variable into another. An intermediate transducer may also be used after primary transducer if a second transduction is desired. However, the sensor should extract a very small amount of energy from the medium because the medium should not be disturbed appreciably when the sensing element is inserted.

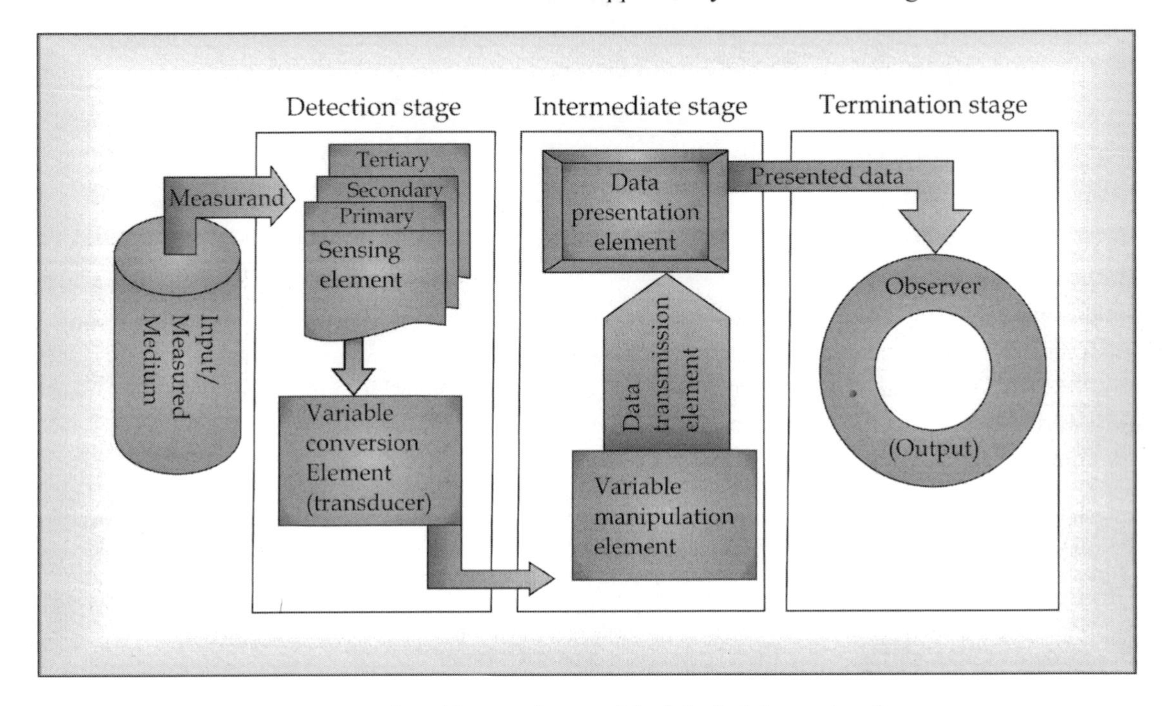

Fig. 1.2 Generalized Input-Output Model of a Measuring System

For example, in measurement of temperature by a clinical thermometer, the mercury bulb senses the temperature when it is in contact with the human body. This temperature sensed will cause the expansion in the mercury. In a uniform cross section column, the liquid (mercury) is allowed to expand which thus becomes a linear measurement i.e., the length of the expansion represents the temperature. Here, the mercury bulb is the primary sensing element and the principle of expansion of liquid is the principle of transduction while the mercury is the transducer. Table 1.1 gives the transduction operations of few physical variables and the examples of systems based on them.

4. **Variable Conversion Element:** After sensing the measurand, it is to be converted to a more suitable variable while preserving its original contents. So, a variable conversion element is employed which acts as an intermediate transducer.

   For example, a nozzle acts as a transducer in pneumatic pressure gauge where displacement is converted into pressure as intermediate transduction. This stage is very important and entire measuring system depends on the principle or the relation between the physical quantities that is adapted at this stage.

5. **Variable Manipulation Element:** This element forms as an intermediate stage in a measuring system. Most of the times, though the sensed signal (input) is converted to

Table 1.1 Transduction Operations of Few Physical Variables

Transduction		Example Systems
From	To	
Temperature	Displacement	Different expansion temperature sensors, bimetallic sensors
	Electric current	Thermo couples, thermopiles
	Pressure	Pressure thermometers
	Temperature	IR pyrometers
	Resistance change	Resistance thermometer.
Pressure	Displacement	Bellows, diaphragms, Bourdon tubes
Flow	Pressure	Orifice plate, venturi, pitot tube
	Displacement	Piston type flow meters
	Temperature change	Hot wire anemometer.
Displacement	Resistance change	Strain gauge
	Voltage	Piezo-electric probes
	Inductance	Rotameters, differential transformers, Inductance strain gauge.

measurable output signal, it may not be in readable/ transmittable format due to noise or too large/small size etc. Therefore, it has to be manipulated to a desired level by modifying or filtering or amplifying or reducing the signal provided the physical nature of variable remains unchanged during this stage. Thus manipulating element performs one or more of the following functions:

(i) Filtering (ii) Amplifying or enlarging (iii) Reducing
(iv) Modifying (v) Fine tuning (vi) Analyzing
(vii) Synthesizing

For example, when you send a captured picture, its size is reduced to certain transmittable level. Similarly, an A.C. amplifier is tuned to the frequency chopper in some spectrophotometers.

6. **Data Transmission Element:** The next step of the measuring system after the signal is manipulated is transmission. If the various functional elements of a measuring system are separated spatially, then it is necessary to transmit signals from one element to another element. The data transmission element carries out this function. It is very important functional element particularly, when a remote controlled operation is desired.
7. **Data Presentation Element:** When the information about the quantity that is measured is transmitted, there should be an element to receive it and communicate to the observer (human) in a desirable form. This information obtained is often used in one or more of the following three ways.

   (i) Monitoring a process or operation (ii) Controlling a process or operation and
   (iii) Analysis of an experiment

Therefore the information is to be presented in a form recognizable by human sense. If the information is to be presented to the computer, the measuring system may be suitably interfaced with the computer. In fact, the purpose of the instrument is known by this functional element. An element which is used for such purpose is called data presentation element.

The data presenting element (often called display unit) performs the following functions

(i) *Transmitting:* To convey the information to a remote point

(ii) *Signaling:* To give signal that desired value is reached

(iii) *Registering:* To indicate by a number or symbol

(iv) *Indicating:* To indicate specific value on a calibrated scale

(v) *Recording:* To produce a record written or kept in memory.

1. **Presented Data:** It contains the information about the quantity that is measured.
2. **Observer/Output:** Based on the presented data the human observer controls, analyzes or monitors the measurand.

1.8.1 Stages of the Measuring System

As shown in Fig. 1.2 the generalized measurement system can also be classified into stages. They are

1. **Detection stage:** This is the first stage of the measuring system which receives a signal or energy or an input from the measuring medium (i.e., measurand). At this stage, the input given is sensed by the device (instrument) and converts into suitable format for which it consists of the respective functional elements. Thus, this stage includes primary sensing element & variable conversion element. From this stage, we can understand the principle of measuring system being designed.
2. **Intermediate stage:** At this stage, the process of measurement is carried out and interpreted in the required form. This stage consists of variable - manipulation element, transmission element & data presentation element. This stage tells us the purpose of the instrument (monitoring, controlling or analyzing) for which it is designed.
3. **Termination stage:** This is the final stage of the measuring system which includes presented data and output/observer. This stage gives information that can be used for intended application.

We shall now observe the above through some illustrations.

More Instruments for Learning Engineers

Polarimeter measures the angle of rotation of polarized light through an optically active substance. Some chemical substances are optically active and polarized (unidirectional) light will rotate either to the left (counter-clockwise) or right (clockwise) when passed through these substances. The amount by which the light is rotated is known as the angle of rotation due to polarization measured by polarimeter.

Polarimeter

Illustration-1.1

A pressure type thermometer is used to measure the temperature of fluid. The thermometer works on the principle of differential expansion of liquid which in turn imparts pressure to the bourdon tube. The displacement of the free end of the bourdon tube is magnified by a differential transformer transducer and its outputs are given to a display device. Prepare a block diagram and identify the functional elements in it.

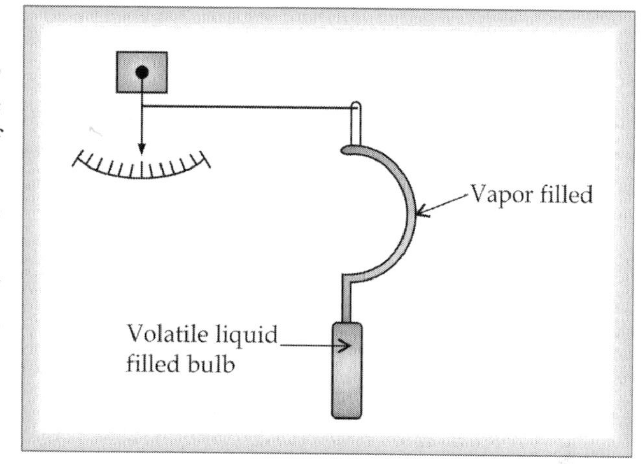

Fig. 1.3 Pressure Thermometer

Solution: The above information pertains to the instrument shown in Fig. 1.3 whose components are as follows:

1. ***Primary sensor:*** The liquid filled bulb.
   It senses the input (temperature) receives the input signal in the form of thermal energy.
2. ***Variable conversion element (primary transducer):*** Liquid in the bulb.
   Liquid bulb is constrained to thermal expansion and the filling fluid results in pressure (mechanical energy) built up in the bulb.
3. ***Data transmission element:*** The pressure tubing. Pressure tube transmits the pressure to the bourdon tube.
4. ***Variable conversion element (secondary transducer):*** The bourdon tube
   The bourdon tube converts the fluid pressure into displacement of its tip.
5. ***Manipulation element:*** Mechanical linkage and gearing.
   The displacement is manipulated by the linkage and gearing to give a larger pointer motion.
6. ***Data presentation element:*** The scale and pointer.

The block diagram is shown in Fig. 1.4.

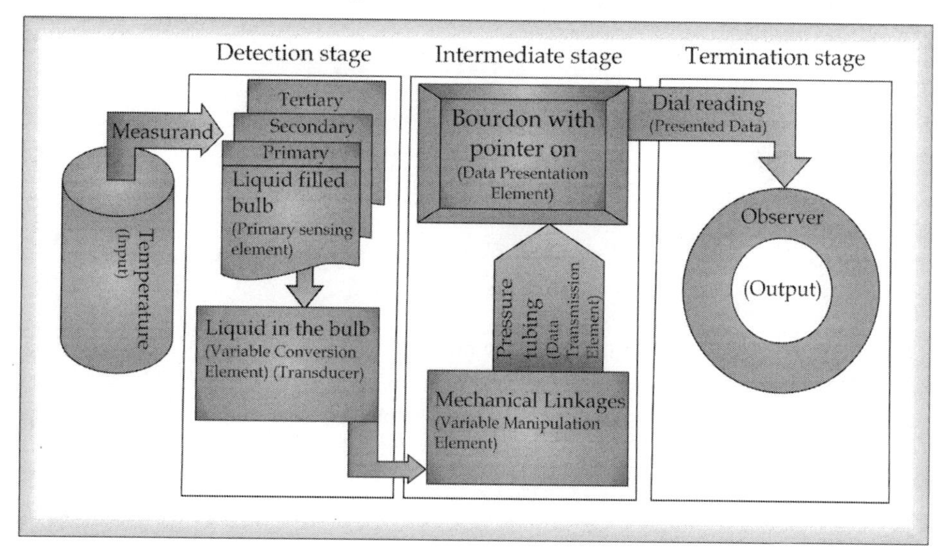

Fig. 1.4 Generalized Input-Output Configuration of Measurement Systems

ILLUSTRATION-1.2

Consider a dial indicator comprising a spindle connected with a rack on which gear-train of three gears, the last of which is fitted with a pointer associated with scale. The linear motion of the spindle is being transmitted and converted into an angular displacement of the pointer by means of the gear train. Identify various functional elements and the stages of measurement system and interpret as generalised. I/O model of measuring system with the help of block diagram.

Solution: The information belongs to the instrument shown in Fig. 1.5 whose components and their stages are as given below.

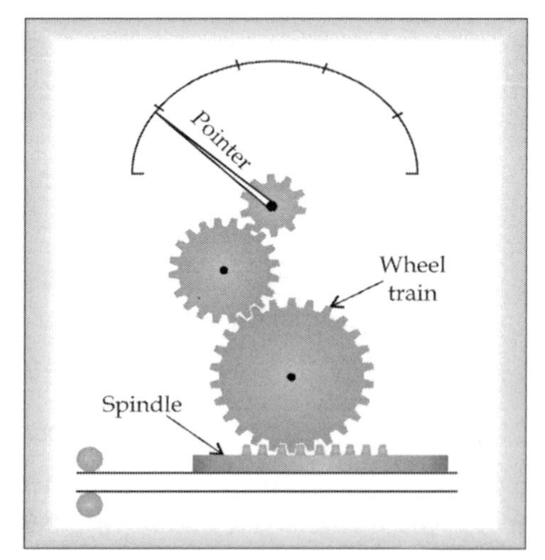

Fig. 1.5 Dial Indicator

1. **Detection Stage:**
   (a) ***Primary sensing element:*** The spindle is sensitive to the linear displacement and acts as the primary sensing element.
   (b) ***Conversion element:*** The spindle and gear train alignment acts as conversion element as it converts translator motion to rotator motion.
2. **Intermediate Stage:** The gear train performs the following different functions and acts as the elements shown against each function:
   (a) ***Transducer element:*** Change in the form of signal from translation to rotation (this may be included in detection stage/intermediate stage)
   (b) ***Manipulation element:*** Amplification (multiplication) of the input signal so that a large output displacement
   (c) ***Transmission element:*** Transmission of input signal from the spindle to the pointer
   (d) ***Data presentation element:*** The pointer and the associated scale comprise the data presentation element
3. **Termination Stage:**
   (a) Data presented is the pointer position on the dial
   (b) Observer notes/records the output

More Instruments for Learning Engineers

An **electricity meter** or **energy meter** measures the amount of electric energy consumed by a house/institution/business/ organization, or an electrically powered device, calibrated in billing units, often *kWh*. With special settings we can measure demand, the maximum use of power in some interval.

Electricity Meter

The above stages are shown in generalised I/O model of meaning system as block diagram in Fig. 1.6.

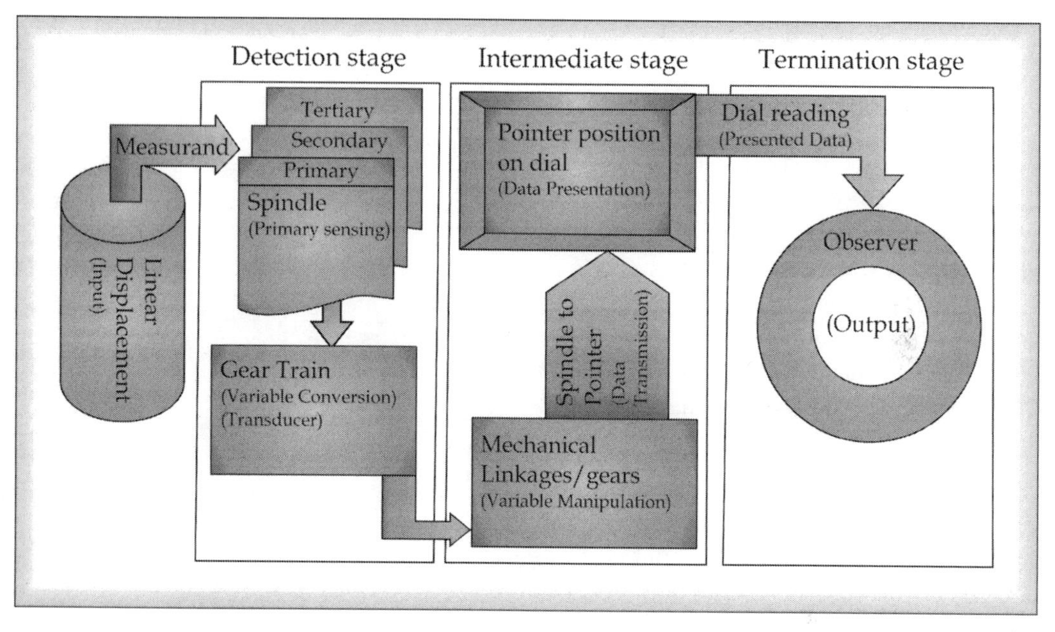

Fig. 1.6 Block Diagram of I/O Model Dial Indicator

Self Assessment Questions-1.3

1. Explain the different stages involved in the measuring system. Give examples.
2. Describe various functional elements used in the detection stage and termination stage of a measuring system. Give suitable examples.
3. Describe the general Input-Output Model of measurement system with a block diagram.
4. What is the function of primary sensing element? How the data is converted into a measuring variable?
5. What do you understand by 'variable conversion element' used in instrumentation? Explain with examples. What is its significance?
6. Describe the following functional elements used in measuring systems. Explain with suitable examples, how they function in the instruments.
   (a) Primary sensing element (b) Variable conversion element
   (c) Data transmission element (d) Data presentation element
7. Discuss the significance and functions of the data manipulation element used in measurement system.
8. List the applications of measuring instruments. Explain each of them.
9. What are the functional elements involved in the measuring system? Explain each of them.
10. What are the uses of Data-Transmission Element and Data Presentation Element used in measuring system? Discuss.

1.9 TYPES OF INPUTS

A generalized configuration in inst'ruments can be understood with a significant input-output relationship present in them as shown in Fig. 1.7. Input quantities are classified into three categories:

(i) Desired Inputs, (ii) Interfering Inputs, (iii) Modifying Inputs

1.9.1 Desired Inputs

Desired inputs are defined as quantities for which the instrument or measurement system is specifically designed to measure and respond. The desired input, r_D produces an output component $C_D = G_D r_D$ in accordance with an input-output relationship symbolized by mathematical operator, G_D, which is defined as Transfer Function.

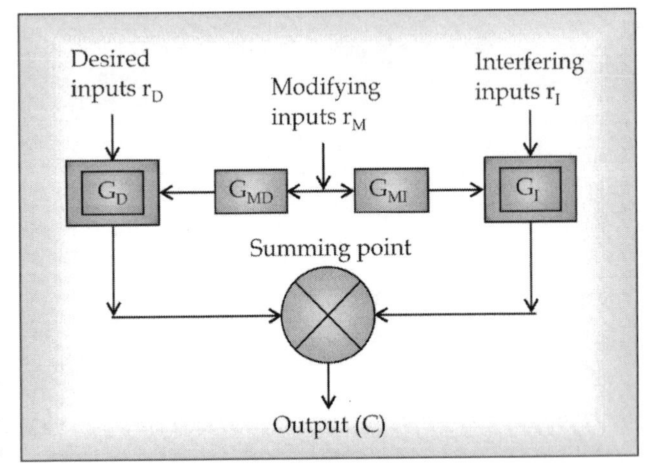

Fig. 1.7 Generalized Input-Output Configuration of Measurement Systems

G_D is necessarily a mathematical operation to get an output from a desired input. Thus if an input 'r' is operated upon by a transfer function 'G', the output is C = Gr. The transfer function may simply be a constant, K, which multiplies the static input, r_D to get an output $CD = Kr_D$ to obtain either an amplified or an attenuated output in linear systems. It should be understood that a constant cannot be used for describing the input-output relationships for non-linear systems. For non-linear systems, the transfer function is represented by either an algebraic or a transcendental function. The input-output relationships for systems subjected to dynamic inputs are represented by differential equations.

In case, a description of the output 'scatter' or dispersion for repeated equal static inputs is desired, a statistical function is needed to represent the input-output relationship

The transfer function, G_D, is therefore representative of a wide range of functions from a constant in the case of linear systems to a statistical function used for statistical measurements.

1.9.2 Interfering Inputs

Interfering inputs represent quantities to which an instrument or a measurement system becomes unintentionally sensitive. The instruments or measurement systems are not desired to respond to interfering inputs but they give an output due to interfering inputs on account of their principle of working, design and many other factors like the environments in which they are placed. The interfering input r_I is operated upon by a transfer function G_I to produce an output in the same manner as a desired input is operated upon by a transfer function, G_D, to produce an output.

1.9.3 Modifying Inputs

This class of inputs can be included among the interfering inputs. However, a separate classification is essential since such a classification is more significant. Modifying Inputs are defined as inputs which cause a change in input-output relationships for either desired inputs or interfering inputs or for both. Thus, a modifying input, r_M is an input that modifies G_D and/or G_I. The symbols G_{MD} and G_{MI} represent the specific manner in which r_M affects G_D and G_I respectively. These symbols G_{MD} and G_{MI} are interpreted in the same general way as G_D and G_I are.

1.10 CLASSIFICATION OF INSTRUMENTS

The classification of instruments may ease the study of the measurement system and can help in developing alternate methods of measurements. Instruments can be classified in several ways depending on the type of input, desirable form of output, the operability, the empirical relationships, principles of transduction, and the complexity of measurement and so forth. The most common classifications and their distinctions are given below.

1.10.1 Manual and Automatic Instruments

The manual instruments need the assistance of the operator while the automatic instruments do not. For instance, a spring balance directly shows the reading (force or weight) on the graduate scale without any assistance of the operator, whereas a deflection magnetometer requires the operators services to set the null position. Similarly, a liquid-in-glass (such as clinical thermometer) is an automatic instrument, but a resistance thermometer is manual type. Owing to the low operation cost, ease of operation and dynamic response, obviously automatic instruments are preferred.

1.10.2 Active (Self-operated) and Passive (Power-operated) Instruments

This classification is based on the power or energy required to operate the instrument. An active instrument is self-operated or self-generating in which the whole energy required for output is supplied by the input signal itself. That is to say, the instrument does not require any outside power in performing its function.

Examples of active instruments:

1. The motion in a Bourdon gauge is caused by pressure in the tube that moves the linkage and hence the pointer
2. Mercury-in-glass thermometer measures temperature due to displacement caused by the thermal expansion of mercury

More Instruments for Learning Engineers

An **electrometer** measures electric charge or potential difference. They are ranging from handmade mechanical to high-precision electronic devices. Modern electrometers use vacuum tube/ solid-state technology to measure voltage or charge with very low leakage currents, up to 1 femtoampere.

An electroscope, works on similar principles but only indicates the relative magnitudes of voltages or charges.

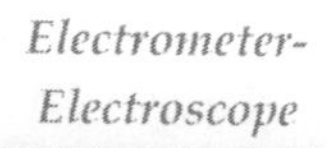

3. Tacho-generator for rotational speed measurement
4. Pitot-tube for the measurement of velocity
5. The dial indicator.

The passive (or power-operated) instruments need some external source of power such as electricity, compressed air, hydraulic supply etc., for the operation. In these devices, the input signal supplies only an insignificant portion of the output power.

Examples of passive instruments:

1. In the digital revolution counter, the power to drive the solenoid comes from the a.c. power lines and not from the rotating shaft
2. LVDT used in the measurement of displacement, force, pressure etc
3. Voltage-dividing potentiometer which converts rotation or displacement into potential difference
4. Strain-gauge load cell using Wheatstone bridge circuit
5. Resistance thermometers.

1.10.3 Self-Contained and Remote Indicating Instruments

This division of instrumentation is based on the operability of the instrument by contact or non-contact mechanisms. In a self-contained instrument all the functional elements are contained in one physical assembly, whereas in a remote indicating instrument, the primary sensing element may be located at a sufficiently long distance from the indicating element. The instrumentation trend in present days moving more toward remote indicating instruments where the important indications can be displayed in the central control rooms. Today almost every television or air conditioner etc., in any town is being operated by remote indications.

1.10.4 Deflection and Null Output Instruments

The null-type instruments are those in which the physical effect caused by the measurand is nullified (deflection maintained at zero) by generating an equivalent opposing effect. The equivalent null causing effect is the measure of the unknown quantity. The deflection type instruments are those in which the physical effect generated by the measuring quantity (measurand) is noted and correlated to the measurand.

The best example for null mode is the working of a physical balance, in which the unknown weight is placed on one pan of the balance and weights of known value are placed in the other pan until a balanced condition is indicated by zero or null position of the pointer.

A spring balance works on the deflection mode. The weight of an object placed on the platform of the scale is indicated by the relative displacement between the pointer and a dial.

Distinction between Null Mode and Deflection Mode of Measurement Systems

1. The accuracy of null type of instruments is higher than that of deflection type. This is because the opposing effect is calibrated with the help of standards which have high degree of accuracy. On the other hand, accuracy of deflection type of instruments is dependent upon their calibration which in turn depends on the instrument constants, normally not known to a high degree of accuracy.

2. In the null type of instruments, the measured quantity is balanced out. This means the detector has to cover a small range around the balance (null) point and therefore highly sensitive. Also the detector need not be calibrated since it has only to detect the presence and direction of unbalance and not the magnitude of unbalance. On the other hand, a deflection type of instrument must be larger in size, more rugged, and thus less sensitive, particularly to measure large magnitude of unknown quantity.
3. Null type of instruments need many manipulations before null conditions are obtained and hence are apparently not suitable for dynamic measurements wherein the measured quantity changes with time. On the other hand, deflection type of instruments can follow the variations of the measured quantity more rapidly and hence are more suitable for dynamic measurements on account of their quicker response. However, there are commercially available automatic control instruments (such as self balancing potentiometers used for measurement of temperature) that maintain a continuous null under rapidly changing conditions and thereby eliminate the need for manipulative operations.

Summarily, deflection instruments are simple in construction and operation, and have good dynamic response. However, they interfere with the state of measurand and as such do not determine its exact state/value/condition. The null-type devices are slow in operation, have poor dynamic response but are more accurate and sensitive, and do not interfere with the state of the quantity being measured.

1.10.5 Analog and Digital Instruments

This classification is based on the output of instruments. In an analog instrument, the signal varies in a continuous pattern and therefore can take on infinite number of values in a given range.

Examples: Wrist watch, speedometer of an automobile, ammeters and voltmeters.

In digital instruments, the signals vary in discrete steps and hence can take a finite number of different values in a given range.

Examples: Timers, counters, odometer of an automobile.

The digital instruments have the merit of high accuracy, high speed and the elimination of human operational errors. However, these instruments are unable to indicate the quantity in between the steps of the instrument. The importance of digital instrumentation is increasing very fast due to the application of digital computers for data handling, cost reduction and in automatic controls.

The output in digital devices may either be a digit (pulse or step) for every successive increment or a coded discrete signal represented by a numeric value of the input. Strictly speaking, this output in digital instruments is a measured analog voltage converted (usually by neon indicator tubes) into digital quantity displayed in numeric values. Thus it is possible to convert analog to digital and vice-versa with the help of analog-to-digital (A/D) and digital-to-analog (D/A) converters and can be interfaced with computers at input-output stages.

1.10.6 Detecting, Recording, Monitoring and Controlling Instruments

Based on the kind of service rendered, the instruments may also be classified as:

(i) **Detecting Instruments:** The detecting instrument senses the signal and gives an output as 'yes or no' type result. Example: Electrical tester, bomb tester, indicator lamps, signal indicators, traffic indicators, leakage dectector etc.

(ii) **Recording Instruments:** These instruments record the output in the hard (written) or soft (computer memory) form. Example: Instrument interfaced with computer/printer, X-ray filming, Electro-cardiogram (ECG) etc.

(iii) **Monitoring Instruments:** These instruments help monitoring the system and regulate the operational parameters. Examples: Flow meters, voltage stabilizers, auto focus instruments in a camera.

(iv) **Controlling Instruments:** The instruments used to control the operations are called controlling instruments. Examples: Thermostats (with bimetallic strip), controllers (with sensors), regulators of electrical signals etc.

1.10.7 Mechanical, Electrical and Electronic Instruments

Based on the technological features, area of operation and the field of application, the instruments may be categorized as

1. **Mechanical Instruments:** These are made up of some mechanical engineering mechanisms such as motion transfer, motion conversion through gears, springs, clutches, pulleys, belts, ropes, chains, fluid properties etc. Examples: Dial gauge, Bourdon tube, liquid-in-glass thermometer, spring balance (platform balance).
2. **Electrical/Electronic Instruments:** These instruments are operated by electrical power or electrical/electronic components such as resistors, capacitors, transistors, transformers are used in them. Examples: Linear Variable Differential Transformer (LVDT), Resistance thermometer, voltmeter, ammeter etc.
3. **Bio-Medical Instruments:** These are the instruments made up of either mechanical parts or electrical/electronic components and so come under one of the above two. However, these have specific fields of application such as medical (hospitals), pharmaceutical field. Examples: clinical thermometer, ECG, pulse meter etc.

Do you know?

Why do electrical/electronic instruments quickly respond?

The basic hindrances to rapid dynamic response are inertia and friction. The mass of the electron and the relatively void space it goes through reduces both these factors to a minimum. Due to this small inertia of electrons, the response time of electronic devices is extremely small. The mechanical movement of the indicator in electric devices also has some inertia which limits the time response to a considerable extent. The frequency response of the majority of electrical instruments is in the range of 0.5 to 25 seconds. The electron beam oscilloscope operates by using the mass of the electron and is capable of following dynamic and transient changes of the order of a few nano-seconds (10^{-9}s).

Mechanical versus electrical/electronic instruments

The mechanical and electrical/electronic instruments are distinguished with their strengths and weaknesses here in the Table 1.2.

Table 1.2 Comparison between Mechanical and Electrical/Electronic Instruments

	Strengths	Weaknesses
Mechanical Instruments	1. long history of development and successful use 2. Simple in design and operability 3. More durable due to rugged construction 4. relatively low cost 5. More reliable and accurate for the measurement of parameters which are stable and non-variant with time. 6. Usually no need of external power supplies for operation	1. Poor frequency response to dynamic and transient measurement. 2. sluggish in operation and not react immediately to the rapid changes of the input signals due to inertia of moving parts 3. incompatibility when remote indication or control is required 4. require large forces to overcome mechanical friction 5. potential source of noise
Electrical/Electronic Instruments	1. Light, compact and more reliable 2. Good frequency and transient response. 3. Feasibility of remote indication and recording 4. Less power consumption and less load on the system being measured 5. Greater amplification than that of a mechanical including hydraulic and pneumatic systems 6. Possibility of mathematical processing of signals like summation, differentiating and integration 7. Possibility of non-contact measurements	1. Power consumption and goes handicapped in absence of power 2. Difficult in fault diagnosis in the event of failure of instrument 3. Hysteresis and other losses 4. Not easy to design and manufacture as compared to that of mechanical 5. Difficult to understand the mechanism and the transduction principles

1.11 OBJECTIVES OF INSTRUMENTATION

The instrumentation plays a vital role in industry particularly, in the area of research and development. Instrumentation engineering is always concerned with the following aspects, which can also be considered as the objectives* of instrumentation and control systems

1. How to measure a given physical quantity?
2. Principle of measurement.
3. Construction of equipment.
4. Working of the instrument.
5. Range through which the instrument can work.
6. Advantages of the instrument
7. Disadvantages of the instrument.
8. Limitations of its utility or operation.
9. Applications, suitability and compatibility.

10. For what other functions or measurements the same instrument can be used.
11. What alternative instrument can be used as substitute?
12. Comparison of various alternatives.
13. Errors those are possible to occur under various conditions/situations.
14. How the errors can be eliminated.
15. Calibrating the instrument(s).
16. Suitable interpretations of response or output.

The readers of this subject are advised to study every chapter with a focus of the above aspects.

SELF ASSESSMENT QUESTIONS-1.4

1. List the objectives of instrumentation.
2. How do you classify the measuring instruments? Briefly describe them.
3. Distinguish between following types of instruments
   (a) Active and passive (b) Null mode and deflection mode
   (c) Mechanical and electrical (d) Analog and digital
4. How do you classify the input quantities? Explain each of them with the help of a block diagram.
5. Distinguish among desired, interfering and modifying inputs with examples.
6. With suitable examples describe instruments used for detecting, monitoring, recording and controlling the operations.
7. Why electrical/electronic instruments are preferred to mechanical instruments? What are their limitations?
8. What are the strengths and weaknesses of mechanical instruments? Under what conditions/ situations are they more suitable to use?

SUMMARY

This subject is concerned with construction and working of various instruments. The process of measurement is concerned with the input compared with a standard resulting in some output. The term *'measurement'* is used to describe the quantity in some numeric value or units of physical entities such as length, weight, temperature etc., or a change in one of these physical entities of a material. The physical quantity or the characteristic condition that is to be measured is named as *measurand, measurement variable, instrumentation variable* or *process variable.* The *output* is readout by the observer after comparing the object with a quantity of same kind, called *standard. The equipment used to measure a physical quantity under the specified conditions and defined procedure is called an 'Instrument'*. Thus Instrumentation is defined as, '*The subject that deals with the study of construction, working principle, and operational procedure of the measuring equipment or instruments under specified conditions'*. Measurement can be made either directly with standard or indirectly by using certain device or a chain of devices called 'transducer' which converts one form of physical quantity to another. The complexity of an instrument mostly depends upon the measurement being made, the characteristics of the measurand and the accuracy level required.

In the primary measurement, a physical parameter is determined by simply comparing it directly with reference standards. Indirect measurements involving one translation are called *secondary measurements,* similarly involving two conversions are called *tertiary measurements.* Measuring instruments are generally applied in three ways monitoring, controlling a process or an operation and experimental analysis and engineering. Generalized measurement instrumentation consists of Measured Medium (the input), Measurand, Primary sensing (receives input), Variable-Conversion (intermediate transducer), Variable-Manipulation, Data-Transmission, Data Presentation and Output. The three stages of generalized measurement system are Detection, Intermediate and Termination stages.

Input quantities are classified into Desired, Interfering and Modifying Inputs. The most common classifications of instruments are Manual/Automatic; Active (Self-operated)/Passive (Power-operated); Self-contained/Remote Indicating; Deflection and Null Output; Analog and Digital; Detecting/Recording/Monitoring and Controlling; Mechanical, Electrical and Electronic Instruments. This book covers the major objectives of instrumentation such as principle of measurement, construction, operation, range, merits, demerits and applications of various instruments.

KEY CONCEPTS

Measurement: The process of obtaining a quantitative comparison between a predetermined standard and an unknown magnitude of the same parameter.

Measurand: The physical variable such as length, temperature, pressure etc., which is the object of measurement of an instrument, also called *measurand or the measured variable or measurement variable or instrumentation variable* or *process variable*.

Independent variables: Variable with spatial independence.

Pervariables or through variables: Variables specified and measured at one point in space.

Transvariables or across variables: Variable require two points (usually one point is the reference) to specify or measure them.

Output: Readout by the observer after comparing the object with a quantity of same kind, called *standard.*

The standard: The reference of the comparison of certain physical quantity.

The Process of measurement: Activity concerned with comparison of input with a standard.

The instrument: The equipment used to measure a physical quantity under the specified conditions and defined procedure.

The instrumentation: The subject that deals with the study of construction, working principle, and operational procedure of the measuring equipment or instruments under specified conditions.

Range: The region between the limits within which an instrument is designed to operate for measuring a physical quantity.

Accuracy: The uniformity of a measured value with an accepted standard value.

Precision: How closely the individual measured values of a quantity agree with each other when the measurements are carried out under identical conditions at a short interval of time (It refers to repeatability).

Calibration: The procedure laid down for making adjusting, or checking a scale, so that readings of an instrument confirm to an accepted standard.

Error: The deviation from the true or exact value.

Characteristics: Properties or features of an instrument which are understood by its response to a given input.

Direct comparison: The measured quantity is compared (directly) with a standard.

Indirect comparison: The comparison is made indirectly by using certain device or a chain of devices called 'transducer' which converts one form of physical quantity to another.

Primary measurement: A physical parameter is determined by simply comparing it directly with reference standards.

Secondary measurement: Indirect measurements involving one translation.

Tertiary measurement: Indirect measurements involving two conversions.

Measured medium: The input to a measurement system that supplies energy to the primary sensing element.

Primary sensing element: Primary sensing element receives energy/input from the medium to be measured and produces a proportional output.

Transducer: Device that converts one physical variable into another.

Variable-conversion element: Converts to a more suitable variable while preserving its original contents.

Variable-manipulation element: Changes to a desired level by modifying or filtering or amplifying or reducing the signal provided the physical nature of variable remains unchanged.

Data-transmission element: Transmits signals from one element to another.

Data presentation element: Receives the information about the quantity that is measured and communicates to the observer (human) in a desirable form.

Presented data: Contains the information about the quantity that is measured.

Detection stage: The first stage of the measuring system which receives a signal or energy or an input from the measuring medium (i.e., measurand).

Intermediate stage: The process of measurement is carried out and interpreted in the required form.

Termination stage: The final stage of the measuring system which includes presented data and output/observer.

Desired inputs: Quantities for which the instrument or measurement system is specifically designed to measure and respond.

Interfering inputs: Quantities to which an instrument or a measurement system becomes unintentionally sensitive.

Modifying inputs: Inputs which cause a change in input-output relationships for either desired inputs or interfering inputs or for both.

Manual instruments: Need the assistance of the operator.

Automatic instruments: Do not need the assistance of the operator.

Active (self-operated) instruments: Whole energy required for output is supplied by the input signal itself.

Passive (power-operated) instruments: Need some external source of power for the operation.

Self-contained instruments: All the functional elements are contained in one physical assembly.

Remote indicating instruments: The primary sensing element may be located at a sufficiently long distance from the indicating element.

Deflection instruments: The physical effect caused by the measurand is nullified (deflection maintained at zero) by generating an equivalent opposing effect.

Null output instruments: The physical effect generated by the measuring quantity (measurand) is noted and correlated to the measurand.

Analog instruments: The signal varies in a continuous pattern and hence can take on infinite number of values in a given range.

Digital instruments: The signals vary in discrete steps and hence can take a finite number of different values in a given range.

Detecting instruments: Senses the signal and gives an output as 'yes or no' type result.

Recording instruments: Record the output in the hard (written) or soft (computer memory) form.

Monitoring instruments: Help monitoring the system and regulating the operational parameters.

Controlling instruments: Used to control the operations.

Mechanical Instruments: Made up of some mechanical engineering mechanisms such as motion transfer, motion conversion through gears, springs, clutches, pulleys, belts, ropes, chains, fluid properties etc.

Electrical/electronic instruments: Operated by electrical power or electrical/ electronic components.

REVIEW QUESTIONS

Short Answer Questions

1. Define the terms:
   (i) Instrumentation (ii) Measurement.
2. Define the following
   (a) Measurand (b) Primary measuring element (c) Calibration
3. Briefly describe the salient terms used while measuring with an instrument?
4. List out different types of variables used in the process of measurement through an instrument.

5. Differentiate between direct measurement and indirect measurement.
6. Define the following terms and briefly describe with examples.
   (a) Range (b) Accuracy (c) Precision (d) Calibration
7. Define the following terms and briefly describe with examples.
   (a) Measurand (b) Measurement (c) Error (d) Instrument
8. Distinguish between precision and accuracy.
9. Distinguish between measurand and measurement.
10. Explain the different stages involved in the measuring system. Give examples.
11. What are the strengths and weaknesses of mechanical instruments? Under what conditions/ situations are they more suitable to use?
12. Why electrical/electronic instruments are preferred to mechanical instruments? What are their limitations?
13. Distinguish between following types of instruments
   (a) Active and passive (b) Null mode and deflection mode
   (c) Mechanical and electrical (d) Analog and digital
14. List the objectives of instrumentation.
15. List the applications of measuring instruments.
16. What do you mean by instrumentation? Write the objectives of instrumentation?

Long Answer Questions

1. Explain different measurement systems.
2. Draw block diagram of generalized measurement system. Explain it with a suitable example.
3. What do you mean by functional elements? Explain the division of a measurement system into functional elements with examples.
4. Differentiate among desired, interfering and modifying inputs to a measurement system with examples.
5. What is the generalized input/output configuration of measurement system? Give example.
6. Differentiate between null mode and deflection of operation of measurement systems with examples.
7. What are primary, secondary and territory measurements? Explain with suitable examples.
8. What are the various methods of measurements? Explain them.
9. Describe various functional elements used in the detection stage and termination stage of a measuring system. Give suitable examples.
10. Describe the general Input-Output Model of measurement system with a block diagram.
11. Describe the following functional elements used in measuring systems. Explain with suitable examples, how they function in the instruments.
   (a) Primary sensing element (b) Variable conversion element
   (c) Data transmission element (d) Data presentation element

12. Discuss the significance and functions of the data manipulation element used in measurement system.
13. What are the functional elements involved in the measuring system? Explain each of them.
14. What are the uses of Data-Transmission Element and Data Presentation Element used in measuring system? Discuss.
15. How do you classify the measuring instruments? Briefly describe them.
16. How do you classify the input quantities? Explain each of them with the help of a block diagram.
17. Consider a clinical thermometer used to measure the temperature of human body. The thermometer works on the principle of expansion of liquid. Prepare a general I/O block diagram and identify the functional elements in it.
18. Describe the general I/O model block diagram and explain using an example of spring balance.
19. Represent the measuring system of float gauge used to indicate the liquid level, as general input-output block diagram and identify the general functional elements in it.

MULTIPLE CHOICE QUESTIONS

1. The term _______ is used to describe the quantity in some numeric value or units of physical entities
   (a) Instrument
   (b) Measurement
   (c) Inspection
   (d) Product
2. __________ have a spatial freedom.
   (a) Independent variables
   (b) Pervariables
   (c) Across variables
   (d) Standard variable
3. The standard is a result of measurement, expressed as a number representing the ratio of the unknown quantity to ________.
   (a) Adopted standard
   (b) Unknown standard
   (c) Adopted variable
   (d) Known standard
4. The equipment used to measure a physical quantity under the specified conditions and defined procedure is called __________.
   (a) Instrument
   (b) Sensing device
   (c) Measurand
   (d) Mechanical device
5. _______ is defined as the nearness of a measured value with an accepted standard value.
   (a) Accuracy (b) Precision
   (c) Standard (d) Calibration
6. The procedure laid down for making adjustments so that readings of an instrument conform to an accepted standard is called _______.
   (a) Instrumentation
   (b) Calibration
   (c) Correction
   (d) Measurement
7. In measurement, the comparison is made indirectly by using certain device or a chain of devices known as _________.
   (a) Instruments (b) Gauges
   (c) Transducers (d) Sensors
8. Which measurements provide subjective (by human senses) information only.
   (a) Secondary (b) Primary
   (c) Direct (d) Indirect

9. "The conversion of pressure into displacement by means of bellows" is the example of ______ measurements.
   (a) Secondary (b) Primary
   (c) Tertiary (d) Direct

10. In monitoring of process ________ is used to indicate the condition or status of the situation or environment.
    (a) Speedometer (b) Water meter
    (c) Thermometer (d) Tachometer

11. Measured quantity is also called ________.
    (a) Lot (b) Measurand
    (c) Standard (d) Output

12. In bimetallic sensors the transduction takes place from temperature to _______.
    (a) Electric current (b) Pressure
    (c) Displacement (d) Volume

13. In desired inputs G_D is defined as _______.
    (a) Transfer function
    (b) Desired function
    (c) Constant
    (d) Output

14. The input-output relationships for systems subjected to dynamic inputs are represented by__
    (a) Algebraic function
    (b) Differential equations
    (c) Logarithmic functions
    (d) Statistical function

15. If a description of the output 'scatter' or dispersion for repeated equal static inputs is desired, a _________ function is needed to represent the input-output relationship.
    (a) Statistical
    (b) Logarithmic
    (c) Differential
    (d) Exponential

16. Measurement provides us with means of describing a natural phenomenon in ______ terms.
    (a) Quantitative (b) Qualitative
    (c) Volumetric (d) Standard

17. The measuring instrument is an essential integral component of _________ system.
    (a) Physical
    (b) Automatic control
    (c) Chemical
    (d) Inspection

18. Depending upon the degree of complexity, the instrumentation systems are categorized as ________ instrumentation systems.
    (a) Primary, secondary, tertiary
    (b) First, second, third
    (c) Third, second, first
    (d) n, n+1, n+2

19. The indirect measurements involving one translation are called ------ measurements.
    (a) Secondary (b) Primary
    (c) Tertiary (d) Mono

20. The unit of a measuring system where translation of a measurand takes place is called _____
    (a) Sensor (b) Gauge
    (c) Transducer (d) Instrument

21. The measurement refers to _______.
    (a) Measured variable (b) Output
    (c) Secondary signal (d) Procedure

22. The purpose of instruments is to _________.
    (a) Change signals
    (b) Allow measurements to be made
    (c) Transmit the information
    (d) Avoid error

23. The temperature measurement by a thermocouple is _________ measurement.
    (a) Primary (b) Secondary
    (c) Tertiary (d) Direct

24. The output stage of a generalized measurement system may comprise
    (a) Detector – transducer
    (b) Manipulator
    (c) Indicating or recording unit
    (d) Sensors and Transducers

25. A high - grade set of slip gages preserved in a factory and not to put into general use would be a
    (a) Primary
    (b) Secondary
    (c) Tertiary standard
    (d) Sometimes primary/secondary.

ANSWERS

1.	(b)	2.	(a)	3.	(a)	4.	(a)	5.	(a)
6.	(b)	7.	(c)	8.	(c)	9.	(a)	10.	(c)
11.	(b)	12.	(c)	13.	(a)	14.	(b)	15.	(a)
16.	(a)	17.	(b)	18.	(a)	19.	(a)	20.	(c)
21.	(b)	22.	(b)	23.	(c)	24.	(c)	25.	(b)

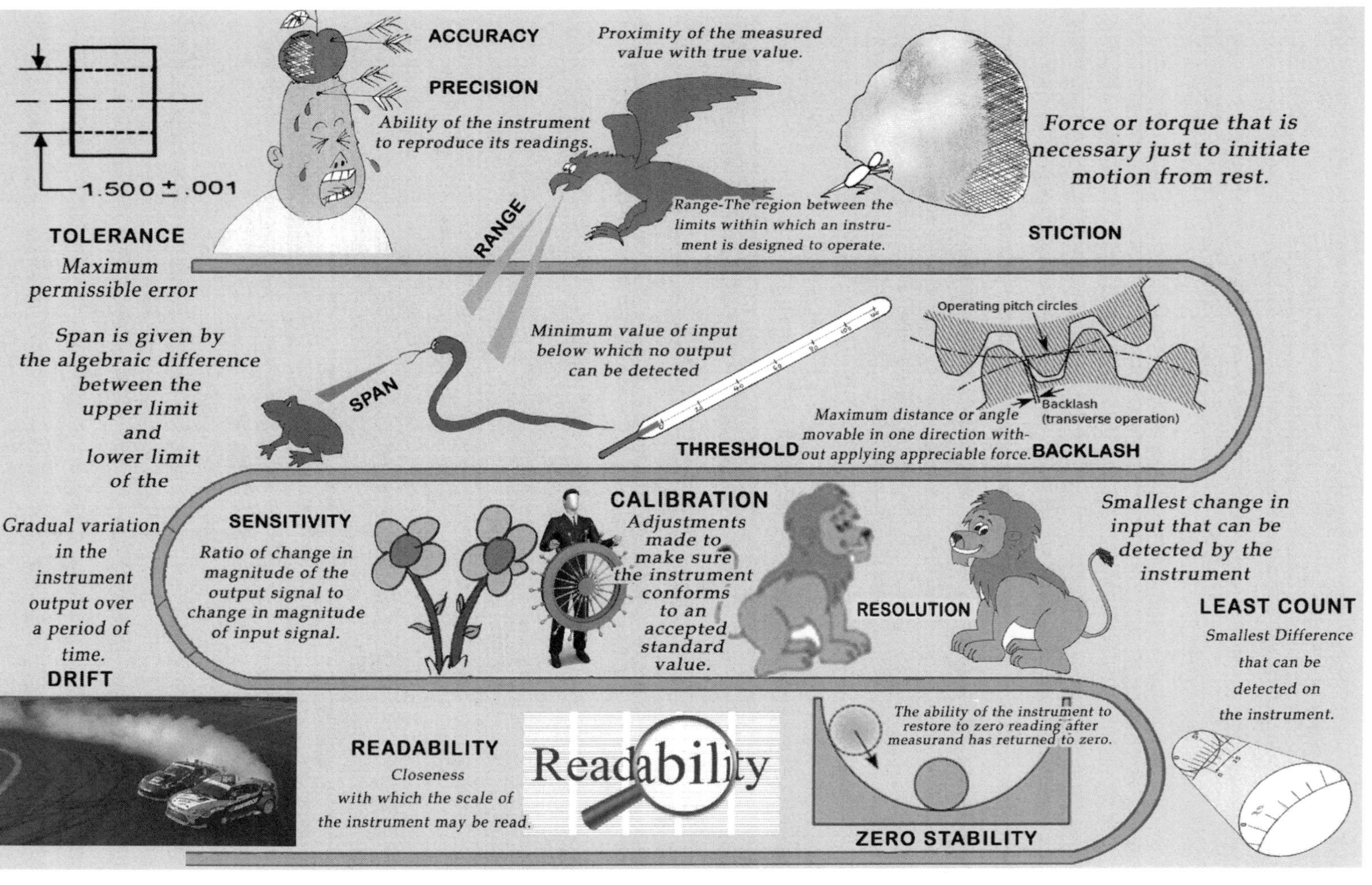

CHAPTER INFOGRAPHIC

2 Glimpses of Characteristics of Instruments

Characteristics or features that are identified ot recognized by its response to a given input usually portray an Instrument.

Characteristics are two types: Static (do not vary in time domain) & Dynamic characteristics (vary in time domain) such as (i) Steady state periodic quantities: Input magnitude has a definite repeating time cycle and Transient quantities : Input mangnitude does not repeat with time.

STATIC PERFORMANCE CHARACTERISTICS

Range: The region between units in which an instrument is designed to operate.
Span: Algebraic difference between the upper limit and lower limit of the Instrument.
Calibration: Procedure laid down for making, adjusting or checking a scale, so that readings of an instrument conform to an accepted standard.
Drift: Gradual variation in the instrument output over a period of time.
Treshold: Minimum value of input below which no output can be detected.
Resolution: Smallest change in input signal or measured value which can be detected by the instrument.
Tolerance: Range of inaccuracy which can be tolerated in measurement.
Readability: Closeness with which the scale of the instrument may be read.
Least count: Smallest difference that can be detected on the instrument scale.
Backlash: Maximum distance/angle in which any part of mechanical system may be moved in one direction without applying appreciable force or motion
Zero stability: To restore to zero reading after the measurand has returned to zero, and other variations are removed.
Stiction: Force or torque that is necessary just to initiate motion from rest.

DYNAMIC PERFORMANCE CHARACTERISTICS

1. Speed of response and measuring lag 2. Fidelity and dynamic error 3. Overshoot 4. Dead time and Dead zone 5. Frequency respone

The dynamic response of instrument is represented by a mathematical model

A mathematical model is a relationship between input and output of the instrument, which is determined taking into account the physical paramenters of its elements. Basic elements of mechanical systems are mass, spring and a damper or dashpot; Basic elements of electrical systems are resistance, inductance and capacitaince.

The order of the system is the order of the differential equation representing

Zero order system

$$\frac{\theta_0}{\theta} = ks$$

First order system

$$\frac{\theta_0}{\theta} = k/(\tau D + 1)$$

Second order system

$$\frac{\theta_0}{\theta} = k/(Dt\omega^2 + 2t\omega_n + 1)$$

Standard Test signals

1. **Step signal** $R(s) = A/S$
2. **Ramp signal** $R(S) = A/S^2$
3. **Parabolic signal** $R(S) = A/S^3$
4. **Simple harmonic input (sinusoidal)** $R(S) = A\frac{\omega}{s^{2+}\omega^2} - \varepsilon$
5. **Impulse signal** $\delta(t)dt = 1$

Dynamic Response of First Order System

1. **Unit step response** $\frac{\theta_0}{\theta_\tau} = 1 - \varepsilon^{\frac{-\tau}{\tau}} = 0.632\theta_{\tau 2}$
2. **Unit step response** $\theta_0 = \Omega\tau - \Omega\tau\varepsilon^{\frac{-\tau}{\tau}}$
3. **Harmonic sinusoidal response** $\frac{A_o}{A_i} = \frac{\omega_V^2}{\sqrt{(\omega_n^2 - \omega) + (2\xi\omega_n\omega)^2}}$

Dynamic Response of Second order System

1. **Unit stpe response** $\theta_o = pe^{-\xi\omega}n^t + \left[\cos\omega_d t + \frac{\xi}{\sqrt{1-\xi^2}}\sin\omega_d t\right]$
2. **Harmonic/sinusoidal response**

$$\theta_0 = pe^{\frac{-1}{\tau}} + \frac{A_i}{\sqrt{1-(\omega t)^2}}\sin(\omega t - \varnothing)$$

Chapter 2

CHARACTERISTICS OF INSTRUMENTS

STARTERS

To study this chapter, you should have awareness on the following concepts. For a better understanding, it is always a good idea to revise these prerequisites.

- An understanding on system approach i.e., input-output models
- A brief idea on system behaviours
- Differentiation, partial differentiation, integration and differential equations
- Basic electronics and principles
- Various fundamental and derived quantities, their denotations, and units
- Standards of measurement and methods of standardizing
- Notations, and units of various quantities
- International standards, National standards and Local standards
- Definitions and basics of various physical quantities
- Conversion factors of one system of units to the other for various quantities
- Basic principles of solid mechanics, fluid mechanics, elasticity, heat, light, sound, magnetism, electricity and semi-conductors

And

The contents discussed in the first chapter of this book.

LEARNING OBJECTIVES

After studying this chapter you should be able to

- Understand and interpret the static and dynamic characteristics of instruments
- Explain various types of input signals and their graphical and mathematical interpretation
- Describe the static and dynamic responses for zeroth, first and second order systems

2.1 INTRODUCTION

Suppose you are asked to describe yourself in an interview. How do you describe? Yes! You would first say about your birth date, birth place, qualification, experiences, strengths, weaknesses, your hobbies or achievements and other soft skills what you have. From this information, the interviewer will be able to estimate your characteristics, what kind of person you are, your problem solving skills, your knowledge, communication skills, your culture, attitude, demography and so forth.

Similarly, if you have to describe about an instrument, what aspects characterize that device, what is range, responsiveness, accuracy, possible errors and so forth. These are often termed as characteristics and/or features of an instrument.

An instrument is usually described by its characteristics or features that are identified or recognized by its response to a given input. The characteristics can depict performance of an instrument by which one can easily understand the behaviour of the instruments for various inputs. These characteristics are generally studied under two heads viz. static and dynamic characteristics. The static characteristics are those which do not vary in time domain while dynamic characteristics are those which vary with time. For example your date of birth is your static characteristic while your age is a dynamic characteristic. Often, the characteristics are expressed in terms of certain physical quantities whose nature also will be either static or dynamic. In this chapter, we shall discuss these characteristics.

2.2 STATIC AND DYNAMIC QUANTITIES

The characteristics of measurement can be categorically described with two kinds of quantities. They are 1. Static quantities 2. Dynamic quantities

1. **Static Quantities:** These are either constant or vary very slowly with time which is negligible
2. **Dynamic Quantities:** These vary with time and based on variation with time, they are further divided into two types (a) Steady state periodic quantities (b) Transient quantities.
   (a) ***Steady State Periodic Quantity:*** It is the dynamic quantity whose input magnitude has a definite repeating time cycle.

More Instruments for Learning Engineers

Electronic tuner is used by musicians to detect/ display the pitch of notes played on musical instruments. The tuner size ranges from a pocket clip to 19" rack-mount units. The simplest tuners use LED lights to indicate if the pitch is low/ high. These can only detect and display the tuning for a single pitch or for a small number of pitches, such as the 6-pitches of standard tuning of a guitar (E,A,D,G,B,E). A complex tuner uses chromatic tuning that allows all the 12 notes. Some electronic tuners have additional features, like adjustable pitch calibration, different scale options, desired pitch through an amplifier and speaker, and adjustable "read-time" settings etc.

Electronic Tuner

(*b*) ***Transient Quantity:*** It is the dynamic quantity whose input magnitude does not repeat with time.

To have a clear understanding about the above quantities, let us consider a cantilever beam, as shown in Fig. 2.1.

Suppose a load "P" is placed at the end "B" of a cantilever fixed at A, which will result in deflection and strain "å" at any point. Now, we can notice the above quantities in the three situations as given below:

(a) Keep a weight on the beam at A and slowly move toward the end B (or slowly release the load at B) as shown in Fig. 2.1(a). We can observe that the beam would be subjected to the deflection gradually. The strain diagram on the time frame can be obtained as straight line parallel to X-axis (time). Thus, a static loading (input) shall result in static deflection and strain (output).

(b) Next, consider the cantilever to be acted upon by a force causing it to vibrate continuously. The input force is time varying and the resulting output that is deflection and strain are also time varying or dynamic in nature as shown in Fig. 2.1(b), which shows a time-varying force causing a harmonically varying strain, changing from tension to compression cyclically.

(c) If the end "B" is subjected to a sudden load or hit by an impact force once, the deflection and strain at the point B shall be transient in nature as shown in Fig. 2.1(c).

Thus from the above experiment, we can understand that the performance of an instrument can be studied under the two headings given below:

1. Static Performance Characteristics.
2. Dynamic Performance Characteristics.

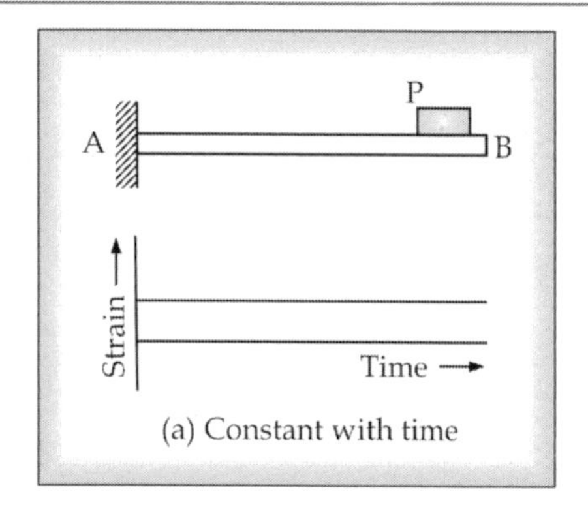

Fig. 2.1(a) Example for Static Quantity

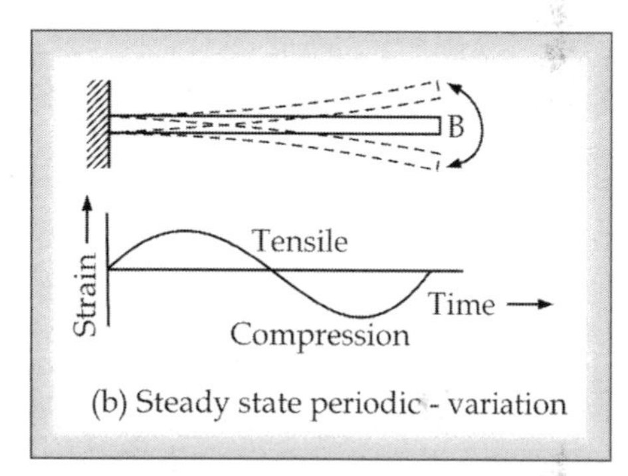

Fig. 2.1(b) Example for Quantity at Steady State (periodic)

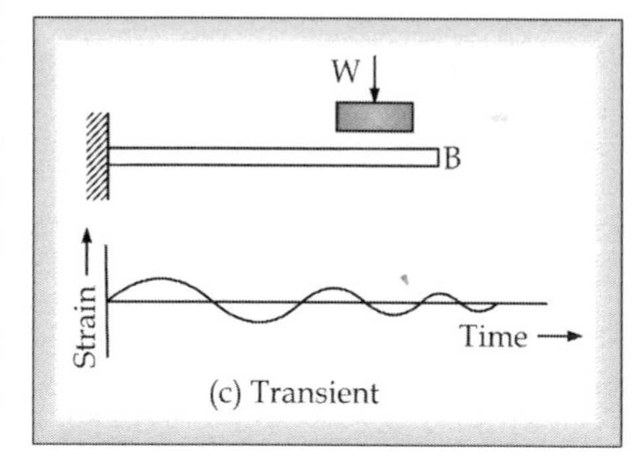

Fig. 2.1(c) Example for Quantity in Transient State

2.3 STATIC CHARACTERISTICS

Static characteristics are those characteristic which do not vary in the time domain. Strictly speaking, there is no 100% static, however, in a time interval either they do not change or the change is negligible even if they change. So, those characteristics of instruments which have negligible change in a given period of time are said to be static. The following are the characteristics that measure the static performance of an instrument.

2.3.1 Range

The region between the units in which an instrument is designed to operate for measuring a physical quantity is called the range. It is expressed by stating Lower and Upper limits

For example:

Range of the thermometer is 0 °C to 100 °C

Range of a pressure gauge 5 N/mm^2 to 100 N/mm^2

Range of an ammeter = 1 A to 10 A

2.3.2 Span

It is given by the algebraic difference between the upper limit and lower limit of the Instrument.

For example:

Span of the above mentioned thermometer 100 – 0 = 100 °C

Span of the above mentioned pressure gauge is 100 – 5 = 95 N/mm^2

Span of the above mentioned ammeter is 10 – 1= 9 A.

2.3.3 Accuracy

Accuracy of a measured value may be designed as conformity with an accepted standard value.

It refers to how closely the measured value agrees with the true value.

The accuracy of an instrument can be specified as given below.

(i) Percentage of true value $= \dfrac{\text{Measured value–True value}}{\text{True value}}$

(ii) Percentage of full scale deflection is $= \dfrac{\text{Measured value–True value}}{\text{Max scale value}}$

2.3.4 Precision

It refers to repeatability or how closely the individual measured values of a quantity agree with each other when the measurements are carried out under identical conditions at a short interval of time (or) It prescribes the ability of the instrument to reproduce its readings over & over again for a constant input signal.

Difference between Accuracy and Precision:

No.	Accuracy	Precision
1.	Def: Accuracy of a measured value may be defined as uniformity with an accepted standard value	Def: It refers to repeatability or how closely the individual measured values of a quantity standard value agree with each other when the measurement are carried out under identical conditions at a short interval of time.
2.	Due to systematic errors accuracy will be low	Due to random errors precision of instruments will be low

The difference between accuracy and precision can be clearly understood by considering the following example.

The arrangement has shown in Fig 2.2 corresponds to the game of darts where one is asked to strike a target represented by center circle. The center circle then represents the true value and the result achieved by the striker has been indicated by the mark "*".

From the above examples it is clear that an accurate measurement may not necessarily be precise & vice versa.

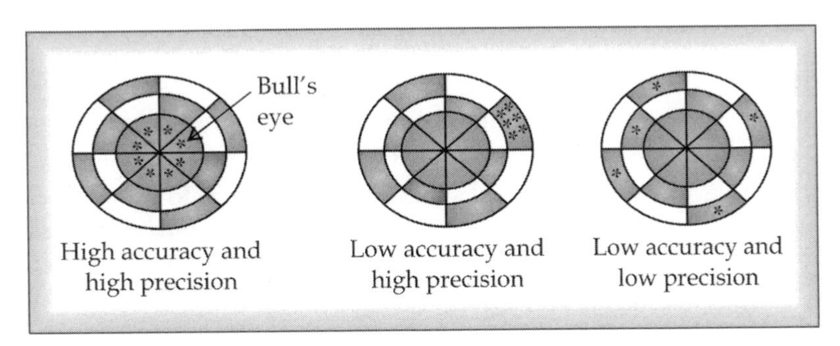

Fig. 2.2 Difference between Accuracy and Precision

2.3.5 Calibration

The value of the error and consequently the correction to be applied is determined by making a periodic comparison of the instrument with standard which are known to be constant. The procedure laid down for making, adjusting or checking a scale, so that readings of an instrument conform to an accepted standard is called the calibration. The graphical representation of the calibration record is called calibration curve which relates standard values of input or measurand to actual values of output throughout the operating range of the instruments.

A comparison of the instrument reading may be made with either

(i) A primary standard
(ii) Secondary standard of accuracy greater than the instrument to be calibrated, or
(iii) A known input source.

More Instruments for Learning Engineers

Mile 2.2

Strobo tuners, basically stroboscopes are most accurate tuning devices, which can tune any instrument, including the beating of steel-pan drums, harp, violin, guitar, bagpipes, accordions, calliopes, bells, the pins in Music Boxes or any audio device much accurate than regular LED/LCD tuners. These are costly and need periodic servicing and so, used by specialists and professionals.

Strobo Tuner

For example, we may calibrate a pressure gauge by comparing it with a standard pressure gauge at the National Bureau of Standards; by comparing it with another pressure gauge which has already been compared with a primary standard or by direct comparison with a primary measurement. The calibration standard should be one step accurate than the instrument being calibrated.

The following points are to be considered while calibrating any instrument.

(i) Calibration of the instrument must be carried out with the instrument in the same position and environmental conditions under which it is to operate while in service.

(ii) The instrument is calibrated with values of input signals both increasing and decreasing. The results are then expressed, graphically, with output as ordinate and input as abscissa.

(iii) Output readings for a series of input signals going up the scale may not agree with the output readings for the same input signals when going down.

(iv) Lines or curves plotted in the graphs mayor may not close to form a loop.

In a typical calibration curve as shown in Fig. 2.3, OPQ represents the readings obtained while RST represents the readings during descent; KLM represents the median and is commonly accepted as the calibration curve. The term median refers to the mean of a series of up and down readings. Quite often, the indicated values are plotted as abscissa and the ordinate represents the variations of the median from the true values.

A properly prepared calibration curve gives information about the absolute static errors of the measuring device, the extent of the instruments linearity or conformity, and the hysteresis and repeatability of the instrument.

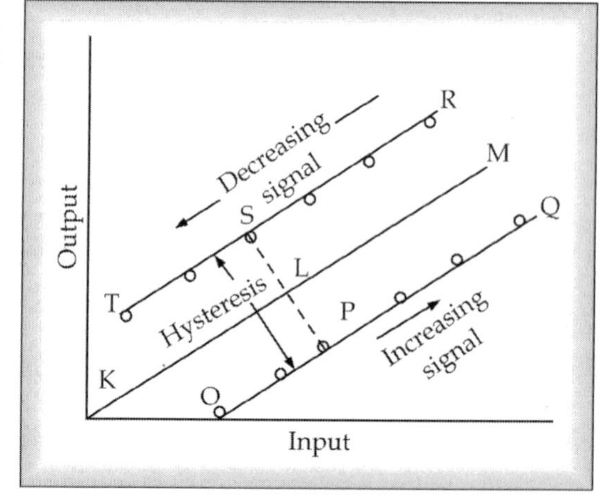

Fig. 2.3 Calibration Curve

2.3.6 Drift

It is a gradual variation in the instrument output over a period of time that is unrelated to changes in input, operating conditions or load. This can be caused by component's instability or temperature changes, wear, erosion, polarization, saturation and aging may also cause drift. It may occur in obstruction flow meters because of wear and erosions of the orifice plate, nozzle, or venturimeter. Drift occurs in thermocouples and resistance thermometers due to the contamination of the metal and a change in its atomic or metallurgical structure. Drift occurs very slowly and can be checked only by periodic inspection and prevented by periodic maintenance of the instrument.

2.3.7 Sensitivity

Sensitivity of an instrument is defined as the ratio of change in magnitude of the output signal to change in magnitude of input signal

$$\text{Static sensitivity, K} = \frac{\text{Change of output signal}}{\text{Change of input signal}}$$

Sensitivity is represented by the slope of the calibration curve if the ordinates are expressed in the actual units. With a linear calibration curve, the sensitivity is constant. However, if the calibration curve is non-linear, the static sensitivity is not constant and must be specified in terms of the input value as illustrated in Fig. 2.4.

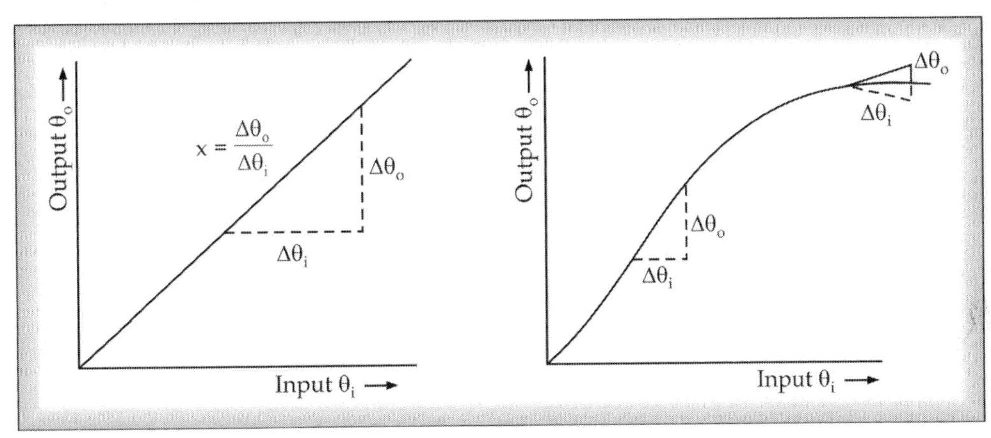

Fig. 2.4 Static Sensitivity for Linear and Non-linear Instruments

Sensitivity has a wide range of units, and these depend upon the instrument being investigated.

For example, the operation of a resistance thermometer depends upon a change in resistance to change in temperature and as such its sensitivity will have units of ohms/°C. It becomes easier to measure (or read) the output if the sensitivity of the instrumentation system is high.

Let the different elements comprising a measurement system, shown in Fig. 2.5 have static sensitivities of K_1 K_2 K_3 ... etc. When these elements are connected in series or cascades, then the overall sensitivity is worked out from the following relations:

θ_i → K_1 → θ_1 → K_2 → θ_2 → K_3 → θ_o = θ_i → $K_1 K_2 K_3$ → θ_o

Fig. 2.5 Overall System Sensitivity

$$K_1 = \frac{\theta_1}{\theta_i};\ K_2 = \frac{\theta_2}{\theta_1};\ K_3 = \frac{\theta_0}{\theta_2};$$

More Instruments for Learning Engineers

Dilatometer is a scientific instrument that measures volume changes caused by a physical or chemical process. A familiar application of a dilatometer is the mercury-in-glass thermometer, in which the change in volume of the liquid column is read from a graduated scale. Because mercury has a fairly constant rate of expansion over normal temperature ranges, the volume changes are directly related to temperature.

Dilatometer

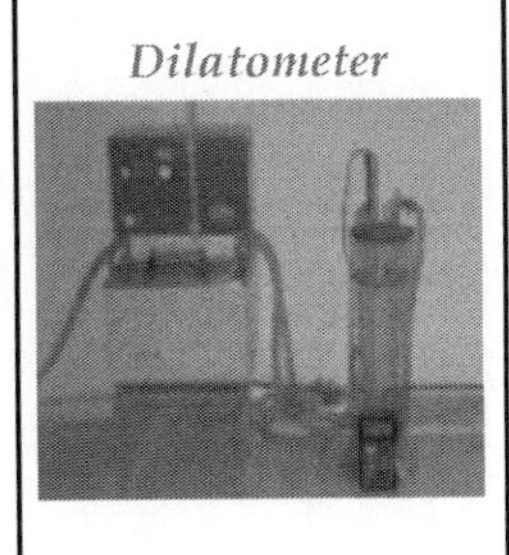

Overall sensitivity, $$K_1 = \frac{\theta_0}{\theta_i} = \frac{\theta_1}{\theta_i} \times \frac{\theta_2}{\theta_1} \times \frac{\theta_0}{\theta_2} = K_1 K_2 K_3$$

Therefore over all sensitivity = Product of individual sensitivities.

The above relation is based upon the assumption that no variation occurs in the value of the individual sensitivities K_1, K_2, K_3 etc., due to loading effects.

2.3.8 Threshold

When the input signal to an instrument is gradually increased from zero, there will be some minimum value of input before which the instrument will not detect any output change. This minimum value is called the threshold of the instrument. Hence it is defined as the minimum value of input below which no output can be detected. It must be caused by backlash or internal noise.

2.3.9 Resolution

It is defined as the smallest change in input signal or measured value which can be detected by the instrument. The least count of an instrument can be taken as the resolution of the instrument.

When the input signal is increased from non-zero value, one observes that the instrument output does exceed. This is termed as resolution or discrimination.

2.3.10 Tolerance

It is the range of inaccuracy which can be tolerated in measurement. In other words, it is the maximum permissible error. For example, the tolerance would be ± 1 % when an inaccuracy of ± 1 bar can be tolerated for 100 bar value of pressure i.e., any value between 99 to 101 bar is acceptable.

2.3.11 Readability and Least Count

The term readability indicates the closeness with which the scale of the instrument may be read. The term least count represents the smallest difference that can be detected on the instrument scale. Both readability and least count are dependent on length scale, spacing of graduations, size of the pointer and parallax effect.

2.3.12 Backlash

The maximum distance or angle through which any part of a mechanical system may be moved in one direction without applying appreciable force or motion to the next part in a mechanical system.

2.3.13 Zero Stability

A measure of the ability of the instrument to restore to zero reading after the measurand has returned to zero, and other variations (temperature, pressure, humidity, vibration etc.) have been removed.

2.3.14 Stiction (Static Friction)

Force or torque that is necessary just to initiate motion from rest.

SELF ASSESSMENT QUESTIONS-2.1

1. Distinguish between
   (a) Static and Dynamic quantities
   (b) Steady state periodic quantities and Transient quantities
   (c) Precision and accuracy
   (d) Range and Span
2. Describe different static characteristics which can characterize an instrument.
3. List out various static characteristics and explain any five in detail with examples.
4. Define and explain the following characteristics in a measurement system
   (a) Calibration (b) Resolution (c) Sensitivity (d) Threshold
   (e) Zero stability
5. How can you describe the nature and behavior of an instrument with reference to the following characteristics?
   (a) Accuracy (b) Precision (c) Drift (d) Threshold
   (e) Tolerance
6. Discuss the importance of the following characteristics in instrumentation
   (a) Sensitivity (b) Readability (c) Stiction (d) Backlash
   (e) Zero stability
7. Explain how the following can characterize a measurement system
   (a) Sensitivity (b) Tolerance (c) Least count (d) Calibration
   (e) Zero stability

2.4 DYNAMIC CHARACTERISTICS

Dynamic characteristics are those which vary in the time domain. The variation can be measured and is a function of time or some parameters. The following are common dynamic characteristics considered in the instrumentation.

2.4.1 SPEED OF RESPONSE AND MEASURING LAG

In a measuring instrument the speed of response or responsiveness is defined as the rapidity with which an instrument responds to a change in the value of the quantity being measured. Measuring

More Instruments for Learning Engineers

Mile 2.4

Eudiometer (in Greek *eúdios* means clear or mild) measures the change in volume of a gas mixture in a physical or chemical change. It consists graduated 50 and 100 ml cylinders closed at the top & bottom immersed in water or Hg. The liquid traps a sample of gas in cylinder and measures gas volume. Two platinum wires are placed in sealed end and an electric spark created to initiate reaction in gas mixture; the graduation on the cylinder reads the change in volume due to reaction. The use of the device is quite similar to the original barometer, except that the gas inside displaces some of the liquid used.

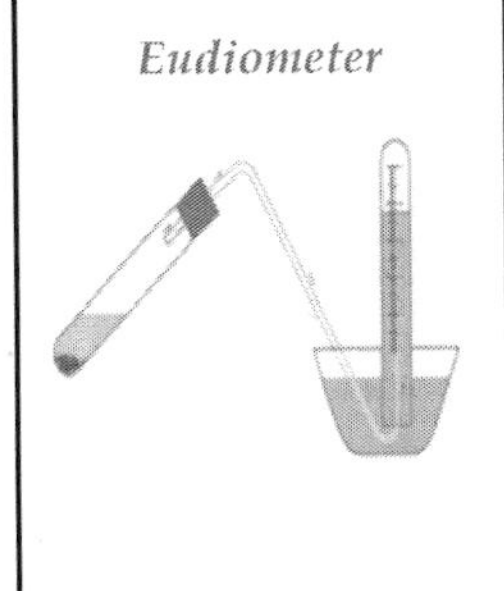

lag refers to retardation or delay in the response of an instrument to a change in the input signal. The lag is caused by conditions such as capacitance inertia or resistance.

2.4.2 Fidelity and Dynamic Error

Fidelity of an instrumentation system is defined as the degree of closeness with which the system indicates or records the signal which is impressed upon it. It refers to the ability of the system to reproduce the output in the same form as the input. If the input is a sine wave then for 100% fidelity, the output should also be a sine wave. The difference between the indicated quantity and the true value of the time varying quantity is the dynamic error; here static error of the instrument is assumed to be zero.

2.4.3 Overshoot

Because of mass and inertia, a moving part, i.e., the pointer of the instrument does not immediately come to rest in the final deflected position. The pointer goes beyond the steady state i.e., it overshoots as shown in Figure 2.6.

The overshoot is defined as the maximum amount, by which the pointer moves beyond the steady state.

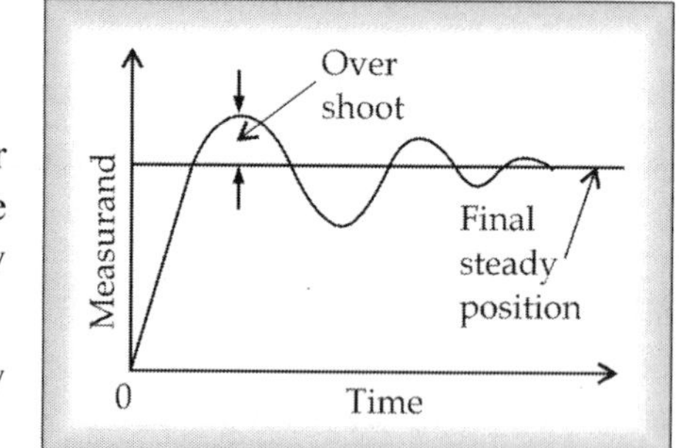

Fig. 2.6 Overshoot

2.4.4 Dead Time and Dead Zone

Dead time is defined as the time required for an instrument to begin respond to a change in the measured quantity. It represents: the time before the instrument begins to respond after the measured quantity has been altered. Dead zone defines the largest change of the measurand to which the instrument does not respond. Dead zone is the result of friction, backlash or hysteresis in the instrument.

Some of the dynamic terms are graphically shown in Figure 2.7 where the measured quantity and the instrument readings are plotted as function of time.

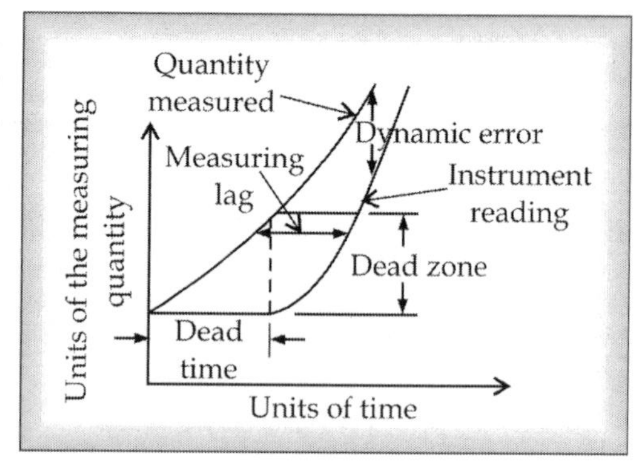

Fig. 2.7 Dynamic Terms

2.4.5 Frequency of Response

Maximum frequency of the measured variable that an instrument is capable of following without error. The usual requirement is that the frequency of measurand should not exceed 60 percent of the natural frequency of the measuring instrument.

2.5 MODELING FOR DYNAMIC PERFORMANCE

The dynamic response of an instrument is represented by a mathematical model. A mathematical model is a relationship between input and output of the instrument which is determined taking into account the physical parameters of its elements. The basic elements of mechanical and electrical systems are given in Table 2.1. There can be thermal, hydraulic, pneumatic and mixed system of

Table 2.1 Basic Elements of Mechanical Electrical Systems

System	Basic Elements
Mechanical	Mass, spring and a damper or dash pot
Electrical	Resistance, Inductance and a Capacitance

instruments. These systems of instruments may have completely different physical appearances, yet they can be analogous. Two systems are said to be analogous to each other if their differential equations or transfer functions are in identical form. The advantage is obvious, if the response of the physical system to a given input is determined, the response of all other analogous systems can be easily determined for the same input.

A generalized relationship for a second order system can be expressed as

$$A_2 \frac{d^2\theta}{dt^2} + A_1 \frac{d\theta_0}{dt} + A_0\theta_0 = B_0\theta_i$$

where θ_i = input signal; θ_0 = output signal

A_2, A_1, A_0, B_0 are constants and represent system parameters

The order of the system is the order of the differential equation representing: Above cited differential equation contains second order derivative, so it is second order system. The above concept can be easily understood by considering a mechanical system involving mass, spring and a damper (dash pot) as shown in Figure 2.8.

Let the above system be subjected to a force

$$F = F(t)$$

and let m = mass (kg)

K = stiffness of the spring (N/M)

C = damping constant $\left(\frac{N}{m/s}\right)$

x = displacement of the mass from the mean or equilibrium position.

More Instruments for Learning Engineers

Graphometer (or **Semicircle**)**,** introduced in Philippe Danfrie's, (Paris, 1597), popularized by French geodesists is a surveying instrument used for angle measurements having a 180° semicircular limb subdivided into minutes. The limb is subtended by the diameter with two sights at ends and a box and needle compass at middle. The alidade with two other sights are fitted and mounted on a staff with a ball and socket joint. Sometimes, another half-circle is graduated 180/360° (another line) on limb.

Graphometer

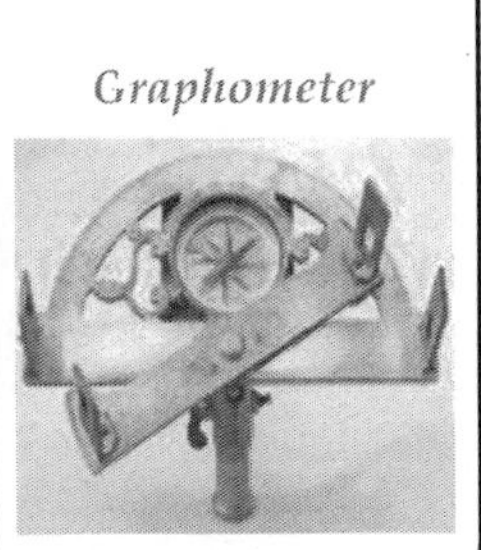

We know that velocity $= \frac{dx}{dt}$ and acceleration $= \frac{dx^2}{dt^2}$

Now, Applying Newton's law of motion,

$$m \frac{dx^2}{dt^2} = F(t) - kx - c \frac{dx}{dt}$$

$$m \frac{dx^2}{dt^2} + c \frac{dx}{dt} + kx = F(t)$$

For a rotating system, a similar equation can be desired as,

$$J \frac{d^2\theta_0}{dt^2} + c \frac{d\theta_0}{dt} + k\theta_0 = T(t)$$

where, J = mass moment of inertia (kg - m²)

K = torsional stiffness $\left(\frac{Nm}{rad}\right)$

c = damping constant $\left(\frac{Nm}{\frac{rad}{s}}\right)$

T (t) = torque

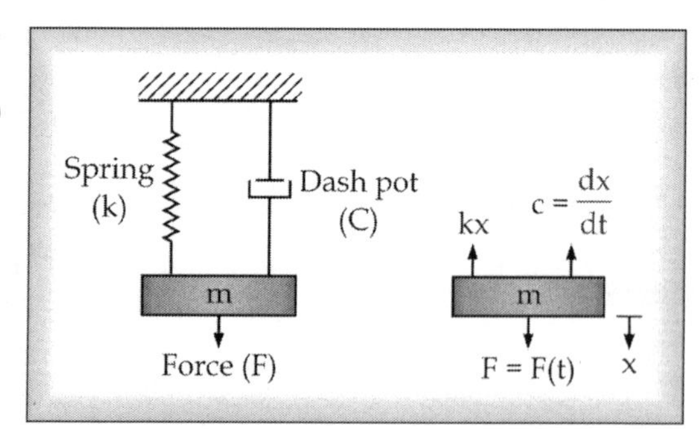

Fig. 2.8 Mechanical Model of a Force Excited Second Order System

Similarly, for electrical, thermal and other systems, equations can be derived using an appropriate law (Kirchhoff's law, Heat conduction / convection relationship etc) that governs the system.

Self Assessment Questions-2.2

1. Distinguish between
   (a) Static and Dynamic characteristics (b) Speed of response and measuring lag
   (c) Precision and accuracy (d) Fidelity and dynamic error
2. Describe different dynamic characteristics which can characterize an instrument.
3. List out various dynamic characteristics and explain any five in detail with examples.
4. Define and explain the following characteristics in a measurement system
   (a) Dead time and dead zone (b) Fidelity and dynamic error
   (c) Speed of response and measuring lag (d) Frequency of response
   (e) Overshoot
5. How can you describe the nature and behavior of an instrument with reference to the following characteristics?
   (a) Overshoot (b) Fidelity
   (c) Drift (d) Dynamic error
   (e) Backlash

6. Discuss the importance of the following characteristics in instrumentation
   (a) Sensitivity (b) Overshoot
   (c) Backlash (d) Zero stability
7. Distinguish between the following terms in the light of instrumentation
   (a) Static error and dynamic error
   (b) Overshoot and zero stability
   (c) Speed of response and frequency of response
   (d) Backlash and measuring lag
8. Explain how the following can characterize a measurement system
   (a) Stiction (b) Speed of response
   (c) Sensitivity (d) Zero stability

2.6 SYSTEMS OF VARIOUS ORDERS

2.6.1 Zero Order System

Zero order system is represented by a differential equation of zero order as,

$$A_0 \theta_0 = B_0 \theta_i$$

or

$$\theta_0 = \frac{B_0}{A_0}\theta_i \quad \text{or} \quad \theta_0 = k\,\theta_i$$

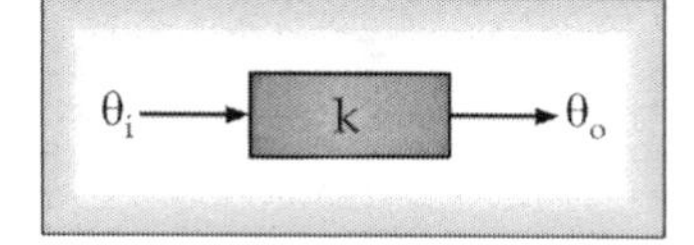

Fig. 2.9 Block Diagram of Zero Order System

where, k is the static sensitivity of the system and is the only parameter which characterizes a zero order system. It is a system where output is directly proportional to the input, no matter how input varies. Output is a true reproduction of the input. Examples of zero order system.

1. A mechanical lever (force is proportional to displacement) and
2. Linear electrical potentiometer (voltage proportional to the displacement of the wiper). Block diagram of zero order system is shown in Figure 2.9.

More Instruments for Learning Engineers

Protractor measures angles in degrees(°) or radians and is made of transparent plastic or glass in rectangular/ square/ circular/ semicircular shape. These are used in variety of engineering related applications, but most commonly in geometry lessons in schools. Some protractors are simple half-discs. More advanced protractors, such as the bevel protractor, have one or two swinging arms, which can be used to help measure the angle.

Protractor

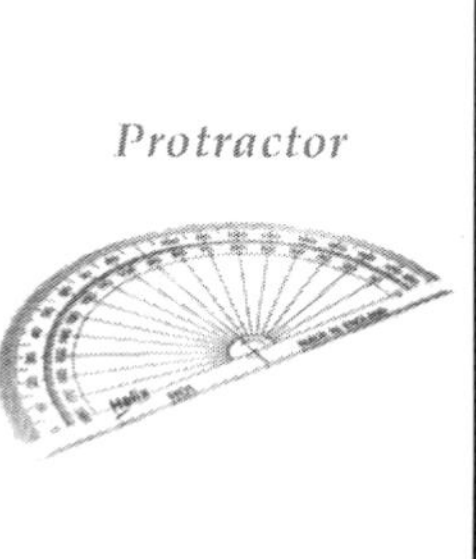

2.6.2 First Order System

It is a system is represented by a first order differential equation of the form

$$A_1 \frac{d\theta_0}{dt} = A_0\theta_0 = B_0\theta_i$$

or

$$\left(\frac{A_1}{A_0}\right)\left(\frac{d\theta_0}{dt}\right) + \theta_0 = \frac{B_0}{A_0}\theta_i$$

$$\tau\frac{d\theta_0}{dt} + \theta_0 = k\ \theta_i$$

where, $\tau = \frac{A_1}{A_0}$ and is called time constant

$K = \frac{B_0}{A_0}$ and is called static sensitivity

In terms of D-operator $\left(D = \frac{d}{dt}\right)$

$$\tau\ D\ \theta_0 + \theta_0 = k\ \theta_i$$

$$(\tau\ D + 1)\theta_0 = k\ \theta_i$$

$$\frac{\theta_0}{\theta_i} = \frac{k}{(\tau D + 1)}$$

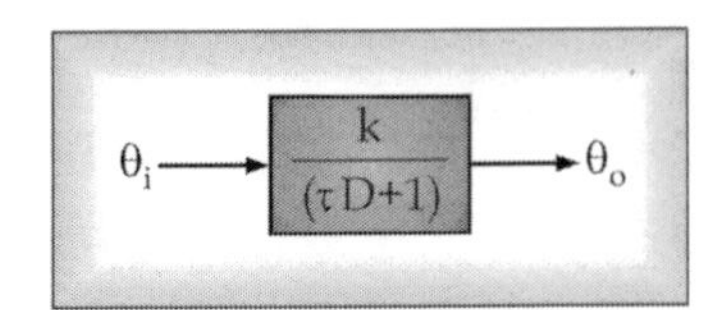

Fig. 2.10 Block Diagram of First Order System

Above equation represents the ***transfer-function*** operator for the first order system. Examples of first order systems are temperature measuring devices like thermocouple, mercury-in-glass thermometer and electrical network of resistance and capacitance.

A first order mechanical system is represented by a spring and a dash pot, assuming the mass to be zero. In this case time constant is

$$\tau = \frac{c}{k} = \frac{\text{Damping constant}}{\text{Stiffness of the spring}} = \frac{N/ms^{-1}}{N} / m = s\ (\text{second})$$

In a resistance capacitance circuit, $\tau = RC$

Block diagram of first order system is shown in Figure 2.10.

2.6.3 Second Order System

A second order system is represented by a differential equation of second order of the form.

$$A_2\frac{d^2\theta_0}{dt^2} + A_1\frac{d\theta_0}{dt} + A_0\theta_0 = B_0\theta_i$$

Taking the operator $D = \frac{d}{dt}$ and dividing the above equation by $A_{0.}$ we have

or $$\left[\left(\frac{A_2}{A_0}\right)D^2 + \left(\frac{A_1}{A_0}\right)D + 1\right]\theta_0 = \frac{B_0}{A_0}\theta_i$$

Above equation can be written as,

$$\left[\frac{D^2}{\omega_n^2} + \frac{2\xi}{\omega_n}D + 1\right]\theta_0 = K\theta_i$$

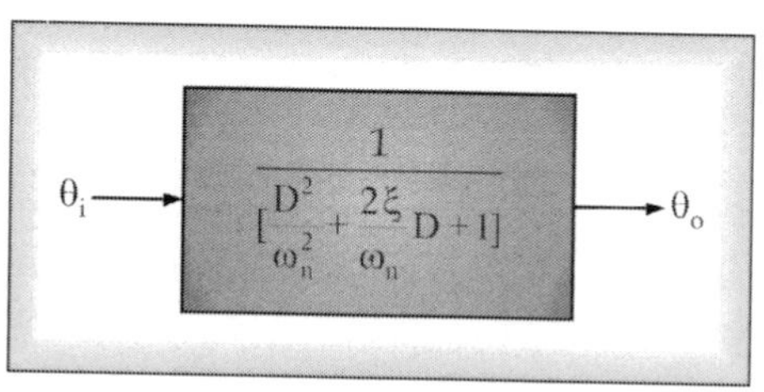

Fig. 2.11 Block Diagram of a Second Order System

The various terms are defined as,

$$\omega_n = \sqrt{\frac{A_0}{A_2}}, \text{ undamped natural frequency (rad/s)}$$

$$\xi = \frac{A_1}{2\sqrt{A_0 A_2}}, \text{ damping ratio} = \frac{\text{Actual damping}}{\text{Critical damping}}$$

$$k = \frac{B_0}{A_0}, \text{ Static sensitivity}$$

To understand the significance of these terms, reconsider the second order mechanical system as shown in Figure 2.8. The corresponding terms can be expressed as,

$$\omega_n = \sqrt{\frac{k}{m}},$$

$$\xi = \frac{c}{c_c} = \frac{c}{2\sqrt{km}}; c_c = 2\sqrt{km} \text{ (critical damping)}$$

Another term often used is

$$\omega_d = \omega_n\sqrt{1-\xi^2}, \qquad (\omega_d = \text{damped natural frequency})$$

More Instruments for Learning Engineers

Mile 2.7

Voltmeter measures potential difference between two points in a circuit and are available as analog & digital. It may be fixed in a panel to monitor or portable to measure current and resistance in the form of multimeter. Further, if any measurement (e.g.pressure, temperature, flow, level) can be converted to voltage, it is suitably calibrated to act as an instrument to measure such quantities.

Voltmeter

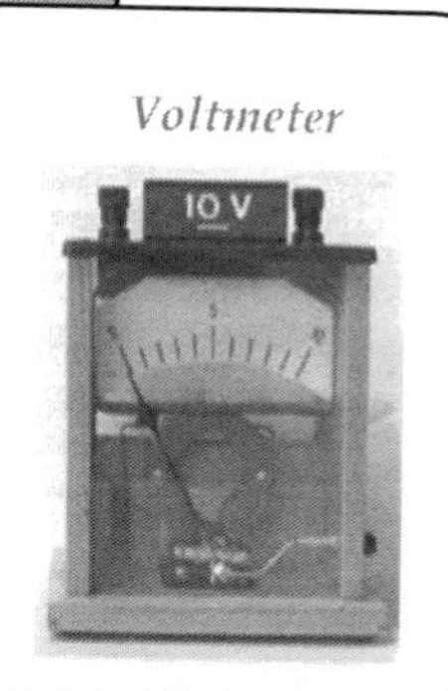

Seismic instruments used for acceleration measurement, pressure measuring devices, U-V galvanometer and R-L-C Circuit can be represented as second order systems.

Block diagram of second order system is shown in Figure 2.11.

2.7 STANDARD TEST SIGNALS

2.7.1 Step Signal

The step is a signal whose value changes from one level (usually zero) to another level A in zero time. The mathematical representation of the step function is

$$r(t) = A\,u(t)$$

where $u(t) = A, \text{ for } t \geq 0$

$= 0, \text{ for } t < 0$

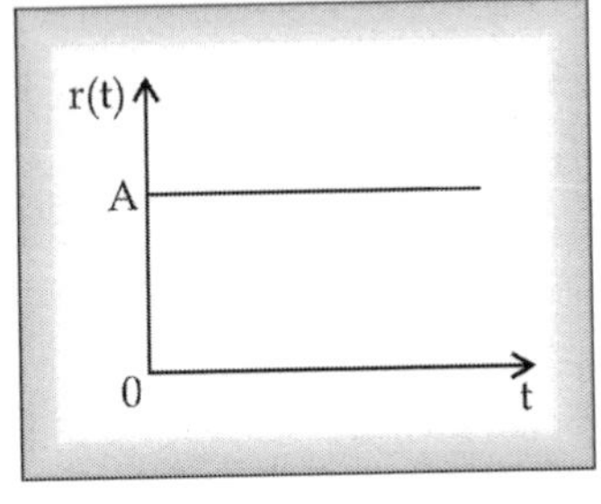

Fig. 2.12 Step Signal

In the form of Laplace transform, we have

$$R(s) = \frac{A}{s}$$

The graphical representation of a step signal is shown in Fig. 2.12.

The signal whose height is unity is called step signal.

2.7.2 Ramp Signal

The ramp is a signal which starts at a value of zero and increases linearly with time.

Mathematically,

$$r(t) = At, \text{ for } t \geq 0$$

$$= 0, \quad \text{for } t < 0$$

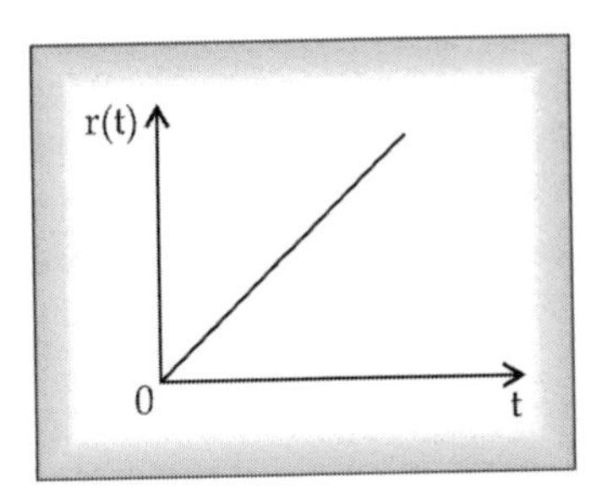

Fig. 2.13 Ramp Signal

In the form of Laplace transform we have

$$R(s) = \frac{A}{s^2}$$

The graphical representation of a ramp signal is shown in Fig. 2.13.

It is to be noted that a ramp signal is integral of a step.

signal (i.e., $\frac{A}{s}$)

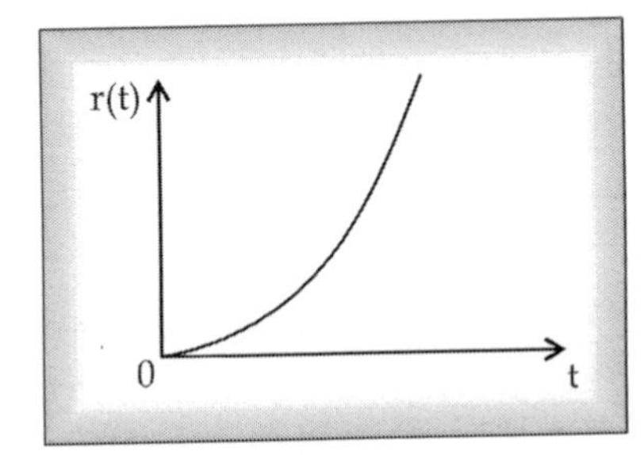

Fig. 2.14 Parabolic Signal

2.7.3 Parabolic Signal

The mathematical representation of this signal is

$$r(t) = \frac{At^2}{2}, \text{ for } t \geq 0$$

$$= 0, \quad \text{for } t < 0$$

In the form of Laplace transform

$$R(s) = \frac{A}{s^2}$$

The graphical representation of a parabolic signal is shown in Fig. 2.14.

It is to be noted that a parabolic signal is integral of a ramp signal.

2.7.4 Simple Harmonic Input (Sinusoidal)

It varies sinusoidal with constant maximum amplitude.

Mathematically

$$R(t) = A \sin \omega t, \quad \text{for } t \geq 0$$
$$= 0, \quad \text{for } t < 0$$

where A = Amplitude of input

ω = Circular frequency (rad/s)

$$R(s) = \frac{A\omega}{s^2 - \omega^2}$$

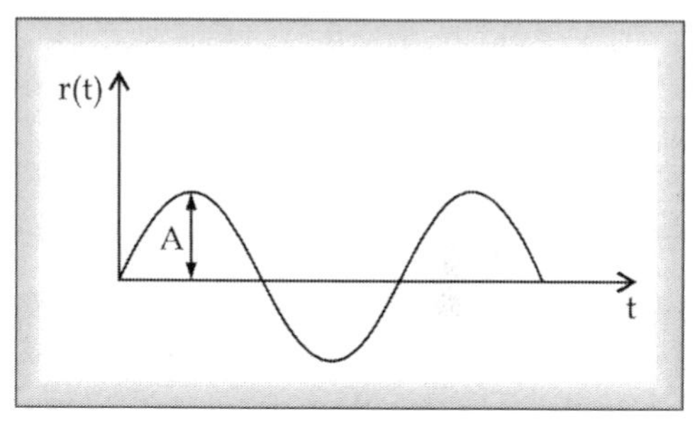

Fig. 2.15 Harmonic (Sinusoidal) Signal

The graphical representation is shown in Fig. 2.15.

If $r(t) = A \cos \omega t$ then $R(s) = \dfrac{A\omega}{s^2 - \omega^2}$

2.7.5 Impulse Signal

A unit-impulse is defined as a signal which has zero value everywhere except at $t = 0$, where its magnitude is infinite. It is generally called δ-function and has the following property.

$$\delta(t) = 0 \; ; \quad t \neq 0$$

$$\int_{-\epsilon}^{+\epsilon} \delta(t)\, dt = 1$$

where ϵ tends to zero.

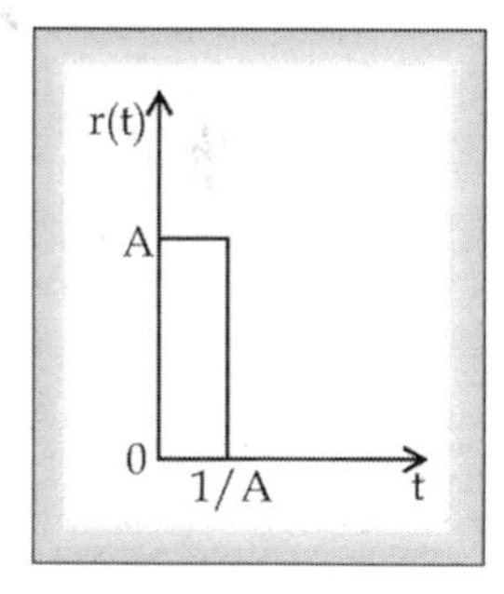

Fig. 2.16 Impulse Signal

More Instruments for Learning Engineers

Mile 2.8

Ammeter measures electric current in amperes (A), are available in micro and milli also. Early ammeters were laboratory instruments which relied on the Earth' s magnetic field for operation.

Multimeter can measure voltage, current and resistance also. All these instruments are based on the basic electric laws such as Ohm's law, Kirchoff's laws etc.

Ammeter

Multimeter

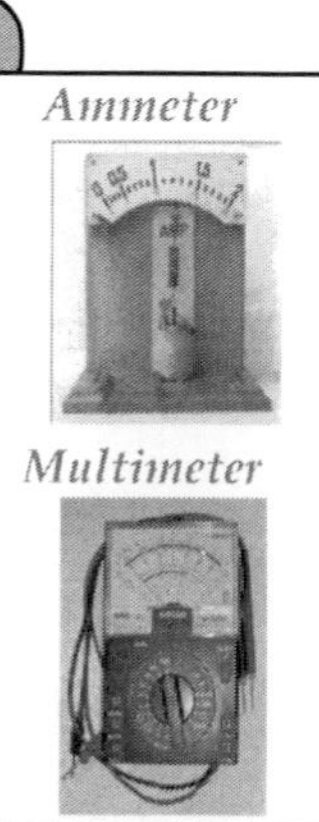

Practically a perfect impulse cannot be achieved. Hence, it is approximated by a pulse of small width of unit area as shown in Figure in 2.16.

The impulse function whose area is equal to unity is called the unit - impulse function of the Dirac delta function.

Mathematically an impulse function is the derivative of a step function i.e.,

$$\delta (t) = u(t)$$

The Laplace transform of a unit-impulse is

$$£ [\delta (t)] = 1$$

SELF ASSESSMENT QUESTIONS-2.3

1. Distinguish between zero, first and second order systems
2. With the help of the governing equation explain first order systems and describe its transfer function
3. With the help of the governing equation explain second order systems and describe its transfer function
4. What is a transfer function? Give the transfer functions of first and second order systems.
5. List out the standard types of signals and explain briefly.
6. Write short notes on
   (a) Step signal
   (b) Ramp signal
   (c) Parabolic signal
   (d) Sinusoidal signal
   (e) Impulse signal
7. Deduce the transfer function of first order and second order systems from their respective governing equations.

2.8 DYNAMIC RESPONSE OF FIRST ORDER SYSTEM

2.8.1 UNIT STEP RESPONSE OF FIRST ORDER SYSTEM (STEP INPUT)

The governing equation for first order system is

$$(\tau D + 1)\theta_0 = k\, \theta_i (t) \qquad .....(2.1)$$

where τ = time constant

k = static sensitivity

θ_0 = output signal (Function of time)

θ_i = input signal.

Assuming k = 1

$$(\theta D + 1)\, \theta_0 = \theta_i (t) = \theta_i \qquad .....(2.2)$$

Step Input $\theta_i = 0$, for $t < 0$

$\theta_i = Q$, for $t \geq 0$

(a) **Complementary functional /Transient response:** It is obtained from the auxiliary equation that is formed by equating the input θ_i; equal to zero.

∴ Substituting $\theta_i = 0$ in eqn. (2.2)

$$(\tau D = 1)\,\theta_0 = 0$$

The solution is assumed as

$$\theta_0 = A\,e^{mt} \quad \text{where m is an algebraic variable}$$

Substituting $\theta_0 = A\,e^{mt}$ in equation (2.2)

$$(t + \tau D)\,A\,e^{mt} = 0$$

$$A\,e^{mt} + \tau \frac{d}{dt}(Ae^{mt}) = 0$$

$$A\,e^{mt} + \tau A\,me^{mt} = 0 \quad \left[\because \frac{d}{dt}e^{mt} = me^{mt}\right]$$

$$A\,e^{mt}(1 + m\,\tau) = 0$$

∴ $1 + m\,\tau = 0$ but $A\,e^{mt} \neq 0$

Hence, $m = -\dfrac{1}{\tau}$

∴ $\theta_0 = A\,e^{-t/\tau}$

It remains the same for different standard inputs.

(b) Particular integral/Steady state response: This is different for different inputs, which is got from

$$(1 + \tau D)\,\theta_0 = \theta_i$$

$$\theta_0 = (1 + \tau D)^{-1}\theta_i$$

If $\theta_0 = A$

Then $\dfrac{d}{dt}\theta_0$ or $D\,\theta_0 = 0$, putting this value in the equation (2.1)

$$(\tau.0 + A) = \theta_i$$

More Instruments for Learning Engineers

Quadrat is a small plot used in ecology and geography to isolate a standard unit of area for study of the distribution of an item over a large area. While originally rectangular, modern quadrats can be rectangular, circular and even irregular. The quadrat is used for sampling plants, slow-moving animals (such as millipedes, snails and insects) and some aquatic organisms. When an ecologist wants to know how many organisms there are in a particular habitat, it would not be feasible to count them all. Instead, sampling of plants or animals that do not move much (such as snails), can be done using a sampling square called a quadrat. A suitable size of a quadrat depends on the size of the organisms being sampled. For example, to count plants growing on a school field, one could use a quadrat with sides 0.5 or 1 meter in length.

Quadrat

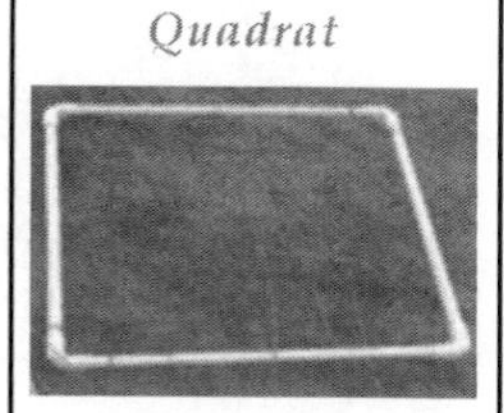

or $A = \theta_i$

or $A = \theta_0 = \theta_i$

Total solution is,

$$\theta_0 = A e^{-t/\tau} + \theta_i \qquad \text{.....(2.3)}$$

Applying the initial conditions, at $t = 0$, $\theta_0 = 0$

$$0 = A e^0 + \theta_i \Rightarrow 0 = A + \theta_i$$

$$A = -\theta_i$$

Substituting $A = -q_i$ in eqn. (2.3)

$$\theta_0 = \left[\theta_i - \theta_i e^{-t/\tau}\right] \quad \text{or} \quad \theta_0 = \theta_i\left[1 - e^{-t/\tau}\right] \qquad \text{.....(2.4)}$$

If input is not zero at $t = 0$ and has an initial value of

θ_A at $t = 0$, then

$$\theta_0 = \theta_i + (\theta_A - \theta_i)\, e^{-t/\tau}$$

Significance of Time Constant:

We have $\theta_0 = \theta_i\left[1 - e^{-t/\tau}\right]$

At, $t = \tau$,

$$\frac{\theta_0}{\theta_i} = 1 - e^{-1} = 1 - 0.368$$

$$\theta_0 = 0.632\,\theta_i$$

Or time constant (t) is defined as the time required reaching 63.2% of the steady state value.

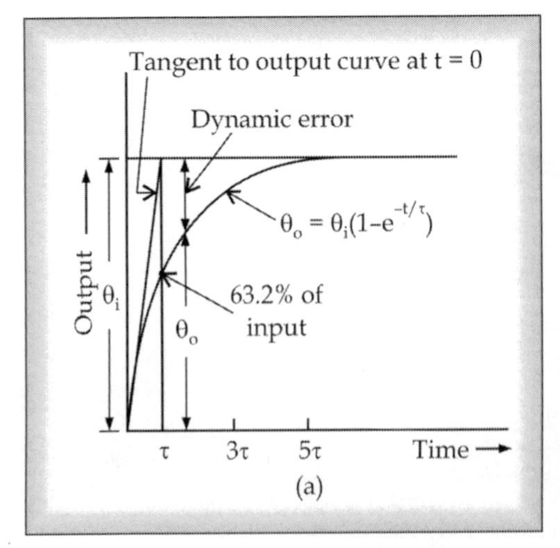

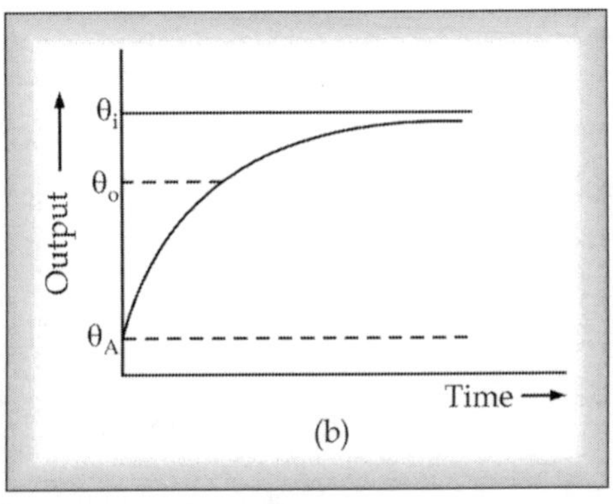

Fig. 2.17 Time Response of a First Order System to a Step Input

Similarly, for $t = 3\tau$,

$\theta_0 = 95\%$ of steady state value and for $t = 5\tau$, $\theta_0 = 99\%$ of steady state value is achieved.

It is often assumed that during five times constant, an instrument attains a steady state value or the process is completed as shown in Figure 2.17(a).

Time constant, indicates the speed of response. A larger time constant corresponds to a slow system response.

Time response of a first order system to a step input: Input $\theta_i = 0$.

In Fig. 2.17 (b), initial value of input is $\theta_i = \theta_A$ Second term $[\theta_i\, e^{-t/\tau}]$ as the RHS of the equation is time dependent and decreases in value as time increases.

$$\text{As} \qquad t \to \infty,\ \theta_0 = \theta_i$$

The vertical difference between input and response is the dynamic error.

The equation may represent either a progressive process, where output increases with time, or a decaying process, where the output decreases with time.

ILLUSTRATION- 2.1

A thermometer has a time constant of 3.44. It is quickly taken from a temperature 0 °C to a water bath having a temperature of 100 °C. What temperature will be indicated after 1.55 sec?

Solution:

We have

Unit step response of first order system (step input)

$$\frac{\theta_0}{\theta_i} = 1 - e^{-t/\tau}$$

where θ_0 = output signal, θ_i = input signal, τ = time constant, t = time

$$\theta_0 = \theta_i\left[1 - e^{-t/\tau}\right] = 100\left[1 - e^{-\frac{1.55}{3.44}}\right] = 36.27\ °C$$

∴ The temperature indicated by thermometer after 1.55 sec is 36.27 °C.

ILLUSTRATION- 2.2

A response test on a thermometer was thrust into temperature controlled bath of water maintained at 100 °C and the time was observed as the indicated temperature reached preselected values giving the following readings.

Time (sec)	0.0	1.2	3.0	5.6	8.0	11.0	15.0	18.0
Temp. (deg C)	20	40	60	80	90	95	98	99

Draw the response curve on a graph paper and show that it follows closely the form of a simple lag with a time constant of 4 seconds.

Solution:

We have

Unit response of first order system (step input)

$$\frac{\theta_0}{\theta_1} = 1 - e^{-t/\tau}$$

where θ_0 = output signal, θ_i = input signal

τ = time constant, $\theta_0 = \theta_i\left(1 - e^{-t/\tau}\right)$

Time constant (τ) = 4 sec

Temperature after

$$t = 0 \text{ sec}, \theta_0 = 100\left(1 - e^{-0/4}\right) = 0\ °C$$

$$t = 1.2 \text{ sec}, \theta_0 = 100\left(1 - e^{-1.20/4}\right) = 25.72\ °C$$

$$t = 3.0 \text{ sec}, \theta_0 = 100\left(1 - e^{-3/4}\right) = 52.76\ °C$$

$$t = 5.6 \text{ sec}, \theta_0 = 100\left(1 - e^{-5.6/4}\right) = 75.34\ °C$$

$$t = 8.0 \text{ sec}, \theta_0 = 100\left(1 - e^{-8/4}\right) = 86.47\ °C$$

$$t = 11.0 \text{ sec}, \theta_0 = 100\left(1 - e^{-11/4}\right) = 93.61\ °C$$

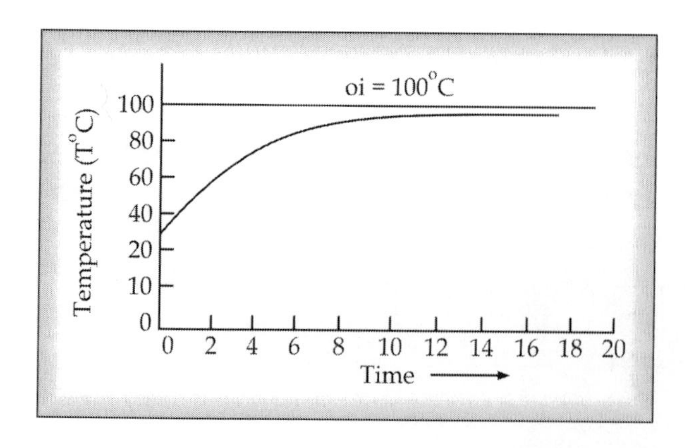

Fig. 2.18 Graphical Representation of Illustration-2.2

$$t = 15.0 \text{ sec}, \theta_0 = 100\left(1 - e^{-15/4}\right) = 97.65\ ^\circ C$$

$$t = 18.0 \text{ sec}, \theta_0 = 100\left(1 - e^{-18/4}\right) = 98.89\ ^\circ C$$

If actual temperature is as indicated by thermometer and calculated values are compared, it can be easily observed that it follows a simple lag (see Figure 2.18).

ILLUSTRATION- 2.3

A thermometer has been suddenly plunged into a steaming water bath whose temperature remains steady at 100 °C. It takes 10 seconds for the thermometer to reach the equilibrium condition, which occurs at five time constant ($t = 5\tau$). Calculate the time constant and the time taken by the thermometer to indicate half of the temperature difference. The initial thermometer temperature can be considered to be zero.

Solution:

We have

Unit step response of first order system (step - input)

$$\frac{\theta_0}{\theta_i} = 1 - e^{-t/\tau}$$

where θ_0 = output signal, θ_i = input signal, t = time constant,

$$\theta_0 = \theta_i\left(1 - e^{-t/\tau}\right)$$

Equilibrium condition is reached after 10 sec's which occurs at 5 time constant.

$$t = 5\,\tau$$

$$\tau = \frac{t}{5} = \frac{10}{5} = 2$$

Now, θ_0 = half of initial temperature

$$= \frac{100}{2} = 50\ ^\circ C$$

More Instruments for Learning Engineers

Caliper (or **calliper**) measures the distance between two opposite sides of an object. A caliper is like a compass with inward/outward-facing points. The tips of the caliper are adjusted to fit on the points to be measured, and is then removed and read on a ruler. It is used in many fields such as metalwork, forestry, woodwork, science, medicine and many more civil/ mechanical engineering applications.

Caliper

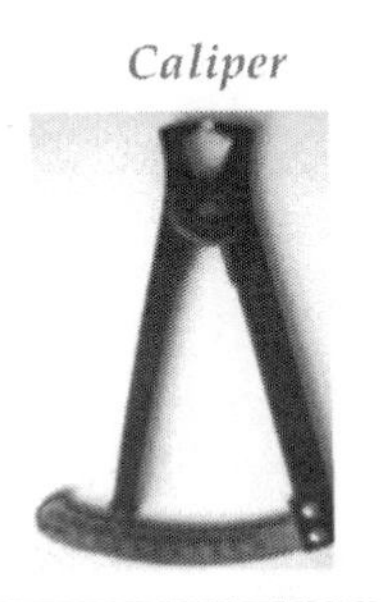

$$50 = 100\left(1 - e^{-t/2}\right)$$

$$0.5 = 1 - e^{-t/2}$$

$$-e^{-t/2} = 0.5 - 1$$

$$+e^{-t/2} = 0.6$$

$$\log_e e^{-t/2} = \log_e 0.5 = -0.653$$

$$-t/2 = -0.653$$

$$t = 1.366 \text{ sec.}$$

2.8.2 Unit Ramp Response of First Order System (Ramp Input)

Ramp input signal is represented as,

$$\theta_i = 0, \quad \text{for } t < 0$$

$$\theta_i = \Omega t, \quad \text{for } t \geq 0,$$

where Ω is the slope of input Vs time relation.

Governing differential equation becomes

$$(\tau D + 1)\,\theta_0 = \Omega t$$

or $$\tau D \theta_0 + \theta_0 = \Omega t \qquad .....(2.5)$$

Complimentary function/Transient part is of the form,

$$\theta_0 = A\, e^{-t/\tau} \qquad .....(2.6)$$

(Remained same as given earlier) (2.6)

Particular interval/steady state response

$$\theta_0 = At + R, \text{ where A and R are constants} \qquad .....(2.7)$$

$$\frac{d}{dt}\theta_0 = A \quad \text{or} \quad D\,\theta_0 = A \qquad .....(2.8)$$

Substituting eqn. (2.7) and eqn. (2.8) in eqn. (2.5)

$$A\tau + At + R = \Omega t \text{ or } Qt + R = \Omega t - Q\tau \qquad .....(2.9)$$

Equating coefficients,

$$A = \Omega$$

and $$R = -A\,\tau = \Omega\,\tau$$

Hence, $$\theta_0 = \Omega t - \Omega \tau \qquad .....(2.10)$$

Total solution is the sum of (2.6) and (2.10),

$$\theta_0 = A\, e^{-t/\tau} + \Omega t - \Omega \tau$$

When, $t = 0$, $\theta_0 = 0$, substituting

$$0 = A\,e^{0} - \Omega\,\tau \text{ hence, } A = \Omega\,\tau$$

So, $$\theta_0 = \Omega\,\tau\,e^{-t/\tau} + \Omega\,t - \Omega\,\tau$$

$$\theta_0 = \Omega\,t - \Omega\,\tau + \Omega\,\tau\,e^{-t/\tau}$$

Output error $$\theta_i - \theta_0 = \Omega\,t - (\Omega\,t - \Omega\,\tau + \Omega\,\tau\,e^{-t/\tau}) \text{ as, } \theta_i = \Omega\,t$$

$\therefore$ Output error $\theta_0 = \Omega\,\tau + \Omega\,\tau\,e^{-t/\tau}$

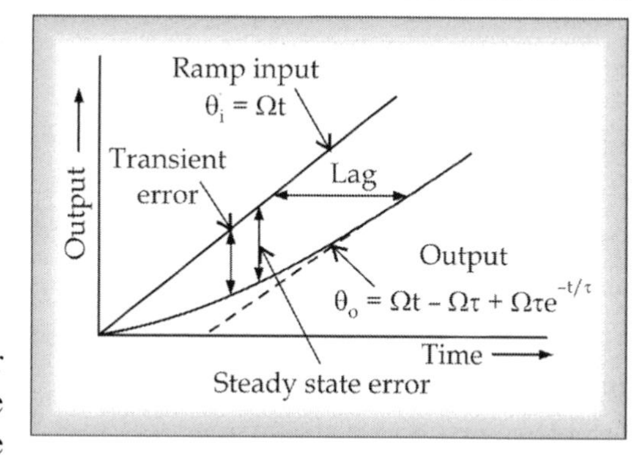

Fig. 2.19 Response for first Order System to Ramp Input

Figure 2.19 shows the response of a first order system to a ramp input. The output is seen to be lagging behind the input. The temperature $\Omega\,t$ of the output increases with time. The term is independent of time and shall continue to exist and is called the steady state error. The term $\Omega\,\tau\,e^{-t/\tau}$ depends on time and decreases with time.

2.8.3 Harmonic Sinusoidal Response of First Order System

Governing equation is

$(1 + \tau\,D) = \theta_i$

Input is of the form

$\theta_i = A_i \sin \omega\,t$ (shown in Fig. 2.20)

where

A_i =Amplitude of input

ω = Circular frequency of the input.

Total solution is of the form

$$\theta_0 = \Pr e^{-t/\tau} + \frac{A_i}{\sqrt{1+(\omega\tau)^2}}\sin(\omega t - \varphi)$$

Steady state solution is,

$$\theta_0 = P\ e^{-t/\tau} + \frac{A_i}{\sqrt{1+(\omega\tau)^2}}\sin(\omega t - \varphi)$$

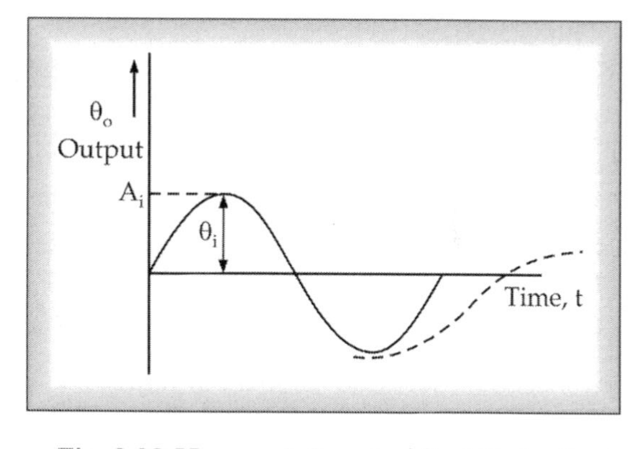

Fig. 2.20 Harmonic Input of first Order System

More Instruments for Learning Engineers

Mile 2.11

Steel square, (or framing square) is tool used by carpenter/ mason/ brick layer to measure if an object is square/right-angle. It is made of steel or aluminum, with graduated 1½"narrow long-arm called **tongue** and a short 2' wide arm (**blade** or base) connected at an angle of 90 degrees in 'L' shape. It is used for laying out common rafters, hip rafters and stairs. The latest framing square has a diagonal scale, board foot scale and an octagonal scale and also degree conversions for different pitches and fractional equivalents.

Steel Square

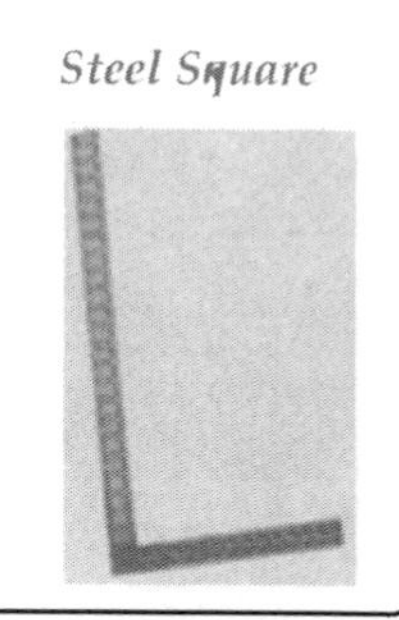

The amplitude of output θ_0 is A_o and is related as

$\theta_0 = A_0 \sin(\omega t - \varphi)$

$$\frac{A_0}{A_i} = \frac{1}{\sqrt{1+(\omega\tau)^2}}$$

For $$\omega = \frac{1}{\tau},\ \frac{A_0}{A_i} = \frac{1}{\sqrt{2}} = 0.707$$

and is called break point frequency.

Above relation is the ratio of amplitudes of output, and input and φ is the phase lag of output given by,

$$\varphi = \tan^{-1}(\omega\,\tau)$$

Figure 2.21 shows the response of first order system to a sinusoidal input. It can be seen that output is also sinusoidal and lags the input by a phase angle 'φ'. Both amplitude and phase lag depend on the product of input frequency or excitation frequency ù and time constant of the system. The response will approach an ideal value if the product ($\omega\,\tau$) is small.

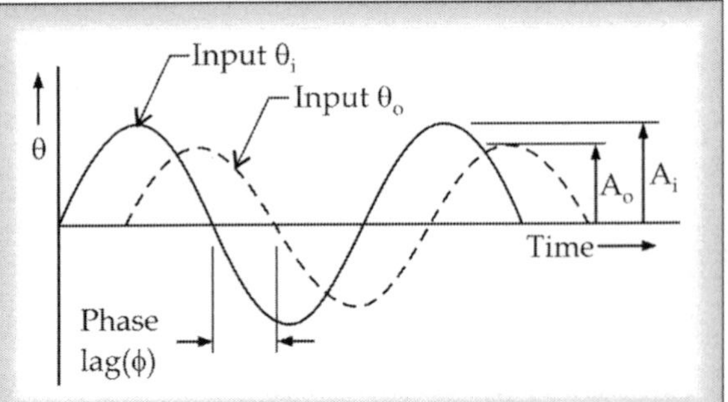

Fig. 2.21 Response to first Order Sinusoidal Input

2.9 DYNAMIC RESPONSE OF SECOND ORDER SYSTEM

Governing equation for the second order system as shown earlier is,

$$A_2 \frac{d^2\theta_0}{dt^2} + A_1 \frac{d\theta_0}{dt} + A_0\theta_0 = B_0\theta_i$$

This can be written as

$$\left[\frac{D^2}{\omega_n^2} + \frac{2\xi}{\omega_n} D + 1\right]\theta_0 = k\theta_i$$

Let "k" assumed to be unity.

The complimentary function or the transient response can be obtained from the auxiliary equation (by setting $\theta_i = 0$) and replacing operator "D" by an algebraic variable "x".

The auxiliary equation is

$$\frac{1}{\omega_n^2} x^2 + \frac{2\xi}{\omega_n} x + 1 = 0$$

The roots of this quadratic equation are,

$$x = -\xi\omega_n \pm \omega_n \sqrt{(\xi^2 - 1)}$$

The transient solution is of the form,

$$(\theta_0)_{transient} = Pe^{x_1 t} + Qe^{x_2 t}$$

where P and Q are arbitrary constants to be determined from the initial conditions and x_1 and x_2 are the roots of the auxiliary equation.

The roots x_1 and x_2 can be (i) real and different (ii) real and equal or (iii) Imaginary depending on the value of damping ratio (ξ).

The transient solution thus depends upon the value of damping ratio ξ with respect to unity. The three cases are possible. They are

$\xi > 1$: Over damped

Roots x_1 and x_2 are real and unequal. The response to the final value is non oscillatory but sluggish.

$\xi = 1$: Critical damped

Roots are real and equal. The response to the final value is smooth, quick and non-oscillatory.

$\xi < 1$: Under damped

Roots are imaginary. The response is oscillatory and the system takes long time to reach final value. Most instruments are designed with this type of damping and are under damped.

The total response or output is the sum of particular integral and complimentary function corresponding to the given damping condition. Most instruments are designed to keep **'the damping ratio' in between 0.65 to 0.75**. Generally, it is the study state response which is of main interest.

2.9.1 Step Response of Second Order System

We already know how to get complimentary functions.

The particular integral for the step input,

$\theta_i = 0$ for $t < 0$

θ_i = constant for $t \geq 0$

If $\theta_0 = \theta_i$

More Instruments for Learning Engineers

A **cathetometer** measures vertical distances if scale cannot be placed close to the points. It consists of a graduated scale and a horizontal telescope movable up/down along a rigid vertical column equipped with Vernier scale. After leveling, the cross hair in the eyepiece of the horizontal telescope is brought into coincidence with the images of points and the positions are noted. The difference is the required distance. It is used to read the liquid levels in a capillary, such as in measuring surface tension and also used to know the changes in liquid level in dilatometer due to a chemical reaction.

Cathetometer

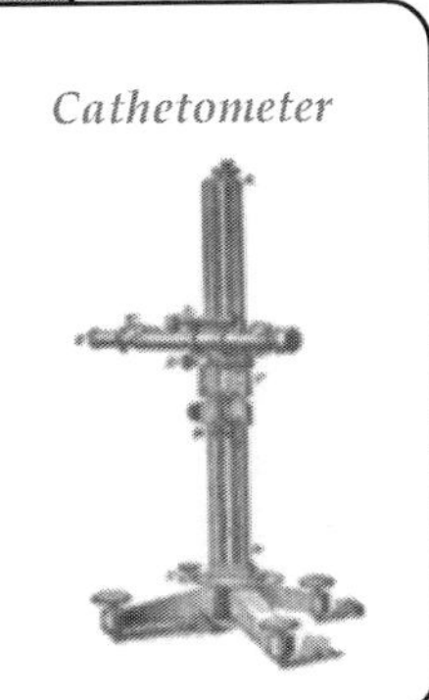

The total solution or output for an under damped system, after applying initial condition can be written as,

$$\theta_0 = k\,\theta_i\left[1 - e^{-\xi\omega_n t} + \left\{\cos\omega_d t + \frac{\xi}{\sqrt{1+\xi^2}}\sin\omega_d t\right\}\right]$$

$$\omega_d = \omega_n\sqrt{(1-\xi^2)}$$

(where ω_d = damped natural frequency)

Figure 2.22 shows the response of a second order system to a step input. For damping ratio less than unity the output is oscillatory and over shoots the step value, and the number of oscillations increases as ξ is reduced. After sufficient time, the output is virtually same as the input and there is no steady state error.

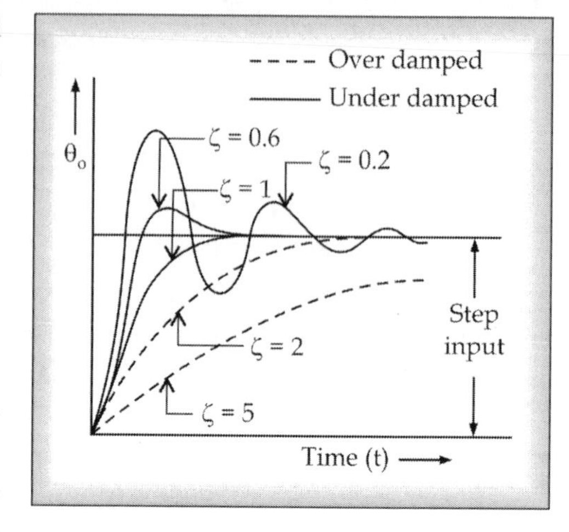

Fig. 2.22 Response of a Second Order System to a Step Input

For damping ratio more than unity, the response is non-oscillatory.

2.9.2 Harmonic/Sinusoidal Response of Second Order System

Harmonic input is expressed as,

$$\theta_i = A_i \sin \omega t$$

where A_i = Amplitude of input; ω = frequency of input or excitation frequency.

Output θ_0 is of the form

$$\theta_0 = A_o \sin(\omega t - \varphi)$$

where

A_o = Amplitude of output

ω = Frequency of excitation

φ = Phase lag between output and input.

ω_n = the natural frequency of the system.

ξ = Damping ratio

Amplitude ratio of output signal to input signal can be expressed as

$$\frac{A_0}{A_i} = \frac{\omega_n^2}{\sqrt{\left(\omega_n^2 - \omega^2\right) + \left(2\xi\omega\omega_n\right)^2}}$$

$$= \frac{1}{\sqrt{\left[1 - \left(\frac{\omega}{\omega_n}\right)^2\right]^2 + \left[2\xi\frac{\omega}{\omega_n}\right]^2}}$$

Let $\frac{\omega}{\omega_n} = \frac{\text{Excitation frequency (input)}}{\text{Natural frequency of system}} = r$

Then $\frac{A_0}{A_i} = k \frac{1}{\sqrt{[1-r^2]^2 + [2\xi r]^2}}$

For sensitivity k, $\frac{A_0}{A_i} = k \frac{1}{\sqrt{[1-r^2]^2 + [2\xi r]^2}}$

Phase lag $\varphi = \tan^{-1}\left[\frac{2\xi r}{1-r^2}\right]$

Figure 2.23(a) shows the variations of amplitude ratio (A_0/A_i) with input frequency at different damping ratio (ξ). It is a measure of system response to frequency. Ideally, it is expected to be constant with frequency or insensitive to frequency of excitation.

Figure 2.23(b) shows the phase lag φ with frequency ratio for various damping ratios. The phase shift should remain linear with frequency ratio, to ensure a proper time relationship between the sinusoidal components of a complex input.

In designing a second order instrument, both the amplitude response and phase response are used. Figure 2.23(a) shows that amplitude ratio reasonably constant only for a limited frequency range and for certain damping ratios. For dynamic measurement, a definite damping ratio must be used and frequency range should be established. If the damping ratio is kept between 65-75% and the frequency of excitation is kept at about 40% of the natural frequency of the system, then amplitude ratio of approximately unity can be expected.

Also phase shift is linear under these conditions. Some salient points of the steady state response, as evident from Fig. 2.23 (a) & (b) are

(i) As $\frac{\omega}{\omega_n} \rightarrow 0$ the amplitude ratio $\rightarrow 1$; then the phase lag $\varphi \rightarrow 0$

More Instruments for Learning Engineers

Mile 2.13

Altimeter or **altitude meter** measures the altitude of an object above a fixed level. The measurement of altitude is called **altimetry**, which is related to the term bathymetry, the measurement of depth underwater. Popular altimeters are

1. **Pressure altimeter:** Used in hiking/ climbing, in aircraft, in ground effect vehicle;
2. **Radar altimeter:** Used in GPS, in Satellites, transport systems;
3. **Sonic altimeter:** Used in air crafts

Altimeter

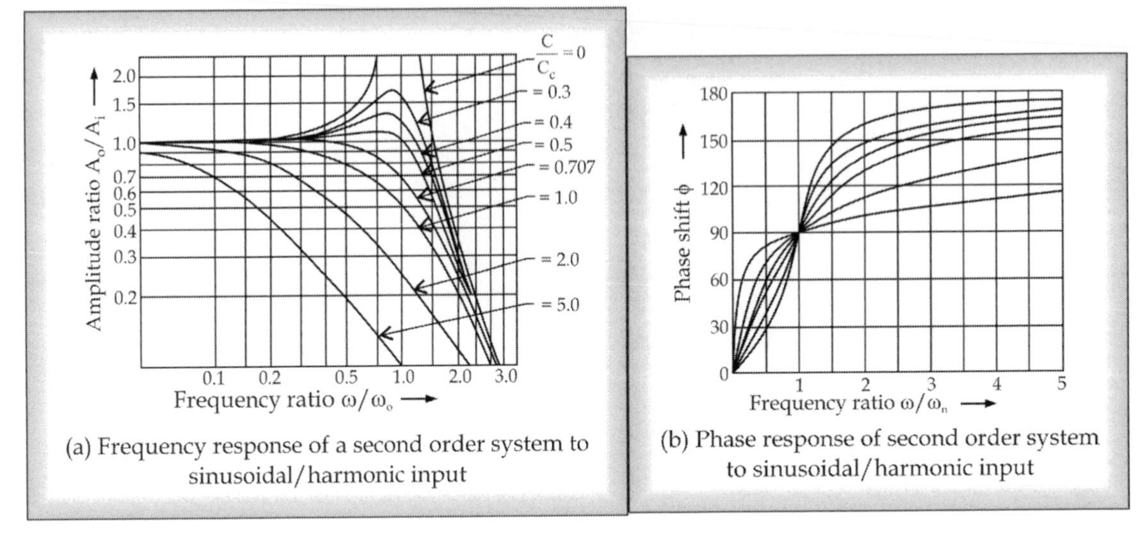

(a) Frequency response of a second order system to sinusoidal/harmonic input

(b) Phase response of second order system to sinusoidal/harmonic input

Fig. 2.23 (a)&(b) Frequency and Phase Responses of Second Order System to Harmonic Input

(ii) As $\frac{\omega}{\omega_n} \to \infty$, amplitude ratio $\to 0$; then phase lag $\varphi \to -180^\circ$

(iii) For un damped system ($\xi = 0$) when $\frac{\omega}{\omega_n} = 1$ amplitude ratio $\to \infty$;

then phase lag $\to$ ($-90°$) in all systems irrespective of the values of damping ratio ξ.

When $\frac{\omega}{\omega_n} = 1$ or $\omega = \omega_n$. This condition is called resonance condition.

(iv) Amplitude ratio is nearly unity for a limited range of exciting frequency ω.

Self Assessment Questions-2.4

1. Describe the response of a first order system for a unit step input signal in
   (i) Transient state (ii) Steady state
2. Discuss the significance of time constant for a step input in a first order system.
3. Explain and deduce the expression for the unit ramp response in a first order system.
4. Explain and deduce the expression for the harmonic response in a first order system.
5. Describe the response of a second order system for a unit step input.
6. Explain and deduce the expression for the unit ramp response in a second order system.
7. Explain and deduce the expression for the harmonic response in a second order system.
8. What is damping ratio? What is its significance with reference to the second order system?

SUMMARY

Characteristics are expressed in terms of certain physical quantities whose nature also will be either static or dynamic. Static characteristics are those characteristic which do not vary in the time domain, which include Range (region in which an instrument is designed to operate), Span (the difference of upper and lower limit), Accuracy (nearness to standard value), Precision (repeatability), Calibration (value of the error and consequently the correction to be applied), Drift (Gradual variation), Sensitivity (the ratio of change in output to that of input), Threshold (minimum value of input), Resolution (smallest detectable change in input), Tolerance (Range of permitted inaccuracy), Readability (closeness with which the scale is read), Least Count (smallest measurement), Backlash (maximum distance or angle movable in one direction without applying appreciable force or motion to the next part), Zero Stability (ability to restore to zero) and Stiction (Static friction).

The dynamic characteristics include Speed of Response (rapidity to respond to a change), Measuring Lag (Retardation or delay in the response), Fidelity (degree of closeness to indicate or record the signal), Dynamic Error (difference between the indicated and true value of time varying quantity), Overshoot (maximum amount, by which pointer moves beyond steady state), Dead Time (time required to begin to respond to a change), Dead Zone (largest change of measurand not responded) and Frequency Response (Maximum frequency of the measured variable without error).

Zero order System (zero order differential equation) is, $A_0\theta_0 = B_0\ \theta_i$ or $\theta_0 = \frac{B_0}{A_0}\ \theta_i$ or $\theta_0 = k\ \theta_i$ where, k is static sensitivity and is the only parameter to characterize zero order system. 1st Order System (1st order differential equation) is $A_1 \frac{d\theta_0}{dt} + A_0\,\theta_0 = B_0\,\theta_i$ (*transfer-function is* $\frac{\theta_0}{\theta_i} = \frac{k}{(\tau D+1)}$ 2nd Order System (2nd order differential equation) is $A_2 \frac{d^2\theta_0}{dt^2} + A_1 \frac{d\theta_0}{dt} + A_0\theta_0 = B_0\theta_i$ (*transfer-function is* $\left[\frac{D^2}{\omega_n^2} + \frac{2\xi}{\omega_n} D + 1\right]\theta_0 = K\ \theta_i$). With reference to these, Step, Ramp, parabolic, Simple Harmonic (Sinusoidal), and Impulse Signals are discussed in this chapter.

KEY CONCEPTS

Static quantities: Constant or vary very slowly with time which is negligible.

Dynamic quantities: Vary with time and based on variation with time.

Steady state periodic quantity: The dynamic quantity whose input magnitude has a definite repeating time cycle.

Transient quantity: The dynamic quantity whose input magnitude does not repeat with time.

Range: The region between the units in which an instrument is designed to operate for measuring a physical quantity, expressed by stating Lower and Upper limits.

Span: The algebraic difference between the upper limit and lower limit of the Instrument.

Accuracy: Conformity with an accepted standard value.

$$\text{Percentage of true value} = \frac{\text{Measured value} - \text{True value}}{\text{True value}}$$

$$\text{Percentage of full scale deflection} = \frac{\text{Measured value} - \text{True value}}{\text{maxscale value}}$$

Precision: The ability of the instrument to reproduce its readings over & over again for a constant input signal.

Calibration: The procedure laid down for making, adjusting or checking a scale, so that readings of an instrument conform to an accepted standard.

Calibration curve: The graphical representation of the calibration record which relates standard values of input or measurand to actual values of output throughout the operating range of the instruments.

Drift: Gradual variation in the instrument output over a period of time that is unrelated to changes in input, operating conditions or load.

Sensitivity: The ratio of change in magnitude of the output signal to that of input signal.

$$\text{Static sensitivity, } K = \frac{\text{Change of output signal}}{\text{Change of input signal}}$$

Threshold: The minimum value of input below which no output can be detected.

Resolution: The smallest change in input signal or measured value which can be detected by the instrument.

Tolerance: Range of inaccuracy which can be tolerated in measurement.

Readability: The closeness with which the scale of the instrument may be read.

Least count: The smallest difference that can be detected on the instrument scale.

Backlash: The maximum distance or angle through which any part of a mechanical system may be moved in one direction without applying appreciable force or motion to the next part in a mechanical system.

Zero stability: A measure of the ability of the instrument to restore to zero reading after the measurand has returned to zero, and other variations are removed.

Stiction (Static friction): Force or torque that is necessary just to initiate motion from rest.

Speed of Response: The rapidity with which an instrument responds to a change in the value of the quantity being measured.

Measuring lag: Retardation or delay in the response of an instrument to a change in the input signal.

Fidelity: The degree of closeness with which the system indicates or records the signal which is impressed upon it.

Dynamic error: The difference between the indicated quantity and the true value of the time varying quantity.

Overshoot: The maximum amount, by which the pointer moves beyond the steady state.

Dead time: The time required for an instrument to begin respond to a change in the measured quantity.

Dead zone: The largest change of the measurand to which the instrument does not respond.

Frequency response: Maximum frequency of the measured variable that an instrument is capable of following without error.

Zero order System: Represented by a differential equation of zero order as, $A_0\ \theta_0 = B_0\ \theta_i$ or $\theta_0 = \frac{B_0}{A_0}\theta_i$ or $\theta_0 = k\theta_i$ where, k is the static sensitivity of the system and is the only parameter which characterizes a zero order system.

First order system: Represented by 1st order differential equation of the form

$$A_1 \frac{d\theta_0}{dt} = A_0\theta_0 = B_0\ \theta_i \text{ (or the transfer-function is } \frac{\theta_0}{\theta_1} = \frac{k}{(\tau D+1)} \text{).}$$

Second order system: Represented by 2nd order differential equation of the form $A_2 \frac{d^2\theta_0}{dt^2} + A_1 + A_0\theta_0$ $= B_0\theta_i$ (or the transfer-function is $\left[\frac{D^2}{\omega_n^2} + \frac{2\xi}{\omega_n}D + 1\right]\theta_0 = K\theta_i$)

Step signal: A signal whose value changes from one level (usually zero) to another level A in zero time.

Ramp signal: A signal which starts at a value of zero and increases linearly with time.

Parabolic signal: Integral of a ramp signal.

Simple harmonic input (sinusoidal): Varies sinusoidal with constant maximum amplitude.

Impulse signal: A signal which has zero value everywhere except at $t = 0$, where its magnitude is infinite.

REVIEW QUESTIONS

SHORT ANSWER QUESTIONS

1. List out various static characteristics.
2. Discuss the following
   (a) Sensitivity (b) Readability (c) Stiction (d) Backlash
   (e) Zero stability

3. Describe different dynamic characteristics which can characterize an instrument.
4. Define the following
   (a) Dead time and dead zone
   (b) Fidelity and dynamic error
   (c) Speed of response and measuring lag
   (d) Frequency of response
   (e) Overshoot
5. List out the standard types of signals and explain briefly.
6. Write short notes on
   (a) Step signal
   (b) Ramp signal
   (c) Parabolic signal
   (d) Sinusoidal signal
   (e) Impulse signal
7. What is damping ratio? What is its significance with reference to the second order system?
8. What do you mean by the term calibration? Explain it.
9. Explain the following terms:
   (a) Speed of response (b) Sensitivity (c) Dead time (d) Dead zone
10. Discuss the relative features of static & dynamic measurements.
11. Explain the term calibration of an instrument?
12. What are the basic concepts in dynamic measurements?

Long Answer Questions

1. Distinguish between
   (a) Static and Dynamic quantities
   (b) Steady state periodic quantities and Transient quantities
   (c) Precision and accuracy
   (d) Range and Span
2. List out various dynamic characteristics and explain any five in detail with examples.
3. With the help of the governing equation explain first order systems and describe its transfer function
4. With the help of the governing equation explain second order systems and describe its transfer function
5. Discuss the significance of time constant for a step input in a first order system.
6. Explain and deduce the expression for the unit ramp response in a first order system.
7. Explain and deduce the expression for the harmonic response in a first order system.
8. Describe the response of a second order system for a unit step input.
9. Explain and deduce the expression for the unit ramp response in a second order system.
10. Explain and deduce the expression for the harmonic response in a second order system.
11. Discuss the significance of calibration in industrial measurement & control.
12. What are the different standard inputs for studying the dynamic response of a system? Define & sketch them.
13. A thermometer has a time constant of 3.44. It is quickly taken from a temperature of 0 °C to a water bath having temperature of 100 °C. What temperature will be indicated after 1.55 seconds?

MULTIPLE CHOICE QUESTIONS

1. The region between the units in which an instrument is designed to operate for measuring a physical quantity is called ________.
   (a) Range (b) Span
   (c) Resolution (d) Least count
2. The algebraic difference between the upper limit and lower limit of the Instrument is ______.
   (a) Range (b) Span
   (c) Resolution (d) Least count
3. ________is defined as the smallest change in input signal or measured value which can be detected by the instrument.
   (a) Range (b) Resolution
   (c) Span (d) Least count
4. _________indicates the closeness with which the scale of the instrument may be read.
   (a) Readability (b) Least count
   (c) Resolution (d) Span
5. ___________ represents the smallest difference that can be detected on the instrument scale
   (a) Span (b) Resolution
   (c) Readability (d) Least count
6. Force or torque that is necessary just to initiate motion from rest is called ___________.
   (a) Stiction
   (b) Zero Stability
   (c) Backlash
   (d) Span
7. For real equal and negative roots, the system response represents damping
   (a) Less than 1
   (b) Greater than 1
   (c) Critical
   (d) Zero
8. The damping ration is defined on the basis of ________ damping to ________.
   (a) Critical, actual
   (b) Actual, critical
   (c) Zero, critical
   (d) Critical, zero
9. The least change of the measured variable, which can be detected at the output of measuring system is
   (a) Sensitivity
   (b) Discrimination
   (c) Least count
   (d) Stability
10. The resolution of the system refers to
    (a) Difference between the true value of the input
    (b) Retardation of response
    (c) Smallest change in the Measurand that can be measured
    (d) Least count
11. The ratio of the output to input change for a given measuring is referred to as
    (a) Sensitivity
    (b) Linearity
    (c) Stability
    (d) Discrimination
12. Dynamic quantities
    (a) Vary rapidly with time
    (b) Remain constant over a period of time
    (c) Are displaced from zero position
    (d) Remain at zero position
13. The largest change in the measured variable, which produces no instrument response is known as
    (a) Dead band
    (b) Dynamic error
    (c) Fidelity
    (d) Threshold

14. A step-input signal
    (a) Starts from zero and raises uniformly with time
    (b) Changes rapidly with time
    (c) Starts from zero, raises to a particular value and returns back to zero
    (d) Starts form zero and raises to a value in zero time

15. The speed of response of a first-order system is judged by
    (a) Time constant
    (b) Transient response
    (c) Steady state value
    (d) All of the above

16. When the driving frequency approaches the system natural frequency, there is a rapid increase in the output amplitude of an undamped second order system. This condition is known as
    (a) Vibration (b) Instability
    (c) Resonance (d) Oscillation

17. The time taken by an under-damped second order system to reach peak output in response to step input is known as
    (a) Settling time
    (b) Response time
    (c) Rise time
    (d) Time constant

18. A damping ratio of 0.7 is described for the measuring instruments because
    (a) it gives minimum possible settling time
    (b) it gives the highest sensitivity
    (c) no resonance occurs at this value
    (d) no overshoot occurs when step input signal is applied

19. The__________characteristics pertaining to the performance of an instrument when the quantities to be measured are _________ or vary slowly with time.
    (a) Constant, static
    (b) Static, constant
    (c) Constant, constant
    (d) Static, static

20. The accuracy of a measuring system is the _______ with which the readings given by the instrument approach the ________ values of the quantities being measured.
    (a) Closeness, true
    (b) Constant, static
    (c) Exactness, static
    (d) Closeness, dynamic

21. Accuracy can be specified in terms of percentage of the value percentage of ____ deflection.
    (a) Error, full
    (b) Significant, full
    (c) True, full scale
    (d) All of the above

22. The act of comparing the performance of the instrument against a precise standard is called________
    (a) Accuracy (b) Precision
    (c) Calibration (d) Sensitivity

23. __________ is the smallest increment of the measurand which can be detected with accuracy by the instrument, and _______ defines the minimum value of input which is necessary to cause a detectable change.
    (a) Sensitivity & threshold
    (b) Linearity & sensitivity
    (c) Resolution and threshold
    (d) threshold & resolution

24. _______ is a retardation or delay in the response of an instrument to change in the measurand and _______ is the degree to which an instrument indicates changes in the measurand without dynamic error.
    (a) Dead time, fidelity
    (b) Fidelity, lag
    (c) Lag, fidelity
    (d) Dead time, speech of time

25. An ideal potentiometer is a _____ order system, temperature measuring instruments are ______ order systems, and accelerometers containing seismic masses generated _______ order equations.
    (a) second, first zero
    (b) Zero, second first
    (c) First, second zero
    (d) Zero, first and second

26. The two solutions of a linear differential equation are known as the _____ -solution and the _____ solution, and the complete solution of the equation is the ______ of the two solutions.
    (a) transient, steady state, difference
    (b) transient, steady state, sum
    (c) transient, steady state, multiplication
    (d) transient, steady state, division

27. The initial response at the output of a system to a sinusoidal input signal will be ______ in nature.
    (a) transient (b) steady state
    (c) constant (d) dynamic

28. The amplitude ratio is defined as the magnitude of the _____ signal divided by the magnitude of _____ signal.
    (a) Output, input
    (b) Half of the input, half of the output
    (c) Input, output
    (d) Twice the output, half of the input

29. With increase in the frequency of input signal, the magnitude of output signal _____ and the phase lag of system
    (a) Increases, decreases
    (b) Decreases, increases
    (c) Remains constant, decreases
    (d) Decreases, remains constant

30. Retardation or delay in the response of an instrument to a change in the input signal is called _________.
    (a) Measuring Lag
    (b) Speed of Response
    (c) Stiction
    (d) Fidelity

ANSWERS

1.	(a)	2.	(b)	3.	(b)	4.	(a)	5.	(d)
6.	(a)	7.	(c)	8.	(b)	9.	(b)	10.	(c)
11.	(a)	12.	(a)	13.	(a)	14.	(d)	15.	(a)
16.	(c)	17.	(c)	18.	(c)	19.	(b)	20.	(a)
21.	(c)	22.	(c)	23.	(c)	24.	(c)	25.	(d)
26.	(b)	27.	(a)	28.	(a)	29.	(b)	30.	(a)

CHAPTER INFOGRAPHIC

3 Glimpses of Inaccuracies in Instruments

Error: Difference a between measured & true value of input signal.

Failure: Failure is a state of equipment or instrument at which it is unable to perform to its rated efficiency at prescribed conditions.

Error = True value – Measured value
Fault is the cause and error is the effect.

Mistake: It is a man-made error that happens during process or measurement
Mistake is cause; defect is effect

Fault: mistake or a wrong that occurs in operation or process that leads end result in error or wrong operation.

Defect: Defect is non–performing element in the process due to which the instrument or equipment fails to work at the rated level.

Sources of Error

Instrument Calibration	• Frequently used Instruments may go out of calibration which reads in error.
Instrument Precision	• Instrument may be error if it is not used under conditions identical to those during calibration
Measuring Arrangement	• Influenced by arrangement, when the comparator law is not followed during measurement of length.
Environmental Conditions	• Environmental conditions such as pressure, temperature, humidity, electrical/ magnetic field also affect the measurements
Nature/properties of Work Piece	• Hardness, roughness, elasticity etc. may produce errors in measurement.
Observer's Skill	• Measurement of a physical quantity varies from one observer to another and even for the same observer it may vary due to his/her physical and mental states.

Terms Used in Assessing Instrument Performance

Static sensitivity

$$\frac{\text{Change of output signal}}{\text{Change of input signal}}$$

Accuracy: Degree of nearness to true value.

Precision: Difference between instruments reported values during repeated measurements of same quantity.

Resolution: Smallest increment of change in measured value that can be determined from the instrument's readout scale.

Sensitivity: Change of instrument's or transducer's output per unit change in measured quantity.

Error of measurement is the inaccuracy in measurement which results in deviation from the expected or standard value.

Measurement of Errors is the determination of uncertainty in a measurement and a way of specifying the uncertainty in an analytical fashion

Classification of Errors:

Random Errors: Variable in magnitude and usually follow a certain statistical law such as a normal distribution law.

Systematic Error: Occur in a specific direction (positive or negative) and have definite magnitude

Illegitimate Errors: Outright mistakes such as incorrectly writing down a number, failing to turn on/off an instrument, or miscalculating.

Mechanical Friction: Variation in measuring systems due to friction

Hysteresis: Detected and corrected by measuring first in increasing order and in decreasing order, and then observing the difference

Reading Error: Occurs while reading a number from the display scale of an instrument.

Chapter 3

INACCURACIES IN INSTRUMENTS

STARTERS

To study this chapter, you should have awareness on the following concepts. For a better understanding, it is always a good idea to revise these prerequisites.

- An understanding on system approach i.e., input-output models.
- A brief idea on system behaviours.
- A basic understanding on errors, faults and failures of systems.
- General idea and methods of measurement, error of measurement and measurement of error.
- Various fundamental and derived quantities, their denotations, and units.
- Standards of measurement and methods of standardizing; International standards, National standards and Local standards.
- Notations, and units of various quantities.
- Definitions and basics of various physical quantities.
- Basic principles of solid mechanics, fluid mechanics, elasticity, heat, light, sound, magnetism, electricity and semi-conductors.

And

The contents discussed in the first and second chapters of this book.

LEARNING OBJECTIVES

After studying this chapter you should be able to

- Understand the types of errors.
- Explain and distinguish between error of measurement and measurement of error.
- Understand and explain systematic and random errors.
- Describe various terms related to inaccuracies in measurement and performance of instruments.

3.1 INTRODUCTION

Everyone wants an instrument to work correctly. Any deviation from this correctness is inaccuracy. The inaccuracy in instrument may be described with several terms such as error, defect, fault, failure, mistake, blunder and so forth. All these terms seem to be same and many times they are used synonymously and interchangeably. However, there is a small difference among one another. We shall first distinguish these terms and then discuss them to understand the causes and effects of these in this chapter.

3.2 TERMINOLOGY

It is always a good idea to know about 'what not to do' before doing a thing. Similarly, when a subject is being learnt, it is good to know what the terms often used in it. So, we shall first know the terminology and then 'what not to do' in instrumentation.

3.2.1 Terms used to Describe Inaccuracy in the Instrument

The following terms are often used to describe the inaccuracies in an instrument.

Error

Error is defined as the difference between the measured value and the true value of the input signal.

$$\text{Error} = \text{True value} - \text{Measured value}$$

Thus error corresponds to the measurement.

Fault

Fault is a mistake or a wrong that occurs in an operation or process that leads the end result in an error or wrong operation.

Fault is the cause and error is the result.

Failure

Failure is a state of equipment or instrument at which it is unable to perform to its rated efficiency at prescribed conditions.

Failure leads to low performance and inability to produce desired output.

Defect

Defect is non–performing element in the process due to which the instrument or equipment fails to work at the rated level.

More Instruments for Learning Engineers

Error has Error

1. It may be noted that the absolute value of error cannot be determined due to the fact that the true value of quantity cannot be determined accurately.
2. Every manufacturer assures that the error in functioning of a device he sells is not greater the limit he sets. This limit of error is known as limiting errors or guarantee error. So error is inevitable, but treated as not an error as long as it is not affecting the functioning within the set values or acceptable to the user.

To err is human;
To forgive is divine;
To measure is instrumentation

Mistake

It is a man-made error that happens during process or measurement

Blunder

It is unnoticed mistake or a gross error by oversight occurs due to unscientific method.

Terms Used in Assessing Instrument Performance

The following terms are often employed to describe the quality of an instrument's readings. They are related to the expected errors of the instrument.

- *Accuracy:* The degree of nearness to the true value. This is represented by difference between the measured and true values. Typically, a manufacturer will specify a *maximum* error as the accuracy; manufacturers often neglect to report the odds that an error will not exceed this maximum value.
- *Precision:* The difference between the instruments reported values during repeated measurements of the same quantity. Typically, this value is determined by statistical analysis of repeated measurements. Simply it is the repeatability of the result of instrument.
- *Resolution:* This is the smallest increment or change in the measured value that can be determined from the instrument's readout scale. The resolution is often on the same order as that of the precision; sometimes it is smaller.
- *Sensitivity:* The change of an instrument or transducer's output per unit change in the measured quantity.

$$\text{Static sensitivity, K} = \frac{\text{Change of output signal}}{\text{Change of input signal}}$$

A *more sensitive* instrument's reading changes significantly in response to smaller changes in the measured quantity. Typically, an instrument with higher sensitivity will also have finer resolution, better precision, and higher accuracy.

3.3 ERROR OF MEASUREMENT AND MEASUREMENT OF ERROR

The two terms often confuse the readers of instrumentation are measurement error (or error of measurement) and error measurement (or measurement of error). We shall discuss these two first to get clarity.

3.3.1 Error of Measurement

Error of measurement is the inaccuracy in measurement which results in deviation from the expected or standard value.

This term is often misunderstood. Here, we should notice and note that the error of measurement does not mean the fault or mistake or blunder in measurement or the errors we commit while taking the measurement. We shall try to clarify with an example.

For example, a pressure indicator in an aircraft must never fail nor show errors during operation. Therefore, it is undoubtedly important to establish the quality, reliability, accuracy and precision of the instrument that a data set or calibrated can be used in an engineering or scientific application with a great confidence. Here, it is to be noted that, how good may be the instrument, the error may still occur in the measurement. Moreover, the error of measurement may become a critical also at times.

Further, sometimes a data may become the foundation for a new theory or the disproving existing theory. Surprisingly, sometimes though the instrument is correct, even the method of measurement is right, still error may exist. Therefore, "How good is the data?" is the fundamental question to put to any experimentalist while drawing a conclusion from a set of measurements.

The answer to the above question depends on the perception and presumption of the meaning we assign to the word 'good'. Most often, we say the data "good" if they match well with theoretically derived results. Of course, theory is a model that replicates the physical system but, there is no guarantee that it exactly represents the actual, real and practical system.

When a physical relation or a law or a hypothesis is drawn or derived, at the end of it, often we come across the words, "within the 'experimental errors' or within the 'measurement errors' this law is true". What does this error mean?

The accuracy of even the most fundamental theories, such as Boyle's law, Ohm's law, Hook's law or Newton's laws, are limited by certain assumptions (such as at constant temperature for Boyle's or Ohm's law; within the elastic limits for Hook's law) used when calculating with it and the accuracy of the data from which the theory was developed. Therefore, measurements compared with theoretical values cannot assess their quality. So, it is always better idea to compare with the *actual or standard or the practical value* of the physical quantity. This difference between the measured value and the true physical value of the quantity is thus defined as the '*error of measurement*'.

Thence, the original question should be reiterated as, "What is the error of the data?" instead of asking "How good is the data?"

Interestingly, on close observation, we can find that the definition of *error* itself has an *error*, because, the error cannot be calculated exactly unless we know the true value of the quantity being measured! Obviously, we can never know the true value of a physical quantity without first measuring because some error is present in every measurement. Therefore, the true value is something we can never know exactly. Hence, we can never know the error exactly, either. It is helpful, but, it suffers from this major flaw. Isn't it?

More Instruments for Learning Engineers

Mile 3.2

Relative Error

Relative Error or Fractional Error is defined as the ratio of the error and the specified magnitude of the quantity. If dA is the error in the magnitude A then Relative Error $= \frac{dA}{A}$. If $A = a_1 + a_2$ then $\frac{dA}{A} = \frac{a_1.da_1}{A.a_1} + \frac{a_2.da_2}{A.a_2}$ i.e., resultant limiting error is equal to the sum of products formed by multiplying the individual relative limiting errors by the ratio of each term to the function; and if $A = a_1.a_2$ then $\frac{dA}{A} = \frac{da_1}{a_1} + \frac{da_2}{a_2}$. So, the resultant error is simply summation of relative **errors in measurement** of terms.

Error is not a mistake because error is inevitable and relative. And so error is admissible but mistake is not.

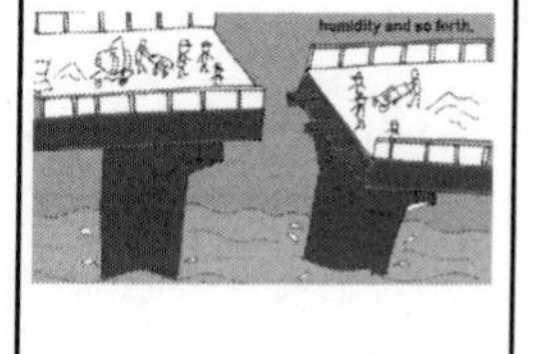

So, the *measurement of error* is an important task in the instrumentation and study of measurement to know the *error of measurement*.

3.3.2 Measurement of Errors

The basic objective of any measurement is to describe some physical property of a system in quantitative terms. However, every measurement of such a quantity has a certain amount of uncertainty, which is already discussed in the above paragraphs. We shall now discuss on two major points i.e., how to measure the error and how much is the error for which consider the following three different practical situations of measurement of the temperature of a given object.

The temperature can be measured by several methods using different instruments. Let us take three different instruments such as a liquid-in-glass thermometer (say, LGT), a solid rod thermometer (say, SRT), and a vapour pressure thermometer (say, VPT). Now conduct the following three experiments.

Experiment – 1: Take all the three instruments, one of each type among LGT, SRT and VPT. Record the temperature using one after the other instrument separately applying on the same point on the object. Each of these instruments may give different measurements for the same temperature.

Experiment – 2: Now, take three instruments of the same type say, LGTs. And now, measure the temperature of the same body at the same point. Again three values of the temperature by the three different instruments may be obtained.

Experiment – 3: Finally, take only one instrument of LGT to measure the temperature of the same body, but three times. Here also we get three values of temperature.

If you observe at the values obtained from the above three experiments, there is no surprise if all these values differ, when one is compared with another. It is evident that repeated measurements by the same instrument may also give different values of the same physical quantity. Therefore, it can be concluded that all measured values are inaccurate to some degree. Thence, it is nearly a fact that it is impossible to find the true value of physical quantity.

So, we have no other option than to compromise to find the most probable value and assign an uncertainty to it (such as the measurement made is 99% true with a 1% uncertainty). Of course, the experimenter need not make the uncertainty of the measurements as small as possible. Even a crude result can serve the purpose as long as the uncertainty in the measurements ascertains or assures to an extent that the conclusions drawn from the result are not affected.

Thus, the determination of uncertainty in a measurement and a way of specifying the uncertainty in an analytical fashion is known as the measurement of error.

Self Assessment Questions-3.1

1. Define and explain the terms accuracy, precision with reference to measuring instruments.
2. Distinguish between error of measurement and measurement of error. Discuss.
3. Distinguish the terms error, failure, fault, mistake, and defect. Also explain their meaning clearly.
4. Differentiate the terms resolution and sensitivity with reference to the measuring instruments.
5. 'The definition of *error* itself has an *error*', do you accept the statement? Justify your answer.
6. 'The error cannot be calculated exactly unless we know the true value of the quantity being measured and true value cannot be measured unless error is known'. Do you agree? If so, how do you solve this riddle?

3.4 ERROR

As discussed at very beginning of this chapter error is defined as the difference between the measured value and the true value. Interestingly, the definition of error itself has an error (discussed in previous section 3.3 under sub-section 3.3.1 of this chapter). So, the definition of *error* is not as clear as it appears because most of the times estimate the likelihood that the deviation (so called error) exceeds some specific value. Let us observe the following example for some more clarity.

At a petrol pump, the vendor claims that the petrol supplied by him is exact by quantity. However, the measurement of quantity can only be done with a specific value or a standard (say, standard liter jar) which we can presume as a true value (Of course, 'how much true these values are' is again a different point of discussion). Now, if you really measure quantity by the standard jar, there will be definitely some difference. Then, the vendor takes a level of significance say 5% (or 95% confidence). Now, his claim is said to be valid if one or both of the following statistical statements must be true.

- 19 times out of 20 (i.e., 95%) of the measurements, the readings measured should be one liter exactly.
- Reading of every measurement [or at least 95 times out of 100 (or 19 of 20)] should be within certain tolerance limits such as greater than or equal to 995 ml (or 995 to1005 ml).

Thus, on close observation, it is evident that in both of the above cases, there is a margin of accepted error with some degree of uncertainty. Thence, is the definition of error valid?

So, we come to a conclusion that the error or uncertainty may be estimated with statistical tools when a large number of measurements are taken. Moreover, the experimenter must also predetermine; be aware of how the instruments perform and of how well they are calibrated in order to establish the possible errors and their probable magnitudes.

3.5 SOURCES/CAUSES OF ERROR

A measurement is concerned with the following aspects. Obviously, any error that occurs in a measurement process must generate from one or more of these only.

(i) The Manufacturer of the instrument
(ii) The Measuring device or the Instrument
(iii) The Measurer or the operator/observer

More Instruments for Learning Engineers

Mile 3.3

A Simple Error can cause a Disaster

In 1950s, the jet, de Havilland Comet, was with many never-before-seen features, such as a pressurized cabin that allowed it to fly higher and faster than other aircraft. But, in 1954, two Comets disintegrated midflight for no apparent reason, killing 56 people total. The Laughably Simple Flaw is that it had square windows. (As per design engineering, the sharp corner (or groove) is stress concentrated point and is a spot where the shape of the object makes it more likely to break under stress)

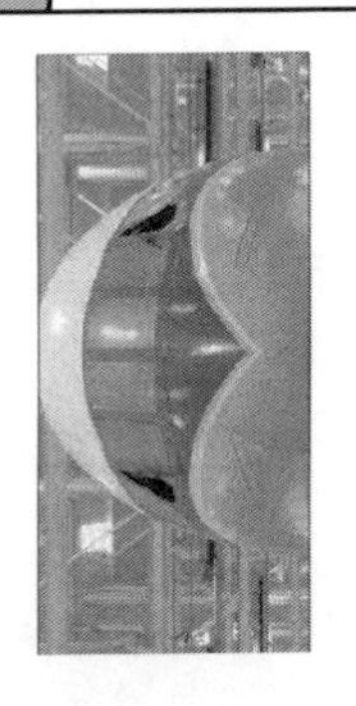

(iv) The Material of the job on test or the work piece
(v) The Measuring environment and the conditions
(vi) The Measuring method and arrangement

Now let us see where and how the errors generate i.e., the source of error with reference to the aspects.

1. **Instrument Calibration:** As the instruments are used frequently, the instrument may go out of calibration. The measurement made with such an instrument reads in error. The error of this kind is usually regular and hence categorized under systematic errors. This error can be corrected by comparing with standard instrument and make necessary adjustment in the instrument or its display which is known as calibration. Any instrument must be calibrated at frequent intervals.
2. **Instrument Precision and Reproducibility:** Even though a measuring device has been calibrated under a set of conditions, the measurements may still be in error if it is not used under conditions identical to those during calibration. Certain mechanical defects such as friction in bearings, backlash in micrometer screw etc, cause error in measurement. These errors may be either accidental or systematic.
3. **Measuring Arrangement:** The measurement is sometimes influenced by its arrangement. Particularly, when the comparator law is not followed strictly during measurement of length. According to this law of comparators, errors of first order are eliminated when the measuring device and scale axes are collinear. This law should be meticulously followed while designing and using opto-mechanical instruments.
4. **Environmental Conditions:** Environmental conditions such as pressure, temperature, humidity, electrical/ magnetic field also affect the measurements if the device is not operated under the same conditions specified during calibration. Suppose, the length measurement where a work piece of length L is measured with an instrument calibrated at room temperature, say 27 °C, the error introduced when the measured at a temperature 't' will be $= (\alpha_2 - \alpha_1). (t - 27)$, where α_1 and α_2 are thermal expansion co-efficient of work piece and of scale of the measuring device respectively.
5. **Nature or Properties of Work Piece:** The nature of work piece such as hardness, roughness, elasticity etc., may produce errors in measurement. Most mechanical and opto-mechanical instruments are operated under constant pressure condition. Obviously, soft and hard work pieces respond differently leading to an error in measurement.
6. **Observer's Skill:** Human interpretation plays a great role particularly, in decision making, even though automation is available, or digital measurement is done. The observer's skill is considered as a part of measurement. It is more than a fact that the measurement of a physical quantity varies from one observer to another and even for the same observer it may vary due to his/her physical and mental states. Such errors may be of systematic or random or both depending on the experimental conditions.

3.6 CLASSIFICATION OF ERRORS

The errors or failures or faults or defects almost used synonymously and situational are broadly classified into two types which have similarity in their respective categorical characteristic behavior and more or less mean same. These are:

No.	Errors	Failures	Faults	Defect	Mistake
1.	Accidental	Sudden	Unknown	Random	Blunder
2.	Systematic	Gradual	Known	Bias	Computational

Though all the above terms are applicable to the instruments, it is more appropriate and customary to use the term 'error' in instrumentation. So whatsoever may be the cause, reason, characteristic, the term 'error' is used here and classified with the same notion.

Usually any error that occurs in an instrument will have two components i.e., a suddenly developed or an accidental error and a gradually developed systematic error due to generic inherent problems. The identification of proportion of these two components can be useful in assigning uncertainty to the measured value. For example, there may be gross blunder that arises due to faulty design circuit etc. However, it may not immediately result in a failure at once, but will gradually develop before it results in failure. The intelligence of experimenter lies in the ability to identify and eliminate these errors.

As described above, categorizing all the possible errors identified into either accidental or systematic error would be most convenient but it is very difficult to classify in certain cases, particularly those categories of error which overlap or ambiguous. Some other errors behave as systematic (or bias) error in one situation and as accidental (or random or precision) error in other situations. Further, some errors do not fit neatly into either category. However, for purposes of discussion, typical errors may be roughly sorted as follows:

1. Accidental or Precision or random error
   (a) Disturbances to the equipment
   (b) Fluctuating experimental conditions
   (c) Errors due to insufficient measuring system sensitivity
2. Systematic or bias error
   (a) Calibration errors
   (b) Human errors
   (c) Equipment Defects or Instrumental Errors
   (d) Loading errors
   (e) Limitations of system resolution
3. Illegitimate error
   During measurement
   (a) Blunders
   (b) Mistakes
   After measurement
   (c) Computational errors (after experiment)
4. Errors that are sometimes bias error and sometimes precision error
   (a) Backlash Errors
   (b) Errors due to friction
   (c) Hysteresis errors
   (d) Drift errors
   (e) Calibration errors

(f) Variation in testing or measuring method
(g) Errors due to environmental conditions
(h) Procedural errors
(i) Parallax error

On close observation of the above listed types of errors, it is evident that most of the errors are either overlapping between the two kinds viz. systematic and accidental or probable to occur in both classes. Therefore, it is better, we discuss the most common and significant of these errors under separate heads.

SELF ASSESSMENT QUESTIONS-3.2

1. What do you understand by the term 'Error'? Discuss in light of measurement systems.
2. What is an error? What are the sources of errors in measurement systems? Give examples.
3. Classify errors? Explain any two types of errors.
4. What are the causes of error? List out the error generating points in instrumentation.
5. What are illegitimate errors?
6. List out random errors and systematic errors. Some errors appear in both types, what are they?

3.7 RANDOM ERRORS (OR PRECISION OR ACCIDENTAL ERROR)

The accidental errors or random errors are so called as they occur randomly. These errors are variable in magnitude and usually follow a certain statistical law such as a normal distribution law. The parameters such as average and standard deviation of the distribution of these variations will be the measure of accidental errors. We know that the standard deviation on either side of maximum of a normal distribution encompasses 68.27% of the area under the curve with increasing number of measurements of the same quantity. The value of standard deviation gets smaller as the value of the quantity approaches its true value.

The accidental or precision or random errors often occur due to
(a) Disturbances in the equipment

More Instruments for Learning Engineers

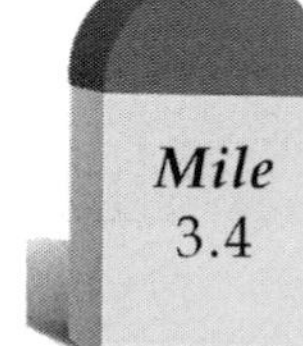

Trying to Avoid Making Mistakes? No...

1. Alfred Nobel's mistakes (killed his brother also), have shown the way how to store explosives such as Trinitro glycerine
2. Penicillin was discovered by Alexander Fleming out of his mistake, finally earned him a Nobel Prize also.
3. Dr. John Kellogg along with his brother Keith discovered the corn flakes by mistake only.
4. Plastics, rubbers (by vulcanization) were the outcomes of a brilliant error of Charles Goodyear.
5. The mistake is so good to build a great company, Dot-Com & many more

Penicillin

Kellogg's

(b) Fluctuating experimental or environmental conditions
(c) Insufficient measuring-system sensitivity

3.7.1 Disturbance in the Equipment

Sometimes the electromechanical type measuring device gets an unknown disturbance in its signal either while receiving or during the processing or while displaying. This may occur due to insufficient amplification or improper modulation signal voltage. In the mechanical instruments, the uncertainty in measurement may be caused due to suddenly developed friction or obstruction due a foreign particles at the mating parts, and the pin or pointer or clogging in filters of flow channel etc. All these errors are due to unknown disturbances in the equipment occur accidentally result as random error and affect the precision of the equipment

3.7.2 Fluctuating Experimental or Environmental Conditions

In spite of utmost care and precautions is taken by experimenter, it is nearly impossible to eliminate all possible errors and certain errors are bound to creep in due to fluctuation in experimental conditions. For instance, even in an apparently simple measurement of flow velocity with a Pitot tube any misalignment of the probe, leakages in the pressure tubing, changes in the bore and surface conditions of the manometer, any fluctuations in the atmospheric and stream pressure are likely to alter the probe readings and give rise to uncertainties.

Environmental conditions such as pressure, temperature, humidity, electrical/ magnetic field also affect the measurements if the device is not operated under the same conditions specified during calibration. Further, these conditions may not be constant all the time and keep on changing regularly. A fluctuation (an unknown or unpredicted) change in the environmental condition may induce the error in instrument output.

Insufficient measuring-system sensitivity

As discussed above in this chapter (and also in chapter -2), the sensitivity of an instrument is defined as the ratio of change in magnitude of the output signal to change in magnitude of input signal.

$$\text{Static sensitivity, K} = \frac{\text{Change of output signal}}{\text{Change of input signal}}$$

Sensitivity is represented by the slope of the calibration curve if the ordinates are expressed in the actual units. With a linear calibration curve, the sensitivity is constant.

Errors and uncertainties are inherent in the process of making any measurement and in the instrument. If the measurement and measuring devices are built with any deficiency or insufficiency, or if the parts are spurious or of sub-standard quality, the instrument tends to induce the error in its output. A pressure measurement device designed to measure 1 atm pressure range may damage if higher pressures are used because of sensitivity.

Observe the following case: An electro-mechanical instrument manufactured in United Kingdom (UK). The instrument runs with the electrical power and is rated as 120 V-5 W. This instrument is bought by an Indian and brought to his home town. When it is operated in India, the instrument burnt off. Can you imagine – why? Yes! The normal household electricity in UK is at 110-120 V while this is 230-240 V in India. So, check the instrument sensitivity sufficiency to withstand before operating it.

Similarly, a clinical thermometer (which can measure to a maximum of 40 °C) cannot be used to measure the temperature of boiling water or hot iron in a forging furnace because of the sensitivity of the material and the range of the thermometer.

In addition, these errors may creep not only while manufacturing but also during assembling the instrument. For example, assembly errors resulting from incorrect fitting of the scale zero with respect to the actual zero position of the pointer, non-uniform division of the scale, and bent or distorted pointers. The assembly errors do not alter with time and arise out suddenly.

3.8 SYSTEMATIC ERROR (OR BIAS ERROR)

The systematic errors or bias errors occur in a specific direction (positive or negative) and have definite magnitude and hence are so called. A simple example of systematic error is zero error of an instrument. These are often more troublesome as repeated measurements need not necessarily reveal them. Even when their existence or nature has been established, it is sometimes very difficult to determine and eliminate them. Therefore, in most of the cases the systematic errors are not used for correcting the measurements but are taken with their full values along with accidental errors as uncertainty.

The experimenters sometimes use theoretical methods to estimate the magnitude of systematic errors. For instance, the error caused in the measurement of temperature due to exposed portion of a mercury column of thermometer can be estimated theoretically. This error is a systematic error and is of a definite magnitude in a specific direction. Systematic or bias errors are further classified as follows based on the source of generation and cause of occurrence.

(a) Calibration errors

(b) Gross errors or Human errors

(c) Defective equipment and Instrumental Errors

3.8.1 Calibration Error

The most common form of bias error is found to occur during calibration. These errors happen if the scale of the instrument is not adjusted to read the measured value properly at regular intervals. Most common calibration errors are *zero-offset* errors, which make all readings to vary (offset)

More Instruments for Learning Engineers

Mile 3.5

A **polygraph**, popularly known as a **lie detector** or **deception detector** measures and records several physiological indices such as BP, pulse, respiration and conductivity while the subject is asked and answers a series of questions. The belief for the use of polygraph is that deceptive answers produce physiological responses that can be differentiated from those associated with non-deceptive answers; the polygraph is one of several devices used for lie detection. It was invented by John Augustus Larson (1921), a medical student and a police officer in Berkeley, California. In some countries polygraphs are used as an interrogation tool with criminal suspects or candidates for sensitive public or private sector employment. Polygraph examiners, or polygraphers, are licensed or regulated in some jurisdictions.

Lie Detecter or Polygraph

by some fixed quantity say, $X_{off\text{-}set}$, or *scale errors or zero-error.* This error usually reads an error by fixed percentage in the ratio of output to input.

Calibration is the process of identifying and eliminating these errors by measuring system's readout scales through a comparison with a standard. Of course, the standards themselves also have uncertainties, though smaller.

[The uniqueness of certain primary standards makes them exceptions to this statement. In particular, the mass standard (the International Prototype Kilogram) has by *definition* a mass of exactly 1 kg. In the sense of practical applications, however, uncertainty will nevertheless occur; even primary standards require the use of ancillary apparatus, which necessarily introduces some uncertainty].

3.8.2 Gross Errors or Human Errors

A classical saying 'To err is human' is applicable to instrumentation and measurements also. Some of such errors are listed below:

- An experimenter, due to an oversight reads the temperature as 31.5 °C while the actual reading may be 81.5 °C
- A person may transpose the reading while recording, 52.8 mm and write 58.2 mm instead
- A person records a number as 1 (one) instead of 7 (seven) or 11 (one-one or eleven) by improperly listening over phone
- A record could be erratic if it is four threes (3333) instead of three fours (444)
- A person may record '44 legs for 20 horses', when you pronounce 'forty forelegs for twenty horses'
- You may read 24s (two four 's'), but write 44

All the examples mentioned above, are 'gross errors' committed by humans. Further, as long as human beings are involved, some gross errors will definitely be committed. Although complete elimination of gross errors is probably impossible, one should try to anticipate and correct them. Some gross errors are easily detected while others may be very difficult to detect.

These errors mainly occur due to human mistakes or blunders in reading instruments, recording and calculating measurement results. The responsibility of the mistake normally lies with the experimenter only because –

A machine does not make mistake, but man does. And even if a machine makes a mistake, it is because of some mistake of man who designs or manufactures or operates it.

Perhaps! With the above notion, there is a saying...

'Instruments are better than the people who use them'

Gross errors may be of any amount and therefore their mathematical analysis is impossible. However, they can be avoided or minimized by the following ways:

1. Take a great care in reading, pronouncing and recording the data.
2. Repeat at least two, three or even more times to read or pronounce

3. Cross check or counter check, preferably by different person
4. Create some check points or make into clusters or groups of convenient size for random inspection if the data is too large
5. Take more readings of measurement at a steady state
6. It is better to take readings by different experimenters
7. Take the readings at different points of similar nature to avoid re-reading with the same error
8. It should be understood that no reliance be placed on a single reading. So, it is always a good idea to take sufficiently large number of readings as a close agreement between readings, assures that no gross error has been committed.

Human errors are often found in two kinds viz.

1. The operational errors and
2. The observational errors

3.8.3 Operational Errors

Quite often, errors are caused by poor operational techniques. Observe the following cases

- Suppose a doctor uses the clinical thermometer on a patient to find the body temperature and if does not allow to fall down the raised mercury level into the bulb before using it on another patient, it gives a wrong reading.
- A thermometer, if its bulb (the sensing element) is immersed insufficiently or improperly, gives an erratic reading.
- The accuracy of a differential type of flow meter is questionable if it is placed immediately after a valve.
- A steam calorimeter shows inaccurate indication of the dryness fraction of steam it is incorrectly exposed to the steam.

In all the above cases, the inaccuracy is due to the wrong operation and can be avoided by correctly operating.

More Instruments for Learning Engineers

Mile 3.6

An Error Meter or **Fraud-o-meter** is a management tool recently developed. It is graduated qualitatively as inadvertent, negligence, gross negligence, reckless disregard, wilful blindness and deceptive intent, or quantitatively with reference to some measurable criteria. Suitable criteria are developed to indicate the level of fraud or errors or mistakes and are quantified to indicate on the scale. The frequency of errors and the loss or cost or effects of errors etc. are usually the criteria chosen. The data collected may be fed to computer and the result is obtained by a programming. In contrast a **Trust-o-meter** is a tool that indicates how much you can believe a person based on the performance. This is similar to the **Reliability-Meter** used to judge about the reliability of an equipment based on the failure analysis.

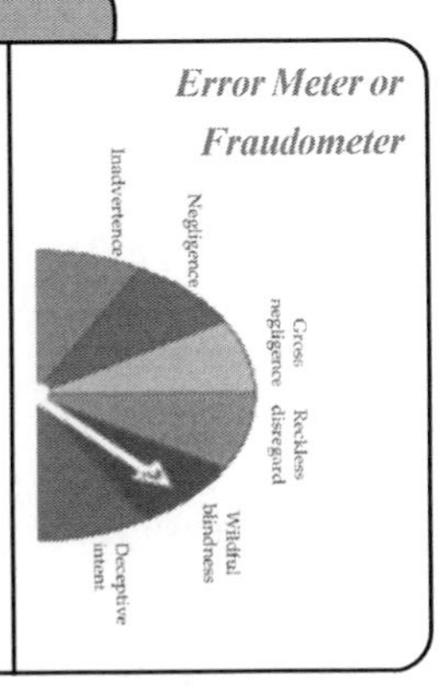

3.8.4 Observation Errors

Even when an instrument is properly selected, carefully installed and faithfully calibrated, shortcomings in the measurement occur due to certain faults on the part of the observer, The observation errors may be due to

- Parallax, i.e., apparent displacement when the line of vision is not normal to the scale.
- Inaccurate estimates of average reading, lack of ability to interpolate properly between graduations.
- Incorrect conversion of units in between consecutive readings, and non-simultaneous observation of interdependent quantities.
- Personal bias, *i.e.,* a tendency to read high or low, or anticipate a signal and read too soon.
- Wrong scale reading, and wrong recording of data.

The poor mistakes resulting from the inexperience and carelessness of the observer are obviously remedied with careful training, and by taking independent readings of each item by two or more observers.

Self Assessment Questions-3.3

1. What are systematic errors? why are they so called?
2. Errors may occur due to disturbance in equipment or environment. Are they random errors or systematic? Discuss.
3. Describe the errors caused due to disturbance in the equipment and the environment.
4. What are operational errors and observational errors? explain with examples.
5. What type of errors are the following – systematic or random? Give examples.
   (a) Operational errors
   (b) Observation errors
   (c) Environmental errors
   (d) Equipment errors
   (e) Calibration errors
6. What are calibration errors? explain with an example.
7. 'Instruments are better than people who use them', elaborate the statement in light of the inaccuracies occur in instruments.
8. List out some examples of observational errors. How do you control or rectify them?
9. "A machine does not make mistake, but man does. And even if a machine makes a mistake, it is because of some mistake of man who manufactures or operates it", Critically appreciate the statement and explore the possible errors that occur.

3.8.5 Equipment Defects and Instrumental Errors

Sometimes, the equipment itself may have built-in error originated from incorrect design, fabrication, or maintenance. These errors usually result from the components, incorrect scale graduations, and so forth. Errors of this type are often consistent in sign and magnitude, and because of their consistency they may sometimes be corrected by calibration. When the input is time varying, however, introducing a correction is more complicated. For example, distortion caused by poor frequency response cannot be corrected by the usual "static calibration" one based on a signal that is constant in time. Such frequency response errors arise in connection with seismic motion detectors.

The instrumental errors are the results of three main reasons:

(a) Inherent defects in the instruments.

(b) Misuse of the instruments, and

(c) Loading effects of instruments.

3.8.6 Inherent Defects

Variations in the actual quantity being measured may also appear as precision error in the results. These errors are not seen openly and are present inherently in instruments because of their mechanical structure. They might originate at the stage of construction or calibration or operation of the instruments.

Sometimes these variations are a result of poor experimental design, such as a measuring system originally designed to run at a constant speed may be modified to run at varying speed. This may inherently get some changes affecting the system performance and hence may show the difference in the output. Truly speaking, this variation in the measurand is not a measurement "error"; how-ever, the same statistical techniques may be applied to estimate the variations treating as errors and the mean value of the measurand may be added to its uncertainty.

Similarly, in bellows gauges or a bourdon tube, lose the property of elasticity gradually after some days. Due to constant and continuous use, it is just possible that after some time, the tube (bellows) may suffer leakage or damage or tube straightened affecting its elastic property. This change in elastic property of the transducer element causes it to show erratic reading of pressure which is exhibited in the form of change in pointer deflection. So, the zero error or an off-set is to be corrected by adjusting the position of pointer in such a way that the changed angle of deflection may be corrected by altering the magnification of sector-pinion arrangement i.e., by moving the point of attachment of the connecting link or readjust the dial so that the read value itself will become correct value. The following are some of such examples of instrument errors possible to occur.

- If the bellows in a bellows gauge lose their elasticity (or any leakage), the instruments goes erratic.
- A defective spring may cause error in the instrument reading if the transduction is carried out by a spring.
- A rubber components if used in an instrument after its shelf-life may lead to incorrect measurement.
- A magnet in an instrument may become weak and show a wrong reading.
- A electromagnet might not be charged fully and so show insufficient magnetism.
- The instrument reads high due to hysteresis or malfunctioning of components.
- Faults of construction due to knife edges.
- Lost motion due to insufficient clearance in gear teeth and bearings.
- Excessive friction at the mating parts and wear, backlash, yielding of supports, pen or pointer drag and hysteresis of elastic members due to aging.
- Improper selection and poor maintenance of the instrument.
- Unavoidable physical phenomenon such as capillary action and imperfect rarefaction, mismatching of parts and components etc.

So, when precision measurements are made, the possibility of such errors have to be observed and eliminated, or at least to be reduced to the maximum possible extent by using the following methods;

- The process of measurement has to be carefully planned and alternative methods or calibration against standards may be used for the purpose.
- Correction factors should be applied after determining the instrumental errors.
- The instrument may be re-calibrated carefully.

3.8.7 Misusing or Mishandling Errors

More than an element of truth in the instrumentation is that, most of the errors occurring in measurements are caused by the fault of the operator than that of the instrument. Even a good instrument used or handled in an improper way gives erroneous results. Some examples of misusing or mishandling of instrument are given below:

- Improper way to adjust the zero of instruments.
- Wrong initial adjustments or wrong connections.
- Connecting the leads of too high or too low resistance or capacitance etc., in the circuit
- Vigorous tapping would lead to delicate bearings being injured and thus increasing friction all the more.
- Giving wrong input voltage.
- Not following input instructions or procedural steps as per the maintenance manual.
- Half knowledge on the system or device mechanism.
- Operator is untrained or unskilled.

No doubt the above improper practices may not always cause a permanent damage to the instrument but they cause errors. However, there are certain ill-practices like using the instrument contrary to manufacturer's instructions and specifications which in addition to producing errors cause permanent damage to the instruments as a result of overloading and overheating that may ultimately result in failure of the instrument and sometimes the system itself.

3.8.8 Loading Errors

The most common errors committed by beginners is due to the improper use of the instrument for measurement which is often referred to as *Loading error*. It is the effect of the measuring procedure on the system on test. The measuring process changes the characteristics of both the source of the measured quantity and the measuring system itself. Therefore, the measured value differs by some quantity with the actual measurement. For example, the sound level sensed by mouth-piece (a microphone) of your mobile phone when it close to your mouth is not the same as that when is a little far. Is this not the effect of loading?

Loading errors generally appear as precision errors. In the instrumentation, these errors may arise from external disturbances that occur while loading to the measuring system, such as noise, distortion, wave interference, temperature variations or mechanical vibrations, humidity and so forth. These disturbances mix up with actual input (load) and cause an error. For example, a sound or music recorded in a recording theatre and in outside environment would be different. Thus the loading errors in a measuring system may also include poorly controlled processes that lead to random variations in the system output.

For instance, a voltmeter may misread voltage when connected across a high resistance while the same may read correctly when connected in a low resistance. These examples illustrate that the voltmeter has a loading effect on the circuit, altering the actual circuit conditions by the measurement process.

3.9 ILLEGITIMATE ERRORS

Illegitimate errors are errors that would not be expected. These include outright mistakes (which can be eliminated through great care or repetition of the measurement), such as incorrectly writing down a number, failing (or forgetting) to turn on/off an instrument, or miscalculating. These errors have no logic and no scientific base for occurrence. Sometimes, it can be found that the data is extremely unlikely to have arisen from precision error by analyzing with statistical tools.

3.9.1 Blunders and Chaotic Errors

The uncertainties or inaccuracy in the measurement may arise due to the following blunders or chaotic situations:

- Improper or unclear signaling.
- Starting the watch before or after the signal is given.
- Stopping the watch a little before or after the signal is given.
- Wrong reading of the tape when noting the distance between the marks.
- Incorrect recording of the observation noted.

These errors can be minimized by repeating the measurement.

3.10 OTHER ERRORS

There are some other errors that occur in the instruments which behave sometimes as precision errors while some other times as systematic. Sometime they even overlap. These errors may be generated due to instrumental or human reasons.

More Instruments for Learning Engineers

Sound-level Meter or Noise-meter measures the intensity of noise, music, and other sounds. It consists of a microphone for picking up the sound and converting it into an electrical signal and calibrated to read in decibels dB; a logarithmic unit used to measure the sound intensity. Threshold of hearing is about '0'db for the avg. young listener and that of pain (extremely loud sounds) is around 120 db, i.e. 10^{12} times greater than '0'db. The frequency-weighting scale (A-Z) describes how complex noises affect a person, which is recognized internationally to prevent deafness from excessive noise in work environments. In 1970s, as a concern to noise pollution, accurate, portable noise-meters were developed. Sound level is not a measure of loudness, as loudness is a subjective factor and depends on the characteristics of the listener's ear. To overcome this, scales are devised to correlate loudness with objective sound measurement such as Fletcher–Munson curve.

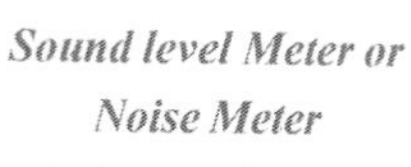

Sound level Meter or Noise Meter

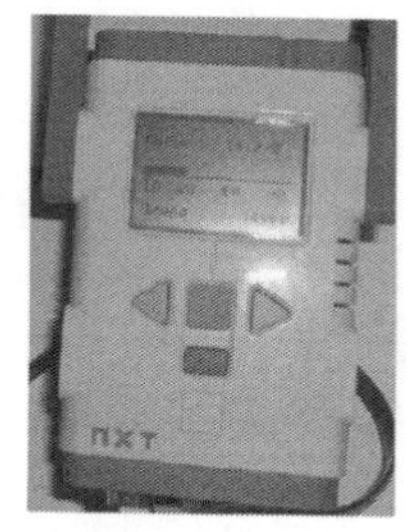

3.10.1 Backlash

This is the maximum distance or angle through which any part of a mechanical system may be moved in one direction without applying appreciable force or motion to the next part in a mechanical system. When there is a gap in gear teeth (due to wear/ tear) this error comes up.

3.10.2 Mechanical Friction

This is yet another important source of variation in measuring systems. For example, friction may cause needle of a galvanometer to lag behind its actual position

3.10.3 Hysteresis

This error depends on how a sequence in measurements is taken and hence it may behave as either a bias error or a precision error. This error is detected and corrected by measuring first in increasing order and in decreasing order, and then observing the difference (called hysteresis loss). This approach is sometimes called the *method of symmetry.*

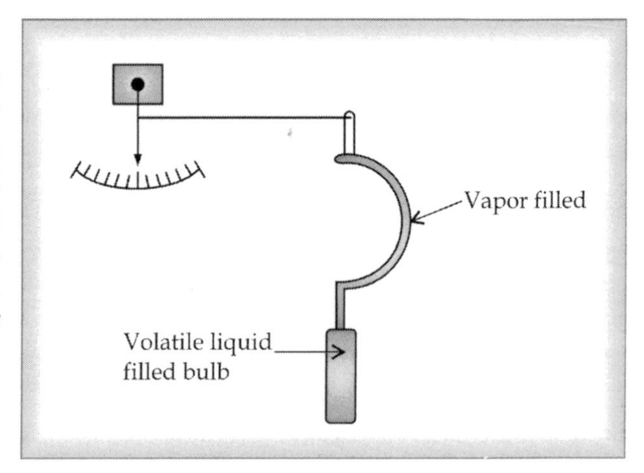

Fig. 3.1 Hysteresis Error

3.10.4 Drift

It is a gradual variation in the instrument output over a period of time that is unrelated to changes in input, operating conditions or load. This can be caused by component's instability or temperature changes, wear, erosion, polarization, saturation and aging may also cause drift. It may occur in obstruction flow meters because of wear and erosions of the orifice plate, nozzle, or venturimeter. Drift occurs in thermocouples and resistance thermometers due to the contamination of the metal and a change in its atomic or metallurgical structure. Drift occurs very slowly and can be checked only by periodic inspection and prevented by periodic maintenance of the instrument.

More Instruments for Learning Engineers

A **ceilometer** determines the height of a cloud base using a laser or other light source. It can also be used to measure the aerosol concentration within the atmosphere. They can sometimes be fatal to birds, as the animals become caught in light-beams and suffer exhaustion and collisions with other birds. In the worst recorded ceilometer non-laser light-beam kill-off, approximately 50,000 birds of 53 different species died at the Warner Robins Air Force Base in the US during one night in 1954.

Ceilometer

When the test duration is long, environmental conditions may fluctuate throughout the test, causing different calibration errors for each successive measurement. In this case, the fluctuations create a precision error.

When the experiment is repeated for sufficient number of times using different equipment or experimenters, the bias errors of successive experiments may occur which can be randomized, and they become another form of precision error in the set of all experiments.

3.10.5 Reading Error

Reading error occurs while reading a number from the display scale of an instrument. This type of error may sometimes be a bias error caused due to truncation or rounding off the actual value to nearest resolution. Reading error may also result out of inadequate instrument sensitivity, i.e., if instrument does not respond to the smallest fluctuations of the measurand.

For example, a digital display may truncate an actual value of 9.4 to 9 if its sensitivity is insufficient to read below ±0.5. This is a bias error in the sense that 9.4 will always be displayed as 9. Now, if several values are read, then the error may cumulate and may become a precision error if the many values have no particular relation to one another.

Self Assessment Questions-3.4

1. What are various equipment (instrumental) errors? Explore the causes or sources of such errors.
2. With suitable examples, describe the effect of the following in causing instrumental errors.
   (a) Inherent design defects (b) Mishandling
   (c) Loading (d) Reading (observation)
3. Explain the effect of the following in causing errors in instruments with examples
   (a) Mechanical friction (b) Drift
   (c) Hysteresis (d) Backlash
4. What is illegitimate error? What are blunders and chaotic errors? when and how do they occur?
5. Explain how misreading causes errors in instruments with examples.

SUMMARY

Error is an inaccuracy, defined as the difference a between the measured value and the true value of the input signal. Fault, failure, defect, mistake, blunder are various synonymous terms used with a small variations of meanings. Accuracy is degree of nearness to the true value while precision is the repeatability. Resolution is the smallest increment of change in the measured value that can be determined from the instrument's readout scale. Sensitivity is the change of an instrument or transducer's output per unit change in the measured quantity. A good instrument is one with higher sensitivity will also have finer resolution, better precision, and higher accuracy.

Error of measurement is the inaccuracy in measurement which results in deviation from the expected or standard value. The determination of uncertainty in a measurement and a way of specifying the uncertainty in an analytical fashion is known as the measurement of error.

An error in a measurement generates at one or more among the manufacturer, measuring device, measurer, material, measuring environment and measuring method. Errors are generally classified as accidental (or precision or random) and systematic (or bias) errors. These may be due to calibration, human, equipment defects, loading, disturbances, experimental conditions, insufficient measuring-system sensitivity etc. The illegitimate errors may occur during measurement in the form of blunders, mistakes and after measurement as computational errors. The miscellaneous errors include backlash, mechanical friction, hysteresis, drift, parallax error and so forth. On close observation of the above listed types of errors, it is evident that most of the errors are either overlapping between the two kinds viz. systematic and accidental or probable to occur in both classes.

Quite often, errors are caused by poor operational techniques. Even when an instrument is properly selected, carefully installed and faithfully calibrated, shortcomings in the measurement occur due to certain faults on the part of the observer. The instrumental errors are the results of inherent defects or misuse or loading effects.

KEY CONCEPTS

Error: The difference a between the measured value and the true value of the input signal. (Error = True value – Measured value).

Fault: Mistake or a wrong that occurs in an operation or process that leads the end result in an error or wrong operation.

Failure: State of equipment or instrument at which it is unable to perform to its rated efficiency at prescribed conditions.

Defect: Non–performing element in the process due to which the instrument or equipment fails to work at the rated level.

Mistake: Man-made error that happens during process or measurement.

Blunder: Unnoticed mistake or a gross error by oversight occurs due to unscientific method.

Accuracy: The degree of nearness to the true value.

Precision: The repetitiveness of producing the results.

Resolution: The smallest increment of change in the measured value that can be determined from the instrument's readout scale.

Error of measurement: The inaccuracy in measurement which results in deviation from the expected or standard value.

Measurement of errors: The determination of uncertainty in a measurement and a way of specifying the uncertainty in an analytical fashion.

Sources of errors: Manufacturer, Measuring device, Measurer, Material, Measuring environment and Measuring method.

Calibration: Comparing with standard instrument and make necessary adjustment in the instrument or its display so as to make the instrument reads correct value.

Random or precision or accidental error: Occur randomly, variable in magnitude and usually follow a certain statistical law such as a normal distribution law.

Disturbance in the equipment: An unknown disturbance in signal either while receiving or during the processing or while displaying.

Environmental conditions: Pressure, temperature, humidity, electrical/ magnetic field etc.

Sensitivity: The ratio of change in magnitude of the output signal to change in magnitude of input signal. Static sensitivity, $K = \frac{\text{Change of output signal}}{\text{Change of input signal}}$

Systematic or bias error: Occur in a specific direction (positive or negative) and have definite magnitude.

Calibration error: Happen if the scale of the instrument is not adjusted to read the measurand properly at regular intervals.

Operational errors: Caused by poor operational techniques.

Observation error: Occur due to certain faults on the part of the observer such as parallax, inaccurate estimates of average reading, lack of ability to interpolate properly between graduations, incorrect conversion of units, personal bias, wrong scale reading, and wrong recording of data etc.

Inherent defects: Originate at the stage of design, construction or calibration or operation of the instruments.

Misusing or mishandling errors: Caused by the fault of the operator than that of the instrument due to handling in an improper way.

Loading errors: Arise from external disturbances that occur while loading to the measuring system, such as noise, distortion, wave interference, temperature variations or mechanical vibrations, humidity and so forth.

Illegitimate errors: Outright mistakes (which can be eliminated through great care or repetition of the measurement), such as incorrectly writing down a number, failing (or forgetting) to turn on/off an instrument, or miscalculating.

Blunders and chaotic errors: Arise due to improper or unclear signaling, starting/stopping the watch a little before or after the signal, Wrong reading of the tape, Incorrect recording of the observation noted.

Backlash: The maximum distance or angle through which any part of a mechanical system may be moved in one direction without applying appreciable force or motion to the next part in a mechanical system.

Mechanical friction: Variation in measuring systems due to friction.

Hysteresis: Detected and corrected by measuring first in increasing order and in decreasing order, and then observing the difference (called hysteresis loss).

Drift: Caused by component's instability or temperature changes, wear, erosion, polarization, satura-tion and aging.

Reading Error: Occurs while reading a number from the display scale of an instrument.

REVIEW QUESTIONS

Short Answer Questions

1. Distinguish between accuracy and precision with reference to measuring instruments.
2. Distinguish between error of measurement and measurement of error. Discuss.
3. Define the terms error, failure, fault, mistake and defect. How do they differ?
4. What are various equipment (instrumental) errors? Explore the causes or sources of such errors.
5. How do the resolution and sensitivity affect the output of an instrument?
6. Explain how misreading causes errors in instruments with examples.
7. What are the causes of error? List out the possible error-generating aspects of an instrument.

Long Answer Questions

1. What is an error? What are the sources of errors in measurement systems? Give examples.
2. Classify errors? Explain any two types of errors.
3. List out random errors and systematic errors. Some errors appear in both types, what are they? Explain how they occur.
4. What are systematic errors? Why are they so called? What are its causes?
5. If errors occur due to disturbance in equipment or environment. Are they random errors or systematic? Discuss.
6. List out the errors caused due to disturbance in the equipment and the environment.
7. Distinguish between operational errors and observational errors? How and where do they generate?
8. What are calibration errors? How do you calibrate an instrument? Explain with examples.
9. 'Instruments are better than people who use them', elaborate the statement in light of the inaccuracies occur in instruments.
10. List out some examples of observational errors. How do you control or rectify them?
11. "Every instrument needs calibration", critically appreciate the statement and explore the possible errors that occur.
12. Explain the impact of the following characteristics of instruments on their performance and accuracy.
    (a) Mechanical friction
    (b) Drift
    (c) Hysteresis
    (d) Backlash

MULTIPLE CHOICE QUESTIONS

1. The difference between true value and measured value is defined as______.
   (a) Fault
   (b) Error
   (c) Mistake
   (d) Defect

2. A wrong in an operation/process leading end result in wrong operation is defined as
   (a) Fault (b) Error
   (c) Mistake (d) Defect
3. Non–performing element in the process due to which the instrument or equipment fails to work at the rated level is defined as______.
   (a) Fault (b) Error
   (c) Mistake (d) Defect
4. Accuracy means
   (a) Near to expected value
   (b) The degree of nearness to the true value
   (c) Repetitiveness of producing the results
   (d) Difference between true and measured values
5. Precision is defined as_____.
   (a) Near to expected value
   (b) The degree of nearness to the true value
   (c) Repetitiveness of producing the results
   (d) Difference between true and measured values
6. Resolution is defined as_______.
   (a) Near to expected value
   (b) The degree of nearness to the true value
   (c) smallest increment of change in measurand that can be determined from readout scale.
   (d) Unnoticed mistake or a gross error by oversight occurs due to unscientific method
7. Which of the following is not a source of instrument error_____.
   (a) Manufacturer
   (b) Measuring Device
   (c) Plant layout
   (d) Material
8. Sensitivity is defined as______.
   (a) Ratio of Change of output signal to Change of input signal.
   (b) Sum of Output signal to input signal
   (c) Difference between Output signal to input signal
   (d) Measure of Error related to accuracy
9. The defect that Originates at the stage of Design is called as______.
   (a) System inaccuracy
   (b) Inherent defect
   (c) Design Error
   (d) Material insufficiency
10. The defect that corrected by increasing and decreasing order is called_____.
    (a) Drift
    (b) Mechanical Error
    (c) Hysteresis
    (d) Reading Error
11. The defect caused by temperature change is called as______.
    (a) Drift
    (b) Mechanical Error
    (c) Hysteresis
    (d) Reading Error
12. Comparing with standard instrument and make necessary adjustment in the instrument or its display so as to make the instrument reads correct value.
    (a) Adjustment (b) Calibration
    (c) Calculation (d) Comparison
13. The ratio of the output to input change for a given measuring system is referred to as
    (a) Sensitivity (b) Linearity
    (c) Stability (d) Discrimination
14. The largest change in the measured variable, which produces no instrument response, is known as
    (a) Dead band (b) Dynamic error
    (c) Fidelity (d) Threshold
15. The speed of response of a first–order system is judged by
    (a) Transient response

(b) Time constant
(c) Steady state value
(d) Settling time

16. Error =
(a) True value + Measured value
(b) True value – Measured value
(c) True value / Measured value
(d) True value × Measured value

17. The smallest increment of change in the measured value that can be determined from the instrument's readout scale.
(a) Sensitivity (b) Resolution
(c) Stability (d) Discrimination

18. _____ is the inaccuracy in measurement which results in deviation from the expected or standard value.
(a) Error of measurement
(b) Measurement of error
(c) Error
(d) Fault

19. Systematic errors are also called as
(a) Human errors (b) Simple errors
(c) Bias errors (d) Common errors

20. Man-made error that happens during process or measurement is called
(a) Error (b) Fault
(c) Defect (d) Mistake

ANSWERS

1.	(b)	2.	(a)	3.	(d)	4.	(b)	5.	(c)
6.	(c)	7.	(c)	8.	(a)	9.	(b)	10.	(c)
11.	(a)	12.	(b)	13.	(a)	14.	(a)	15.	(b)
16.	(b)	17.	(b)	18.	(a)	19.	(c)	20.	(d)

CHAPTER INFOGRAPHIC

4 Glimpses of Transducers and Measurement of Displacement

Transducer is an energy converting device that receives signal from a physical situation or condition that is object of measurement and is converted into a definitely associated signal that is more appropriate to use as the output to a measurement system. A sensor can be defined as the element, which first detects the measurand and it is the element in contact with the process.

Classification of Transducers

Some important characteristics of transducers are Ruggedness, Repeatability, Good dynamic response, Linearity, Stability, reliability and cost effectiveness.

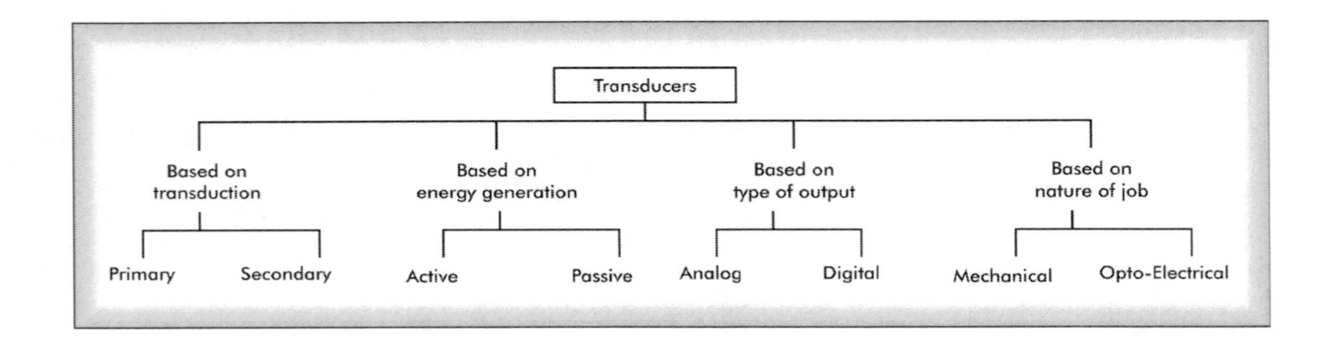

Some important characteristics of transducers are Ruggedness, Repeatability, Good dynamic response, Linearity, Stability, reliability and cost effectiveness.

Hall Effect: When an electric field and magnetic field are applied perpendicularly to the two perpendicular surface of the material, a voltage will be generated on the other surfaces, which is perpendicular to both. This effect is called HALL EFFECT.

Chapter 4

Transducers and Measurement of Displacement

STARTERS

To study this chapter, you should have awareness on the following concepts. For a better understanding, it is always a good idea to revise these prerequisites.

- Definitions and basics of various physical quantities.
- An understanding on various physical relations and expressions.
- A brief idea on system behaviours.
- Basics of mechanical engineering and laws of mechanics.
- Basics of electrical engineering and laws.
- Basic of electronics and principles.
- Fundamentals and laws of physics studied till 10+2 and engineering physics.
- Various fundamental and derived quantities, their denotations, and units.
- Standards of measurement and methods of standardizing.
- Notations, and units of various quantities.
- Principles governing the conversion of one form of physical quantity into another.
- Conversion factors of one system of units to the other for various quantities.
- Basic principles of solid mechanics, fluid mechanics, elasticity, heat, light, sound, magnetism, electricity and semi-conductors etc.

And

The contents discussed in the previous chapters of this book.

LEARNING OBJECTIVES

After studying this chapter you should be able to

- Understand the terms transducer and sensor.
- Classify the transducers and,
- Explain the various types of transducers.
- Explain the applications of transducers.
- Discuss the measurement of displacement.

4.1 INTRODUCTION

Transducer forms an important element of a measuring system. It is energy converting device that receives signal from a physical situation or condition i.e., object of measurement (the measurand) and is converted into a definitely associated signal (more appropriate to use as the output) to a measurement system.

It is essentially based on 'cause and effect' relationship. A spring, when subjected to force (input-cause), changes the length of the spring (output-effect) and thus converts force into displacement and forms the basis of a spring balance a force measuring device. In general, a transducer provides a usable output in response to a specific input, which may be a physical quantity, property or a condition. It may involve a conversion of one form of energy to another according to a specified relationship.

Another important term often used is sensor and is defined as given below:

Sensor: A sensor can be termed as the element, which first detects the measurand and it is the element in contact with the process. Contacting spindle of a dial gauge indicator, acts as a detector or sensing element for displacement sensors, may also serve to transduce the measurand and the unit is called detector transducer. Strictly speaking, mercury thermometer, thermocouple and a photodiode are sensors as well as transducers.

4.2 CLASSIFICATION OF TRANSDUCERS

Transducers can be classified as:

1. Based on stage of transduction
   (a) Primary Transducers (b) Secondary Transducers
2. Based on energy generation
   (a) Active Transducers (b) Passive Transducers
3. Based on type of output
   (a) Analog Transducers (b) Digital Transducers
4. Based on nature of job
   (a) Mechanical Transducers (b) Electromechanical Transducers
   (c) Opto-electrical transducers.

More Instruments for Learning Engineers

Vernier Calipers, Micrometers (also called **screw gauge)** measure smaller lengths and precise measurements of components with the help of a calibrated screw. Micrometers are usually in the form of calipers i.e. opposing ends joined by a frame, (hence called **micrometer caliper**) and has a spindle, accurately machined in the form of screw. The object to be measured is placed between spindle and anvil. The spindle is moved by turning the ratchet knob or thimble until the object to be measured is lightly touched by both spindle and the anvil. Micrometers can also used in telescopes (or microscopes) to measure the apparent diameter of celestial (or tiny) bodies. Other similar metrological instruments are dial, Vernier and digital calipers.

Outside/Inside/Depth Micrometers-Vernier/ Dial/Digital Calipers

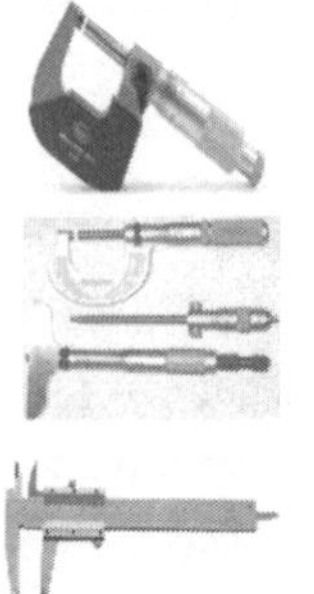

The classification of transducers is shown in Figure 4.1.

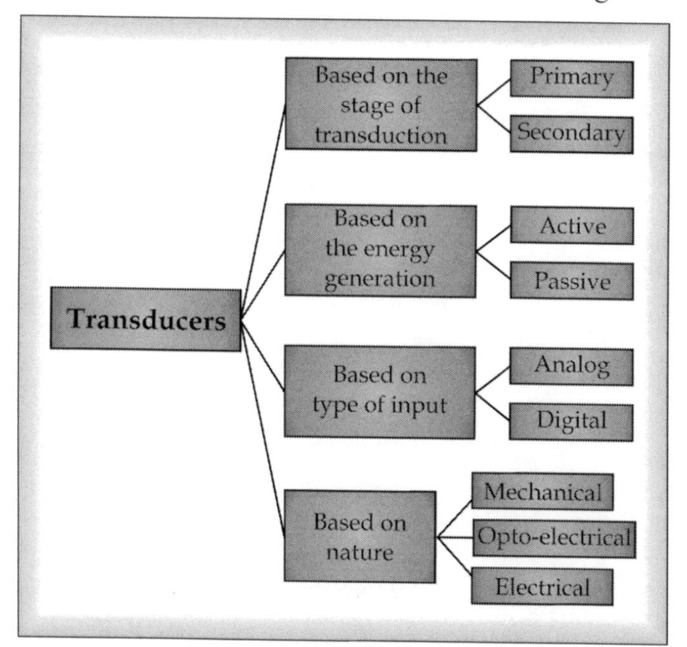

Fig. 4.1 Classification of Transducers

1. **Primary Transducers:** These transducers detect or sense or react to the change in the physical quantity to be measured directly without any mediation in between the process. For example, the liquid-in-glass thermometer senses the heat energy due to which the liquid expands proportionately and the elongation measures the temperature.
   Some primary detector traducer elements and operations they perform are given in Table 4.1.
2. **Secondary Transducers:** These transducers take input from the output of a primary transducer and change it to another form usually to an electrical output. For example, consider an arrangement to measure force/ weight in the Figure 4.2. The displacement is caused by the weight W at the point A due to the spring (Primary Transducer). This becomes an input in the form of movable contact to cause a change in the resistance R of a potentiometer. Thus it converts the displacement due to force/ weight into varying voltage drop and forms a secondary transducer.
3. **Active Transducers (Internally Powered Transducers):** These transducers do not require any auxiliary source of external power and are self generating type. They absorb energy from the physical variable to be measured to produce their output. They operate under the energy conversion principle. Example: piezoelectric and photo voltaic transducer.

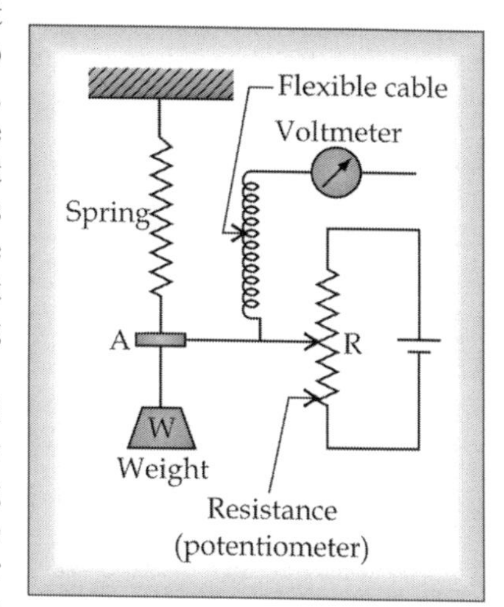

Fig. 4.2 Secondary Transducer

4. **Passive Transducers (Externally Powered Transducers):** These transducers require excitation or energy from the external source. They also absorb some energy from the physical variable to be measured. For example, capacitive transducers, strain gauges come under this category.

5. **Analog Transducers:** In analog transducers, with the variation of input, there is a continuous variation of the output. Most transducers are of analog type.
6. **Digital Transducers:** In these transducers, the outputs are discrete in nature or vary in steps with '0' or '1' type output i.e., "yes" or "no" form such as conducting or not conducting, light or no light.

The following Table 4.1 gives a list of elements and operation they perform.

Table 4.1 Some Primary Detector Transducer Elements and their Operations

Element	Operation
1. Mechanical	
A. Contacting spindle, pin or plunger	Displacement to displacement
B. Elastic member	
(i) Load cells (a) Tension / compression	Force to linear displacement
(b) Bending	Torque to angular displacement
(c) Torsion	Force to linear displacement
(ii) Proving ring	Pressure to displacement
(iii) Bourdon tube	Pressure to displacement
(iv) Bellows	Pressure to displacement
(v) Diaphragm	Pressure to displacement
(vi) Helical spring	Force to linear displacement
C. Mass	
(i) Seismic mass	Forcing function to relative displacement
(ii) Pendulum	Gravitational acceleration to frequency or period
(iii) Liquid column	Force to displacement
D. Thermal	
(i) Thermocouple	Temperature to electric potential
(ii) Biomaterial (including mercury in glass)	Temperature to displacement
(iii) Thermostat	Temperature to resistance change
(iv) Chemical phase	Temperature to phase change
(v) Pressure thermometer	Temperature to pressure
E. Hydro pneumatic	
(i) Static	
(a) Float	Fluid level to displacement
(b) Hydrometer	Specific gravity to relative displacement
(ii) Dynamic	
(a) Orifice	Fluid velocity to pressure change
(b) Venturi	Fluid velocity to pressure change
(c) Pitot tube	Fluid velocity to pressure change
(d) Vanes	Velocity to force
(e) Turbines	Linear to angular velocity

Table 4.1 *Contd....*

Element	Operation
2. Electrical (or Electromechanical)	
A. Resistive	
(i) Contacting type	Displacement to resistance change
(ii) Variable- length conductor	Displacement to resistance change
(iii) Variable- area conductor	Displacement to resistance change
(iv) Variable-dimensions of conductor	Strain to resistance change
(v) Variable resistivity to conductor	Temperature to resistance change
B. Inductive	
(i) Variable coil dimensions	Displacement to change of inductance
(ii) Variable air gap	Displacement to change of inductance
(iii) Changing core material	Displacement to change of inductance
(iv) Changing core position	Displacement to change of inductance
(v) Changing coil positions	Displacement to change of inductance
(vi) Moving coil	Velocity to change in induced voltage
(vii) Moving permanent magnet	Velocity to change in induced voltage
(viii) Moving core	Velocity to change in induced voltage
C. Capacitive	
(i) Changing air gap	Displacement to change in capacitance
(ii) Changing plate areas	Displacement to change in capacitance
(iii) Changing dielectric constant	Displacement to change in capacitance
D. Piezoelectric	Displacement to voltage and/ or voltage to displacement
E. Semiconductor junction	
(i) Junction threshold voltage	Temperature to voltage change
(ii) Photodiode current	Light intensity to current
F. Photovoltaic	
(i) Photovoltaic	Light intensity to voltage
(ii) Photoconductive	Light intensity to resistance change
(iii) Photoemissive	Light intensity to current
Hall Effect	Displacement to voltage

Self Assessment Questions-4.1

1. What is a transducer?
2. What is a sensor?
3. Give the classification of the Transducers.
4. Discuss the following
   (a) Primary Transducers
   (b) Secondary Transducer
   (c) Active Transducers
   (d) Passive Transducers
5. Discuss the following
   (a) Analog Transducers
   (b) Digital Transducers
6. List out Some Primary Detector Transducer Elements with their Operations.

4.3 MECHANICAL TRANSDUCERS

There are several principles available to convert input quantities (to be measured), into analog output. In most of these methods, the output is shown as linear or angular displacement. This displacement can be indicated directly by a pointer on a scale like in the case of a spring balance and or after amplification like in a bourdon tube pressure gauge. Most often, mechanical transducer acts as primary transducer and then a secondary transducer is used to convert its output into a desired electrical quantity. The most common mechanical transducers are listed below. [The principle, construction, operation and applications of these transducers in various instruments are discussed in detail in the forthcoming chapters of this book. Therefore, brief explanations only are given here].

- ***Mass and seismic mass transducers:*** Consider a simple pendulum shown in Figure 4.3(a) with which we can determine time period, frequency, angular velocity and acceleration due to gravity, force or weight etc. Thus the mass of the bob can behave as a transducer to measure these physical quantities. Similarly, seismic mass, mass of liquid column can also be used as transducers. [refer chapter – 12 of this book for details]

> ***Characteristics of Transducers***
>
> Ideally, a transducer should have:
> 1. Ruggedness
> 2. Repeatability
> 3. Good Dynamic Response
> 4. Linearity
> 5. Stability and Reliability
> 6. Cost effectiveness

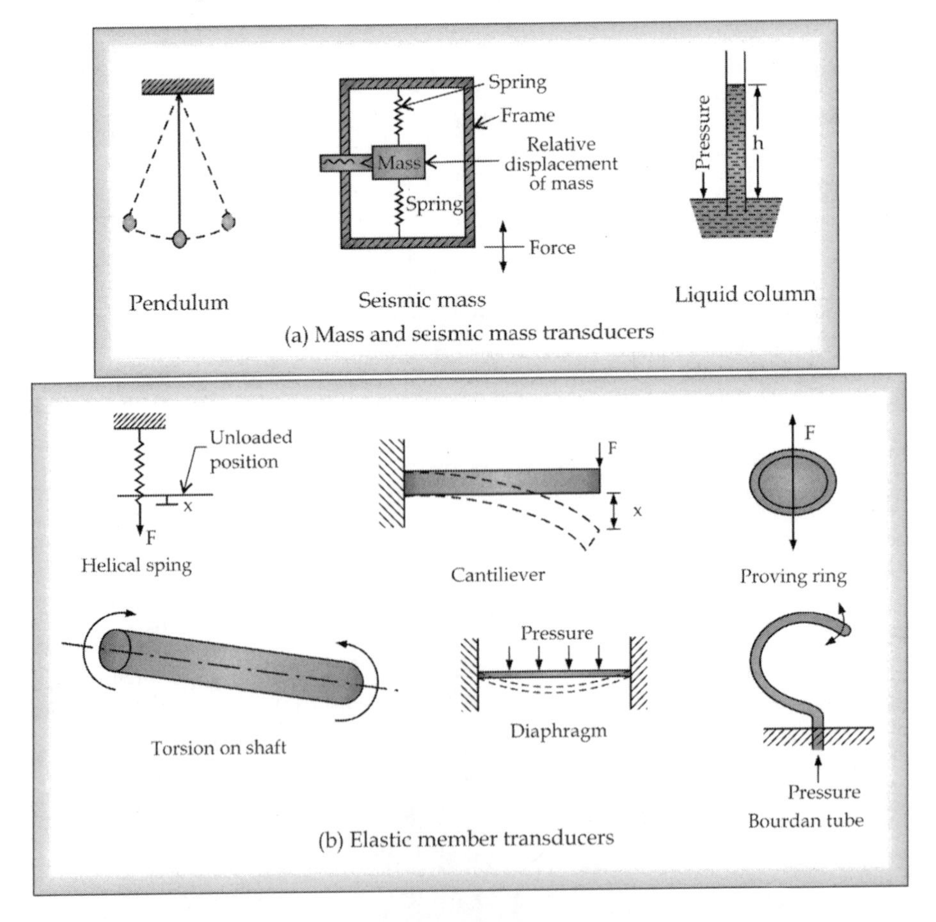

(a) Mass and seismic mass transducers

(b) Elastic member transducers

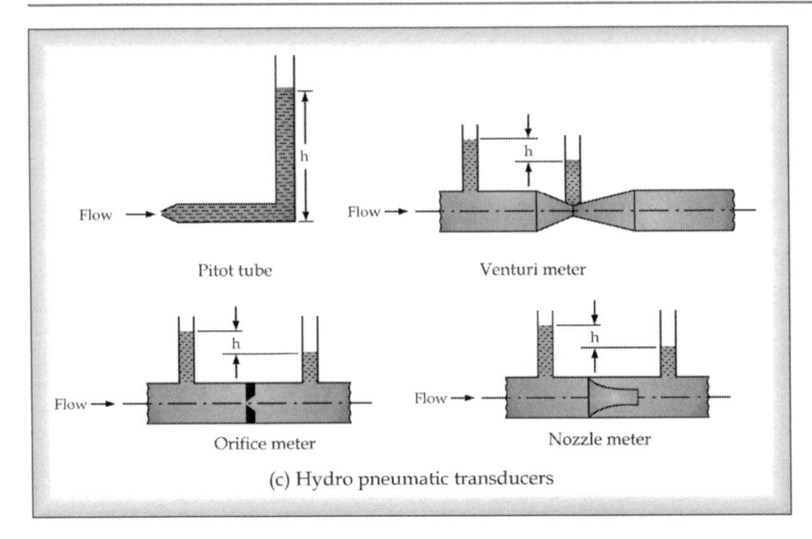

(c) Hydro pneumatic transducers

Fig. 4.3 Mechanical Transducers

- *Elastic member transducers:* A spring, a cantilever, a shaft, a diaphragm, a proving ring and a bourdon tube are various examples of simple elastic members shown in Figure 4.3 (b), can be used to measure the physical quantities such as velocity, acceleration, strain, stress, force, torque, power, pressure and so forth. Here, these transducers use the principle that the deformation or displacement of elastic members is related to the quantity to be measured through a proven physical relationships such as Hook's law, Newton's laws etc. [Respect chapter 12 for more details]
- *Thermal transducers:* The thermal transducers use the simple thermal and heat principles such as expansion of solids and liquids. The bimetallic strip and mercury-in-glass thermometer are some of the examples of thermal transducers used for temperature measurements and other fields of instrumentation. [Refer chapter – 5 for details]
- *Hydraulic transducers:* The hydraulic transducers include pitot tube, orifice meter, flow nozzle, venturi meter used for flow measurement and convert velocity to pressure as shown in Figure 4.3(c). [Refer chapters – 7 & 8 for more details]
- *Pneumatic transducers:* Nozzle and flapper is a simple example of pneumatic transducer.

ILLUSTRATION-4.1

Illustrate the functioning of a Nozzle and Flapper to act as hydraulic/pneumatic control transducer with the help of a neat sketch. Mention the principle, transduction, construction, operation and applications.

Solution:

Nozzle and Flapper Transducer

Principle of transduction

The displacement of the flapper causes a proportional change in the back pressure (P_b), if the supply pressure is kept constant. This linear relationship, is valid over a limited range of displacement as shown in Figure 4.4(b).

Transduction

Displacement to Pressure.

Construction

It consists of two orifices, one is a fixed flow restriction orifice (O_1) and a variable flow restriction orifice (O_2) controlled by a nozzle and a flapper as shown in Figure 4.4(a).

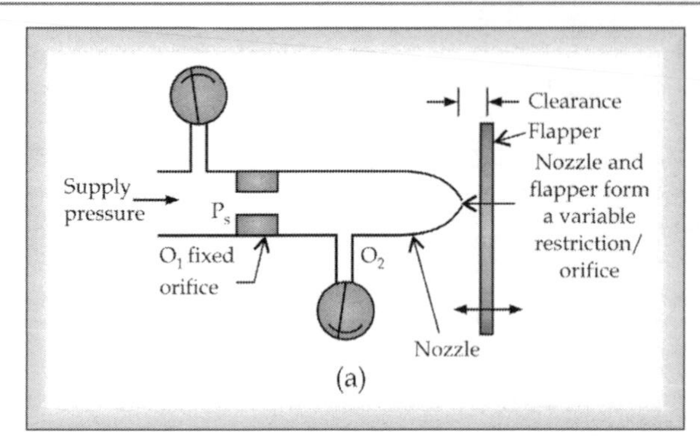

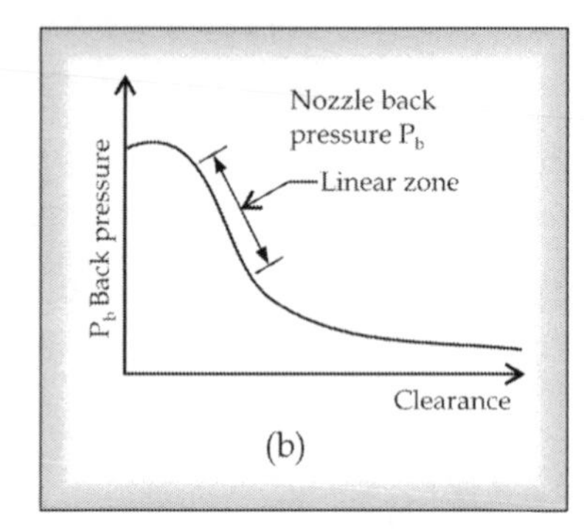

Fig. 4.4 Nozzle and Flapper Transducers

Why electrical transducers are preferred?

Though mechanical transducers are rugged, stable and cost effective, still electrical transducers are preferred by instrumentation designers. This is due to below listed main merits of an electrical transducer.

1. The output can be amplified to any desired level.
2. The output can be indicated and recorded remotely at distance from the sensing medium
3. More than one modulator can be actuated simultaneously.
4. The output can be modified
5. The signal can be conditioned or mixed to obtain any combination with outputs of similar transducers or control signals.
6. The size and shape of the transducer can be suitably designed to achieve the optimum weight and volume,
7. The contour design and dimensions can be so chosen as not to disturb the measured phenomena.

Operation

The variable flow restriction orifice (O_2) is varied by changing the distance of the flapper with respect to the nozzle. The nozzle is set with pressurized air (P_s) through the fixed orifice (O_1). As flapper approaches the nozzle, the resistance to flow of air through the nozzle increases and thence the back pressure (P_b) increases. Similarly, when the flapper is moved away from the nozzle, more air escapes from the nozzle and back pressure drops. Thence, the displacement (of flapper) measures the pressure when calibrated.

Application

Thus, it is a basic component used in pneumatic and hydraulic control and transmission, which converts displacement to pressure. It is also used in precision pneumatic gauges for dimensional measurements.

Self Assessment Questions-4.2

1. List out the various mechanical transducers.
2. Discuss the mass and seismic mass transducers.
3. Explain the elastic member transducers.
4. Describe the thermal transducers.
5. Explicate the hydraulic transducers.
6. Discuss the Pneumatic transducer.
7. Explain Nozzle and Flapper Transducers.

4.4 ELECTROMECHANICAL TRANSDUCERS

These transducers convert mechanical quantities (such as displacement, velocity, force, torque, pressure and temperature) into electrical quantities (such as a change of resistance and capacitance or generate an emf/electrical charge).The change in resistance, inductance or capacitance can then be measured by several devices depending on the magnitude of change. Generally Wheatstone bridge (dc or ac) circuit is used, the output voltage of which becomes an indication of the quantity being measured. The Table 4.2 shows change in parameter and the device employing the change.

Table 4.2 Devices Employing the Change of the Parameters

Change in Parameter	Devices Employing the Change
1. **Length:** Length can be varied by moving a sliding contact, used for the measurement of mechanical linear and angular displacement	Resistance potentiometer (linear and rotary type)
2. **Dimension:** Mechanical strain changes the dimensions of the conductor used for strain measurement	Electrical resistance strain gauges.
3. **Resistivity:** Resistivity changes with temperature used for temperature measurement.	Thermistors

4.4.1 Merits

The merits of electromechanical transducers are:

1. Effect of friction is minimum.
2. Amplification or attenuation can be done easily.
3. Inertia effects are minimum.
4. Minimum size (miniaturization).
5. Indication or recording at a distance from sensing point is facilitated, by using suitable cables.
6. Electrical signals can be made compatible for computer applications (Interfacing to PC is easy).
7. The output power of any magnitude can be provided.

Various electro-mechanical transducers are dealt in the following sections.

More Instruments for Learning Engineers

Fathom is a unit of water depth, and so named, *Fathometer* measures water depth and also used for navigation and safety. Fish finder is an another instrument derived from fathometer work the same way. Now-a-days both are merged as both use similar frequencies and can detect both the bottom and fish. This works on the principle of **Echo sounding,** a type of SONAR used to determine the depth of water by transmitting sound pulses into water and recording the time interval between emission and return of a pulse.

Fathometer & Echo sounding

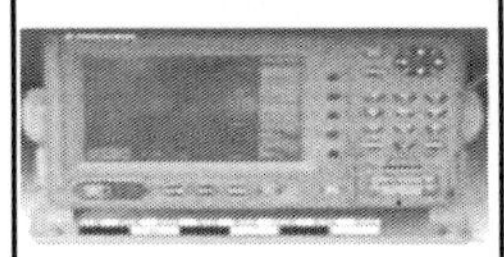

4.4.2 Classification of Electromechanical Transducers

Various electromechanical transducers are classified as give below.

(i) Based on Resistance
   1. Variable Resistance Transducers
   2. Sliding Contact Devices (Potentiometers) or Pots
   3. Resistance Strain Gauges
   4. Thermistors

(ii) Based on Inductance/Induction
   1. Inductive Type Transducers
      (a) Non-self generating type [Variable self inductance (Single/two coil) and Variable Mutual Inductance (two/three coil)]
      (b) Self generating type
      (c) Eddy current transducers
   2. Magnetometer Search Coil Transducers

(iii) Based on Capacitance
   Capacitive transducers

(iv) Piezoelectric Transducers

(v) Photoelectric Transducers
   1. Photo emissive type
   2. Photo conductive type
   3. Photo voltaic type

(vi) Hall effect Transducers

(vii) Ionization Transducers

(viii) Digital Displacement Transducers

We shall now discuss about the above transducers.

4.5 RESISTIVE TYPE TRANSDUCERS

There are four main types of transducers which use the change in resistance (fixed or variable) proportionally to various physical quantities or dimensions as the core principle of transduction. The circuit symbols for fixed and variable resistance are given in Figure 4.5(a).

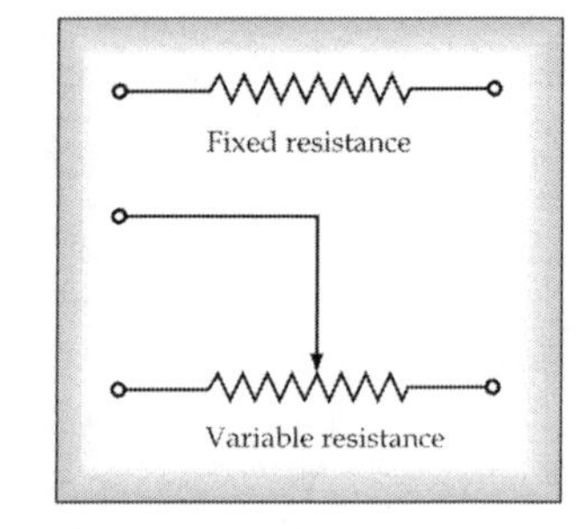

Fig. 4.5(a) Circuit Symbols of Fixed and Variable Resistances

Principle

This transducer works on the principle that resistance varies directly with length and inversely with the area of cross section of an electrical conductor, which is given by the expression, $R = \rho L/A$

where

R = Resistance in ohm

L = length of the conductor in m

A = Cross sectional area of the conductor in m^2

ρ = Resistivity of the material in ohm-m.

Thus the electrical resistance can be varied by altering (i) length (ii) cross sectional area (iii) resistivity or a combination of these. Hence, a physical variable which causes a change in any of the above factors can be used as sensors for a given resistance change [Refer table 4.2].

4.5.1 Variable Resistance Transducers

Transduction

Input Physical quantity → Physical dimensions → change in resistance.

Principle

This transducer operates using the same principle as above i.e., the resistance varies directly with length and inversely with the area of cross-section of an electrical conductor and if this change is made to occur due a desired physical quantity, then the change in resistance becomes the measure of the desired physical quantity when calibrated.

Construction

This transducer resembles the meter bridge circuit which consists of a long wire (acts as resistor) tied between two ends of a bridge as shown in Figure 4.5 (b). A slider or a brush is arranged to glide on a guide rod whose tip moves on the wire. One electrical lead is connected to one of the ends of the bridge while the other is connected to the slider. These leads are connected to a device that measures the resistance or voltage.

Operation

This circuit arrangement is kept the instrument which has to measure a desired physical quantity. When the slider moves on the wire the resistance varies proportionally according to the length of the wire. The resistance can be measured by a device to read the required physical quantity that causes this resistance when calibrated accordingly. This arrangement can be used in a null mode or deflection mode as required. The schematic, circuitry and graphic representations are given in Figure 4.5(b).

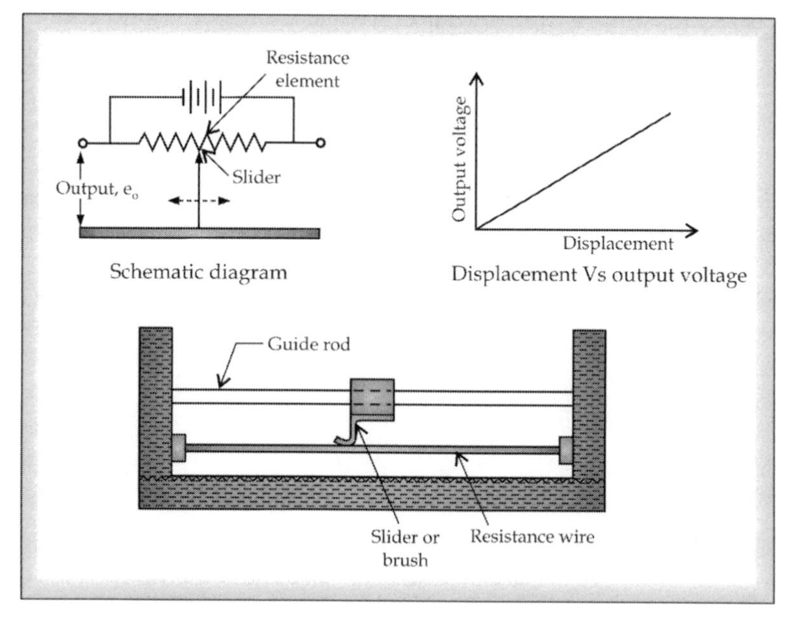

Fig. 4.5(b) Variable Resistance Type Transducer

Applications

This transducer can be used in several forms, some of which are given below.

- Probably the simplest mechanical-to-electrical transducer is the ordinary switch in which resistance is either zero or infinity. It is a yes-no (i.e., conducting-non conducting) device that can be used to operate an indicator. In such transducer, we do not need to keep a variable resistor; instead we can use a fixed resistance, such as a miniature bulb also like in an electrical tester used to check if current is passing or not.
- In its simplest form, the switch may be used as a limiting device operated by direct mechanical contact or it may be used as a position indicator.
- When actuated by a diaphragm or bellows, it becomes a pressure-limit indicator.
- If controlled by a bimetallic strip, it can be used as a temperature limit indicator.
- It may also be combined with a proving ring to serve as either an overload warning device or a device actually limiting load-carrying, such as safety device for a crane.

In the above cases, the slider may be connected to bellows or diaphragm or such part of the instrument in such a way that the part (such as bellows or diaphragm) due to some physical quantity (like pressure or load) causes movement to the slider to vary the resistance.

4.5.2 Sliding Contact Devices – The Pots

These devices are commonly called *resistance potentiometers or simply pots.*

Transduction

Physical dimensions to resistance change.

Principle

Sliding contact resistive transducers also work on the same principle that the resistance changes by changing the effective length *(l)* of the conductor. They convert a mechanical displacement input into an electrical output in the form of voltage or current which are proportional to the resistance.

Construction

In these transducers, some kind of electrical resistance element is used, with which a contactor (slider or pointer or brush) maintains electrical contact as it moves or slides. In its simplest form, the device may consist of a stretched resistance wire and slides, as shown in Figure 4.6 (left-top sub-figure). The effective resistance between either of the ends of the wire and the contactor (slider or pointer or brush) thereby becomes a measure of mechanical displacement.

We can find two types of resistance elements in the sliding contact devices. These are

(a) Wire wound resistance element

(b) Conductive film (or cermet) resistance element

Most usually, the resistance element is made by wrapping a resistance wire around a card or some form. The turns are spaced or separated by insulators to prevent shorting and the brush slides across the turns (from one turn to the next). In actual practice, either the arrangement may be wound for a rectilinear movement or the resistance element may be made as an arc for angular movement as shown in right-top sub-figure of Figure 4.6.

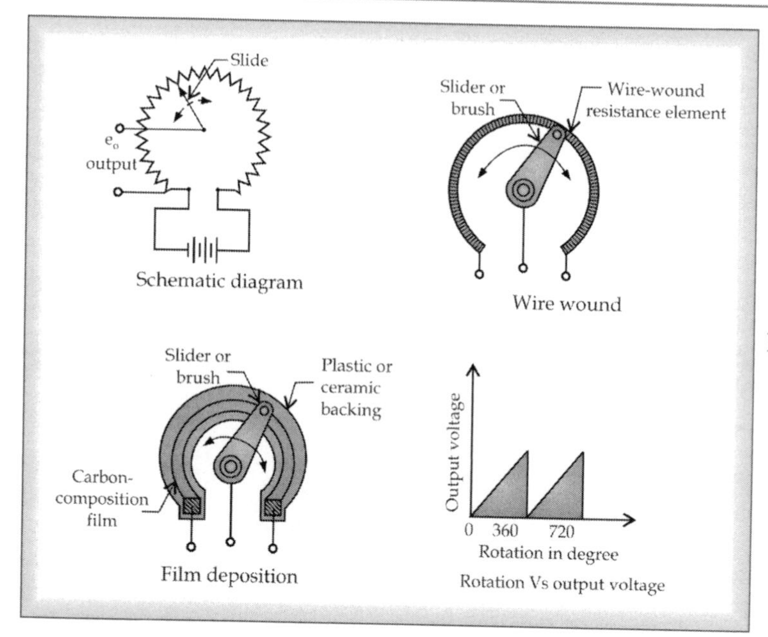

Fig. 4.6 Angular Potentiometer

Sliding contact devices are also made up of conductive films as the variable resistance elements, rather than wires as shown in Figure 4.6 (left-bottom sub-figure). Most commonly used materials are the carbon composition films or cermets, in which graphite or carbon particles are suspended in an epoxy or polyester binder, and ceramic metal. Composition films or cermets (ceramics+metals) are in which ceramic and precious metal powders are combined. In any case, the thin film is backed by a ceramic or plastic.

Potentiometer Resolution

Resistance change in a sliding-contact traversing over a wire-wound resistance element is not a continuous function of contact movement. The smallest increment into which the whole may be divided, gives the resolution. In the case of wire-wound resistance, the limiting resolution is equal to the reciprocal of the number of turns. Suppose 100 turns of wire are used and the winding is linear, the resolution is 1/100, or 0.01%. No matter how refined the remainder of the system may be, it is impossible to divide or resolve the input into parts smaller than 1/100 of the total potentiometer range.

In case of conductive-film potentiometers, the resolution is negligibly small and variation of this slider-contact-resistance is a more significant limitation.

Potentiometer Linearity

Generally, a linear potentiometer is needed when used as measurement transducers. The term 'linear' means here that the resistance measured between one of the ends of the elements and the contactor is a direct linear function of the contactor position in relation to that end. Strictly speaking, the linearity is never completely achieved; however, deviation limits are usually supplied by the manufacturers.

Merits of conductive film devices over wire wound devices

- Conductive film devices are less expensive than wire wound devices.
- The carbon film devices, in particular, have outstanding wear characteristics and long life, although they are more susceptible to temperature drift and humidity effects.

Demerit

- Resolution is negligibly small and variation of this slider-contact-resistance is a significant demerit.

4.5.3 The Resistance Strain Gauges

Transduction

Strain → Change in physical dimensions → Change in resistance

Principle

These are based on the principle that the electrical resistance changes when a resistive element is mechanically strained.

Construction and Operation

The resistive element is bonded to the surface of the member to be strained and as it elongates with applications of strain (such as a tensile strain) its cross-sectional area reduces. Thus it results in a longer length of smaller cross-section element and hence the change in resistance due to this dimensional variation can be measured and calibrated. [However, simply accounting for this dimensional change does not complete l y explain the behavior of the gauge; there is also a change in resistivity with strain and temperature also. This is more clearly discussed in the chapter 12 of this book].

The construction and operation of various instruments by the applications of these transducers is well explained in chapter 12 of this course.

4.5.4 Thermistors

Thermistors are variable thermal resistors made of ceramic like semiconducting materials. Oxides of manganese, nickel and cobalt whose resistivity values are 1000 to 4500000 Ωm are used in these transducers.

These devices have two basic applications:

- As temperature detecting and measuring elements used for the purpose of measurement or control [This is discussed more clearly in chapter – 5, the next chapter of this book].
- As electric power sensing devices wherein the thermistor temperature and hence resistance are a function of the power being dissipated by the device. The second application is particularly useful for measuring radio-frequency power.

Self Assessment Questions-4.3

1. What is the basic function of electromechanical transducers?
2. How do the devices employing the change of the parameters?
3. What are the merits of electromechanical transducers?
4. List out the various electromechanical transducers.
5. With a neat sketch explain resistive type transducers.
6. Describe variable resistance transducers with a neat diagram.
7. Explain sliding contact devices with a neat diagram.
8. Discuss the resistance strain gauges.
9. What is a thermistor? What are its applications?

4.6 INDUCTIVE TYPE TRANSDUCERS

Principle

These transducers work on the principle that inductive reactance (X_L) in ohm of an inductance (L) is given by $X_L = 2\pi f L$

where f = applied frequency in Hz

and L = Inductance in Henry

The total impedance of coil Z is given by $Z = \sqrt{X_L^2 + R^2}$

where R is its d.c resistance

Using the above principles, the inductive transducers operated can be classified into the following three categories.

1. Non-self-generating transducers
   - (a) Variable Self Inductance type
     - (i) Single coil type
     - (ii) Two coil (Single coil with center-tap) type
   - (b) Variable Mutual Inductance type
     - (i) Two coil type
     - (ii) Three coil type
2. Self-generating transducers and
3. Eddy current type transducers

We shall now discuss about these transducers in detail.

Non Self Generating Type

These may consist one or more coil and require an external source of power to energize the coils. They are generally used for the measurement of displacement and can be further categorized as variable self inductance type and variable mutual inductance type. Depending on the number of coils, these may be further divided into two types in each category.

Variable Self Inductance type

(a) Single coil type (b) Two coil (Single coil with centre-tap)

More Instruments for Learning Engineers

A **taximeter** is a mechanical or electronic device installed in taxicabs and auto rickshaws that calculates passenger fares based on a combination of distance travelled and waiting time. It is the shortened form of this word that gives the "taxi" its name.

Taximeter

Variable Mutual Inductance

(a) Two coil type

(b) Three coil type.

4.6.1 Non-Self-Generating Variable Self Inductance-Single Coil Type Transducer

Principle

The inductive reactance varies proportionally with the inductance which can be varied by displacement of armature thus changing the air gap.

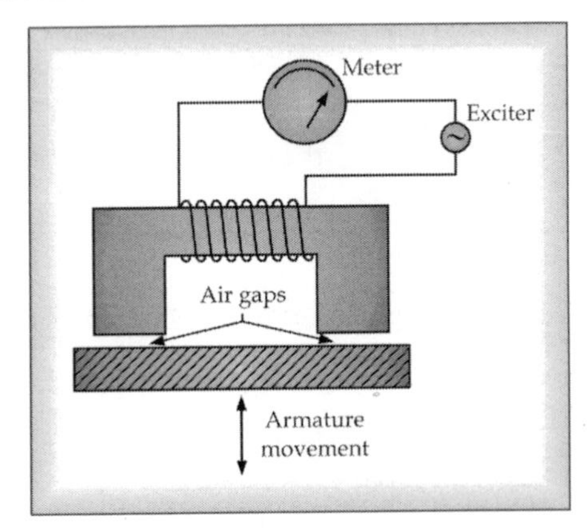

Fig. 4.7 Single Coil Variable Self Inductive Transducer

Construction and Operation

A coil with a core of magnetic material (used for magnetic flux path) acts as an inductor on a.c. exciter as shown in Figure 4.7. This arrangement is connected to a calibrated electric meter (measuring device calibrated with respect to inductance or voltage or current). The inductance can be varied by varying the turns in the coil and the permeability of the flux path. The flux path includes the path through magnetic material as well as the air gap. Since the permeability of magnetic material is very high, the more effective way of varying the inductance is by varying air gap.

The variation in air gap is often caused by the displacement of the armature, resulting in the change of inductance. This change of inductance can be measured by a.c. bridge circuit or by a change of current in the circuit.

The extension to the idea of the above discussed single coil inductive transducer gave a scope to construct the transducers with the following actions.

- Transformer action between two or more coils,
- Generator action in a coil moving in a magnetic field and
- Generation of eddy currents in a conducting object.

4.6.2 Non-Self-Generating Variable Self Inductance Two Coil (or Single Coil with Center Tap) Type Transducer

Principle

When movement of the core changes, the relative inductance of the two portions of the coil (L_1 and L_2) changes i.e., the inductance of one portion increases and the other decreases.

Construction and Operation

The two inductances, L_1 and L_2 can form the two arms of a Wheatstone bridge as shown in Figure 4.8 and the output voltage of the bridge circuit becomes the measure of the motion or displacement of the core [observe that this is in fact like a single coil transducer as discussed under section 4.6.1 and because of the center tap, the coil is split into two. If these two coils are separated and kept as shown in Figure 4.9, it can be made as a two coil mutual inductance type of transducer].

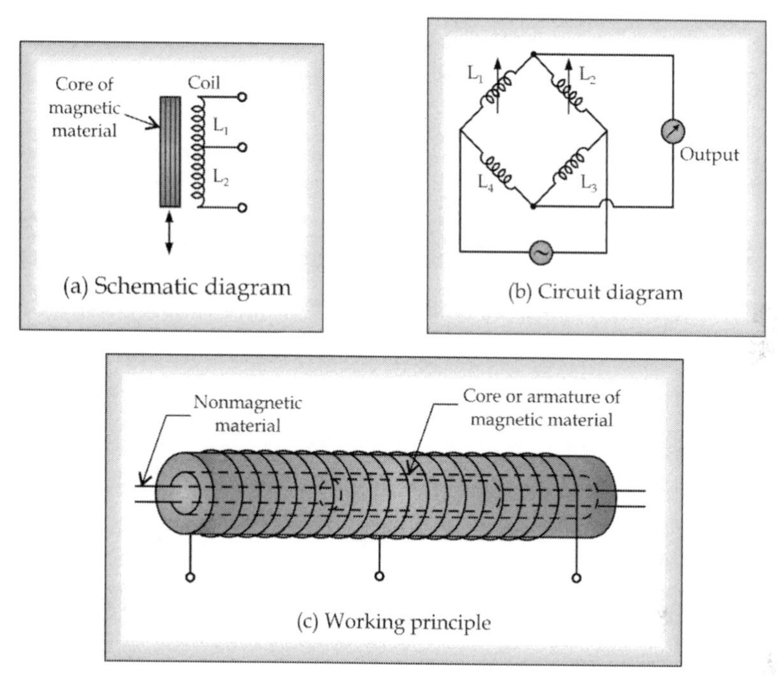

Fig. 4.8 Two Coil Self Inductance or Center Tap-Single Coil Transducer

4.6.3 Non-Self-Generating Variable Mutual Inductance Two Coil Transducer

Principle

The air gap varied by the motion of the armature will change the output voltage of the coil-2 which becomes the measure of displacement when calibrated.

Construction

Two coils are wound to the sides of the core, one connected to the external source of a.c. and the second to the output. This set up is kept at an air gap with the armature as shown in Figure 4.9.

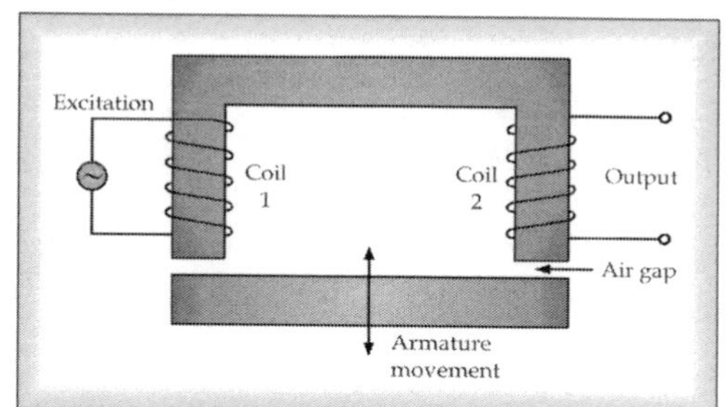

Fig. 4.9 Two Coil Mutual Inductance

Operation

The coil-1 is excited by an external source of alternating current. The flux coupling the output coil-2 depends upon the air gap. The air gap, when varied by the motion of the armature will change the output voltage of the coil-2 as shown in Figure 4.9.

Functionally, coils 1 and 2 act as the two windings of a transformer. Thus the output voltage of coil-2 becomes a measure of the displacement of the armature.

4.6.4 Non-Self-Generating Variable Mutual Inductance Three Coil Transducer

Linear Variable Differential Transformer (LVDT)

Principle

The induced voltage E_0 varies linearly with displacement of the core from the null or central position in a central range.

Construction

The idea of extending the two coils set up to three coils set up paves the way to construct the LVDT, which is a three coil variable mutual inductance transducer.

This transducer consists of a primary coil (P) excited by input ac power source (E_i) and two identical secondary coils S_1 and S_2 placed symmetrically with respect to the primary coil as shown in Figure 4.10. The two secondary coils are connected in series opposition. The magnetic coupling between primary and secondary coils is provided by a core of magnetic material. When the core is in a central position, the voltages (produced by transformer action), in two secondary coils are equal. But the output is zero, as they are connected in series opposition. Motion of the core up and down varies the relative flux linking the two secondary coils. The voltage induced in one coil increases and in the other decreases. The output voltage E_0 varies linearly with displacement of the core from the null or central position in a central range. The construction of the device is shown in Figure. 4.11.

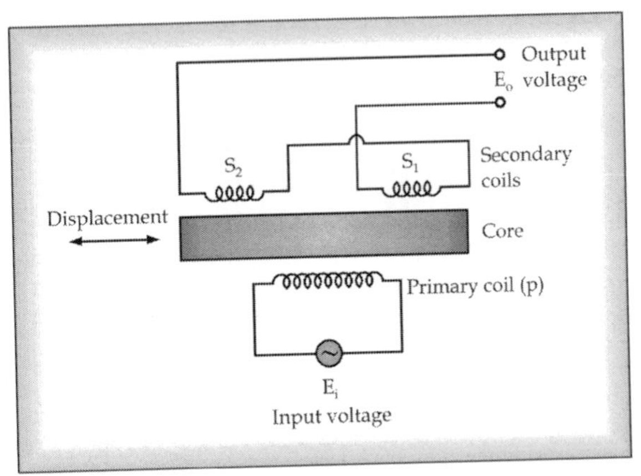

Fig. 4.10 Schematic of a Linear Variable Differential Transformer

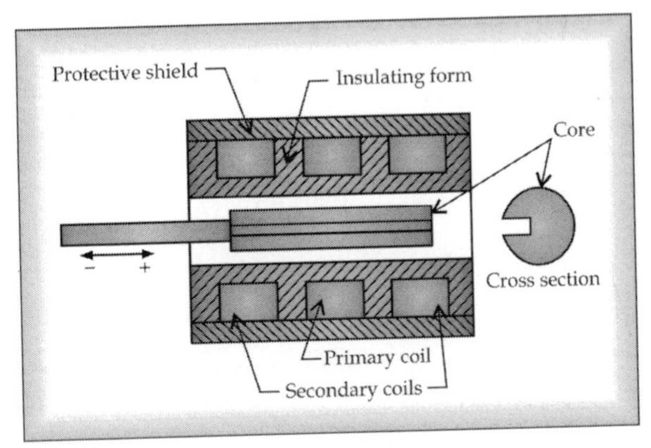

Fig. 4.11 Construction of LVDT

Operation

As long as the core remains near the centre of the coil arrangement, the output is very nearly linear, as indicated in Figure 4.12. Commercial differential transformers are operated in linear range.

When operating in the linear range, the device is called Linear Variable Differential Transformer (LVDT). Near the null position a slight non linearity condition is encountered, as illustrated in Figure 4.13. It will be noted that Figure 4.12 considers the phase relationship of the output voltage, while the "V" graph in Figure 4.13 indicates the absolute magnitude of the output. There is 180^0 phase shift from one side of the null position to the other.

The frequency response of LVDT is primarily limited by the inertia characteristics of the device. In general, the frequency of the applied voltage should be 10 times the desired frequency response.

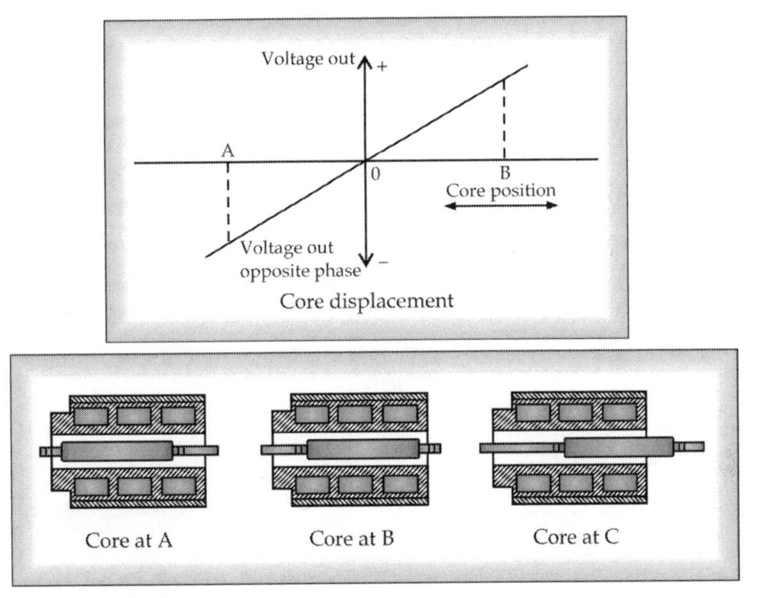

Fig. 4.12 Output Characteristics of a LVDT

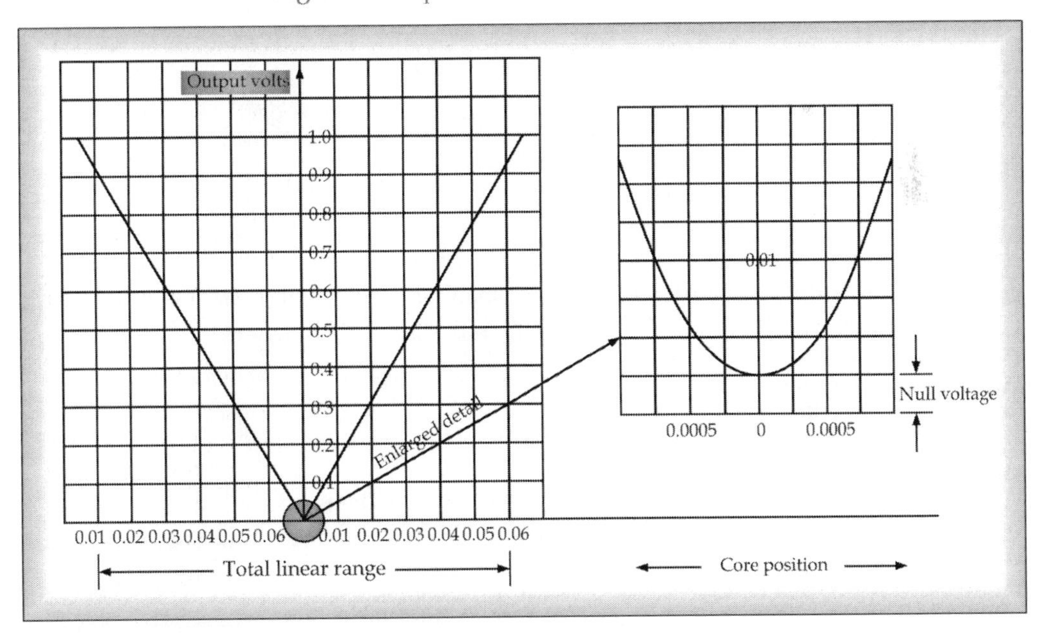

Fig. 4.13 V-Graph for LVDT

4.6.5 Self Generating Type Inductive Transducers

In this type of transducer a voltage signal is generated. The basic components of such a transducer are a conductor and a magnet.

Principle

In these transducers, a rate of change of flux, caused by the motion (of magnet or coil), generates a voltage in the coil or conductor (generator action).

This can be achieved by,

1. Keeping the magnet fixed and moving the coil relative to it
2. Keeping the coil fixed and moving the magnet relative to it
3. Keeping both the magnet and the coil fixed but varying the magnetic flux linking the two by some moving object of magnetic material. These types of transducers are known by various names as electrodynamics, electromagnetic, proximity pickup, etc. and are basically sensitive to velocity. Figure 4.14 shows such type of transducers.

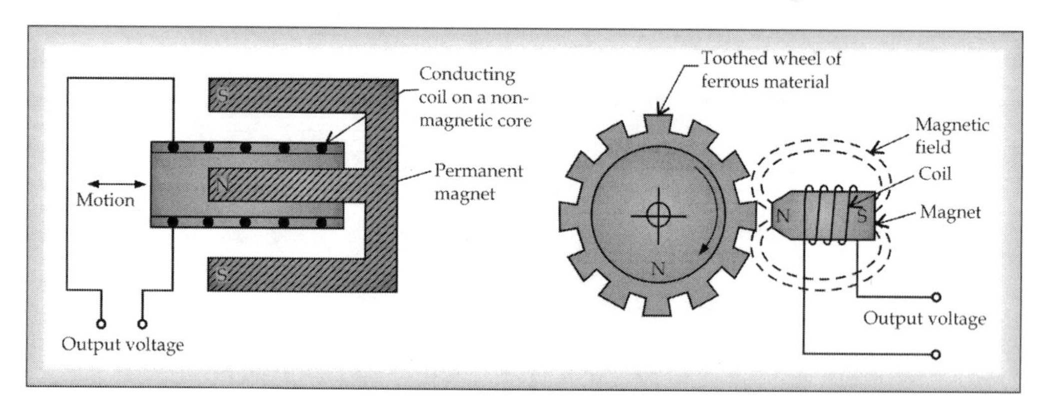

Fig. 4.14 Self Generating Inductive Transducer

4.6.6 Eddy Current Transducers

Principle

A conductor in the form a piece of plate of disc, when moves through a magnetic field, produces eddy currents which when calibrated becomes the measure of displacement.

Construction and Operation

When a conducting material moves through a magnetic field, a voltage is produced; if the conductor is in the form of a wire; it results in a generator action as discussed in the previous sections. If the conductor is a piece of plate or disc, the local currents will be produced, called eddy currents. These currents flow in short circulated paths within the plate and result in heat dissipation. The operation is explained below.

Figure 4.15 shows a non- contact type of motion measuring transducer. The object whose motion is to be measured, acts as a conduction surface. An active coil, powered by alternating current, is placed near a conducting surface, whose motion is to be measured. The magnetic flux from the active coil passes through this conducting surface and eddy currents become stronger and change the imped ance of the active coil. Thus, the inductance of the coil alters with the variations of distance between the active coil and the conducting surface. A bridge-circuit is used measuring change of inductance of the active coil. A balancing coil forms one of the arms of this bridge-circuit.

Eddy current devices are used for torque (dynamometer) and for motion measurement.

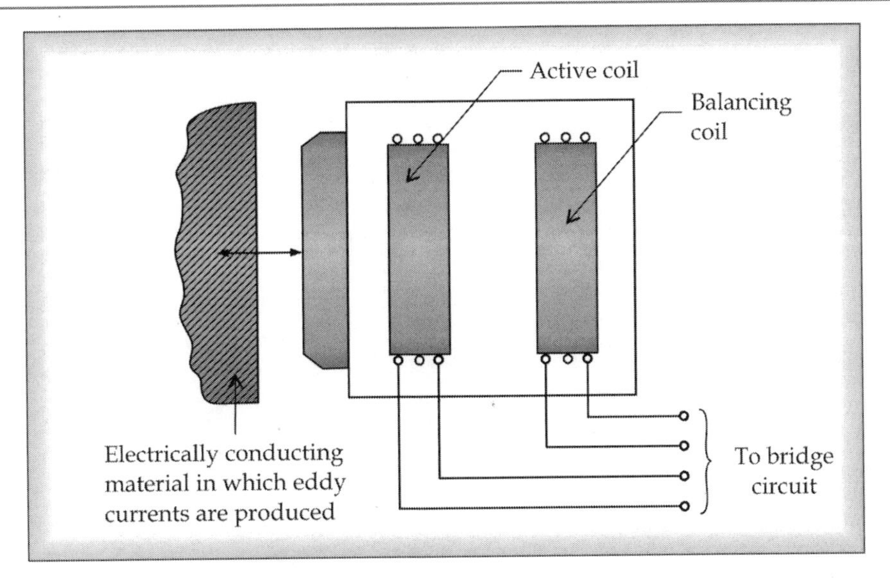

Fig. 4.15 Eddy Current Transducer

4.6.7 Magnetometer Search Coil

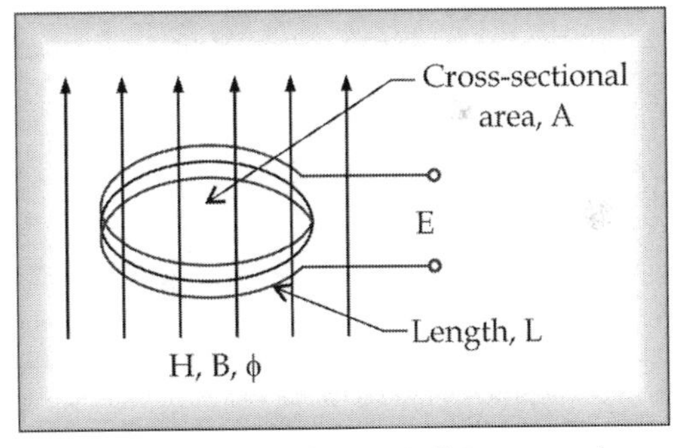

Fig. 4.16(a) Schematic Diagram of Magnetometer Search Coil

Transduction

Magnetic field signal to voltage.

Principle

A flat coil with N turns is placed in the magnetic field as shown in Figure 4.16. The length of the coil L, the cross section area A, the magnetic field strength H and the magnetic flux density B is in the direction as shown, are related as

$B = \mu H$, where μ is the magnetic permeability, and

The voltage output of the coil E is given by

$$E = NA \cos\alpha \, dB/dt$$

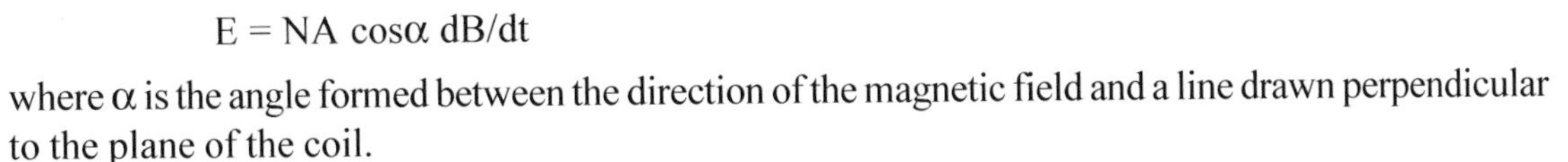

where α is the angle formed between the direction of the magnetic field and a line drawn perpendicular to the plane of the coil.

The total flux through the loop is

$$\Phi = A \cos\alpha \, B$$

So that $E = N \, d\Phi/dt$

Construction and Operation

A schematic of the magnetometer search coil is shown in Figure 4.16. The voltage output of the device is dependent on the rate of change of the magnetic field and that a stationery coil placed in a steady magnetic field will produce a zero voltage output. The search coil is thus a transducer that transforms a magnetic field signal into a voltage.

In order to perform a measurement of a steady magnetic field, it is necessary to provide some movement of the search coil. A typical method is to use a rotating coil, as shown in Figure 4.16 (a)&(b). The rms value of the output voltage for such a device is $E_{rms} = [1/\sqrt{2}]NAB\omega$ where ω is the angular velocity of rotation.

Oscillating coils are also used. The accuracy of the search coil device depends on the accuracy with which the dimensions of the coil are known. The coil should be small enough so that the magnetic field is constant over its area.

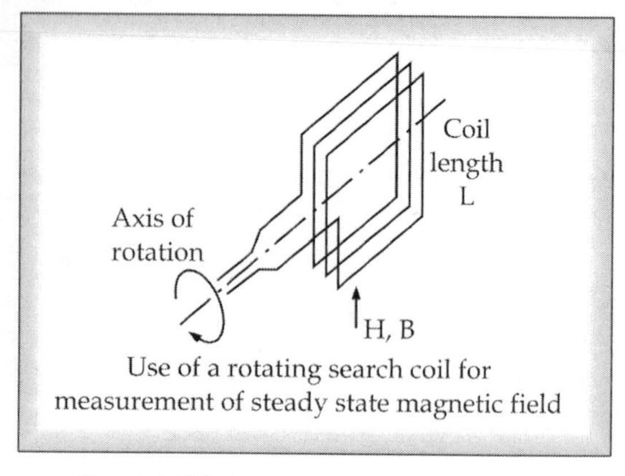

Fig. 4.16(b) Magnetometer Search Coil

[In the above equations the area is in square meters, the time is in seconds, the magnetic flux is in webers, the magnetic flux density is expressed in webers per square meter, the magnetic field strength (magnetic intensity) is in amperes per meter, and the magnetic permeability for free space is $4\pi \times 10^{-7}$ H/m].

4.7 CAPACITIVE TRANSDUCERS

The capacitance of a parallel plate capacitor is given by,

$$C = (1/3.6\pi)\,\epsilon\,A/d = 0.0885 \times 10^{-2}\,\epsilon A/d$$

where, C = capacitance, in μμF

A = Area of plates, in m^2

d = distance between the plates, in m

ϵ = dielectric constant of the medium between the plates (ϵ = 1 for air; ϵ = 3 for plastics)

Hence the capacitance can be varied by varying anyone of the following.

- The dielectric constant (ϵ)
- The effective area of the plates (A)
- Distance between the plates or air gap (d)

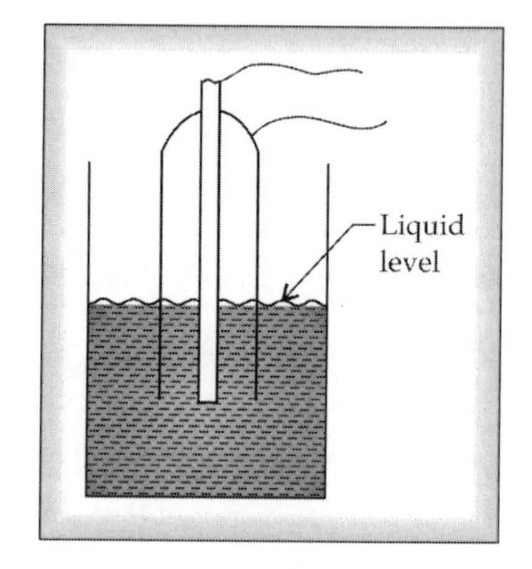

Fig. 4.17 Capacitance Transducer (for Liquid Level)

1. **Di electric Constant:** Figure 4.17 shows a device used for the *measurement of level* in a container. The capacitance, formed between the central rod and the surrounding tube, varies as the changing liquid level varies the dielectric constant.
2. **Area:** A change of capacitance by varying the area finds application in ***torque measurement***. A capacitor is formed between the teeth cut axially on a sleeve fixed around the shaft, and the matching teeth on the shaft. Torque, carried by the shaft, twists it and causes a shift in the relative positions of the teeth, thereby changing the effective area as shown in Figure 4.18 (a) & (b).

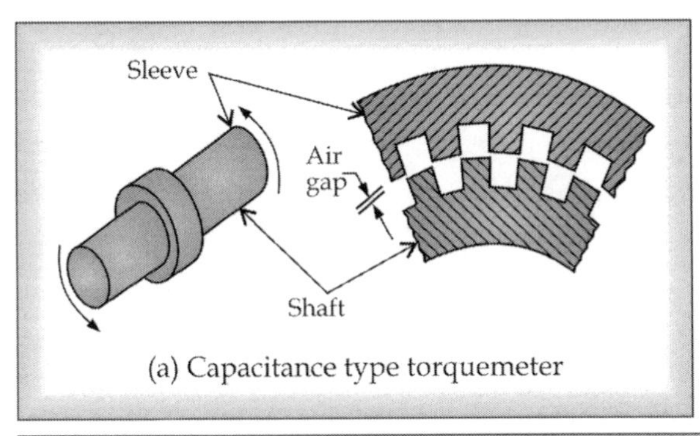

(a) Capacitance type torquemeter

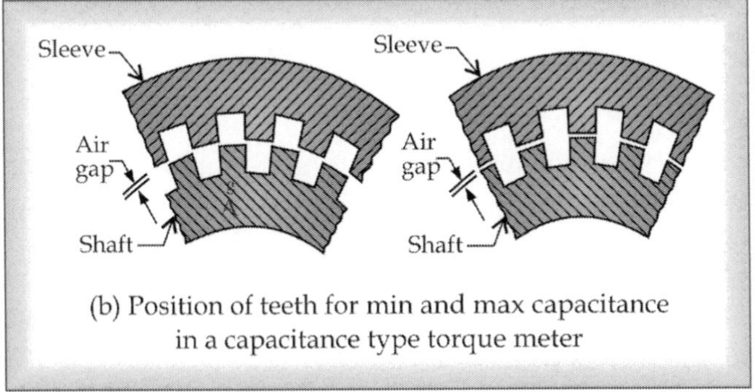

(b) Position of teeth for min and max capacitance in a capacitance type torque meter

Fig. 4.18(a) & (b) Capacitance Type Torque Meter

3. **Distance:** The most common application of varying distance is for ***displacement and motion measurement.*** In a pressure transducer, a fixed electrode and a diaphragm act as a capacitor.
   Application of pressure on the diaphragm changes the distance or air gap, thereby changing the capacitance, which becomes a measure of pressure as shown in Fig. 4.19.

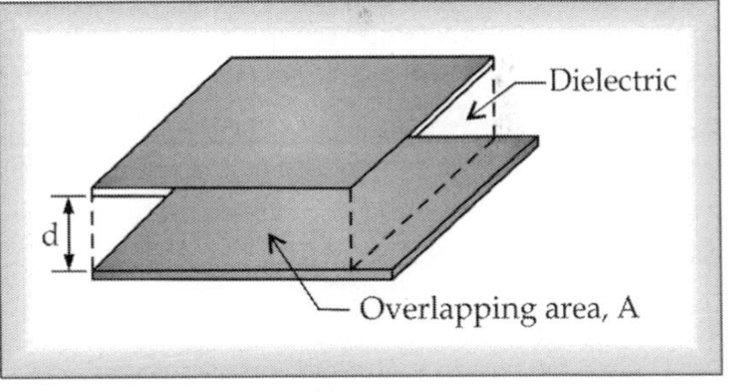

Fig. 4.19 Capacitance Transducer by Varying Distance

Mechanical loading effect is minimum in the case of capacitance transducers used for motion measurement. The capacitance may be measured with bridge circuits.

More Instruments for Learning Engineers

Mile 4.4

Odometer or **odograph** indicates distance traveled by a vehicle (such as bicycle or automobile). The name is derived from the Greek words *hodós* (path or gateway) and *métron* (measure). In some countries it is called **mileometer** or **milometer** or **tripometer**. The device may be electronic, mechanical, or electromechanical or mechatronic.

Odometer

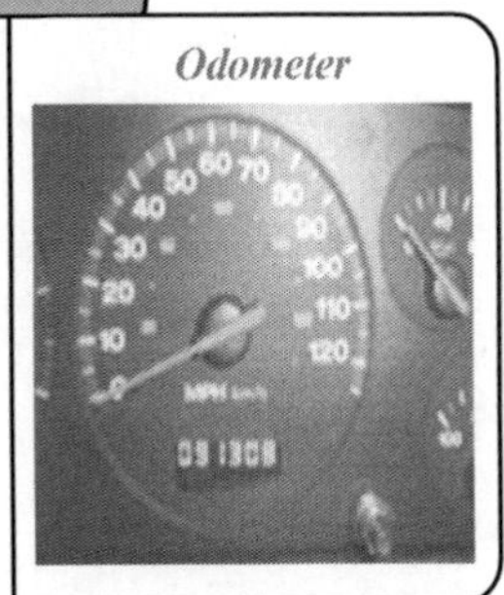

The output independence of a capacitor is given by

$$Z = 1/(2\pi fc)$$

Z = impedance, Ω; f = frequency, Hz; c = Capacitance, F

In general, the output impedance of a capacitive transducer is high, this fact may call for careful design of the output circuit.

Sensitivity of the capacitive transducer is given by $\frac{\partial C}{\partial d}$ is

$$S = \frac{\partial c}{\partial d} = \frac{\partial}{\partial d}\left[\frac{0.0885 \in A}{d^2}\right] = \frac{-2 \times 0.0885 \in A}{d^3} = \frac{0.177 \in A}{d^3}$$

ILLUSTRATION-4.2

A capacitive transducer is constructed by two 25 mm² plates separated by a 25 mm distance in air. Calculate the displacement sensitivity of such an arrangement. The dielectric constant for air is 1.0006.

Solution: The capacitance of a capacitor is given by

$$C = 0.0885 \in A/d^2$$

where, $\in$ = dielectric constant

d = distance between the plates in cm

A = over lapping area in cm²

$$\text{Sensitivity } (S) = \frac{\partial c}{\partial d}$$

$$S = \frac{\partial}{\partial d}\left[\frac{0.0885 \in A}{d^2}\right] = \frac{-2 \times 0.0885 \in A}{d^3} = \frac{-2 \times 0.0885 \times 1.0006 \times \frac{25}{10^2}}{\left(\frac{25}{10}\right)^3}$$

$$= -2.8 \times 10^{-3} \text{ pF/cm}$$

SELF ASSESSMENT QUESTIONS-4.4

1. What is the principle of inductive type transducer?
2. Classify the inductive type transducer.
3. Explain Non-Self-Generating Variable Self Inductance-Single Coil Type Transducer with a neat sketch.
4. Explain Non-Self-Generating Variable Self Inductance Two Coil (or Single Coil with Center Tap) Type Transducer with a neat sketch.
5. With a neat sketch discuss Non-Self-Generating Variable Mutual Inductance Two Coil Transducer.
6. Describe Non-Self-Generating Variable Mutual Inductance Three Coil Transducer Linear Variable Differential Transformer (LVDT) with neat sketches.

7. With a neat sketch explain Self Generating Type Inductive Transducer.
8. What is the principle of Eddy Current Transducer? Explain its Construction and Operation with a neat sketch.
9. Write the principle of Magnetometer Search Coil and discuss its Construction and Operation with neat sketches.
10. By varying which elements the capacitance will be varied? Explain with neat sketches.

4.8 PIEZOELECTRIC TRANSDUCERS

Consider the arrangement shown in figure 4.20 where a piezoelectric crystal is placed between two plate electrodes. When a force is applied to the plates, a stress is produced in the crystal and a corresponding deformation in it. This deformation will produce a potential difference at the surface of the crystal, and the effect is called the *piezoelectric effect*. The induced charge on the crystal is proportional to the impressed force and is given by

$$Q = dF$$

where Q is in coulombs, F is in Newtons, and the proportionality constant d is called the piezoelectric constant.

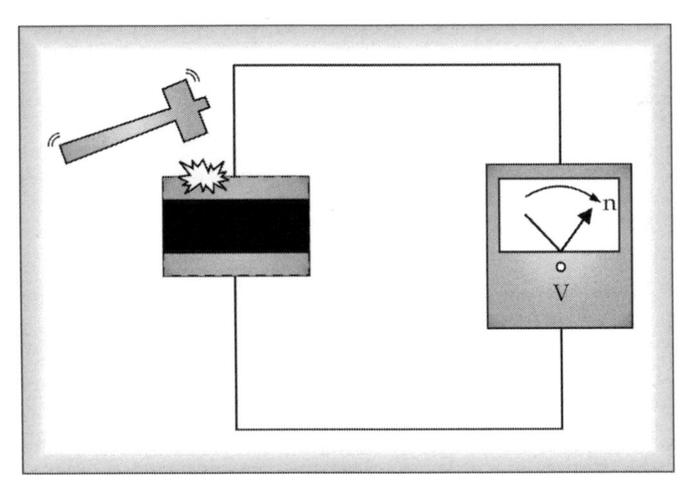

Fig. 4.20 Piezoelectric Effect

The output voltage of the crystal is given by $E = gtp$

where t is the crystal thickness in meters,

p is the impressed pressure in Newton per square meter, and

g is called voltage sensitivity and is given by $g = d/\epsilon$.

Values of the piezoelectric constant and voltage sensitivity for various common piezoelectric materials are given in Table. 4.3.

The voltage output depends on the direction in which the crystal slab is cut in respect to the crystal axes. In Table 4.3 an X (or Y) cut means that a perpendicular to the largest face of the cut is in the direction of the X-axis (or Y-axis) of the crystal.

Table 4.3 Piezoelectric Constants

Material	Orientation	Piezoelectric constant $(C/m^2)/(N/m^2)$	Voltage Sensitivity $(gV/m)/(N/m^2)$
Quartz	X cut; length along *y* length longitudinal	2.25×10^{-12}	0.055
	X cut; thickness longitudinal	-2.04	- 0.050
	Y cut; thickness shear	4.4	- 0.108
Rochelle salt	X cut 45°; Length longitudinal	435.0	0.098
	Y cut 45°; Length longitudinal	- 78.4	- 0.29
Ammonium phosphate dihydrogen	Z cut 0°; face shear	48.0	0.354
	Z cut 45°; length longitudinal	24.0	0.177
Commercial barium titanate	Parallel to polarization	86-130	0.011
	Perpendicular to polarization	- 56	0.005
Lead zinconate titanate titanate	Parallel to polarization	190-580	0.02 - 0.03
Lead meta niobate	Parallel	80	0.036

Piezoelectric crystals may also be subjected to various types of shear stresses instead of the simple compression stress, but the output voltage is a complicated function of the exact crystal orientation. Piezoelectric crystals are used as pressure transducers for dynamic measurements.

4.9 PHOTOELECTRIC EFFECTS

A photoelectric transducer converts a light beam into an electric signal. The electrical voltage signal can be in the form of a voltage generated a change of resistance or electron flow in a circuit. They can be electronic or solid state and can be classified based on their principle of operation as,

(i) photo emissive (ii) photo conductive and (iii) photo voltaic

4.9.1 Photo Emissive

Consider the circuit shown in Figure 4.21. Light strikes the photo emissive cathode and releases electrons, which are attracted toward the anode, thereby producing an electric current in the external circuit. The cathode and anode are enclosed in a glass or quartz envelope, which is either evacuated or filled with an inert gas. The photoelectric sensitivity is defined by

$$I = S\,\Phi$$

where I = photoelectric current Φ = Illumination of the cathode S = Sensitivity

The sensitivity is usually explained in amperes per watt or amperes per lumen.

Photoelectric tube response to different wavelengths of light influenced by two factors

1. Transmission characteristics of the glass tube envelops and
2. The photo emissive characteristics of the cathode material.

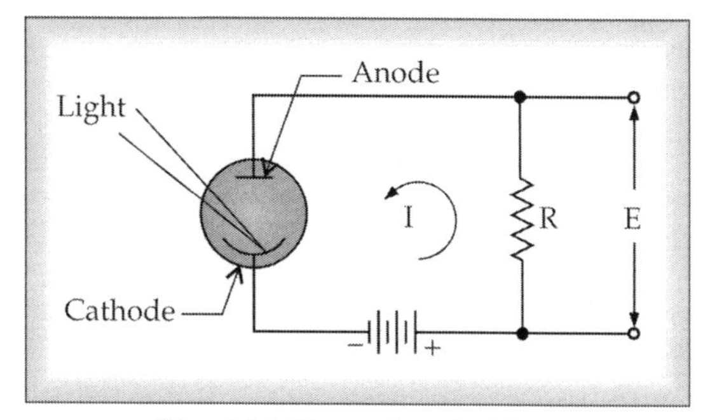

Fig. 4.21 Photoelectric Effect

Photo emissive materials are available which will respond to light over a range to 0.2 to 0.8 μm. Most glasses transmit light in the upper portion of this range, but many do not transmit below about 0.4 μm. Quartz however, can transmit down to 0.2 μm. Various noise effects are present in photoelectric tubes. Photoelectric tubes are quite useful for measurement of light intensity. Inexpensive devices can be utilized for counting purposes through periodic interruption of a light source.

4.9.2 Photo Conductive Transducers

The principle of the photo conductive transducer is shown in Figure 4.22. A voltage is impressed on the semiconductor material as shown. When light strikes the semiconductor material, there is a decrease in the resistance, thereby producing an increase in the current indicated by the meter, A variety of substances are used for photoconductive materials.

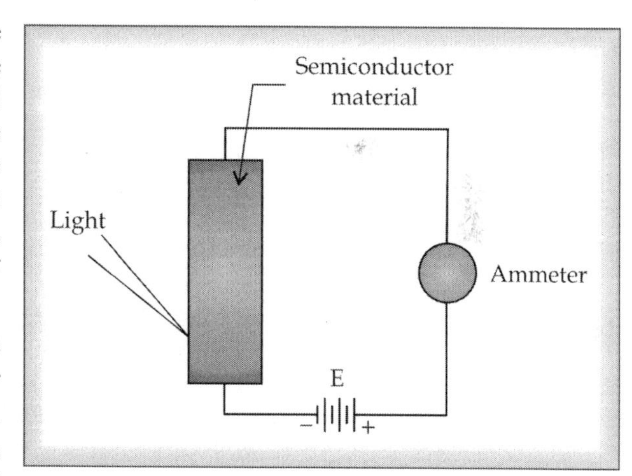

Fig. 4.22 Schematic Diagram of Photoconductive Transducer

Photo conductive transducers enjoy a wide range of applications and are useful for measurement of radiation at all wavelengths. It must be noted, however, that extreme experimental difficulties may be encountered when operating with long wavelength radiation. The responsivity R_v of a detector is defined as

R_v = rms output voltage/rms power incident upon the detector

The Noise Equivalent Power (NEP) is defined as the minimum radiation input that will produce a S/N ratio of unity. The detectivity is defined as

D = R_v /rms noise voltage output of cell.

NEP = rms noise voltage of cell/ R_v.

The detectivity is the reciprocal of NEP.

A normalized detectivity D* is defined as

$$D^* = (A\ \Delta F)^{1/2} D$$

where A is the area of the detector and "F is a noise equivalent bandwidth. The units of D* are usually cmxHZ½/W and the term is used in describing the performance of detectors so that the particular surface area and bandwidth will not effect the result. Figure 4.23 and 4.24. illustrate the performance of several photo conductive detectors over a range of wavelengths. In these figures the wavelength is expressed in micrometers, where 1 μm = 10^{-6}m. The symbols represent lead sulphide (PbS), lead selenide (PbSe), lead telluride (PbTe), indium antimonide (InSb), and gold-and antimony doped germanium (Ge:Au, Sb) for these figures D* is a monochromatic detectivity for an incident radiation, which is chopped at 900 Hz and I-Hz bandwidth.

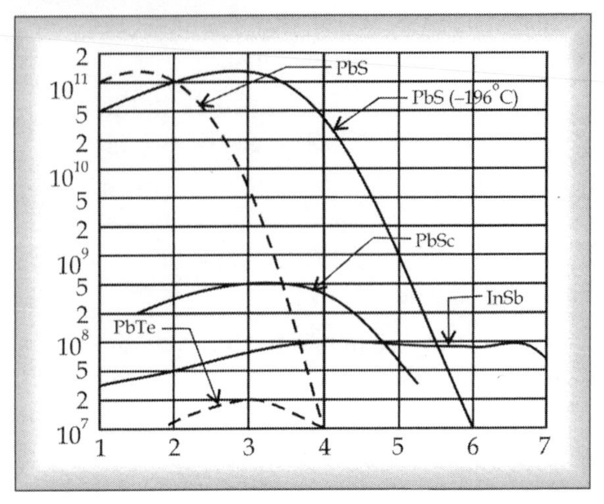

Fig. 4.23 Absolute Spectral Response of Typical Detectors at Room Temperature

The lead-sulphide cell is very widely used for detection of thermal radiation in the wavelength band of 1 to 3 μm. By cooling the detector more favorable response at higher wavelength can be achieved up to about 4 or 5 μm. For measurement at longer wavelengths the indium-antimonide detector is preferred, but it has a lower detectivity than lead sulphide.

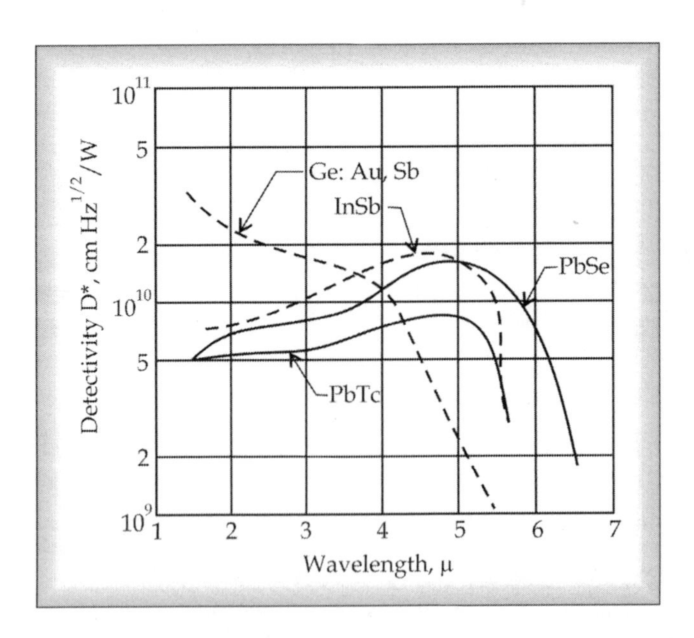

Fig. 4.24 Absolute Spectral Response of Typical Detector Cooled to Liquid Nitrogen Temperature

4.9.3 Photo Voltaic Cells

The photovoltaic-cell principle is illustrated in Figure 4.25. The sandwich construction consists of a metal base plate, a semiconductor material, and a thin transparent metallic layer. This transparent layer may be in the form of a sprayed, conducting layer. When light strikes the barrier between the transparent metal layer and a semiconductor material, a voltage is generated as shown. The output of the device is strongly dependent on the load resistance R. The open circuit voltage approximates a logarithmic function, but more linear behaviour may be approximated by decreasing the load resistance.

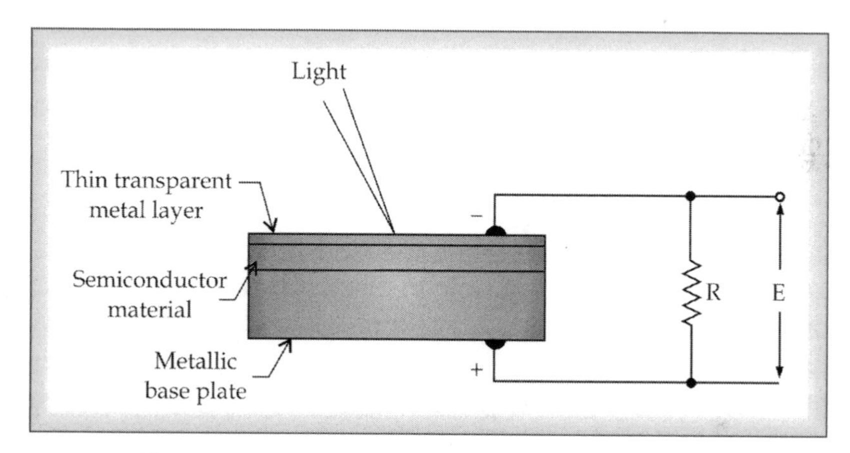

Fig. 4.25 Schematic Diagram of Photo Voltaic Cell

Perhaps the most widely used application of the photovoltaic cell is the light exposure meter in photographic work. The logarithmic behaviour of the cell is a decided advantage in such applications because of its sensitivity over a broad range of light intensities.

4.10 HALL EFFECT

When an electric field and magnetic field are applied perpendicularly to the two perpendicular surfaces of the material a voltage will be generated on the other surface, which is perpendicular to both. This effect is called *Hall Effect.*

Hall Effect Transducers: The principle of the Hall Effect is shown in Figure 4.26. A semiconductor plate of thickness 't' is connected as shown so that an external current passes through the material. When a magnetic field is impressed on a plate in a direction perpendicular to the surface of the plate, there will be a potential E_H generated as shown. This potential is called the hall voltage and is given by

$$E_H = K_H \, (IB)/t$$

where I is in amperes, B is in gauss, and t is in centimeters. The proportionality constant is called the Hall co-efficient and has the units of volt- centimeters per ampere gauss. Typical values of K_H for several materials are given in Table 4.4.

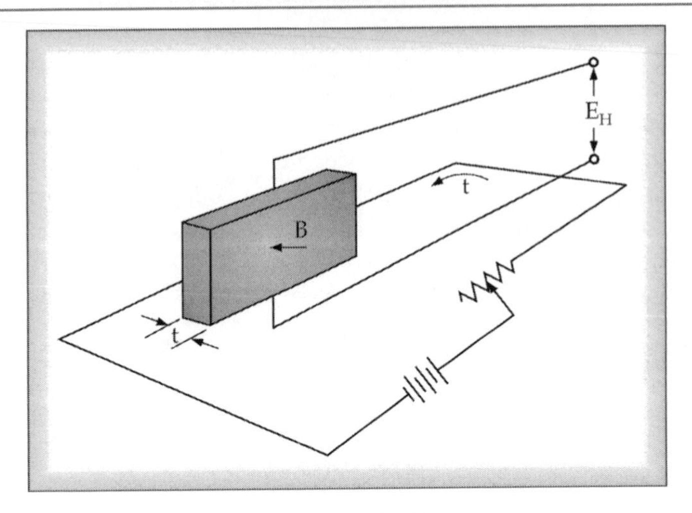

Fig. 4.26 The Hall Effect

Table 4.4 Hall Co-efficient for Different Materials

Material	Field strength, G	Temperature °C	K_H(V/AG)
As	4000-8000	20	4.52×10^{-11}
C	9000-11000	Room	-1.73×10^{-10}
Bi	1130	20	-1×10^{-8}
Cu	8000-22000	20	-5.2×10^{-13}
Fe	17000	22	1.1×10^{-11}
N- Ge	100-8000	25	-8.0×10^{-5}
Si	20000	23	4.1×10^{-8}
Sn	4000	Room	-2.0×10^{-14}
Te	3000 - 9000	20	5.3×10^{-7}

ILLUSTRATION-4.3

A Hall Effect transducer is used for the measurement of a magnetic field of 5000 G. A 3 mm slab of bismuth is used with a current of 5 A. Calculate the voltage output of the device.

Solution:

We have for Bismuth $K_H = -1 \times 10^{-8}$

$$E_H = K_H\,(IB)/t = (-1 \times 10^{-8})(5)(5000)/(3 \times 10^{-1})$$

$$= -\,8.3 \times 10^{-4}\ \text{V}.$$

4.11 MAGNETOSTRICTIVE TRANSDUCER

When a ferro magnetic material (e.g., Iron, Cobalt, Nickel) is subjected to a magnetic field, its dimensions are altered. This change is a function of the strength of magnetic field in which the material is placed and the effect is reversible. Transducers, based on magnetostrictive effect, find use in ultrasonic devices.

4.12 IONIZATION TRANSDUCERS

A schematic diagram of the ionization transducer is shown in Figure 4.27. The tube contains a gas at low pressure while the *RF* generator impresses a field on this gas. As a result of the *RF* field, a glow discharge is created in the gas, and the two electrodes 1 and 2 detect a potential difference in the gas plasma. The potential difference is dependent on the electrode spacing and the capacitive coupling between the *RF* plates, and the gas. When the tube is located at the central position between the plates, the potentials in the electrodes are the same; but when the tube is displaced from this central position, a d.c. potential difference will be created. Thus, the ionization transducer is a useful device for measuring displacement.

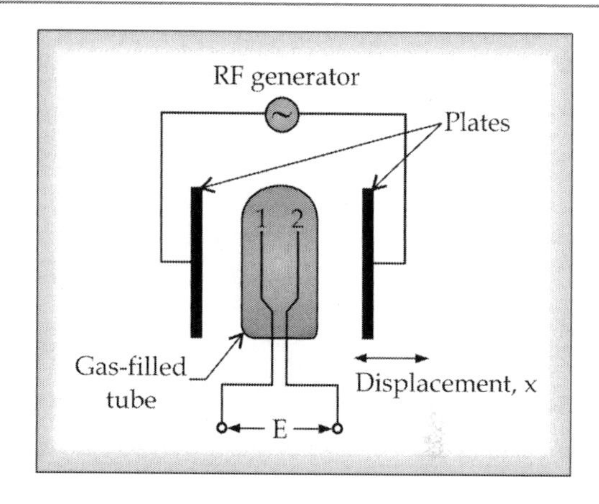

Fig. 4.27 Schematic Diagram of an Ionization

4.13 DIGITAL DISPLACEMENT TRANSDUCERS

A digital displacement transducer can be used for both angular and linear measurement. In Figure 4.28 an angular measurement device is shown. As the wheel rotates, light from the source is alternately transmitted and interrupted, thereby submitting a digital signal to the photo detector. The signal is the number of counts, which are proportional to angular displacement. The frequency of the signal is proportional to angular velocity. Sensitivity of the device may be improved by increasing the number of cut outs. *Example:* transduction by mouse.

A linear transducer which operates on a reflection principle is shown in Figure 4.28 small reflecting strips are installed on the moving device. Light from the source is then alternately reflected and absorbed with linear motion, thereby presenting a digital signal to the photo detector. Read out is the same as with the angular instrument. Calibration with a known displacement standard must be performed.

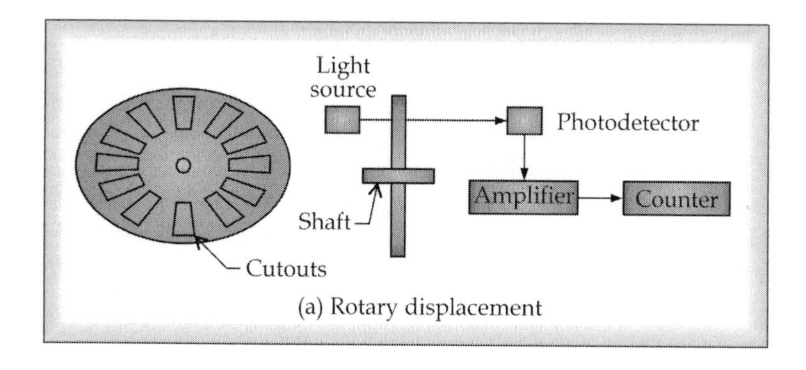

(a) Rotary displacement

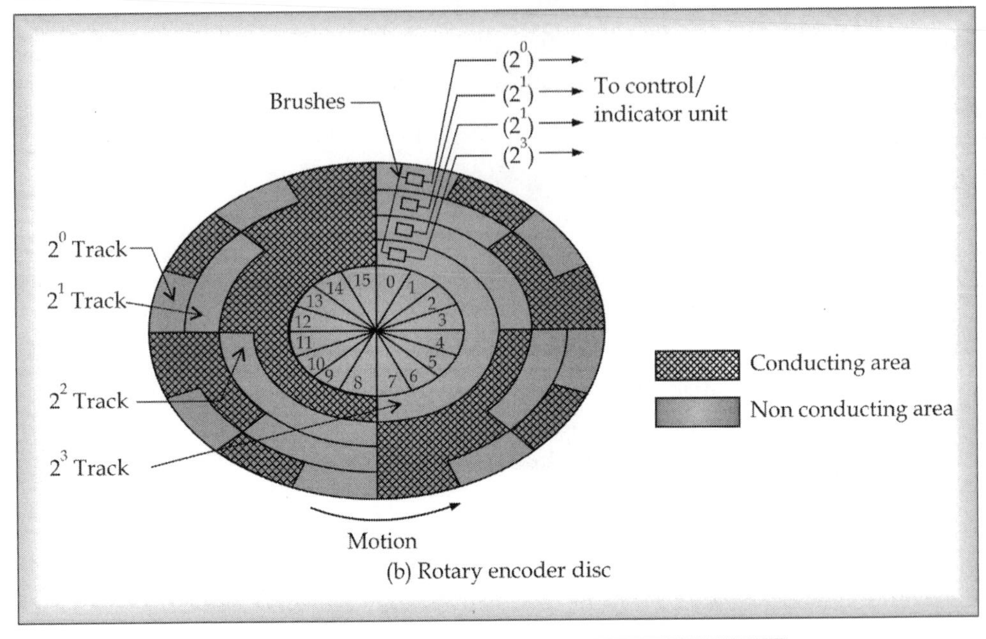

(b) Rotary encoder disc

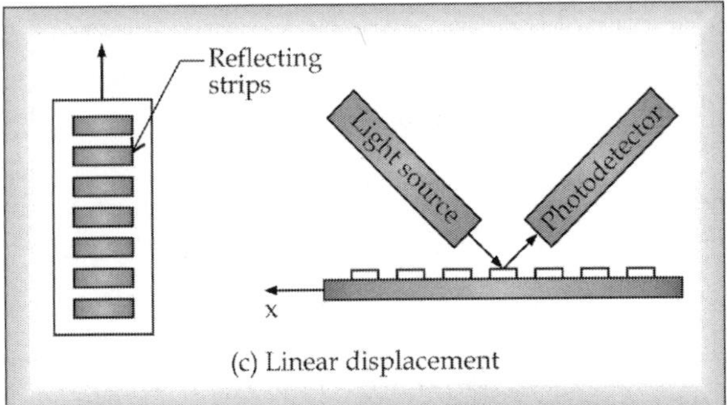

(c) Linear displacement

Fig. 4.28 Digital Displacement Transducers

SELF ASSESSMENT QUESTIONS-4.5

1. What do you mean by the piezoelectric effect?
2. What is piezoelectric constant?
3. Explain about piezoelectric transducers.
4. What is photoelectric effect? What are its types?
5. Explain photo emissive principle.
6. With neat sketches discuss photo conductive transducers.
7. Describe photo voltaic cell with a neat sketch.
8. What is Hall Effect?
9. What is the principle of Hall Effect? Explain Hall Effect.
10. What is magnetostrictive transducer?
11. With a neat sketch discuss about ionization transducers.
12. Explain about digital displacement transducers with a neat sketch.

SUMMARY

Transducer is signal converting device that receives signal from a physical situation or condition (at measurand) and converts into a signal output. A sensor element detects the measurand and contacts with the process. Transducers can be classified as Primary/Secondary based on stage of transduction, Active/ Passive based on energy generation, Analog/Digital based on type of output, and Mechanical/ Electromechanical and Opto-electrical transducers based on nature of job. Mechanical transducers are rugged, stable and cost effective, but electrical transducers are preferred due to merits that the output can be amplified, recorded remotely, modified, besides being quick responsive.

Variable Resistance Transducers can be varied by length, cross section, and resistivity or a combination of these. Sliding Contact Devices convert mechanical displacement input to electrical output by voltage or current. Resistance Strain Gauges are based on electrical resistance changes when a resistive element is mechanically strained. Thermistors are thermally sensitive variable resistors made of ceramic like semi conductors. In inductive type transducer a coil with a core of magnetic material acts on inductor on alternating current.

When a conducting material moves through a magnetic field, a voltage is produced if the conductor is in the form of a wire resulting in a generator action and if the conductor is a piece of plate of disc, the local currents produced are called eddy currents. Eddy current devices are used for torque (dynamometer) and for motion measurement.

Magnetometer search coil transducer transforms a magnetic field signal into a voltage. Capacitive transducers use properties due to changes in dielectric substance, area of parallel plates and separated distance. Piezoelectric transducers use peizo-electric effect i.e., the deformation is proportional to the p.d. of the crystal. A photoelectric transducer converts a light beam into an electric signal. Light strikes the photo emissive cathode to release electrons, attracted toward the anode, thereby produces an electric current in photo emissive transducers. When light strikes the semiconductor material, there is a decrease in the resistance, ammeter thereby produces an increase in the current in a photoconductive transducer. When light strikes the barrier between the transparent metal layer and a semiconductor material, a voltage is generated in photovoltaic-cell.

When an electric field and magnetic field are applied perpendicularly to the two perpendicular surfaces of the material, a voltage will be generated on the other surface in Hall Effect Transducers. In Magnetostrictive Transducers, when a ferromagnetic (Fe, Co, Ni) is subjected to magnetic field, its dimensions vary as a function of the strength of magnetic field and the effect is reversible (used in ultrasonic devices). The potential difference is dependent on the electrode spacing and the capacitive coupling in ionization transducers. A digital displacement transducer can be used for both angular and linear measurement.

KEY CONCEPTS

Primary Transducer: Reacts directly to the change in the physical quantity.

Secondary Transducer: Converts input into output in two steps.

Active (or Internally Powered) Transducers: Self generating type, operate under the energy conversion principle, and do not require an auxiliary source of power to produce output.

Passive (or Externally Powered) Transducers: Operation requires excitation or energy from the external source and absorb some energy from the physical variable to be measured.

Analog Transducers: With the variation of input, there is a continuous variation of the output.

Digital Transducers: The outputs are discrete in nature or vary in steps.

Mass and seismic mass transducers: Based on mass characteristics and used to measure time period, velocity and acceleration.

Mechanical transducer with elastic members: Deformation of elastic members is related to the quantity to be measured and used to measure force, torque and pressure.

Thermal transducers: Based on expansion of solids and liquids and used for temperature measurements.

Hydraulic transducers: Based on hydraulic principles and used for flow measurement and convert velocity to pressure.

Pneumatic transducers: Based on pneumatic principles and used for flow measurement and convert velocity to pressure.

Electro-mechanical transducers: Convert mechanical quantities (such as displacement, velocity, force, torque, pressure and temperature) into electrical quantities (such as a change of resistance and capacitance or generate an emf/electrical charge).

Variable Resistance Transducers: Operated by varying length, cross sectional area and/or resistivity or a combination of these.

Sliding Contact Devices or Resistance Potentiometers or the Pots: Convert mechanical displacement input into an electrical output, either voltage or current.

Resistance Strain Gauges: Based on the principle that the electrical resistance changes when a resistive element is mechanically strained.

Thermistors: Thermally sensitive variable resistors made of ceramic like semi conducting matrerials.

Inductive Type Transducer: A coil with a core of magnetic material (used for magnetic flux path) acts as inductor on alternating current whose inductance can be varied by varying the turns in the coil and the permeability of the flux path.

Single Coil Inductive Type Transducer: Inductance can be varied by the displacement of armature thus changing the air gap.

Two Coil (Self Inductance) Type Transducer: Movement of the core changes the relative inductance of the two portions.

LVDT: Linear Variable Differential Transformer.

Eddy Current Transducers: When the conductor in the form of a piece of plate or disc, moves through magnetic field, the local currents produced are called eddy currents. [used for torque (dynamometer) and for motion measurement].

Magnetometer search coil transducer: Transforms a magnetic field signal into a voltage.

Capacitive transducers: Use properties due to changes in dielectric substance, area of parallel plates and separated distance.

Piezoelectric transducers: Use peizo-electric effect i.e., the deformation is proportional to the p.d. of the crystal.

Photoelectric transducer: Converts a light beam into an electric signal.

Photo emissive transducer: Light strikes the photo emissive cathode to release electrons, attracted toward the anode, thereby producing an electric current.

Photoconductive transducer: When light strikes the semiconductor material, there is a decrease in the resistance, ammeter thereby producing an increase in the current.

Photovoltaic-cell transducer: When light strikes the barrier between the transparent metal layer and a semiconductor material, a voltage is generated.

Hall Effect: When an electric field and magnetic field are applied perpendicularly to the two perpendicular surfaces of the material a voltage will be generated on the other surface, which is perpendicular to both.

Hall Effect Transducers: Operate based on Hall effect.

Magnetostrictive Transducer: When a ferromagnetic (Fe, Co, Ni) is subjected to magnetic field, its dimensions vary as a function of the strength of magnetic field and the effect is reversible (used in ultrasonic devices).

Ionization transducers: The potential difference is dependent on the electrode spacing and the capacitive coupling.

Digital displacement transducer: As the wheel rotates, light from the source is alternately transmitted and interrupted, thereby submitting a digital signal to the photo detector and the signal is the number of counts proportional to angular displacement.

REVIEW QUESTIONS

Short Answer Questions

1. Classify different transducers based on operation: with examples.
2. Compare the following.
   (i) Primary transducers versus secondary transducers.
   (ii) Active transducers versus passive transducers.
3. Describe the construction and principle of
   (i) LVDT (ii) Variable reluctance displacement transducer.

Long Answer Questions

1. What is a transducer? Explain the working of LVDT type transducer. Mention its limitations and applications.
2. Explain the different transduction principles of measuring instruments. Give examples of each.
3. Briefly give the characteristics of the following transducers.
   (i) Electric (ii) Mechanical (iii) Photo-electric.
4. Discuss the principle and operation of ionization transducer & capacitance transducer with neat sketches.
5. Explain the principle and working of a variable capacitance transducer when used for measurement, of displacement.
6. Explain the difference in principle of operation of a photo-emissive cell, a photo-conductive cell & a photo voltaic cell. Give the applications of each of these cells.
7. Explain operation of ionization transducer with a neat sketch and write the applications.

MULTIPLE CHOICE QUESTIONS

1. A sensor can be termed as the element, which ______ the measurand.
   (a) detects (b) measures
   (c) moves (d) enhances
2. Based on nature of input and output, transducers are classified as _____ transducers.
   (a) Analog / digital
   (b) Mechanical / Electromechanical
   (c) Primary / Secondary
   (d) Active / Passive
3. If the sensor senses the element and directly converts it into an output, then the transducer is called as ________.
   (a) Primary transducer
   (b) Secondary transducer
   (c) Mechanical transducer
   (d) Active transducer
4. Which of the following transducers is based on the energy conversion principle.
   (a) Primary transducer
   (b) Secondary transducer
   (c) Mechanical transducer
   (d) Active transducer
5. In analog transducers, with the variation of input, there is _____ variation of the output.
   (a) No (b) Minimum
   (c) Continuous (d) Discontinuous
6. Given two lists, List-I (Element) and List-II (Operation)
   List-I
   (A) Tension / Compression
   (B) Thermocouple
   (C) Liquid column
   (D) Photo emissive
   List-II
   1. Pressure to Displacement
   2. Light intensity to Current
   3. Temperature to Electrical potential
   4. Force to linear displacement
   Now, the correct matching is
   (a) A3, B4, C1, D2 (b) A4, B1, C3, D2
   (c) A4, B3, C1, D2 (d) A3, B1, C2, D4
7. Generally mechanical transducer serves as ________ transducer.
   (a) Primary (b) Secondary
   (c) Tertiary (d) Quaternary
8. Nozzle and Flapper is an example of ________ transducer.
   (a) Electronic

(b) Electromechanical
(c) Mechanical
(d) Pneumatic

9. Which of the following instrument is not used in nozzle and flapper transducer
(a) Orifice
(b) Nozzle
(c) Pressure gauge
(d) Venturimeter

10. In sliding contact device, _______ existing between either end of the wire and the brush thereby becomes a measure of mechanical displacement.
(a) E.M.F
(b) Effective resistance
(c) Friction
(d) Pressure

11. In the sliding contact device, the turns are spaced to prevent ______.
(a) Shorting
(b) Friction
(c) Movement of brush
(d) Elongation of wire

12. In wire-wound resistance, the limiting resolution is equal to the ____ of number of turns.
(a) Minimum (b) Maximum
(c) Reciprocal (d) Square

13. In inductive transducer, ______ can be varied by varying the turns in the coil and the permeability of the flux path.
(a) Inductance (b) Resistance
(c) Voltage (d) Current

14. Three Coil Variable Mutual Inductance is also called ______.
(a) LVDT (b) Potentiometer
(c) Voltmeter (d) Anemometer

15. In LVDT, there is ____ phase shift from one side of the null position to the other.
(a) 0^0 (b) 90^0
(c) 180^{0*} (d) 120^0

16. In eddy current transducer, if the conductor is in the form of wire, it results in ______ action.
(a) Instant (b) Generator
(c) Slow (d) Negative

17. The eddy currents flow in short circulated paths within the plate and get dissipated as
(a) Voltage (b) Heat
(c) Resistance (d) Mechanical

18. In eddy current transducer, the inductance of the coil alters with the variations of distance between the active coil and________.
(a) Passive coil
(b) Bridge circuit
(c) Conducting surface
(d) Balancing coil

19. A spring is a mechanical transducer converting force to________.
(a) Velocity (b) Acceleration
(c) Displacement (d) Torque

20. For an ideal potentiometer, there is a _____ relationship between the output voltage and input displacement.
(a) Straight line
(b) Quadratic curve
(c) Cubic curve
(d) Fourth order curve

21. The maximum excitation voltage is determined by _____ of the power dissipation of the five wires of the potentiometer winding
(a) Minimum (b) Maximum
(c) Zero (d) Infinite

22. The capacitance of a parallel plate capacitor can be varied by changing the ________ the overlapping area or the distance between the plates.
(a) Density
(b) Thickness of the plates
(c) Relative permittivity
(d) Orientation of plates

23. Photo electric transducers produce electrical signals in response to changes in the intensity of
(a) Illumination
(b) Power – electric
(c) Magnetic flux produced by photoelectric
(d) Incident light

24. When one gauge in a symmetrical bridge circuit increases its resistance and an adjacent gauge simultaneously decreases, its resistance will vary by
    (a) One fold (b) Two fold
    (c) Threefold (d) Four fold
25. The active transducer, used for linear or angular velocity measurements, depends upon
    (a) Generation of force
    (b) Variation in mutual inductance of the coils
    (c) Movement of conductor through a magnetic field
    (d) Variation in capacitance of a capacitor
26. The LVDT is an inductive transducer, which functions due to
    (a) Change in the air gap
    (b) Change in the amount of core material
    (c) Mutual inductance variation
    (d) Variation in the position of the core
27. When certain natural or artificial crystal are deformed, and electric change is generated. The characteristics is referred to as
    (a) Thermo-electric effect
    (b) Capacitive effect
    (c) Electromagnetic effect
    (d) Piezoelectric effect
28. Specify the variable in a capacitive transducer that does not necessitate a physical contact between the transducer and the measurand
    (a) Effective or overlapping area of plates
    (b) distance between plates
    (c) Dielectric constant of the insulator
    (d) Capacitance of the plates
29. Specify the transducer, which is generally used for dynamic measurements
    (a) Capacitive (b) Resistive
    (c) Piezoelectric (d) Inductive
30. A potentiometer produces large variation is resistance by
    (a) Moving a conductor through a magnetic field
    (b) Moving a slider across a resistor
    (c) Stretching a metal wire
    (d) Thermally expanding a conductor

ANSWERS

1.	(a)	2.	(b)	3.	(a)	4.	(d)	5.	(c)
6.	(c)	7.	(a)	8.	(d)	9.	(d)	10.	(b)
11.	(a)	12.	(c)	13.	(a)	14.	(a)	15.	(c)
16.	(b)	17.	(b)	18.	(c)	19.	(c)	20.	(a)
21.	(b)	22.	(c)	23.	(d)	24.	(b)	25.	(c)
26.	(d)	27.	(d)	28.	(c)	29.	(c)	30.	(b)

Centigrade, Fahrenheit, Reaumur or Rankine and Kelvin scales related as : $\frac{C-0}{5} = \frac{F-32}{9} = \frac{R-491.69}{9} = \frac{K-273}{5}$

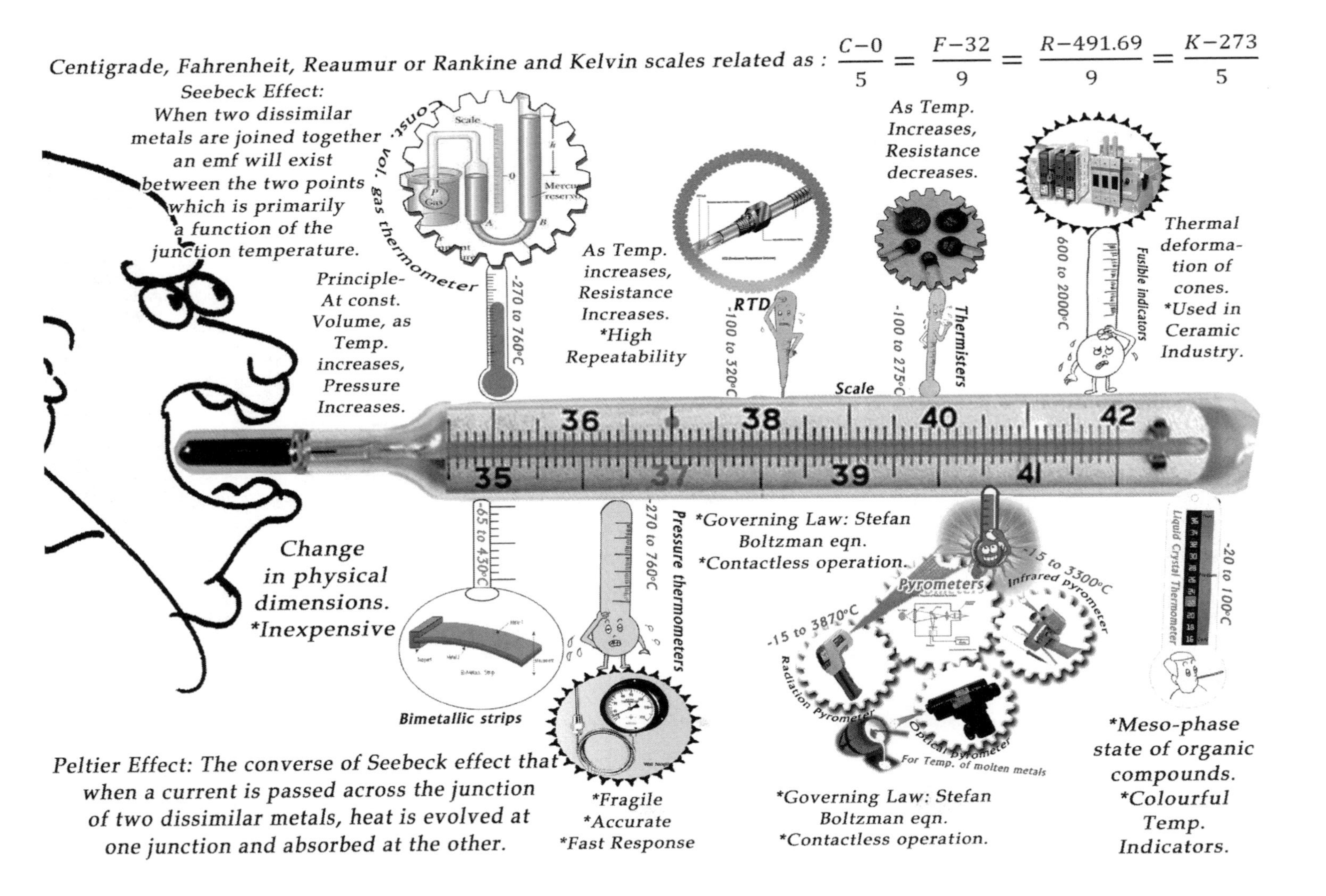

CHAPTER INFOGRAPHIC

Temperature is the degree of hotness or coldness of a body. Temperature Scate Conversion equation is (C-0)/5 = (F-32)/9 = (K-273)/5; The absolute Celsius sclae is called Kelvin scale. The absolute Fahrenheit scale is called Rankine scale

S.No	Instrument	Transducer Type and Principle	Construction	Operation	Application	Merits	Limitations
Temperature measurement by changes in physical dimension							
1a.	Liquid-in-glass thermometers	Change in physical dimensions	Glass tube with relatively large bulb at lower portion of thermometers which holds major portion of liquid.	When temperature merease liquid expands more than grass and rises in capillary and thereby measures temperature.	– 37 to 320 °C	Low cost	Remote reading is not possible
Desirable properties of thermometer liquids are linear relationship, temperature renge, large coefficients of expansion, visibility, and low adhering properties.							
1b.	Bimetallic strips	Change in physical dimensions	It cosists of two pieces of metals with different coefficients of thermal expansions are bounded together to form a strip	When subjected to high temp metals with different ± a expand differently, so bends indicating temp in terms of length.	– 65 to 430 °C	Rugged & Inexpensive	Used up to 430 °C
Temperature measurement by changes in gas pressure or vapour pressure							
2a.	Const. vol. gas thermometer		-	-	-	-	-
2b.	Pressure thermometers	Changes in gas pressure or vapour pressure	Consists of metal bulb, with expanding fluid, a flexible capilary tube, pressure sensitive gauge bourdon tube connected to device to indicate/record signal related to measured temp	When bulb is subjected to increasing temperature the fluid increases, volume /pressure in bulb increases and is transmitted to bourdon tube from capillary which converts signal to a more useful form.	Range of Gas filled guage – 270 to 760 °C; Hg and Vapor Press. – 90 to 370 ° C.	Very accurate, Fast response	Fragile, compensa-tion is required for ambient temperature and long capillary tubes.
Temperature measurement by electrical effects							
3a.	Electrical Resistance Thermometer or Resis-tance Temperature Detector (RTD)	Made of metals whose resistance increases with temp	Platinum resistance thermometer are manufatured open wire/closed wire/ open film /closed film	Modified versions of conventional 4-arm Wheat stone's bride ckts are used in temp measurement, Modified forms are Siemen's 3-lead arrangement; Callender 4-lead arrangement; Floating potential arrangement	– 100 to 320 °C	High repeatability remote applications	Non Linear
3b.		Electrical Resistance thermometers	Mixing of metallic oxide in appropriate proportions & pressing into desired shapes with binders & finally sintering	Negative temperature coefficient (Ressistance decreases with temperature)	100 to 275 °C	Resistance change is large	Less time-stable, reproducibility is not good
1.	Total Radiation Pyrometer	Stefan-Boltzmann Law	One end of a tube (fitted with thermocouple inside) has sight hole with adjustable eyepiece. Other end is opened for thermal radiation	Thermal radiations are directed on to thermocouple which generates emf proportional to temp.	– 15 to 3870 °C	Contact not required; good respo-nse speed and can apply to moving body	Environmental effects: cooling is reqd.

S.No	Instrument	Transducer Type and Principle	Construction	Operation	Application	Merits	Limitations
2.	Optical pyrometer	Brightness comparison	Consists of eyepiece & objective lens on either ends. A power source, rheostat, & multi-voltmeter connected to a reference temp. bulb	Radiation from source is focused onto filament of reference temp. lamp using objective lens. Observer controls lamp current & temp is measured wrt appearance of filament.	For temperature of molten metals	No contact is reqd. easy to operate	1. More than 700 °C only 2. Continuous monitoring not possible
3.	Infrared pyrometer	Uses photocell	Consists of photocell	Output of photocell is amplified and displayed thus temperature measured	– 15 to 3300 °C	Less Response time	Not more than 3300 °C
Temperature measurement by changes in chemical phase							
4a.	Fusible indicators	Melting	Small cones made of an oxide and glass are used to measure temperature	When predetermined temp. is reached, tip of cone softens and curls over, providing an indication that temp is reached.	600 to 2000 °C	Ceramic industry as a means of checking temperatures	Temperature more than – 2000 °C cannot be measured.
4b.	Liquid crystals	Meso-phase state of organic compounds	Organic compounds coated thin film over blackened plastic sheet makes scattered light-more visible; sheet is then affixed to the surface whose temp. is measured.	As temperature increases, liquid crystals successively scatter reds, yellows, greens, blues and violets	– 20 to 100 °C	Colorful temperature indicators	Not more than 100 °C
Laws of thermo electric	**Law of Intermediate Metals or Successive Contracts**: If at given temp., a no. of metals are in successive contact so as to form chain of elements connected in series, emf between extreme elements, if placed in direct contact, is sum of emf between successive adjacent elements				**Law of Intermediate Temperature**: EMF for a couple with junction at T_1 and T_3 is sum of emf of two couples of same metals one with junctions at T_1 and T_2 and other with junctions at T_2 and T_3.		
Thermocouples	**Thermocouple materials**: two unlike conducting materials could be used to form a thermocouple				**Base metal thermocouple**: made up of combinations of pure metal and alloys of Fe, Cu, Ni		
	Rare metal thermocouple: made of combination of pure metal and alloys of Pt, Rh (for up to 1725 °C); W, Rh and Mo for (up to 2625 °C)				**Non-metallic thermo couples**: made of combination of semi conductor material like Si-Ge.		
	Thermopile: Thermocouples connected in series, is called a thermopile, Total output from 'n' thermocouples connected to form a thermopile will be equal to sum of individual emfs and if thermocouples are identical, total output will equal n times output of single couple. Purpose of using thermopile rather than single thermocouples is to obtain more sensitive element.						
Semi Conductor Junction Temperature Sensors: Junction between differently doped regions of semi conductors has a voltage-current curve that depends on temperature. This dependence can be harnessed in two types: Diode sensors and monolithic integrated circuit sensors							
Linear Quartz Thermometer: Uses relationship between temperature and resonating frequency of a quartz crystal is nonlinear					**Stefan-Boltzmann Law:** Total emissive power of black surface is directly proportional to fourth power of temp. of surface in °K		
Pyrometer: Pyrometers are photo detectors, specifically designed for temperature measurement. Pyrometers are of two general types: Thermal detectors and Photon detectors.							
Seebeck effect: When two dissimilar metals are joined together emf will exist between two points P_1 and P_2, which is primarily a function of junction temp.				**Peltier effect:** The converse of Seebeck effect.	**Thomson effect:** When current flows through a thermocouple, heat is evolved/absorbed not only at junctions but also absorbed/evolved all along one or both conductors due to difference of temp. between two ends of same element.		

Contd...

<table>
<tr><td>Laws of thermo electric</td><td colspan="2">Law of Intermediate Metals or Successive Contracts: If at given temp., a no. of metals are in successive contact so as to form chain of elements connected in series, emf between extreme elements, if placed in direct contact, is sum of emf between successive adjacent elements</td><td colspan="2">Law of Intermediate Temperature: EMF for a couple with junction at T_1 and T_3 is sum of emf of two couples of same metals one with junctions at T_1 and T_2 and other with junctions at T_2 and T_3.</td></tr>
<tr><td rowspan="3">Thermocouples</td><td colspan="2">Thermocouple materials: Two unlike conducting materials could be used to form a thermocouple</td><td colspan="2">Base metal thermocouple: made up of combinations of pure metal and alloys of Fe, Cu, Ni</td></tr>
<tr><td colspan="2">Rare metal thermocouple: Made of combination of pure metal and alloys of Pt, Rh (for up to 1725 °C); W, Rh and Mo for (up to 2625 °C)</td><td colspan="2">Non-metallic thermo couples: made of combination of semi conductor material like Si-Ge.</td></tr>
<tr><td colspan="4">Thermopile: Thermocouples connected in series, is called a thermopile, Total output from 'n' thermocouples connected to form a thermopile will be equal to sum of individual emfs and if thermocouples are identical, total output will equal n times output of single couple. Purpose of using thermopile rather than single thermocouples is to obtain more sensitive element.</td></tr>
<tr><td colspan="5">Semi Conductor Junction Temperature Sensors: Junction between differently doped regions of semi conductors has a voltage-current curve that depends on temperature. This dependence can be harnessed in two types: Diode sensors and monolithic integrated circuit sensors</td></tr>
<tr><td colspan="4">Linear Quartz Thermometer: Uses relationship between temperature and resonating frequency of a quartz crystal is nonlinear</td><td>Stefan-Boltzmann Law: Total emissive power of black surface is directly proportional to fourth power of temp. of surface in °K</td></tr>
<tr><td colspan="5">Pyrometer: Pyrometers are photo detectors, specifically designed for temperature measurement. Pyrometers are of two general types: Thermal detectors and Photon detectors.</td></tr>
<tr><td colspan="3">Seebeck effect: When two dissimilar metals are joined together emf will exist between two points P_1 and P_2, which is primarily a function of junction temp.</td><td>Peltier effect: The converse of Seebeck effect.</td><td>Thomson effect: When current flows through a thermocouple, heat is evolved/absorbed not only at junctions but also absorbed/evolved all along one or both conductors due to difference of temp. between two ends of same element.</td></tr>
</table>

Chapter 5

MEASUREMENT OF TEMPERATURE

STARTERS

To study this chapter, you should have awareness on the following concepts. For a better understanding, it is always a good idea to revise these prerequisites.

- Distinction between Temperature and Heat.
- Relation between temperature and heat.
- Effects of temperature on physical dimensions i.e. expansions of solids, liquids and gases with temperature.
- Sensible heat and latent heat.
- Conversion of sensible heat to other forms such as light, sound, electrical energy etc. and vice-versa.
- The relation between Volume and Pressure of gases i.e. Gas laws, particularly Charles laws ($V \alpha T$, $P \alpha T$) and Ideal gas equation.
- Kinetic gas equation.
- Thermo-electric effects (Seebeck's, Peltier's and Thomson's effects).
- Conduction, convection and radiation.
- Relation between electrical resistance and temperature.
- Relation between light and temperature.
- Newton's law of radiation.
- Stefan-Boltzmann law.

LEARNING OBJECTIVE

After studying this chapter you should be able to

- Understand the basic concepts and measurement of temperature,
- Describe Electrical Effects on Temperature.
- Explain Measurement of Temperature by Electrical Methods such as RTD, thermistors and thermocouples,
- Describe Semiconductor Junction Temperature Sensors and the Linear Quartz Thermometer and
- Understand Pyrometers and its types.

5.1 INTRODUCTION

In the early days, the temperature was measured by sense of touch. Even today, for a certain range of touchable temperatures, the measurement is often done by sense of touch. For example, if your friend is suffering from fever, you just touch the forehead or neck and sense the hotness. Many times in our daily life, we assess the temperature just by common sense even without touching also like coldness of an ice cream, hotness of tea or the temperature of water you use for bathing and so on. So, we often think that for ordinary bodies, we do not need any instrument particularly, to measure normal range temperatures. But sometimes even our senses mislead and deceive us in estimating the temperature of the bodies.

Here is one small experiment to prove how we are deceived.

Take three bowls of water of different temperatures, the first one as hot, the second one normal and third one cold. Keep your left finger in hot water bowl and right finger in cold water bowl for a minute or two. Now, remove both fingers and keep both in the normal water bowl. Check what you sense. You would feel the normal water is cold with the left finger while the same water hot with right finger. Is your sense of touch, measuring or at least assessing the temperature of the normal water? Why?

Whatsoever may be the reason, we come to an understanding that temperature measurement is not that simple as we think. After all, we are unable to measure the ordinary temperatures, then how about very high or very low temperatures, particularly for those we come across engineering applications. So, there is a dire necessity to learn about such devices to measure the degree of hotness or coldness of the bodies. To learn about these, we first need to know about temperature, and its relation with various other physical quantities. This chapter deals with those relations, and also the construction, operations and applications of the instruments to measure temperature.

5.2 BASIC CONCEPTS OF TEMPERATURE

Temperature is a surface phenomenon (while heat is substantial phenomenon) and an intrinsic/intensive property of the system i.e., it is independent of the mass or size of the system. Its measurement depends upon the establishment of thermodynamic equilibrium between the system and the device used to measure it. (Thermodynamic equilibrium is said to be established between two bodies in contact if there is no change in temperature in first body as well as that of second).

More Instruments for Learning Engineers

Thermometer measures temperature using variety of principles. It has two important elements: the temperature sensor in which some physical change occurs due to temp. and some means of converting this physical change into a numerical value. (e.g., the visible scale that is marked on a mercury-in-glass thermometer).

Thermometer

Definitions of Temperature

The degree of hotness or coldness of a body is called temperature or

Driving force or potential, causing the flow of heat energy or

Measure of the mean K. E. of the molecules of a substance

Heat and Temperature

Often people get confused between the two terms temperature and heat. To resolve the confusion, these two are distinguished through the Table 5.1.

Table 5.1 Distinction between Heat and Temperature

No.	Heat	Temperature
1.	Heat is form of energy caused due movement of the particles in the body.	Temperature is the degree of hotness or coldness of the body i.e. the measure of mean KE of the molecules
2.	Heat is in two forms internal (latent heat) and external (sensible heat) phenomenon	Temperature is external phenomenon that indicates the sensible heat level.
3.	Heat is spread out all over substance	Temperature is felt on surface
4.	Heat is cause	Temperature is effect
5.	Heat is measure by calorimeter	Temperature is measured by thermometer
6.	Heat is measured in calories or joules	Temperature is measured in degree Centigrade, Fahrenheit, Kelvin etc.

Temperature Scales

There are many temperature scales such as Centigrade, Fahrenheit, Reaumur or Rankine and Kelvin scale. These scales are based on a specification of the number of increments between the freezing point and boiling point of water at standard atmospheric pressure.

The equation used to convert the temperature from one scale to another scale is:

Do You Remember?

- *The absolute* Celsius scale *is called* Kelvin scale.
- *The absolute* Fahrenheit scale *is called* Rankine Scale.
- *The lower fixed point on Kelvin is 273.16° which is the* triple point *(The state at which solid, liquid, and vapour phases are in equilibrium) of water.*

$$\frac{C-0}{100}=\frac{F-32}{180}=\frac{R-491.69}{180}=\frac{K-273}{100}$$

or

$$\frac{C-0}{5}=\frac{F-32}{9}=\frac{R-491.69}{9}=\frac{K-273}{5}$$

In any scale,

$$\frac{\text{Reading} - \text{lower fixed point}}{\text{Fundamental interval}} = \text{constant}$$

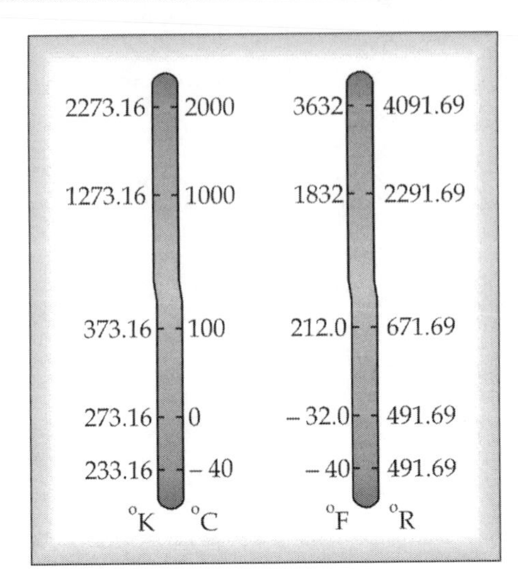

Fig. 5.1 Relationship between Different Temperature Scales

Table 5.2 Different Types of Temperature Scales with their LFP, UFP and FI

Temperature Scale	Lower Fixed Point (LFP) (Freezing Point of Water)	Upper Fixed Point (UFP) (Boiling Point of Water)	Fundamental Interval (FI)
Centigrade (°C)	0 °C	100 °C	100
Fahrenheit (°F)	32 °F	212 °F	180
Kelvin (°K)	273.16 °K	373.16 °K	100
Rankine (°R)	491.69 °R	671.69 °R	180

5.3 MEASUREMENT OF TEMPERATURE

Temperature change is usually measured by observing the change in another temperature dependent property. Since pressure, volume, electrical resistance, expansion coefficient etc, are related to temperature through the fundamental molecular structure, they change with temperature, and these changes can be cleverly used to measure it.

As stated in previous chapter, we can use the sensors which have certain physical characteristics which change with temperature and this effect can be taken as a measure of the temperature. The physical characteristics which can be used are:

1. A change in dimension i.e., either expansion or contraction of material in the form of solid, liquid and gas.
2. A change in electrical resistance of metals and semi conductors.
3. A change in the intensity and colour of radiation emitted by the hot body.
4. Fusion of materials when exposed to the temperature under investigation.
5. Thermo-electric e.m.f. for two different metals and alloys joined together.

5.3.1 Temperature Measuring Instruments

Temperature measuring instruments may be classified either according to the range of temperature measurement or according to the nature of change produced in the temperature sensing element. According to the latter, we have the following classification

1. Changes in physical dimensions: Under this category, the physical dimensions may change due to the following three principles
   (a) Expansion of solids
      (i) Solid rod thermostats/ thermometers
      (ii) Bimetallic thermostats/ thermometers
   (b) Expansion of liquids
      (i) Liquid-in-glass thermometers
      (ii) Liquid in metal thermometers
   (c) Expansion of gases
      (i) Gas thermometers
2. Changes in gas pressure or vapour pressure
   (a) Constant-volume gas thermometers.
   (b) Pressure thermometer (gas, vapour & liquid filled)
3. Changes in electrical properties.
   (a) Resistance thermometer *(RTD, PRT)*
   (b) Thermistors
   (c) Thermocouples
   (d) Semi conductor - junction sensors
      (i) Diodes
      (ii) Integrated circuits
4. Changes in emitted thermal radiation
   (a) Thermal and photon sensors
   (b) Total radiation pyrometers
   (c) Optical and two color pyrometers.
   (d) Infrared pyrometers.
5. Changes in chemical phase
   (a) Fusible indicators
   (b) Liquid crystals

Of the above methods, when automatic or remote recording is desired and when temperature sensors are incorporated into control systems; electrical sensors are mostly used. Bimetallic elements are used in various low-accuracy, low-cost applications. Radiant sensors are used for non-contact temperature sensing, either in higher temperature applications like combustors or for infrared sensing at lower temperatures. Since they are optical in nature, they are also adaptable to whole-field temperature measurement (thermal aging).

The most familiar liquid-in-glass thermometer is widely used for both laboratory and household applications, mainly because of its ease of use and low cost. Changes in chemical phase are somewhat rarely applied in engineering work.

The Table 5.3 outlines approximate ranges and uncertainties of various temperature measuring devices. The values listed in the table are only approximate.

Table 5.3 Characteristics of Various Temperature-measuring Elements and Devices

Type	Useful Range	Accuracy	Comments
Liquid in Glass Mercury filled	-37to 320 °C	0.3 °C	Low cost, Remote reading not practical
Pressurized Mercury	-37to 320 °C		Lower limit of mercury - filled thermometers determined by freezing point of mercury
Alcohol	-75 to 129 °C	0.6 °C	Upper limit determined by boiling point.
Bimetal	-65 to 430 °C	0.5 to 12 °C	Rugged, inexpensive
Pressure systems Gas (Laboratory)	-270 to 100 °C	0.002 to 0.2 °C	Very accurate, Quite fragile, Not easily used.
Gas (Industrial) Liquid(except mercury)	-270 to 760 °C	0.5 to 2 %	Bourdon pressure gauge used for readout. Rugged, with wide range
Liquid (Mercury) Vapour pressure	-90 to 370 °C -37 to 630 °C	1 °C ½ to 2%	Relative elevations/readout and sensing bulb are critical. Smallest bulb up to 3m capillary
		½ to 2%	Fast response, nonlinear lowest cost
Thermocouples General			Extreme ranges - all types.
Type B Pt30%Rh(+)vs.Pt6%Rh(-)	-250 to2400 °C	0.6 °C ±½%	Not for reducing atmosphere or vacuum Generates high emf per degree
Type-E Chromel(+)vs.Constanta(-)	-184 to870 °C		Highest output of common thermocouples
Type J Fe (+) vs Alumel (-)		+ ½%	For reducing or neutral atmosphere, popular and inexpensive
Type K Chromel (+)vs.Alumel (-)	0 to 1260 °C	1 to 6 °C	For oxidizing or neutral atmosphere.
Type R Pt(+)vs.Pt,13% Rh(-)	0 to 1480 °C	+ 3/4%	Attacked by sulfur Most linear of all thermocouples
Type S Pt (+) vs. Pt10% Rh(-)	0 to 1480 °C	½ %	Requires protection in all atmospheres. Higher output than Type S. Linearity poor (below)

Table 5.3 *Contd...*

Type	Useful Range	Accuracy	Comments
Type T Cu(+) vs.Constanta (-) W.5%Rh(+)vs.W26%Rh(-)	-250 to 340 °C -270 to 2310 °C	½ % 0.6 °C ---	Requires protection in all atmospheres. Under proper conditions yields highest precision. May be used in either oxidizing or reducing atmospheres. Good stability. No standards. Reducing or neutral atmospheres. Highest temperature limit of all thermocouples
Resistance Platinum	-260 to 980 °C	0.02 to 0.2 °C	High repeatability, Linear and can be used as far as 1500m from readout.
Nickel Thermistor (Metal Oxide)	-180 to 320 °C -100 to 315 °C	0.2 °C	High repeatability. Nonlinear produces greater resistance change per degree than does Pt. Sensor can be as far as 1500 m from readout.
Thermistor (Doped Ge)	-273 to-173 °C	0.03 °C	Negative temperature coefficient. Highly non linear. Less stable than metal types.
Thermistor (Carbon- Glass)	-272 to 50 °C	0.05 °C	High repeatability. Non linear, Negative temperature coefficient. Cryogenic sensor.
Semi conductor Junction Diode (Silicon, GaAIAs) Linear Integrated Circuit	-272 to 50 °C -50 to 150 °C	0.05 °C 0.5 °C	Non linear, High accuracy requires calibration. Cryogenic sensor. Inexpensive Linear. Easily Integrated into electronics. Limited temperature range.
Pyrometers Optical Total Radiation Infrared	-15 to 3870 °C -15 to 3300 °C	½ to 2% ½ to 2% ½ to 2%	Used only for high temperatures. Requires manual manipulation by operator. Can measure "spot" or average temperatures, Portable, Self - contained.

SELF ASSESSMENT QUESTIONS-5.1

1. Define temperature. Describe various approaches by which temperature can be measured.
2. Distinguish between temperature and heat. Describe various scales of temperature measurement.
3. With a sketch compare Kelvin, Rankine, Celsius, Fahrenheit and Reaumur scales of temperature.
4. List out at least ten temperature measurement instruments and mention their applications.
5. List out any eight temperature measuring devices along with their useful range, accuracy level, applications, merits and demerits.

6. Give exemplary instruments to the following cases and explain briefly.
   (a) Temperature measurement by the change in physical dimension
   (b) Temperature measurement by the change in electrical properties
   (c) Temperature measurement by changes in thermal radiation
7. Discuss the following physical characteristics with a focus on temperature measurement. Also establish their relationships with temperature.
   (a) Physical dimensions and Change of state (b) Electrical properties
   (c) Pressure and volume (d) Radiation

5.4 MEASUREMENT OF TEMPERATURE BY CHANGES IN PHYSICAL DIMENSIONS

Perhaps! The simplest method of measuring temperature is by converting the change in temperature into a known physical dimension such as length and sensing this change on a scale. Thus the principle of transduction here is *"expansion of solids or liquids (to change physical dimension such as length or volume) or gases by increasing temperature"*.

5.4.1 MEASUREMENT OF TEMPERATURE BY EXPANSION OF SOLIDS

Solid thermostat/ thermometers are based on the differential expansion of materials i.e., one material expands more than the other when subjected to same change in temperature.

(a) Solid Rod Thermostats/Thermometers

> ***Transducer:*** *Solid rod in a tube*
> ***Transduction:*** *Temperature to length*
> ***Principle:*** *Differential expansion of materials*

Construction and Operation: The most common solid rod thermostat consists of invar steel rod ($\alpha = 2.7 \times 10^{-6}$ °C^{-1}) inside a brass tube ($\alpha = 34.2 \times 10^{-6}$ °C^{-1}) having one end of the rod and tube hard soldered. On the other end there will be a micro switch and a micrometer. When this set up is kept in temperature bath (measurand), the brass tube will expand more than invar rod. The difference of expansion can be measured by the micrometer. Thus it works as a thermometer.

This type of set up can be used in two ways i.e., as a thermostat as well as a thermometer. If it is used as a thermostat, the set-up is pre-set to a point and when the desired temperature is attained, the difference of expansion reaches to the point, while a mechanism may be built in so as to cut-off the circuit by a micro switch. This kind of thermostat is employed to cut-off supply to electric heaters, ovens and refrigerators etc.

(b) Bimetallic Thermostats/Thermometers (Using Bimetal Temperature Sensing Elements or Bimetallic Strips)

> ***Transducer:*** *Bimetallic Sensing Element*
> ***Transduction:*** *Temperature to Length*
> ***Principle:*** *Due to differential expansion in two dissimilar solid metals (bound together), the bimetallic strip bends*

Construction and Operation: A very widely used method of temperature measurement is by bimetallic strip. Two pieces of metal with different coefficients of thermal expansion are

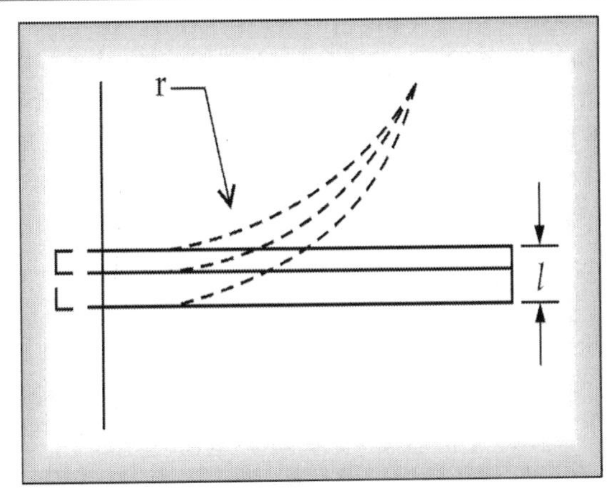

Fig. 5.2 The Bimetallic Strip

bound together to form the device shown in Figure 5.2 when the strip is subjected to a temperature higher than the bonding temperature, it will bend in one direction, when it is subjected to a temperature lower than the bonding temperature, it will bend in the other direction.

Curvature and Deflection of Bimetallic Strip: The radius of curvature 'r' may be calculated as

$$r = \frac{t\{3(1+m)^2 + (1+mn)[m^2 + (1/mn)]\}}{6(\alpha_2 - \alpha_1)(T - T_0)(1+m_2)}$$

where

t = combined thickness of the bonded strip, m

m = ratio of thicknesses of low-to-high expansion materials.

n = ratio of modulus of elasticity of low-to-high-expansion materials.

α_1 = lower coefficient of expansion, per °C

α_2 = higher coefficient of expansion, per °C

T = temperature, °C

T_0 = initial bonding temperature, °C

The thermal expansion coefficients for some commonly used materials are given in Table 5.4.

Table 5.4 Mechanical Properties of Some Commonly used Thermal Materials

Material	Thermal coefficient of expansion per °C	Modulus of elasticity GN/m²
Invar	1.7×10^{-6}	147
Yellow brass	2.02×10^{-5}	96.5
Monel 400	1.35×10^{-5} (thermostats).	179
	1.35×10^{-5}	
Inconel 702	1.25×10^{-5}	217
Stainless steel type 316	1.6×10^{-5}	193

Bimetallic strips are frequently used in simple on-off temperature control devices. Movement of the strip has sufficient force to trip control switches for various devices. The bimetallic strip has the advantages of low-cost, negligible maintenance, and stable operation over extended periods of time. Alternate methods of construction can use a coiled strip to drive a dial indicator for temperatures.

ILLUSTRATION-5.1

A bimetallic strip is constructed of strips of yellow brass and Invar bonded together at 300. Each has a thickness of 0.5 mm. Calculate the radius of curvature when 8.0 cm strip is subjected to a temperature of 100 °C.

Solution:

Temperature (T) = 100 °C

Initial Temperature (T_o) = 30 °C

$$T - T_o = 100° - 30° = 70 \text{ °C}$$

$$\text{Ratio of thickness } m = \frac{0.5}{0.5} = 1$$

Ratio of modulus of elasticity of low-to-high expansion materials (n) 147/96.5 = 1.52 lower coefficient of expansion, per °C

$$\alpha_1 = 1.7 \times 10^{-6} \text{ °C}^{-1} = 0.17 \times 10^{-5} \text{ °C}^{-1}$$

Higher coefficient of expansion, per °C

$$\alpha_2 = 2.02 \times 10^{-5} \text{ °C}^{-1}$$

Combined thickness of the bounded strip

$$t = 2\,(0.5 \times 10^{-3}) = 1 \times 10^{-3} \text{ m}$$

Thus
$$r = \frac{(1\times10^{-3})[(3)(2)^2 + (1+1.52)\,(1 + 1/1.52)]}{6(2.02 - 0.17)(10^{-5})(70)(2)^2}$$

$$= 0.39 \text{ m}$$

More Instruments for Learning Engineers

A **calorimeter** (in Latin *calor* means heat) measures the heat capacity and heat of chemical/physical changes. (The types are differential scanning~, bomb~, accelerated rate~, titration~ and isothermal micro-calorimeters). It consists of thermometer in metal container full of water suspended above combustion chamber. To find enthalpy change/mole of 'A' (in a reaction between A&B), add A&B to calorimeter and note initial & final temp. Multiply the temp. change by the mass & specific heat capacities of the substances to get value of energy released/ absorbed. Divide energy change by no. of moles of A to get enthalpy change. It doesn't account for heat loss through container or heat capacity of thermometer and container itself.

Calorimeter

From the above Figure 5.3 we observe that the angle through which the strip is deflected is related to the strip length L and radius of curvature (r) by $L = r\theta$

Where we assume the increase in length due to thermal expansion is small.

Thus, $\theta = \frac{0.06}{0.39} = 0.154$ rad

As indicated in the accompanying figure, the straight line segment joining the ends of the strip has the length y or

$$r = \frac{t\{3(1+m)^2 + (1+mn)[m^2 + (1/mn)]\}}{6(\alpha_2 - \alpha_1)(T - T_0)(1 + m_2)}$$

$$y = 2\ r\ \sin q = (2)\ (0.390)\ \sin\ (0.154°)$$
$$= 0.78 \times 0.153 = 0.1196$$

The deflection d is related to y by

$$d = y\sin\ \theta = (0.1196)$$
$$\sin\ (0.154°) = 0.018\ \text{m}$$

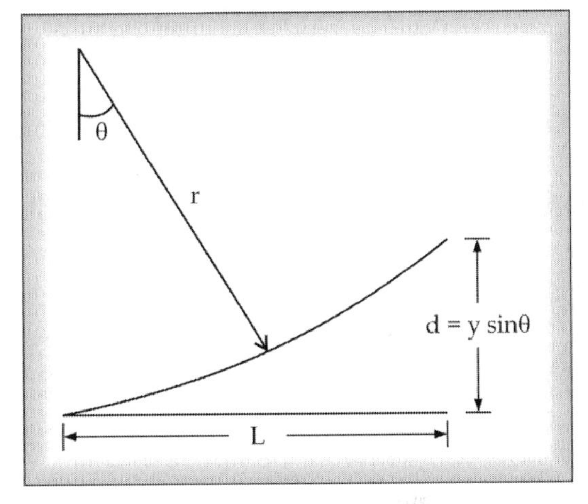

Fig. 5.3 Relation between Arc Length, Radius and Angle in Radians

5.4.2 Measurement of Temperature by using Expansion of Liquids

The most common and easy method is to use liquids (often termed as thermometric liquids such as water, mercury and alcohol) in a glass tube of uniform cross section.

Desirable Properties of Thermometric Liquid

Several desirable properties for the liquid used in a glass thermometer are as follows:

1. *Linear Relationship:* The temperature-dimensional relationship should be linear, permitting a linear instrument scale.
2. *Temperature Range:* The liquid should accommodate a reasonable temperature range without change of phase. Mercury is limited at the low-temperature end by its freezing point – 38.87°C, and spirits are limited at the high-temperature end by their boiling points.
3. *Larger Coefficient of Expansion:* The liquid should have as large a coefficient of expansion as possible. For this reason, alcohol is better than Hg. Its larger expansion makes possible large capillary boxes, and hence provides easier reading.
4. *Visibility:* The liquid should be clearly visible when drawn into a fine thread. Mercury is obviously acceptable in this regard, where as alcohol is usable only if dye is added.
5. *Low Adhering Property:* Preferably, the liquid should not adhere to the capillary walls. When rapid temperature drops occur, any film remaining on the wall of the tube will cause a misreading that is 'low'. In this respect, mercury is better than alcohol.

Within its temperature range, mercury is undoubtedly the best liquid for liquid-in-glass thermometers and is generally used in the higher grade instruments. Alcohol is usually satisfactory. Other liquids are also used, primarily for the purpose of extending the useful ranges to lower temperatures.

Transducer: *Liquid column*
Transduction: *Temperature to length*
Principle: *The property of expansion of liquids*

Liquid-in-glass Thermometers: The ordinary thermometer is an example of the liquid-in-glass type. The construction details of this instrument are shown Fig. 5.4. A relatively large bulb at the lower portion of the thermometer holds the major portion of the liquid, which expands when heated and rises in the capillary tube, upon which are etched appropriate scale markings. At the top of the capillary tube another bulb is placed to provide a safety feature in case the temperature range of the thermometer is exceeded. As the temperature is raised the greater expansion of the liquid compared with that of the glass causes it to rise in the capillary or stem of the thermometer and the height of rise is used as a measure of the temperature. The volume enclosed in the stem above the liquid may either contain a vacuum or filled with air or another gas. For the temperature ranges, an inert gas at a carefully controlled initial pressure is introduced in this volume, thereby raising the boiling point of the liquid and increasing the total useful range. In addition, it is that such pressure minimizes the potential for column separation.

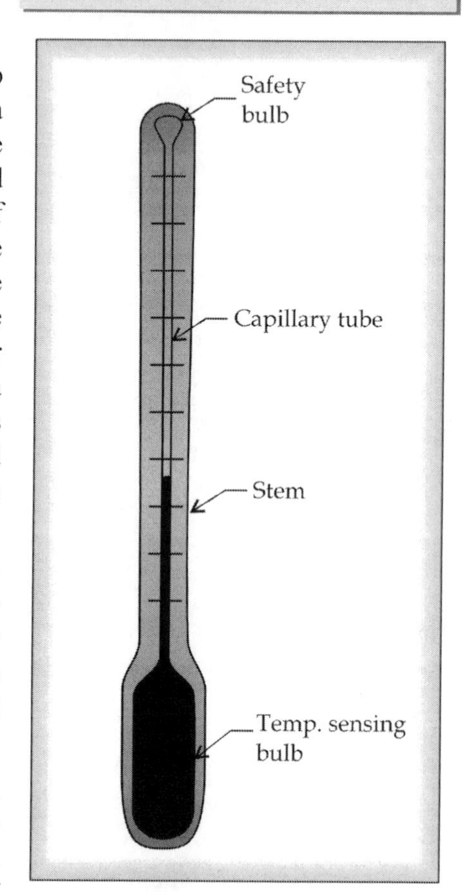

Fig. 5.4 Mercury -in-glass Thermometer

Calibration and Stem Correction: High grade liquid-in-glass thermometer is made with the scale etched directly in the thermometer stem, thereby making it mechanically impossible to shift the scale relative to the stem. The care with which the scale is laid out depends on the intended accuracy of the instrument.

The process of establishing bench marks from which a scale is determined is known as pointing, and two or more marks or points are required. Dimension characteristics of liquid or glass are by the non uniformity of the bore of the column that exists some degree of non-linearity. Two points, freezing and boiling points of water are established and equal divisions are used to interpolate and/or extrapolate the complete scale. For a more accurate scale, additional points–sometimes as many as five are used.

Greatest sensitivity to temperature is at the bulb, where the largest volume of liquid is contained, however all portions of a glass thermometer are temperature-sensitive.

When the thermometer is subjected to temperature variation, the stem and upper bulb will also change dimension, which affects the thermometer reading. For this reason, if maximum accuracy is to be attained, it is necessary to prescribe how a glass thermometer is to be subjected to the temperature.

Full control can be achieved by immersing the entire thermometer into uniform temperature medium. Often this is not possible, especially when the medium is liquid. A common practice therefore is to calibrate the thermometer for a given partial immersion, with the proper depth of immersion indicated by a scribed line around the stem. This does not ensure absolute uniformity because the upper portion of the stem is still subjected to some variation in ambient conditions.

When the immersion employed is different from that used for calibration, an estimate of the correct reading can be done from the following relation.

$$T = T_1 + KT (T_1 - T_2)$$

[For Hg-in-glass thermometers]

where T = The correct temperature in degrees.

T_1 = The actual temperature reading, in degrees.

T_2 = The ambient temperature surrounding the emergent stem

[This may be determined by attaching a second thermometer to the stem of the main thermometer], in degrees.

K = The differential expansion coefficient between liquid and glass (0.9×10^{-4})

T' = degrees of Hg thread emergence to be corrected.

Determination of T: For a total immersion thermometer, *T'* should be the actual length of the thread of mercury that is emerging, measured in degrees. For the partial immersion thermometer *T'* should be the number of scale degrees between the scribed calibration immersion line and the actual point of emergence. When the thermometer is too deeply immersed, the value of *T'* will be negative.

Another factor influencing liquid-in-glass thermometer calibration is the pressure of system in which it is immersed.

5.5 PRESSURE THERMOMETERS

Construction: The essential parts of this thermometer are a metal bulb (A), with expanding fluid (liquid or gas), a flexible capillary tube (B), a pressure sensitive gauge (C) Bourdon tube and a device for indicating or recording a signal (D) related to the measured temperature as shown in Fig. 5.5.

(a) Constant volume thermometer (b) Constant pressure thermometer

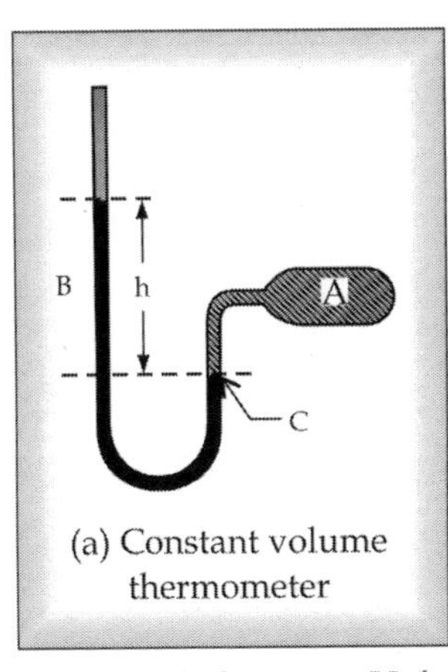

(a) Constant volume thermometer

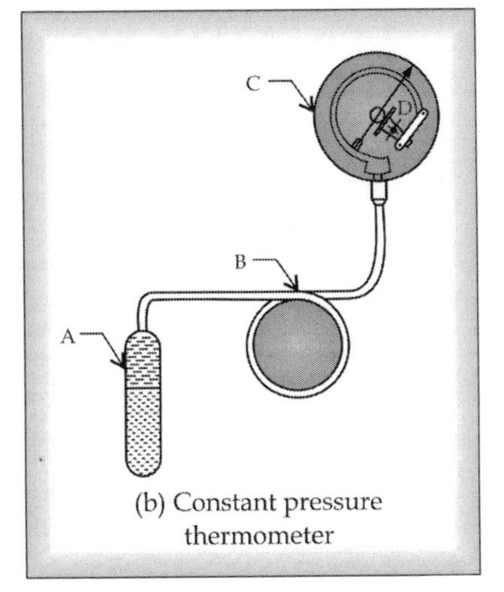

(b) Constant pressure thermometer

Fig. 5.5 Constant Volume and Constant Pressure Thermometers

Operation: When bulb (A) is subjected to increasing temperature, the volume/pressure of fluid in the bulb increases. This increased pressure is transmitted to Bourdon tube from capillary tube, which converts the signal to a more useful form. This signal can be coupled to a mechanical linkage to drive a pointer or pen or may be coupled to a pneumatic or electrical circuit to transmit the signal to a larger distance.

The fluid-filled systems may be divided into three classes based on the fluid used in the bulb. They are

1. Liquid filled (including Hg) 2. Gas filled 3. Vapour-pressure filled.

1. **Liquid Filled Thermal System:** These operate on the principle of thermal expansion similar to that of Hg-in-glass thermometer. The volumetric expansion of the liquid caused by temperature change provides the force required to operate the spiral and thus indicate temperature as shown in Fig. 5.6(a).

   The relation between volume and temperature is given by cubical law of expansion

   $$V_t = V_o(1 + \alpha T + \beta T^2 + \gamma T^3)$$

   $$= V_o(1 + BT)$$

   where V_t = final volume
   V_o = initial volume
   B = Mean coefficient of volumetric expansion
   α, β, γ = coefficients of volumetric expansion

   > ***Transducer:*** *Bulb filled with Liquid*
   > ***Transduction:*** *Temperature to length*
   > ***Principle:*** *The volumetric expansion of the liquid*

2. **Gas-filled Thermal System:** In this, the bulb is filled with a gas and when the temperature of the gas is increased its pressure increases which is communicated to Bourdon tube similar to liquid-filled thermal system as shown in Fig. 5.6(b).

   The expansion of gas is governed by ideal gas law.

   as $PV = RT$

   $$\Rightarrow \quad P = \frac{RT}{V}$$

   $P \propto T$ (keeping V constant)

   > ***Transducer:*** *Bulb filled with Gas*
   > ***Transduction:*** *Temperature to length*
   > ***Principle:*** *Ideal gas equation*
   > *PV = nRT*

   The relationship between pressure and temperature is linear, provided volume remains constant. Commonly used gases in these systems are mostly inert gases and nitrogen. The temperature limits of these thermometers are -130° to 540°C. The accuracy is of the order of ± 1% of the range of the instrument. The volume of the gas required in the bulb is determined from gas expansion and from the temperature range of the instrument

   $$V = R\left(\frac{T_2 - T_1}{P_2 - P_1}\right)$$

   where 1,2 (subscripts) refer to the condition at the lowest & highest points of the scale.

3. **Vapour-pressure Filled Thermal System:** The fluid used in these systems is a very volatile liquid. The bulb is partially filled with a volatile liquid and the rest of the system contains vapour as shown in Figure 5.6(c).

If the bulb is subjected to increasing temperature the volatile liquid vapourizes, which results in the increase of vapour pressure and is measured by pressure measuring device. The problem with this thermometer is that when the bulb temperature is less than die capillary tube and bourdon tube, liquid will be there in the bulb, but in capillary tube and Bourdon tube vapour is present. If bulb temperature is more than the capillary tube and bourdon tube, there will be vapour in the bulb but liquid in capillary and bourdon tube as shown in Fig. 5.6(a) to (c)

> ***Transducer:*** *Bulb filled with vapour*
> ***Transduction:*** *Temperature to length*
> **Principle:** Volatility of liquid & Ideal gas equation PV = nRT

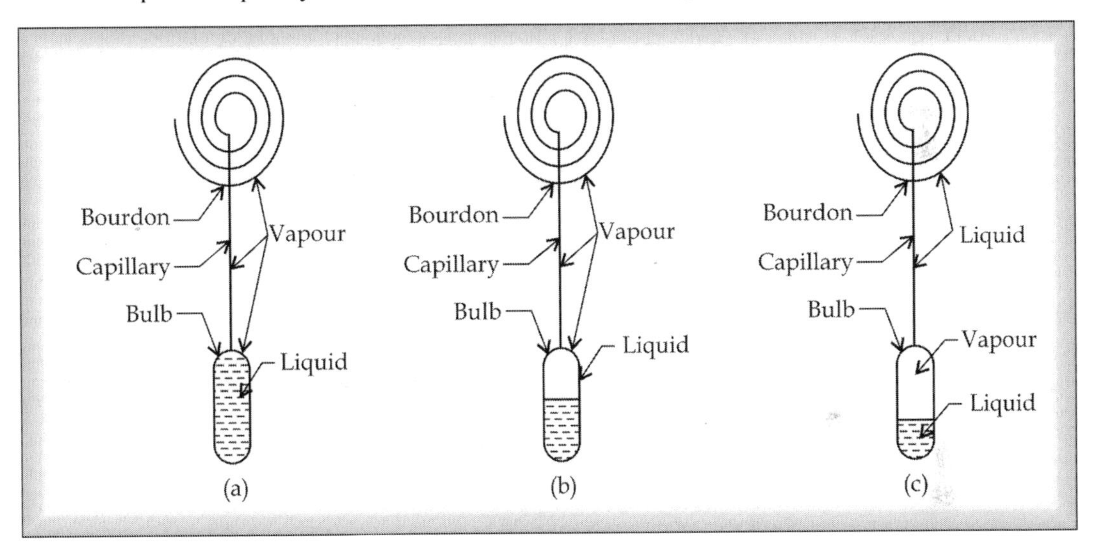

Fig. 5.6 Vapour Pressure Thermal Systems

The above problem can be solved by using dual- filled system as shown in Fig. 5.7 in which a vapourizing and a non-vapourizing liquids are used. The temperature sensitive fluid which vapourizes is called the actuating liquid. This vapour-pressure acts on the other liquid which does not vapourize and is called transmitting liquid and also prevents the volatile fluid from entering the capillary and

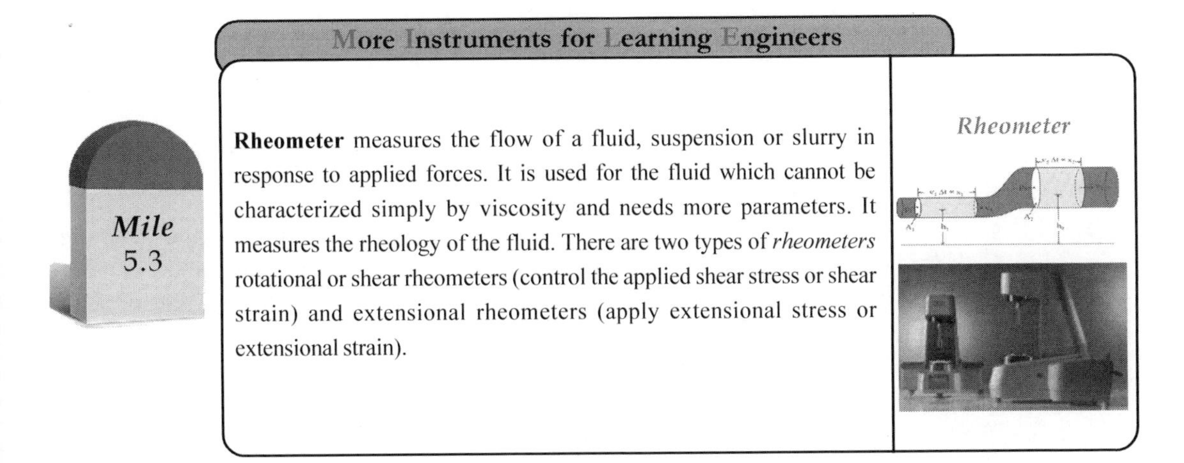

Rheometer measures the flow of a fluid, suspension or slurry in response to applied forces. It is used for the fluid which cannot be characterized simply by viscosity and needs more parameters. It measures the rheology of the fluid. There are two types of *rheometers* rotational or shear rheometers (control the applied shear stress or shear strain) and extensional rheometers (apply extensional stress or extensional strain).

Bourdon tube. The relation between temperature and pressure is $\log_e = P = a - b/T$, which is a logarithmic function end hence the scale of these thermometers is non-linear.

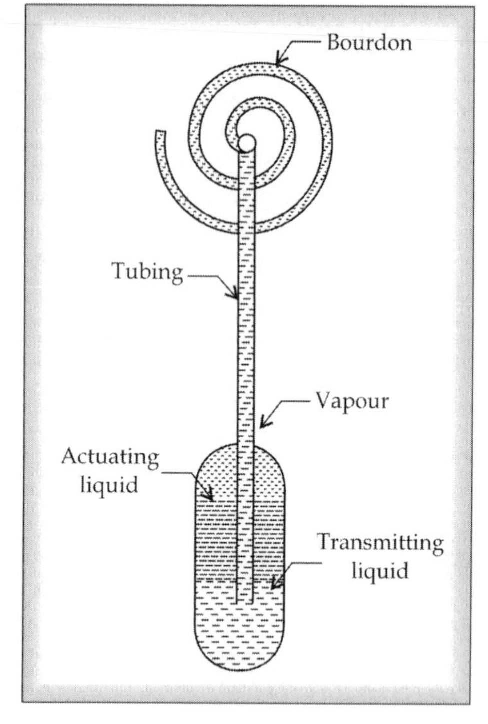

Fig. 5.7 Dual Filled Thermal System

Advantages (Merits)

- Inexpensive design, simple and rugged in construction
- Minimum damage in shipment, installation and use
- Good response, accuracy and sensitivity.
- These can be used in remote applications with capillary tube of length 100 m
- Self operated i.e., absence of auxiliary power unless electrical transmission systems are used.

Limitations (Demerits)

- Large time lag and stem conduction error.
- Compensation is required for ambient temperature and long capillary tubes.
- Maximum temperature measured is less as compared to electrical temperature measuring systems.

We know that the main parts of a pressure thermometer are bulb, capillary tube and Burdon tube. If there is a change in ambient temperature where the receiving element is located, it will produce an indication of the receiving element even though there is no change temperature of the bulb. This is because receiving elements respond to changes in temperature and these elements have no way of differentiating between pressure changes in ambient temperature and the pressure changes that are caused by change temperature of the process being monitored. This effect can be compensated by two methods

1. Case compensation method and 2. Full compensation method

Compensation Methods

1. *Case Compensation Method:* It is accomplished by using a bimetallic strip in the mechanism as shown in Fig. 5.8(a)

   (a) Bimetal case compensation (b) Bourdon case compensation

   The action of the bimetallic strip is in opposition to that of the receiving element (Bourdon tube) thereby negating the effect of changes in ambient temperature.

2. *Full Compensation Method:* In situations where the receiving element is not mounted close to the bulb, a capillary tube of appreciable length is used. In such cases, it is necessary to compensate for the effects produced by changes in tempera-ture in the neighborhood of the tube. This is done by using capillary compensation, which involves the use of the dummy capillary tube, which runs along the active -capillary tube but is not connected to the temperature sensing bulb thereby cancelling the effects produced by change in ambient temperature along the length of the active capillary tube.

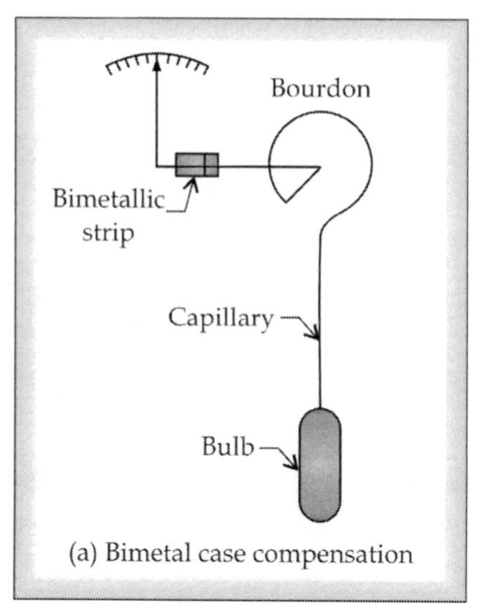

(a) Bimetal case compensation

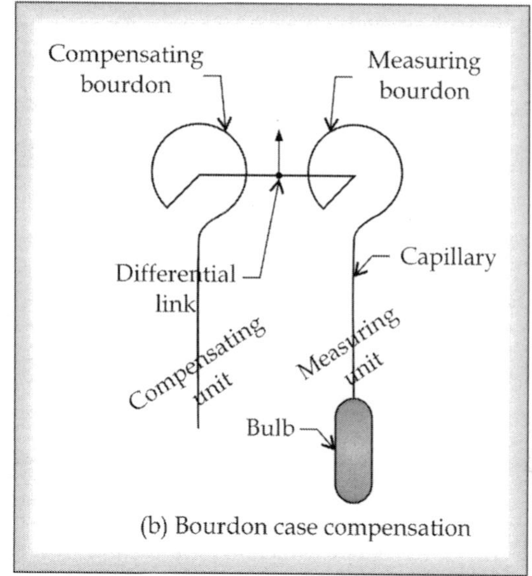

(b) Bourdon case compensation

Fig. 5.8 Case Compensation

Full compensation refers to a situation where in effects of changes in ambient temperature both within the case along the capillary tube are compensated for. This is illustrated in Fig. 5.9. There are two receiving elements, one active, Bourdon tube and the other a dummy or compensating Bourdon type. When the temperature increases, the active spring uncoils and the compensating Bourdon tube coils thereby providing case compensation.

The active capillary tube is connected to active Bourdon tube while the dummy (compensating) capillary tube is connected to the compensating Bourdon tube. The actions of the two negate each other thereby providing capillary compensation.

The liquid vapour filled systems do not require any ambient temperature compensation since the pressure in these systems is solely determined by the temperatures at the free surface of the liquid.

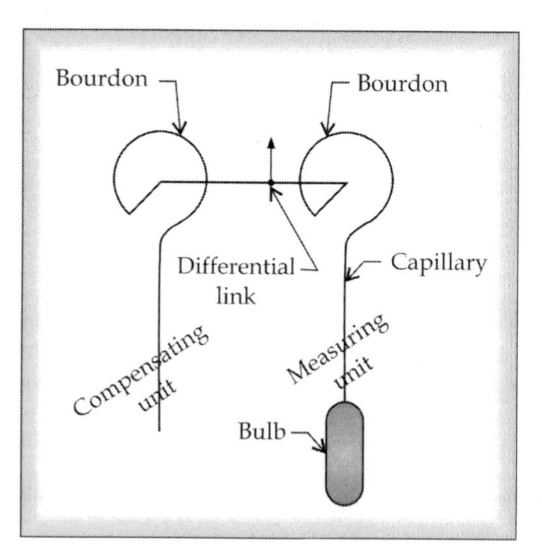

Fig. 5.9 Full Compensation of Filled-in System

SELF ASSESSMENT QUESTIONS-5.2

1. Explain how temperature can be measured with help of the changes in physical dimensions.
2. Explain the construction and operation of solid rod thermostats/ thermometers.
3. Explain the construction and operation of bimetallic thermostats/ thermometers.
4. Explain the transducer and principle of transduction in the following instruments
   (a) Solid rod thermostat (b) Bimetallic thermostat
   (c) Liquid in glass thermometer (d) Pressure thermometer
5. What is meant by thermometric liquid? Give examples. Also list out the desirable properties of thermometric liquids.
6. Why is mercury considered as superior to water and alcohol as a thermometric liquid?
7. Describe the construction and operation of the following instruments.
   (a) Liquid in glass thermometer (b) Pressure thermometer
8. Describe the construction, operation, merits and demerits of pressure thermometers.
9. With neat sketches describe (a) liquid filled (b) gas filled (c) vapour filled pressure thermometers.
10. What is compensation in pressure thermometers? Why is required? Explain the two compensation methods employed in pressure thermometers.

5.6 ELECTRICAL EFFECTS ON TEMPERATURE

To understand the electrical methods of measuring temperature, one needs to know clearly about the fundamentals of thermo-electric elements, thermo-electric materials and thermo-electricity. We shall first discuss about these fundamentals and governing principles which are pre-requisites to understand the measurement of temperature by electrical methods.

5.6.1 THERMO RESISTIVE ELEMENTS

Thermo resistive elements are classified into two types based on the material used for making the elements. We know that electrical resistance of most materials varies with temperature. If the resistive elements are made of materials like metals whose resistance increases with temperature they are called RTD's (Resistance Temperature Detectors) or resistance thermometers and if the resistive elements are made of materials like semi conductors whose resistance decreases with temperature they are called thermistors.

Precisely,

Thermo-resistive element – made of metals – resistance increases with temperature – RTD

Thermo-resistive element – made of semiconductors – resistance decreases with temperature – Thermistor

5.6.2 FUNDAMENTALS OF THERMO ELECTRICITY

The most common method of measuring temperature is by using thermocouples. In these instruments the measurements of temperature is based on thermo electricity, which was discovered by Seebeck in 1821.

When two dissimilar metals are joined the following three effects take place. They are

1. Seebeck effect
2. Peltier effect
3. Thomson effect

1. **Seebeck Effect:** When two dissimilar metals are joined together as shown in Fig. 5.10 an e.m.f will exist between the two points P_1 and P_2, which is primarily a function of the junction temperature. Its phenomenon is called the Seebeck effect.

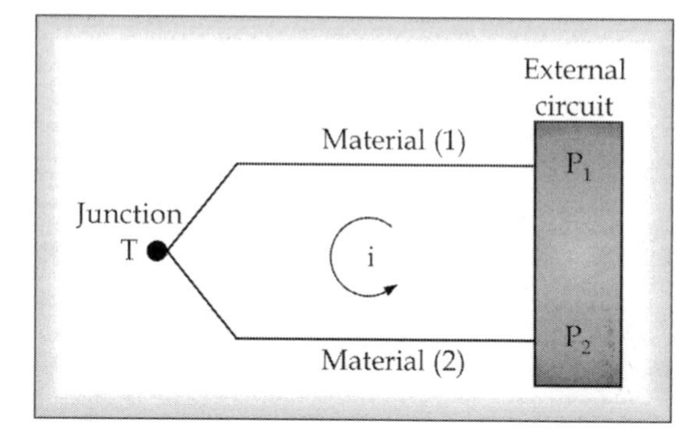

Fig. 5.10 Junction of Two Dissimilar Metals Indicating Thermoelectric Effect

2. **Peltier Effect:** The converse of Seebeck effect is Peltier effect, which was discovered in 1834 by Peltier. He found that when a current is passed across the junction of two dissimilar metals, heat is evolved at one junction and absorbed at the other, i.e., one junction is heated and other is cooled. This effect is known as Peltier effect. For a given pair of metals, the cooling or heating of the junction depends on the direction of the current.

   In Sb-Bi (Antimony -Bismuth) couple when the junction B is heated and A is kept cold, current flows from Sb to Bi at the junction A and from Bi to Sb at the junction B (Seebeck effect) Fig. 5.11(a). When a battery is placed in the circuit as shown in Fig. 5.11(b) heat is evolved at the junction A where the current flows from Sb to Bi and absorbed at the junction B where the current flows from Bi to Sb. If the direction of the current is reversed heat will be evolved at the junction B and absorbed at the junction A i.e., Peltier effect is reversed.

   Explanation of Peltier Effect: The explanation of Peltier effect lies in the fact that when two dissimilar metals are joined together, a potential difference is set up at each junction due to different free electron densities in the two metals.

In case of Iron-Copper couple there is a contact potential difference as shown in Fig. 5.11, the iron being above the copper.

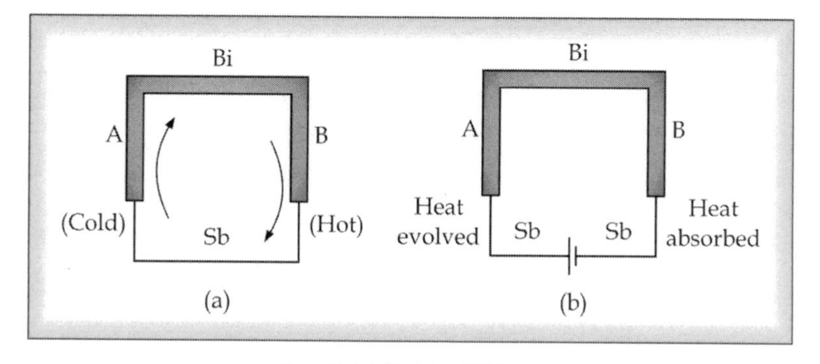

Fig. 5.11 Peltier Effect

3. **Thomson Effect:** Thomson in 1856 suggested that when a current flows through a thermo couple, the heat energy is evolved or absorbed not only at the junctions but it is absorbed or evolved all along one or both the conductors by virtue of the difference of temperatures between the two ends of the same element. Thus the Thomson effect is the absorption or evolution of energy due to the flow of current in an unequally heated conductor. This effect is reversible. In the case of copper the hotter parts are at a higher potentials than the colder ones and it is opposite in the case of iron.

   In case of copper, if the current is passed as indicated in Fig. 5.12(a) heat will be absorbed in the part *AB* and evolved in the part *BC*. Thus heat is absorbed when the current flows from the cold to the hot end, and is evolved when the current flows from the hot to the cold end. The Thomson effect is said to be positive.

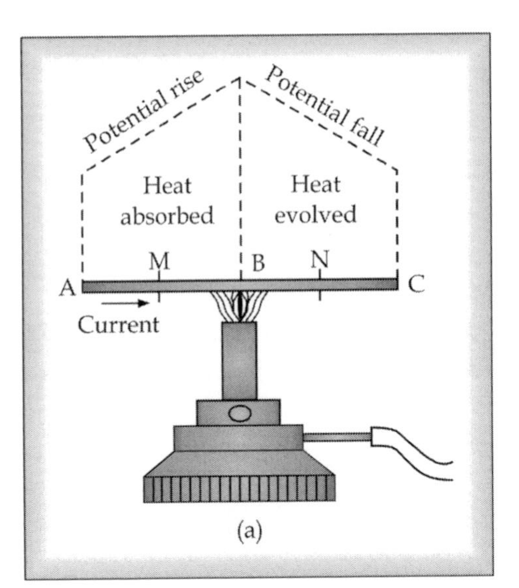

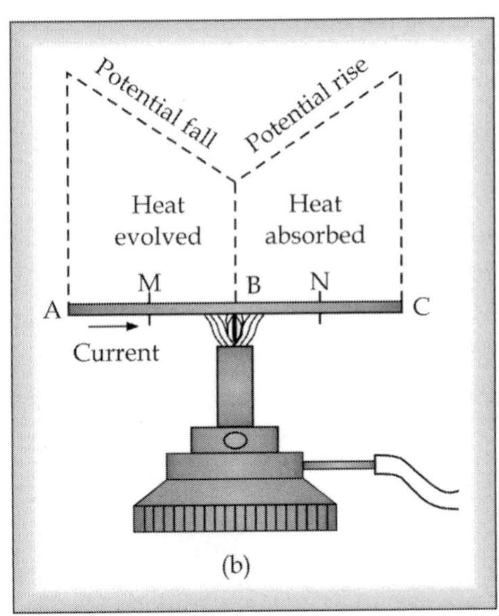

Fig. 5.12 Thomson Effect

More Instruments for Learning Engineers

Pyrheliometer measures direct beam solar irradiance. Sunlight enters the instrument through a window onto a thermopile which converts heat to electrical voltage signal calibrated to read in watts/m^2. It is used with a solar tracking system so as to keep faced at sun. Its measurement specs are subject to ISO and WMO standards. These meters are compared regularly for inter-calibration, to measure the amount of solar energy received. Pyrheliometer is often used in the same setup with a pyranometer.

Pyrheliometer

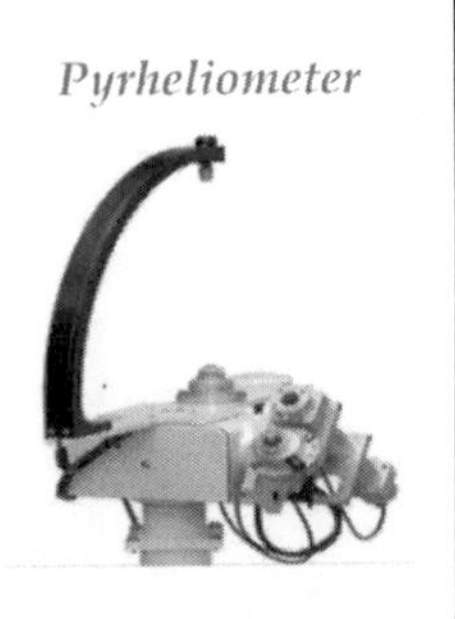

In case of Iron, if the current is passed as indicated in Fig. 5.12(b), heat will be evolved in the part *AB* and absorbed in the part BC. Thus the heat is evolved when the current flows from the cold to the hot end, and vice versa, and the Thomson effect is negative

There are, then, three e.m.fs present in a thermo electric circuit. The Seebeck emf, caused by the junction of dissimilar metals, the Peltier e.m.f, caused by a current flow in the circuit; and the Thomson e.m.f, which results from a temperature gradient in the materials. The Seebeck e.m.f is of prime since it is dependent on junction temperature. If the e.m.f generated at the junction of two dissimilar metals is carefully measured as a function of temperature, then such a junction may be utilized for the measurement of temperature.

5.6.3 Laws of Thermo-Electric Power

The following two laws are quite useful in studying electrical effects on temperature.

1. Law of intermediate metals or successive contacts
2. Law of intermediate temperatures

We shall briefly describe these in the paragraphs to follow.

1. **Law of Intermediate Metals or Successive Contacts:** This law states that if at a given temperature, a number of metals are in successive contact so as to form a chain of elements connected in series, the electromotive force between the extreme elements, if placed in direct contact, is the sum of the electromotive forces between successive adjacent elements. That is, for metals A, B, C, D ... N

   In successive contact, $E_A^N = E_A^B + E_D^C + E_C^D \ldots \ldots \ldots E_M^N$ provided all the junctions in series are at the same temperature as shown in Fig. 5.13.

   Suppose the metals A, B, C, are In contact

$$E_A^C = E_A^B + E_B^C$$

Where E_A^B; E_B^C and E_A^C denotes the e.m.f.s for circuits with metals A & B, B & C and A & C respectively for given fixed junction temperatures.

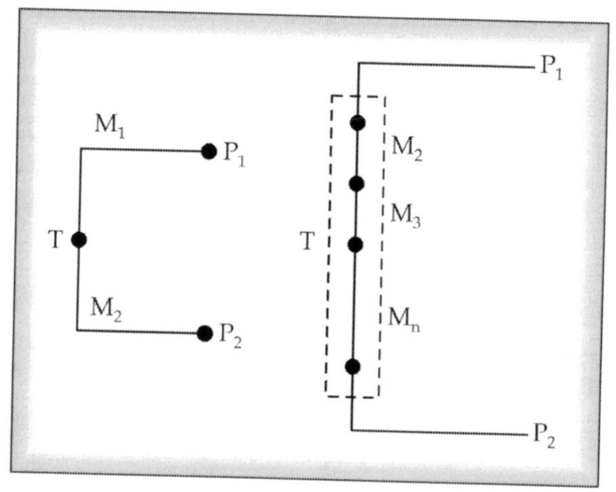

Fig. 5.13 Circuit Illustrating Law of Intermediate Metals

2. **Law of Intermediate Temperatures:** The electromotive force for a couple with junction at T_1 and T_3 is the sum of the electromotive forces of two couples of the same metals one with junctions at T_1 and T_2 and the other with junction at T_2 and T_3 i.e.,

$$E_3^1 = E_2^1 + E_3^2$$

where E_3^1 = e.m.f. when the junctions are at T_1 and T_3

E_2^1 = e.m.f. when the junctions are at T_1 and T_2

E_3^2 = e.m.f. when the junction at T_2 and T_3

After understanding the above concepts and principles, we now can easily discuss about the measurement of temperature using electrical methods as explained below.

5.7 MEASUREMENT OF TEMPERATURE BY ELECTRICAL METHODS

Temperature can conveniently be measured by electrical methods since these methods give a signal that can be easily detected, amplified, or used for control purposes and are quite accurate when properly calibrated and compensated. The following are the most widely applied electrical methods to measure temperature.

1. Resistance Temperature Detector (RTD) or Electrical-Resistance Thermometer
2. Thermistor (Thermal Resistors)
3. Thermocouples
4. Quartz-Crystal Thermometer.
5. Liquid Crystal Thermography.

We shall now discuss these in detail in the following sections.

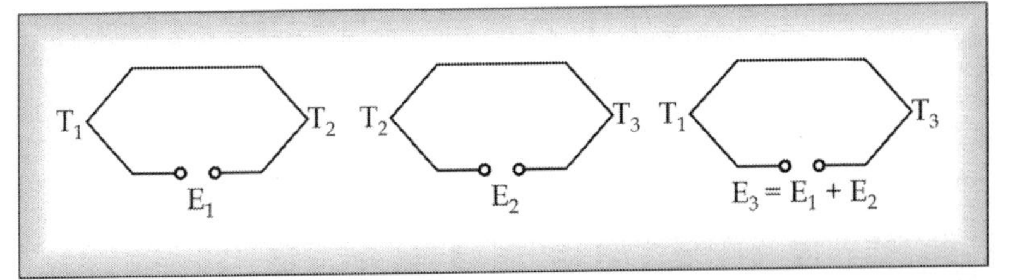

Fig. 5.14 Circuit Illustrations the Law of Intermediate Temperatures

5.7.1 Resistance Temperature Detector (RTD) or Electrical-Resistance Thermometer

As described in the above section 5.6.1, the Resistance Temperature Detector, often called in short as RTD, is a thermo-resistive element made up of metals whose resistance increases with temperature. RTDs are also known as resistance thermometers. One who wishes to design or construct a resistance thermometer should first select right material for the thermo-resistive element.

Desirable properties for the resistance thermometer element

The materials used as resistance thermometer elements should have following properties.

(a) Its resistivity should be in such a way that it allows fabricating in convenient sizes without excessive bulk, which would degrade time response.

(b) Its thermal coefficient of resistivity should be high and as constant as possible, thereby providing an approximately linear output of reasonable magnitude.

(c) The material should be corrosion resistant.

(d) It should not undergo phase changes in the temperature range of interest.

It should be available in a condition providing reproducible and consistent results. For this minimum residual strains must be there in the material.

Ideal or universal resistance thermometer elements are difficult to find. Platinum, Nickel, Copper, Tungsten, Silver and Iron are mostly used.

The temperature resistance relation of an RTD is represented by

$$R(T) = R_o[1 + A(T - T_0) + B(T - T_0)^2]$$

where $R(T)$ = The resistance of temperature T, R_o = The resistance at a reference temperature T_o

A & B = temperature coefficients of resistance depending on material. Over a limited temperature interval, if linear approximation is assumed

Then, $R(T) = R_o[1 + A(T - T_0)]$

Resistance temperature coefficient and resistivity of some material are given in Table 5.5 at 20 °C.

Table 5.5

Substance	α ($^0C^{-1}$)	S(M Ω.m)
Nickel	0.0067	0.06 × 85
Iron (alloy)	0.002 to 0.006	0.1000
Tungsten	0.0048	0.0565
Aluminum	0.0045	0.0265
Copper	0.0043	0.0167
Lead	0.0042	0.206
Silver	0.0041	0.0159
Gold	0.004	0.235
Platinum	0.00392	0.1050
Mercury	0.00099	0.9840
Magnanin	± 0.00002	0.4400
Carbon	- 0.0007	14000
Electrolytes	- 0.02 to -0.09	Variable
Semi conductor (Thermistors)	- 0.068 to + 0.14	10^7

1. **Construction of Platinum Resistance Thermometer Elements:** Several forms of resistance thermometers have been developed for temperature measurement depending upon their requirement such as speed of response, environments condition and ability to withstand corrosion or vibration

   One common construction technique involves winding the platinum on a glass or ceramic bobbin followed by sealing with molten glass. This technique protects the platinum RTD element but is subject to stress varying over wide temperature ranges. The platinum resistance thermometers are manufactured in the following types

   (i) Open wire type (ii) Closed wire type
   (iii) Open film type (iv) Closed film type

Fig. 5.15 shows the common open wire element in which the platinum wire is wound in the form of free spiral or held in place by an insulated carrier such as mica or ceramic in the form of a perforated coil former. The wire is in direct contact with the gas or liquid whose temperature is to be measured. Such an element has an excellent response time, small conduction and self heating errors.

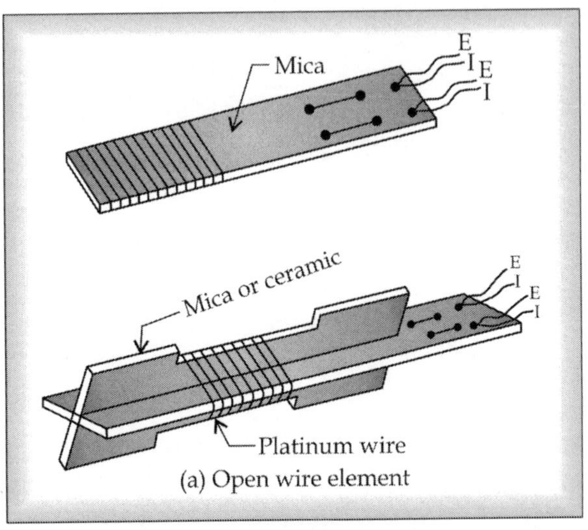

(a) Open wire element

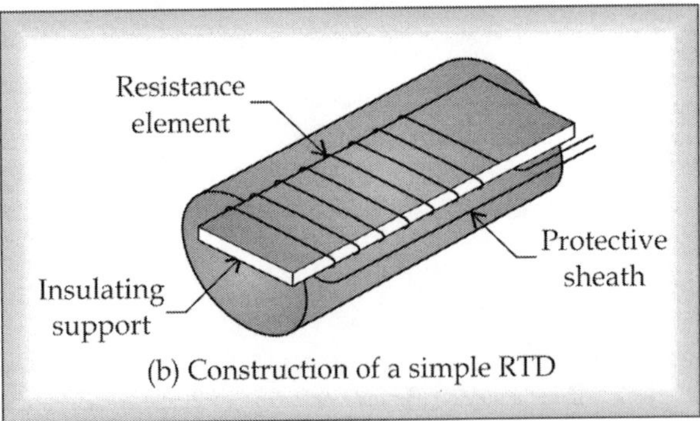

(b) Construction of a simple RTD

Fig. 5.15 Wire Type Platinum Resistance Thermometer

2. **Construction of Film Type Platinum Resistance Thermometers:** RTD sensors may also be constructed by depositing platinum or metal glass slurry on a ceramic substrate. This process is less expensive than the mechanical-winding ceramic but its accuracy is less. The thin film sensor does offer the advantage of low mass and therefore more rapid thermal response and less chance if conduction error.

> ***Transducer:*** *RTD element*
> ***Transduction:*** *Temperature to Resistance*
> **Principle:** $R(T) = R_0[1 + A(T - T_0)]$

3. **Electrical Resistance Thermometer Circuits:** Modified versions of the conventional four arm whetstone's bridge circuits are widely employed for temperature measurements using platinum resistance thermometers. For steady-state measure-ments a null condition will suffice, while transient measurement will usually require the use of a deflection bridge.

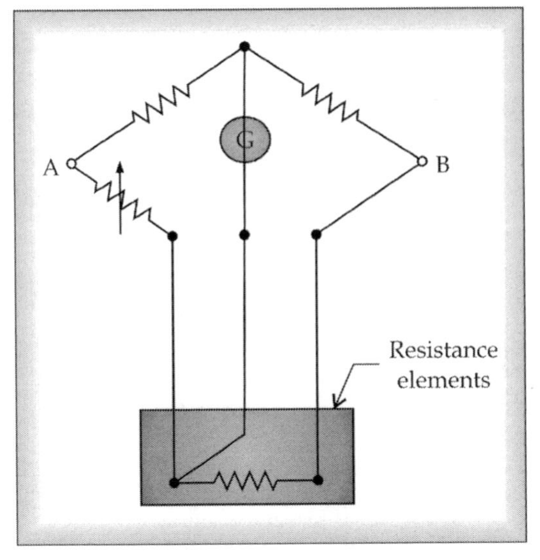

Fig. 5.16 Siemens' Three-lead Arrangement

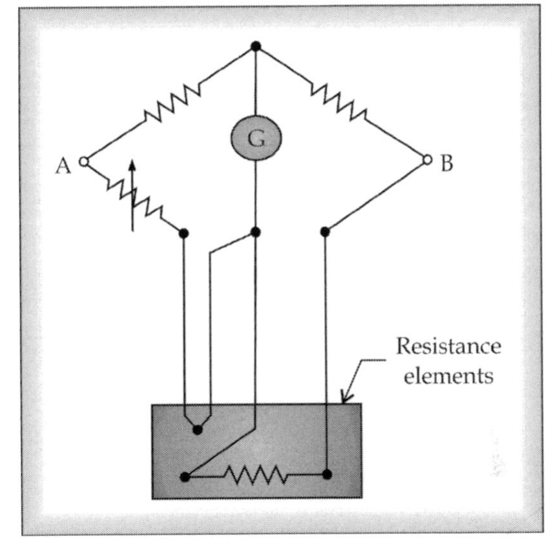

Fig. 5.17 Callender Four-Lead Arragement

One of the primary sources of error in the electrical resistance thermometers is the feet of the resistance of the leads, which connect the element to the bridge circuit. Several modified forms of the Wheatstone bridge circuit are

(i) Siemen's Three - lead arrangement

(ii) Callendar Four - lead arrangement

(iii) Floating - Potential arrangement

(i) ***Siemen's Three- Lead Arrangement:*** It is the simplest type of corrective circuit at balance conditions (Fig. 5.16). The center lead carries no current, and the effect of the resistance of the other lead is cancelled out.

(ii) ***Callendar Four Lead Arrangement:*** This arrangement (Fig. 5.17) solves the problem by inserting two additional lead wires in the adjustable leg of the bridge so that the effect of the lead wires on the resistance thermometer is cancelled out.

More Instruments for Learning Engineers

Mile 5.5

Pyranometer [in Greek, pyr(ðõñ) means fire and ano(íù) means above or sky] is a type of actinometer that measures broadband solar irradiance on a planar surface. It does not need any power to operate and has a sensor that measures flux density (W/m^2) of the solar radiation (spectrum spread 300-2800nm) and covers with a spectral sensitivity (as flat as possible). It has a "directional or cosine response" close to the ideal cosine characteristic.

Pyranometer

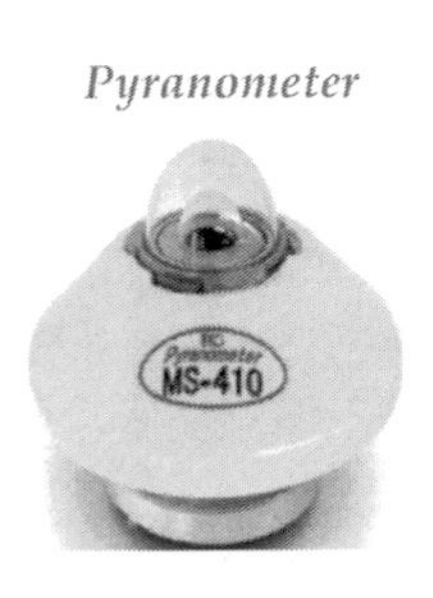

(iii) *Floating-Potential Arrangement:* This type of arrangement is same as Siemen's connection, but an extra lead is inserted. This extra lead is used to check the equality of lead resistance. The reading may be taken in the position shown in Fig. 5.18, followed by additional readings with the two right and left leads inter changed respectively through this interchange procedure, the best average reading may be obtained and the lead error is minimized.

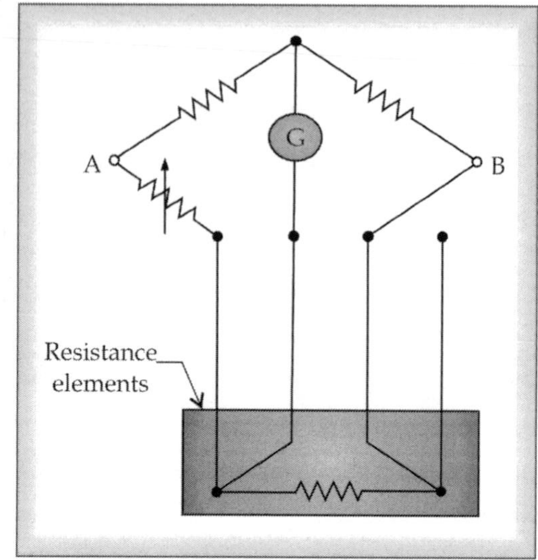

Fig. 5.18 Floating Potential Arragement

ILLUSTRATION - 5.2

A Platinum resistance thermometer is used at room temperature (20°). Assuming a linear temperature variation with resistance, calculate the sensitivity of the thermometer is ohms/degrees centigrade. [Given Resistance Temperature Coefficient (α) = 0.00392°C^{-1}]

Solution: For linear variation of resistance with temperature is

$$R = R_0[1 + \alpha(T - T_0)]$$

$$R = R_0 + \alpha R_0(T - T_0)$$

where R_0 is the resistance at the reference temperature T_0.

The sensitivity is

$$S = \frac{dR}{dT} = \frac{d}{dT}[R_0 + \alpha R_0(T - T_0)] \quad [\because R = R_0 + R_0\alpha(T - T_0)]$$

$$= 0 + \frac{d}{dT}[\alpha R_0(T - T_0)]$$

$$= \alpha R_0\left(\frac{dT}{dT} - \frac{dT_0}{dT}\right) = \alpha R_0$$

R_0 depends on the length and size of the resistance wire.

At room temperature α = 0.00392°C^{-1}

Hence sensitivity = 0.00392 Ω/°C

5.7.2 THERMISTORS (THERMAL RESISTORS)

The thermistor is the short form of "***Thermal Resistors***". They are thermally sensitive resistors made of semi conducting material like oxides of Manganese, Nickel, Cobalt, Copper, Iron and Uranus. These maternal have negative temperature coefficient.

Further, the resistance follows an exponential variation with temperature, which is given by

$$R = R \ \text{exp.}\left[\beta\left(\frac{1}{T} - \frac{1}{T_0}\right)\right]$$

where R_o is the resistance at the reference temperature, T_0 & β is an experimentally determined constant. The numerical value varies between 3500 to 4600 °K depending on the thermistor material and temperature. The Figure 5.19 shows typical temperature resistance relation of Platinum or backed afterwards. Thermistors may be shaped in the form of beads, disks, washers, rods and these standard forms are shown Fig. 5.20.

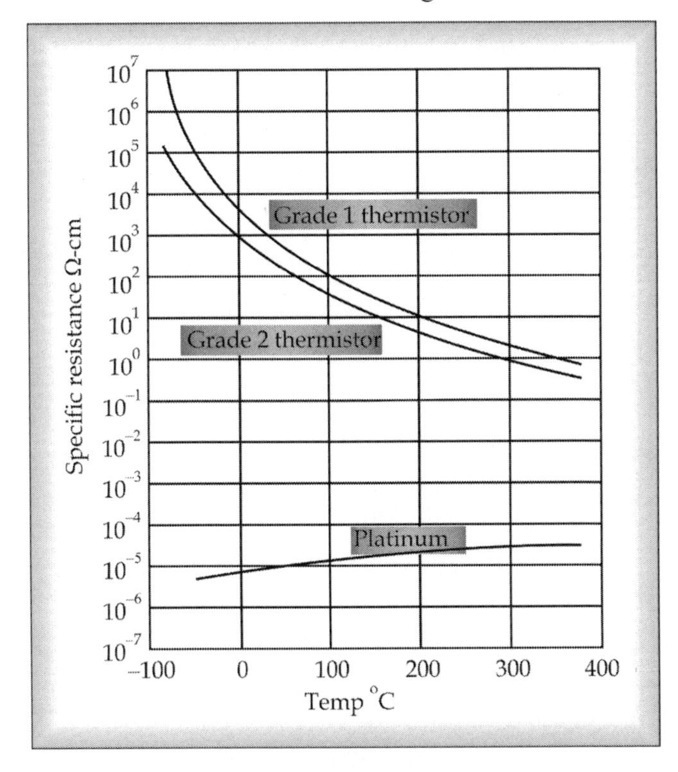

Fig. 5.19 Thermistor Temperature Resistance Relation

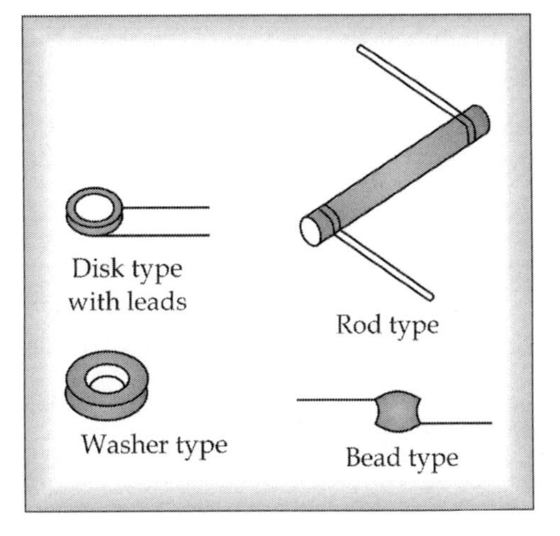

Fig. 5.20 Various Forms of Thermistors

Through proper application of thermistor and electrical circuit characteristics, the devices may be used for temperature measurement or control. Also, time delay actions over large ranges are possible through proper balancing of electrical and heat transfer conditions.

Differences between Resistance Temperature Detectors (RTDs) and Thermistors: The distinction is shown through the following Table 5.6.

Table 5.6 Distinction betweem RTD and Thermistors

	RTD	Thermistors
1.	A temperature measuring device using resistance elements made of metals like nickel, copper, platinum and silver are referred to as resistance thermometer or resistance temperature detector.	A temperature measuring device using resistance elements made of semi conducting materials like metallic oxides of cobalt, manganese and nickel are referred to as thermistors.
2.	Resistance change is small and positive	Resistance change is large and negative
3.	Linear temperature resistance relation	Non linear temperature resistance relation.
4.	The operating range lies between -260 °C to 1000 °C	The operating range lies between 100 °C to 275 °C
5.	More time-stable i.e., better reproducibility.	Less time-stable i.e. reproducibility is not good.

SELF ASSESSMENT QUESTIONS-5.3

1. Describe the relation between the electrical resistance and temperature. How can you make use of the relations to measure the temperature?
2. What is thermo-electricity? State and explain the laws of thermo-electric power.
3. Explicate the following and discuss how they are useful in measurement of temperature.
   (a) Seebeck effect (b) Peltier effect (c) Thomson's effect
4. What is 'RTD'? Explain its construction and operation to act as a measuring instrument.
5. What are the desirable properties of resistance thermometer element?
6. Discuss the transducer and the principle of transduction in the case of RTD.
7. Describe the construction of various platinum resistance thermometer elements with the help of sketches.
8. Describe the various modified Wheatstone bridge arrangements of platinum resistance thermometer elements with the help of circuit diagrams.
9. Sketch and explain the construction of following types of platinum resistance thermometer elements.
   (a) Wire type (b) Film type
10. Explain the following arrangements of platinum resistance thermometer elements with the help of circuit diagrams.
    (a) Siemen's three-lead arrangement (b) Callendar's four-lead arrangement
    (c) Floating-potential arrangement
11. What is 'Thermistor'? Explain its construction and operation to act as a measuring instrument.
12. Distinguish between RTD and Thermistor.

5.7.3 THERMOCOUPLES

Theoretically, any two unlike conducting materials could be used to form a thermocouple. Certain materials and combinations are better than others and some have practically become standard for given temperature ranges.

Based on the composition of metals used the thermocouples are grouped into the following broad categories.

1. Base metal thermocouples
2. Rare metal thermocouples
3. Non-metallic thermocouples

1. ***Base Metal Thermocouples:*** These are made by the combination of pure metal and alloys of iron, copper and nickel. Mostly used to measuring lower ranges of temperature up to 1375 °C.
2. ***Rare Metal Thermocouples***: These are made by the combination of pure metal and alloys of
   (a) Platinum and Rhodium for temperature up to 1725 °C and
   (b) Tungsten, Rhodium and Molybdenum for temperature up to 2625 °C.
3. ***Non-metallic Thermo Couples:*** These are made by the combination of semi conductors materials such as silicon-germanium. These are used in the power packs of the Viking 1 and 2 space explosers. There is a heat conversion from the thermal generator into about 450 W of electrical power, which is sufficient to run all the electrical systems of the module.

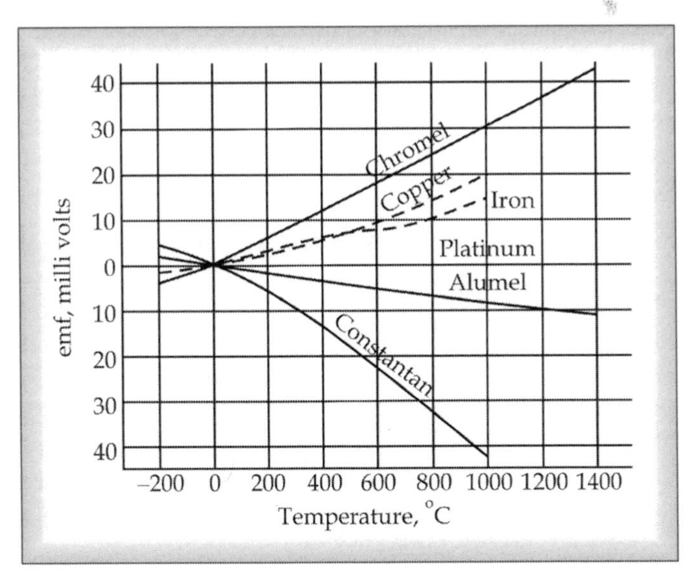

Fig. 5.21 Thermal EMF of Some Thermocouple Materials Related to Platinum

The Desirable Characteristics of Thermocouple Materials

(i) The emf produced per degree of temperature change must be sufficient to facilitate detection and measurement. The e.m.f. of a simple thermo couple circuit is generally given by the equation:

$$E = At + \frac{1}{2}Bt^2 + \frac{1}{3}Ct^3$$

where T is the temperature in °C and A, B & C are constants dependent on the thermocouple material.

The value of E in the above equation is based on the reference junction temperature of 0°C.

The Fig. 5.21 shows the thermal e.m.f of some thermocouple materials related to platinum. The greatest output for a thermocouple pair will be obtained when one material is positive and the other negative with respect to platinum. The sensitivity will be more if, value of potential produced is more. The sensitivity of a thermocouple is given by

$$S = +\frac{dE}{dt} = A + Bt + Ct^2$$

(ii) The emf temperature relation must be reasonably linear and reproducible. This helps in reading the scale more easily.

(iii) The thermocouple should maintain its' calibration over a long period of time i.e., it should not drift.

(iv) The thermocouple materials should be highly resistant to oxidation, corrosion and contamination so that it will have long life.

(v) The material should physically be able to withstand high and rapidly fluctuating temperatures.

(vi) The material must be in such a way that it can be produced in successive batches with the same thermo electric characteristics. Then replacement of thermocou-ples without recalibrating each time is possible.

Preparation of Thermocouples

Thermocouples may be prepared by twisting the two wires together brazing or preferably by welding as shown in Fig. 5.22(a). Low-temperature couples are often used bare; however, for higher temperature some form of protection is generally required. Fig. 5.22(b) illustrates common methods for separating the wires.

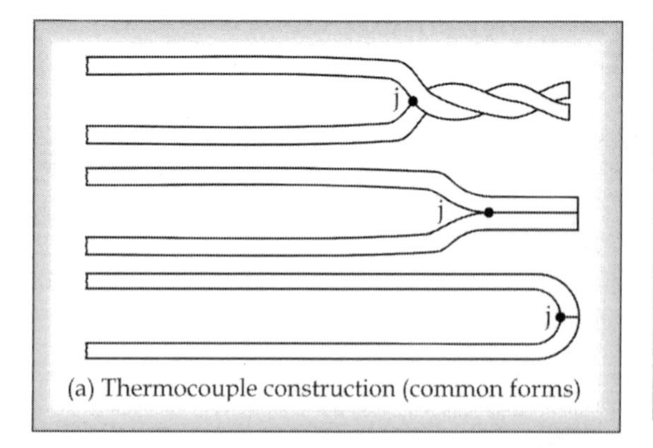

(a) Thermocouple construction (common forms)

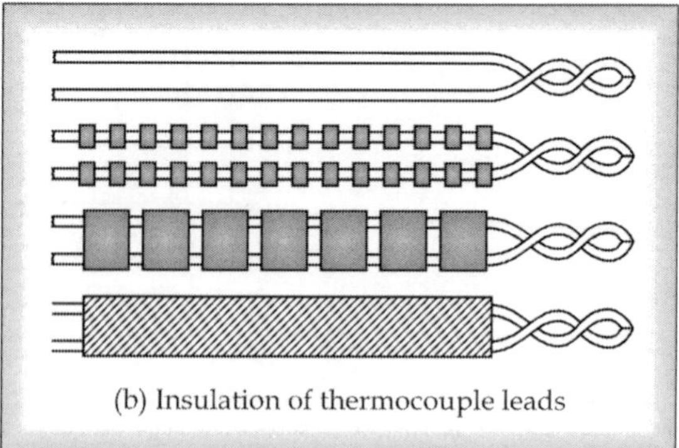

(b) Insulation of thermocouple leads

Fig. 5.22 Thermocouple

Measurement of Thermal EMF

The actual magnitude of electrical potential developed by thermocouples is quite small. Table 5.7 gives an idea of the range of values to be expected.

Table 5.7

Temperature °C	Thermocouple Type				
	Copper vs. Constantan (T)	Chromel vs. Constantan (*E*)	Iron vs. Constantan (*J*)	Chromel vs. Alumel (K)	Platinum vs, Platinum, 10%Rhodium(*S*)
-184.4	-5.341	-8.404	-7.519	-5.632	-
-128.9	-4.149	-6.471	-5.760	-4.381	-
-73.7	-2.581	-3.976	-3.492	-2.699	-
-17.8	-0.674	-1.026	-0.885	-0.692	-0.092
37.8	1.518	2.281	1.942	1.520	0.221
93.3	3.967	5.869	4.906	3.819	0.597
148.9	6.647	9.708	7.947	6.092	1.020
204.4	9.523	13.748	11.023	8.314	1.478
260.0	12.572	17.942	14.108	10.560	1.962
371.1	19.095	26.637	20.253	15.178	2.985
537.8	-	40.056	29.515	22.251	4.609
815.6	-	62.240	-	33.913	7.514
1093.3		-	-	44.856	10.675
1371.1	-	-	-	54.845	14.018
1648.9	-	-	-	-	17.347

Table 5.8 Values of EMF in Absolute Mill Volts for Selected Metal Combinations Based on Reference Junction Temperature at 0 °C

°C	0	5	10	15	20
-200	-5.603	-5.522	-5.439	-5.351	-5.261
-175	-5.167	-5.069	-4.969	-4.865	-4.758
-150	-4.648	-4.535	-4.419	-4.299	-4.177
-125	-4.051	-3.923	-3.791	-3.656	-3.519
-100	-3.378	-3.235	-3.089	-2.939	-2.788
-75	-2.633	-2.475	-2.315	-2.152	-1.987
-50	-1.819	-1.648	-1.475	-1.299	-1.121
-25	-0.94	-0.757	-0.571	-0.383	-0.193
0	0	0.195	0.391	0.589	0.789
25	0.992	1.196	1.403	1.611	1.822
50	2.035	2.250	2.467	2.687	2.908
75	3.131	3.357	3.584	3.813	4.044
100	4.277	4.512	4.749	4.987	5.227

Table 5.8 ***Contd...***

°C	0	5	10	15	20
125	5.469	5.712	5.957	6.204	6.452
150	6.702	6.954	7.207	7.462	7.718
175	7.975	8.235	8.495	8.757	9.021
200	9.286	9.553	9.820	10.090	10.360
225	10.632	10.905	11.180	11.456	11.733
250	12.011	12.291	12.572	12.854	13.137
275	13.421	13.707	13.993	14.261	14.570
300	14.860	15.151	15.443	15.736	16.030
325	16.325	16.621	19.919	17.217	17.516
350	17.816	18.118	18.420	18.723	19.027
375	19.332	19.638	19.945	20.252	20.560

At a given temperature type E thermocouples (Refer Fig. 5.23) have the highest output voltage among common types, but this voltage is still a value measured in milli-volts. The sensitivity of thermocouples is also relatively low. For example, type E voltage increases from 2.281 to 5.869 mV

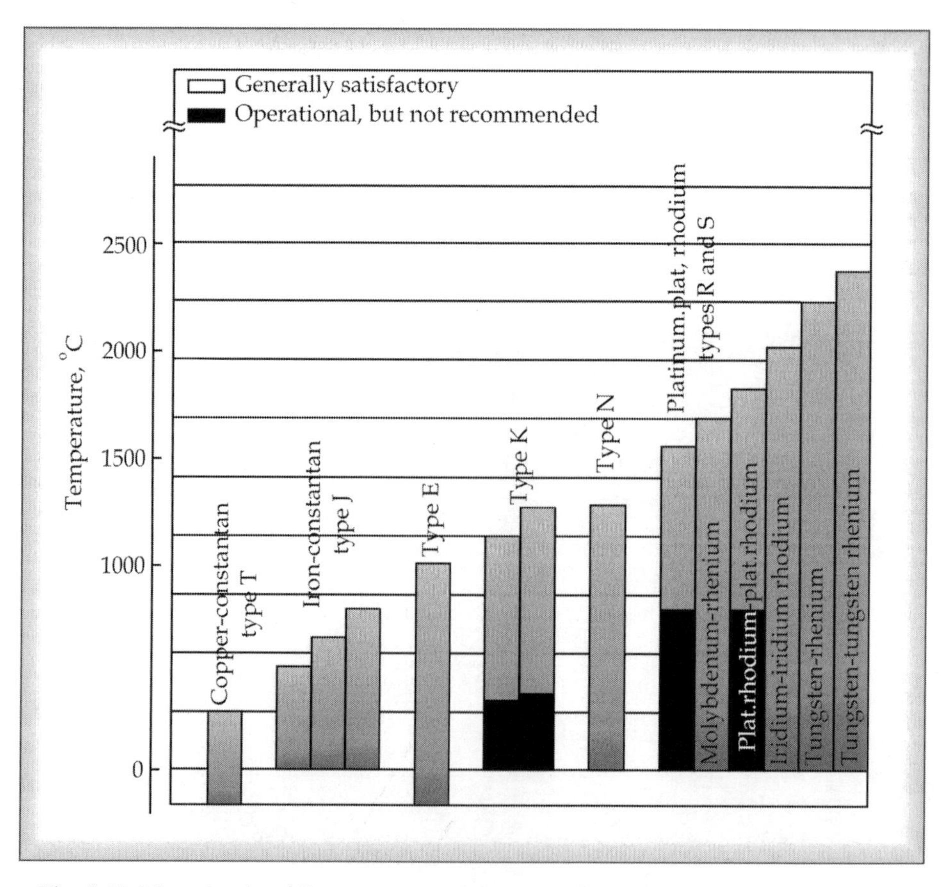

Fig. 5.23 Magnitude of Temperatures Measured by Different Thermocouples

as temperature increases from 37.8 °C to 93.3 °C; the average change per degree Celsius is only about 36μV. Because of these factors, thermocouples, require accurate and sensitive voltage measurement and in practice, cannot be used reliably for temperature changes of less than about 0.1 °C.

Traditionally, thermocouple output has been measured through use of voltage balancing potentiometer. Today, a high quality digital voltmeter is sufficient. Moreover, solid state or integrated-circuit devices have largely replaced manual methods in most applications.

Solid state temperature measurement instrumentation includes digital read-out "Thermometers" and recorders and controllers.

Fig. 5.24 shows a simple temperature measuring system using a thermocouple as the sensing element and potentiometer for indication. In this illustration, the thermoelectric circuit consists of a measuring junction p, and a reference junction. q, at the potentiometer.

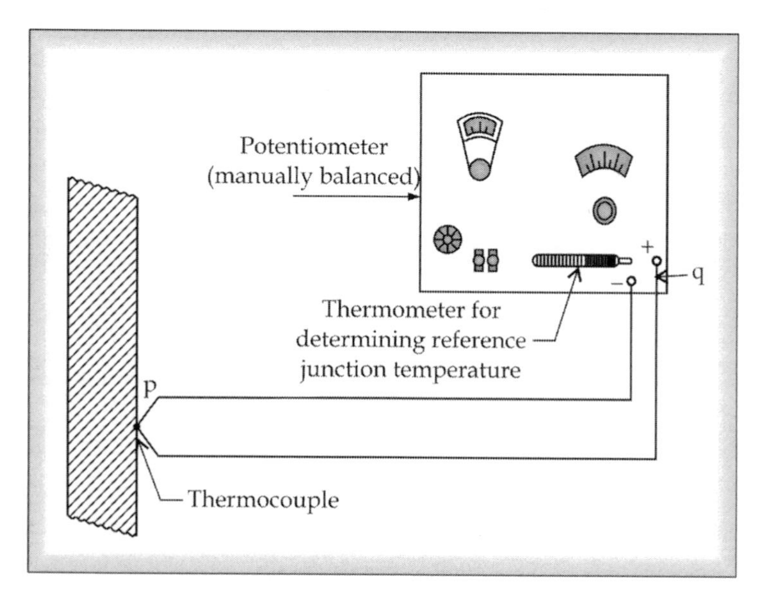

Fig. 5.24 Potentiometer Terminals as a Reference Junction

More Instruments for Learning Engineers

Mile 5.6

Pyrometer measures high temperatures, is a non-contacting device that intercepts and measures thermal radiation, a process known as pyrometry. It is originally coined to denote a device measuring temperatures of objects above incandescence (i.e. objects bright to the human eye). Pyrometer operates on Stefan–Boltzmann law.

Pyrometer

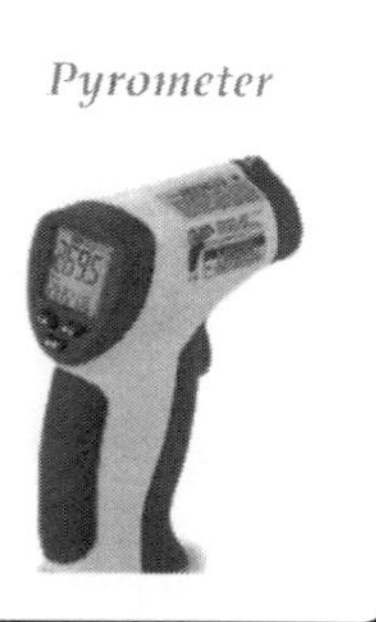

Thermocouple Compensation

Thermocouples will give a value of temperature which is the resultant of the entire circuit. They measure an E.M.F

$$E = E_1 - E_2$$

where E_1 = e.m.f generated at measuring/ hot junction

E_2 = e.m.f generate at reference/ cold function

Thermocouple tables give e.m.fs for a range of temperatures of the measuring function h, with reference junction 1_2 at 0 °C. An instrument calibrated to the tabulated figures will therefore read correctly only if the reference 'junction is at 0 °C. If it is at any other temperature the reading will be an error and hence would required compensation.

The correction may be by simply applying an arithmetic correction to the readings or automatic or manual means.

1. **Arithmetic Correction:** Suppose T_1& T_2 are the temperatures at junctions 1_1 and h with e.m.f s E_1 & E_2 respectively

   By knowing E_1 & E_2 we cannot know the measuring junction temperature because those values are valid only when reference junction is at 0 °C. Temperature T_1 may be found only if $E_1 - E_0$ is known, where E_0 is the e.m.f. produced at the reference junction with its temperature at 0°C. This requires the knowledge of $E_2 - E_0$ as given below.

   $$E_1 - E_2 = (E_1 - E_0) - (E_2 - E_0)$$

   $$E_1 - Eo = (E_1 - E_2) + (E_2 - E_0)$$

   It is clearly understood from the equation that when the reference junction is not at 0 °C, the observed value $(E_1 - E_0)$ must be corrected by adding to it a voltage of $(E_2 - E_0)$ that would have resulted from a temperature difference equal to the amount by which the reference junction is above 0 °C.

2. **Automatic Compensation by Thermistor:** Cold junction or reference junction compensation can be done as shown in Fig. 5.25.

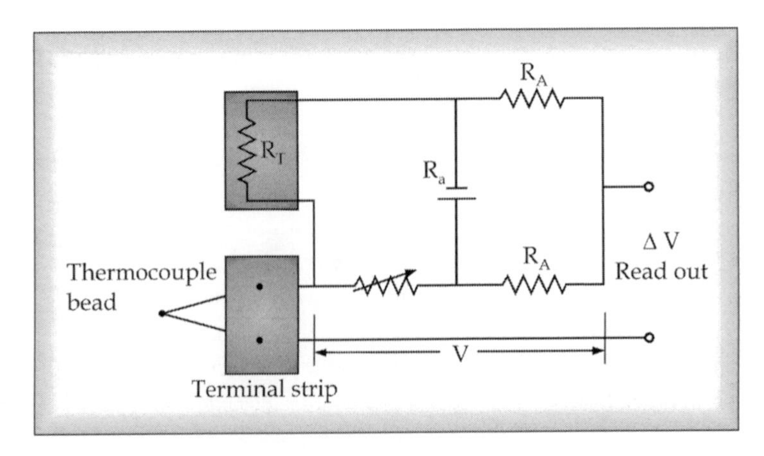

Fig. 5.25 Reference Junction Compensation Circuit using Thermistor

In this a thermistor is placed in thermal contact with the terminal strip to which the thermocouple wires are attached. The voltage V_a and temperature coefficient of the thermistor must be adjusted so that V will match the thermocouple temperature coefficient *in mv/°C*. The value of R_v is adjusted so that the voltage output ΔV is zero at 0 °C; convenient value of R_A is about 1 $k\Omega$. This is called hardware - compensation device; others based on RTD or of solid - state temperature sensors can also be used.

Thermocouples Connected in Series (Thermopile) and in Parallel

Thermocouples may be connected electrically in series or parallel, as shown in Fig 5.26. When connected in series, the combination is generally called a thermopile, where as parallel connected thermocouples have no specific name.

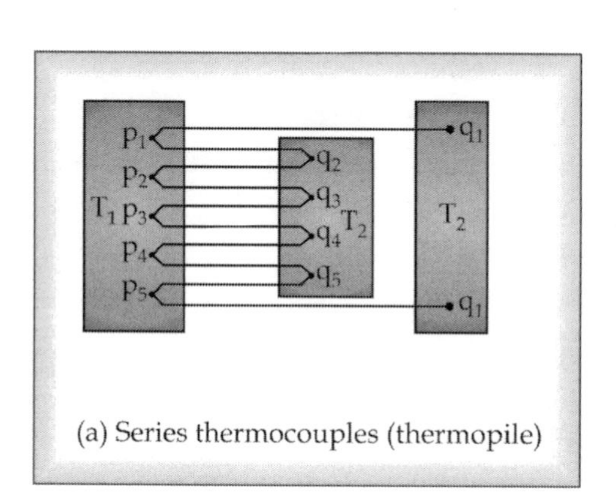

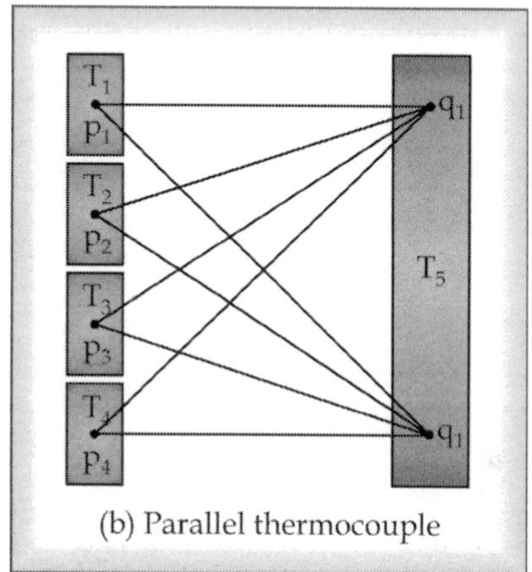

Fig. 5.26 Thermocouples in Series and Parallel

The total output from 'n' thermocouples connected to form a thermopile will be equal to the sum of the individual emfs and if the thermocouples are identical, the total output will equal 'n' times the output of the single couple. The purpose of using a thermopile rather than a single thermocouple is to obtain a more sensitive element.

When the couples are combined in the form of a thermopile, it is usually desirable to cluster them together as closely as possible in order to measure the temperature at an approximate point source. It is obvious, however, that when source of thermocouples are combined in series, the law for intermediate metals, as illustrated in Fig. 5.13, cannot be applied to combinations of thermocouples, for the individual thermocouple e.m.fs would be shorted. Care must therefore be used to ensure that the individual couples are electrically insulated one from the other.

Parallel connection of thermocouples provides on averaging, which in certain cases may be advantageous. The form of combination is not usually referred to as a thermopile.

Comparison between Resistance Thermometer (RTD) and Thermocouple

1. Thermocouples are less accurate as compared to RTDs. RTDs are accurate up to ± 0.3 °C while there is a possibility inaccuracies about ± 1.0 °C in thermocouples owing to the changes in reference junction temperature.
2. Thermocouples are less sensitive than RTDs. However, the response of modern resistance elements is almost same as that of thermocouples.

3. Thermocouples need frequent replacement, but cheaper than RTD elements. Thus overall cost is same for both. However, the inconvenience in replacement of thermocouple is accountable.
4. In the early days, the response in RTD element was somewhat slower than that of thermocouple. However, modern RTD elements are almost same as those of thermocouples.
5. In fact, RTD is preferred to thermocouples in view of convenience and ease of operation, other limitations on temperature measurement.

SELF ASSESSMENT QUESTIONS-5.4

1. What is 'Thermocouple'? Explain its significance in instrumentation.
2. Explain about the thermocouples used in the instruments for measurement of temperature.
3. Explicate the following thermocouple materials and discuss how they are useful in measurement of temperature.
   (a) Base metals (b) Rare metals (c) Non-metals
4. What is 'Thermocouple'? Explain its construction and operation to act as a measuring instrument.
5. What are the desirable properties of thermocouple materials?
6. Discuss the transducer and the principle of transduction in the case of thermocouple.
7. What are the merits and demerits of thermocouples used for temperature measurement?
8. Explicate the following with reference to thermocouples.
   (a) Arithmetic Compensation (b) Automatic Compensation
9. What do you understand by the term 'compensation' with reference to the thermocouple materials? Describe various compensation methods.
10. What is a 'Thermopile'? Explain.
11. Discuss the series and parallel connections of thermocouples and their applications.
12. Distinguish between RTD and Thermocouples.

5.8 SEMICONDUCTOR JUNCTION TEMPERATURE SENSORS

The junction between differently doped regions of a semiconductor has a voltage-current curve that depends strongly, on temperature. This dependence can be harnessed in two types of temperature sensors; diode sensors and monolithic integrated circuit sensors, like many semi conductor sensors, these devices have maximum operating temperatures of 100 to 150 °C. Both types can be small, having dimensions of a few millimeters.

They are more accurate when properly calibrated. The diode is powered with a fixed forward current, of about 10 μA and the resulting forward voltage is measured with a four-wire constant-current circuit. The diode's forward voltage is a decreasing function of temperature, known from the calibration. Typical diodes are either silicon or gallium aluminum arsenide, and they are often applied in cryogenic temperature measurements. Precision diodes are relatively expensive.

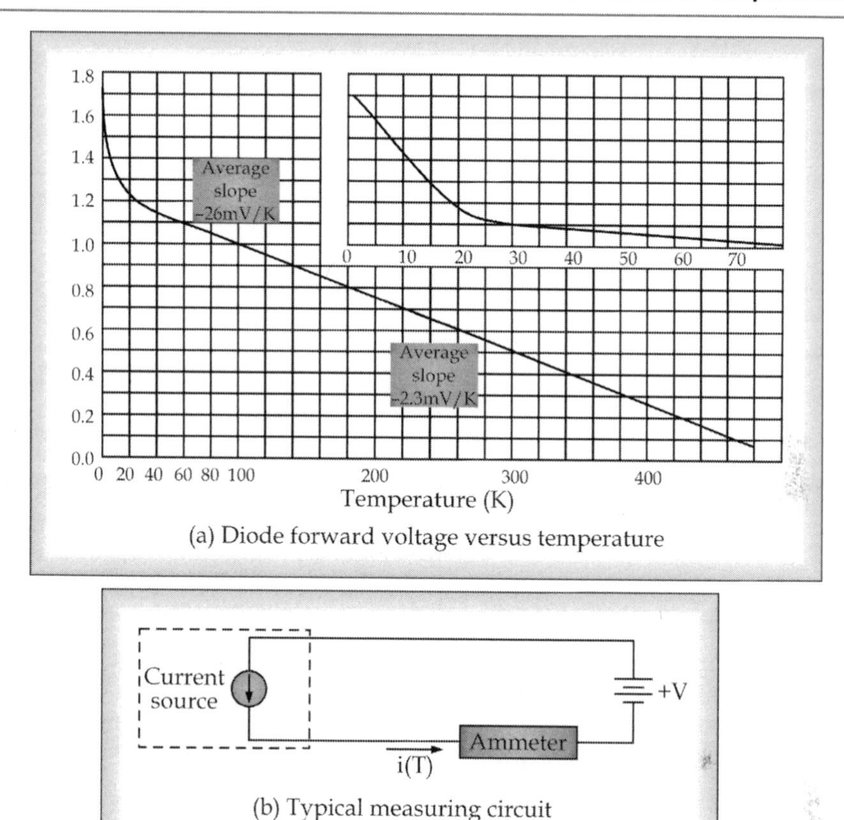

(a) Diode forward voltage versus temperature

(b) Typical measuring circuit

Fig. 5.27 Semiconductor Junction Sensor

Monolithic integrated circuit devices use silicon transistors to generate an output current proportional to absolute temperature. A modest voltage (4 to 30 V) is applied to the sensor and the current through the circuit is monitored with an ammeter as shown in Fig. 5.27.

5.9 THE LINEAR QUARTZ THERMOMETER

The relationship between temperature and the resonating frequency of a quartz crystal has long been recognized. In general, the relationship is non linear and for many applications very considerable efforts has been spent in attempts to minimize the frequency drift caused by temperature variation. Hammond discovered new crystal orientation called the "LC" or "linear cut", which provides a temperature frequency relationship of 1000 Hz/°C with a deviation from the best straight line of less than 0.05% over a range of –40°C to 230°C. This linearity may be compared with a value of 0.55% for the platinum resistance thermometer.

5.10 THERMOMETERS USING CHANGES IN EMITTED THERMAL RADIATION - PYROMETERS

The term pyrometry is derived from the Greek words Pyros, means "fire" and metron, means "to measure". Literally speaking, the term means general temperature measurement only. However, in engineering usage, this word refers to the measurement of temperatures in the high range, extending upward from about 500 °C. Pyrometry generally implies themal-radiation measurement of temperature.

Electromagnetic radiation extends over a wide range of wavelengths (or frequencies), as illustrated in Figure. 5.28. Pyrometry is based on sampling the energies in certain bandwidths of this spectrum. At any given wavelength, a body radiates energy of an intensity that depends on the body's temperature, by evaluating the emitted energy at known wavelengths. Thus temperature of the body can be found.

Classification of Pyrometers

Pyrometers are essentially photo detectors which are specifically designed for temperature measurement. Like Photo detectors, Pyrometers are of two general types:

(i) Thermal detectors and
(ii) Photon detectors.

Thermal detectors are based on the temperature rise produced when the energy radiated from a body is focused on to a target, heating it. The target temperature may be sensed with a thermopile, a thermistor or RTD, or a pyroelectric element.

Photon detectors use semi-conductors of either the photo-conductive or proto-diode type.

Pyrometers may also be classified by the set of wavelengths measured.

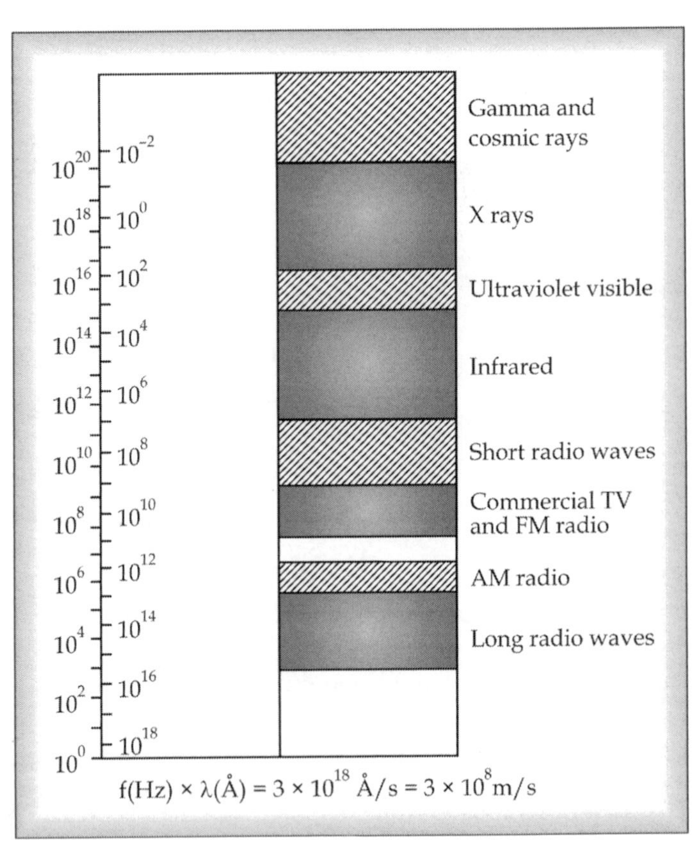

Fig. 5.28 The Electromagnetic Radiation Spectrum

- A *total-radiation pyrometer* absorbs energy at all wavelengths or, at least, over a broad range of wavelengths.
- An *optical pyrometer* measures energy at only one specific wavelength while the two-colour pyrometer compares the energy at two specific wavelengths.
- Perhaps the most commonly used type is the *infrared pyrometer,* which determines temperature from measurements over a range of infrared wavelengths.

Because bodies near room temperature radiate most of their energy at infrared wavelengths, the infrared pyrometer has over turned the traditional perception of pyrometry as a strictly high temperature technique. In fact, the Greek meaning of pyrometry mentioned before, is no longer valid. However, pyrometers retain the distinguishing feature of finding an object's temperature without having direct contact with it.

Pyrometer Theory

All bodies above absolute zero temperature radiate energy. Not only do they radiate or emit energy, but also receive and absorb it from other sources. Radiation striking the surface of a material *is* partially absorbed, reflected & transmitted.

These portions are measured in terms of absorptivity a, reflectivity r, and transmissivity, (t), Where $\alpha + r + t = 1$.

- For an ideal reflector, a condition approached by a highly polished surface $\rho \to 1$.
- Many gases represent substances of high transmissivity, for which $\tau \to 1$.
- A small opening into a large cavity approaches an ideal absorber, or blackbody for which $\alpha \to 1$.

A body in radiative equilibrium with its surroundings emits as much energy as it absorbs. It shows therefore, that a good absorber is also a good radiator, and hence an ideal radiator is one for which the value of ⟨ is equal to unity. When we refer to emitted radiations as distinguished from absorption, the term emissivity (e) is used rather than absorptivity (α). However, the two are directly related by Kirchhoff's law, $\alpha = \varepsilon$

According to the Stefan- Boltzmann law, the net rate of exchange of energy between two ideal radiators A and B that view only each other is

$$q = \sigma\left(T_A^4 - T_B^4\right)$$

When a non-ideal object 'A' radiates to a perfectly absorbing object 'B', the expression is modified as...

$$q = \varepsilon_A F_{BA}\left(T_A^4 - T_B^4\right)$$

where, q = net radiant heat transfer or heat flux from A to B, in W/m^2.

ε_A = The emissivity of object A,

F_{BA} = Shape factor to allow for relative position and geometry of bodies,

T_A and T_B = Absolute temperatures of objects *A* and *B* in *K,*

The Stefan -Boltzmann constant, 5.67×10^{-8} W/m^2.

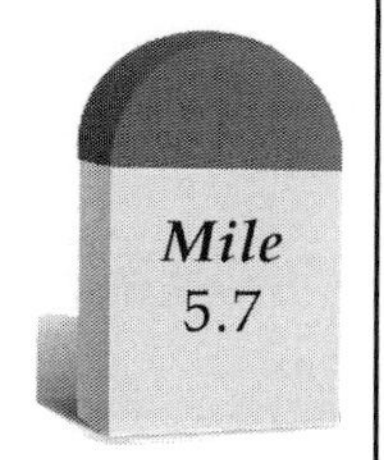

More Instruments for Learning Engineers

Osmometer measures osmotic strength of a solution, colloid, or compound. There are several different methods employed in osmometry:

- **Vapor pressure depression osmometer** measures the concentration of osmotically active particles that reduce the vapor pressure of a solution.
- **Membrane osmometer** measures the osmotic pressure of a solution separated from pure solvent by a semi-permeable membrane.
- **Freezing point depression osmometer** measures the osmotic strength of a solution, as osmotically active compounds depress the freezing point of a solution.

Osmometers are used to measure the concentration of dissolved salts or sugars in blood or urine samples and also to find the molecular weight of unknown compounds and polymers.

Osmometer

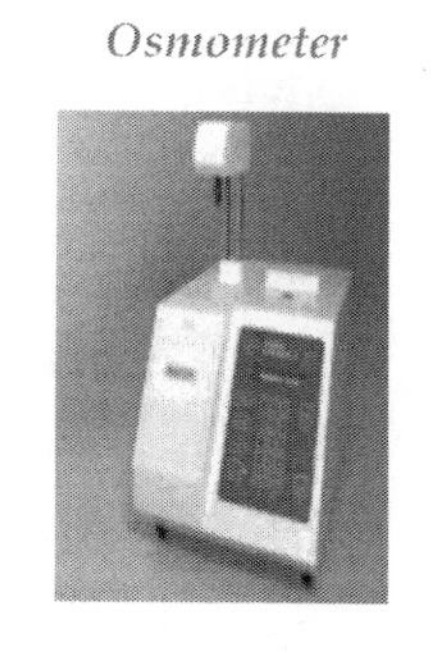

5.11 TYPES OF PYROMETERS

Pyrometers are of three types. They are

1. Total radiation pyrometer
2. Optical pyrometer or disappearing filament pyrometer or monochromatic brightness radiation thermometers
3. Infrared pyrometer

5.11.1 Total Radiation Pyrometer

Principle: The radiation from the hot metal whose temperature is to be measured is focused on a radiation receiving element, such as the hot Junction of a thermocouple, thermopile, or a photo-electric transducer and the e.m.f developed becomes a measure of the temperature of the hot body when calibrated i.e., radiation pyrometer is used to measure the total energy of radiation from a heated body, which is based on "Stefan-Boltzmann law. *This law states that the total emissive power of a black surface (ideal) is directly proportional to the forth power of the temperature of the surface expressed in ^{o}K"*

$$E = \varepsilon \sigma T^4$$

where ε = emissivity of the grey body and is defined as the ratio of the emissive power of a grey body to the emissive power of a black body.

Description: The radiation pyrometer consists of tube. One end of this tube has a sighting hole with and adjustable eye piece. The other end of the housing tube is opened for thermal radiation from hot body. Inside this tube there is a concave mirror whose position can be adjusted with the help of a rack and pinion arrangement. The radiations from the hot body are focused by this mirror on to a thermocouple as shown in Fig. 5.29. The radiation shield will be protecting the thermocouple by avoiding radiations to fall directly onto it. This thermocouple is connected to a calibrated milli-voltmeter to indicate temperature directly.

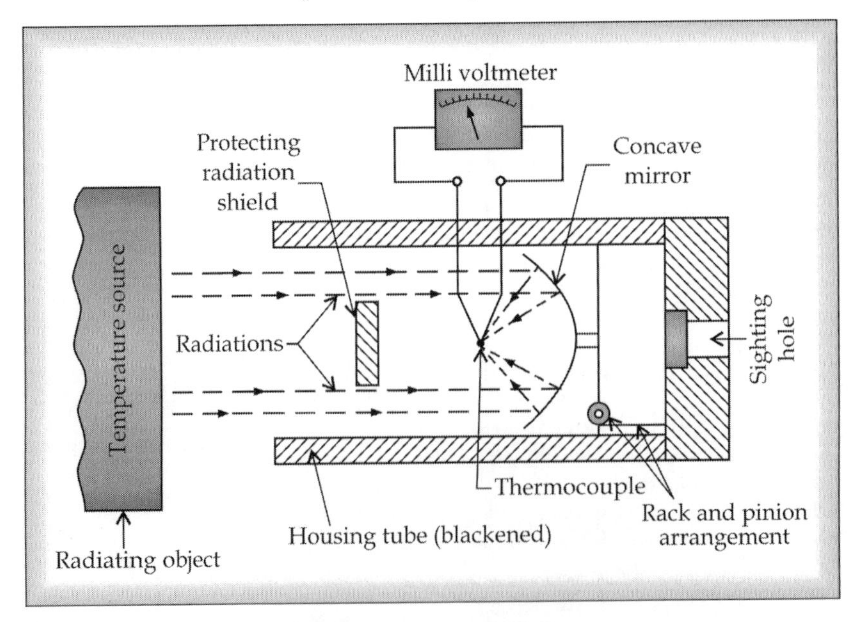

Fig. 5.29 Radiation Pyrometer

Operation: The open end of the pyrometer is focused on to the radiating object such that the thermal radiations enter into the tube. The position of the concave mirror is adjusted by rack & pinion arrangement such that the radiations fall on the receiving element. The focusing of thermal radiations can be done in two ways by using a lens or a parabolic reflector as shown in Fig. 5.30(a) & (b).

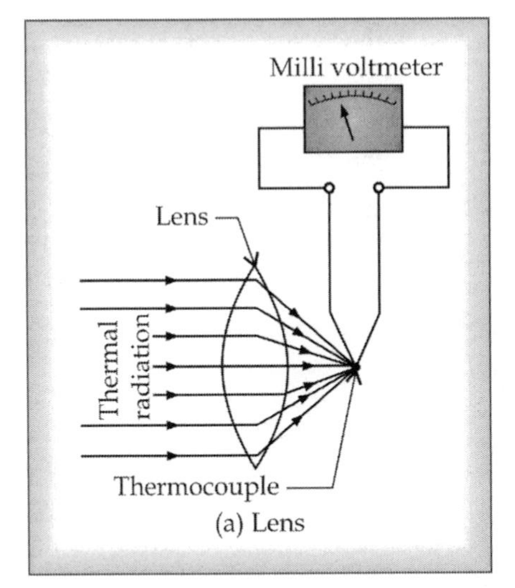

(a) Lens

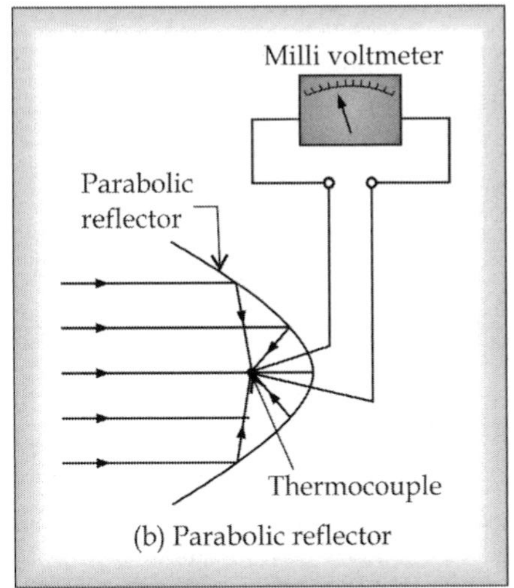

(b) Parabolic reflector

Fig. 5.30 Focusing of Thermal Radiation

An emf will be setup in the thermocouple, which is measured using a mill voltmeter. By measuring voltage, temperature of the hot body can be measured.

Application

The radiation pyrometers are used to measure temperatures ranging from –15 to 3870 °C.

Advantages

- Physical contact of the hot body is not required.
- The speed of response is more.
- Used to measure temperatures of stationary as well as moving objects.
- Their accuracy is of the order of ± 2 % of the range.

Limitations

- The presence of dust, smoke and overheated gases between the radiating object and the concave mirror will induce errors in temperature measurement.
- Sensitivity is low at lower temperatures.
- Cooling of the instrument is required to avoid overheating.

5.11.2 Optical Pyrometer or Disappearing Filament Pyrometer or Monochromatic Brightness Radiation Thermometer

Principle: The principle of temperature measurement by brightness comparison is used in this optical pyrometer. A colour variation with growth in temperature is taken as an index of temperature.

This compares the brightness of the image produced by the temperature source with that of reference temperature lamp. The current in the lamp is adjusted until the brightness of the lamp is equal to the brightness of the image produced by the temperature source.

Since the intensity of light of any wave length depends on the temperature of the radiating objects, the current passing through the lamp becomes a measure of the temperature source when calibrated.

Description: The optical pyrometer consists of an eye piece at one end and an objective lens at the other end. A power source (battery), rheostat: and multi-voltmeter connected to a reference temperature bulb. An absorption screen is placed in between the objective lens and reference temperature lamp. The absorption screen is used to increase the range of temperature, which can be measured by the instrument as shown in Fig. 5.31. The red filter between the eye piece and the lamp allows only a narrow band of wave length.

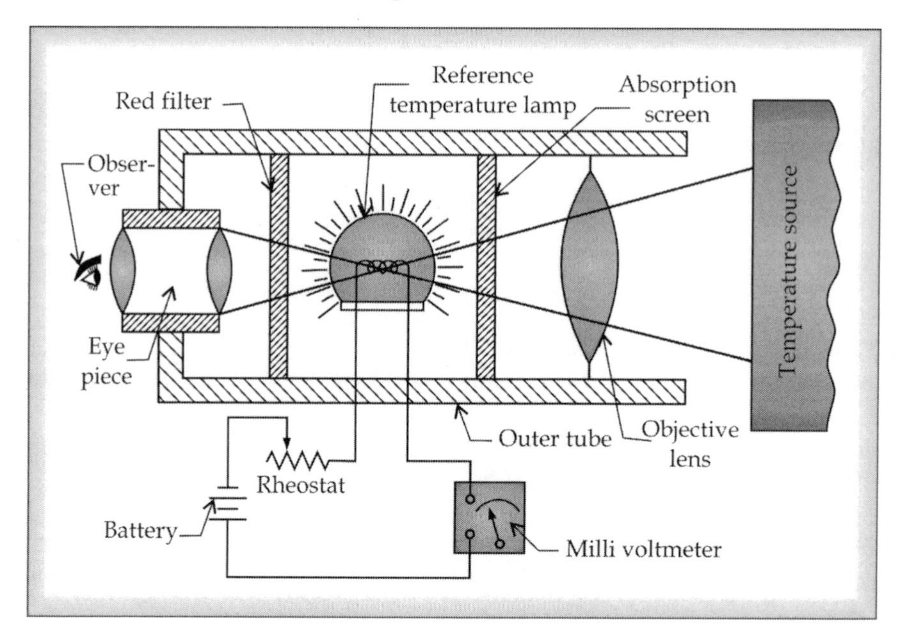

Fig. 5.31 Optical Pyrometer

Operation: The radiation from the source is focused onto the filament of the reference temperature lamp using the objective lens. Now the eye piece is adjusted so that the filament of the reference temperature lamp is in sharp focus and the filament is seen super imposed on the image of the temperature source. Now the observer starts controlling the lamp current and if the filament is cooler than the temperature source, the filament will appear dark as in Fig. 5.32(a). If the filament is hotter than the temperature source, the filament will appear bright as in Fig. 5.32(b). If the filament and temperature source are at the same temperature, the filament will not be seen (disappears) as shown in Fig. 5.32(c). Hence the observer should control the lamp current until the filament and temperature source have the same brightness which will be noticed when the filament disappears as in Fig. 5.32(c) in the super imposed image of the temperature source. At this instance, the current flowing through the lamp which is indicated by the multi-voltmeter connected to the lamp becomes a measure of a temperature source when calibrated.

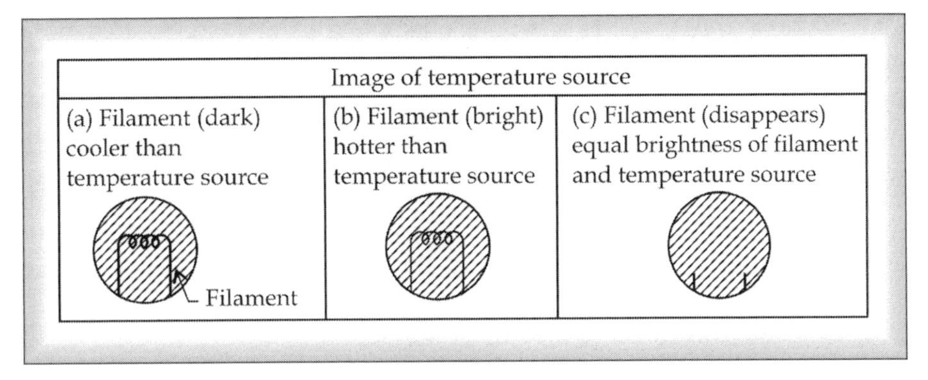

Fig. 5.32 Different Views of Filament

Applications

These are used to measure temperature of molten metals and temperature of furnaces.

Advantages

(i) Physical contact of the temperature source is not required.

(ii) Accuracy is high ± 5 °C of the range.

(iii) Provided a proper sized image of the temperature source is obtained in the instrument, the distance between the instrument and the temperature source does not matter.

(iv) Easy to operate.

***Do You Know: Actinometers** are instruments used to measure the heating power of radiation. An actinometer may be a chemical system or physical device that detects the number of photons in a beam integrally or per unit time. This name is commonly applied to devices used in the UV and visible wavelength ranges. For example, solutions of iron (III) oxalate can be used as a chemical actinometer, while bolometers, thermopiles, and photodiodes are physical devices giving a reading that can be correlated to the number of photons detected. They are used in meteorology to measure solar radiation as pyrheliometers.*

Limitations

(i) Temperature of more than 700 °C can only be measured since illumination of the temperature source is a must for measurement.

(ii) Since it is manually operated, it cannot be used for continuous monitoring and controlling purpose.

Two-colour Pyrometer: It is an optical technique that minimizes the influence of the emissivity, specifically, it measures source intensity at two wave lengths λ_1 & λ_2. If the emissivity is independent of wavelength or if the wavelengths are nearly equal, then the ratio of measured intensities depends only on temperature.

5.11.3 Infrared Pyrometer

This pyrometer is similar to radiation pyrometer. In both, a sample of radiation is detected and its energy level is measured. In total radiation pyrometer a thermal detector is used whereas, Infrared pyrometers employ a photo-cell. The output of this cell is amplified and displayed from which temperature can be measured. The response time is very less and is used in the range of – 150 °C to 3300 °C.

5.12 TEMPERATURE MEASUREMENT BY CHANGES IN CHEMICAL PHASE

Fusible Indicators: Among these Seger Cones or pyrometeric cones are first to be mentioned, which are being used in the ceramic industry as a means of checking temperatures. These are small cones made of an oxide and glass. When a predetermined temperature is reached, the tip of the cone softens and curls over, thereby providing the indication that the temperature has been reached as shown in Fig. 5.33(a)&(b). The range of temperature measurement is 600 °C to 2000 °C.

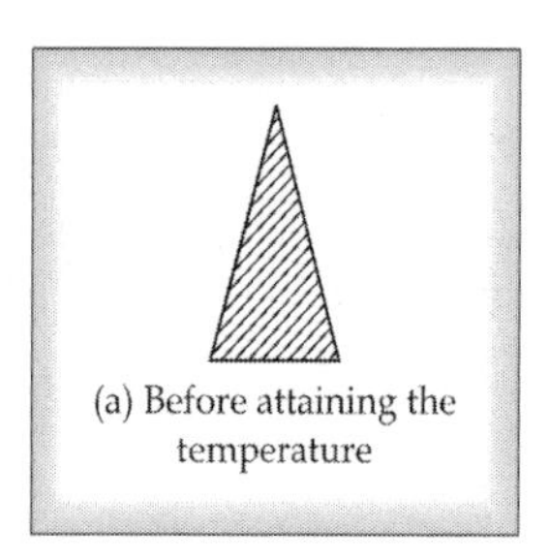

(a) Before attaining the temperature

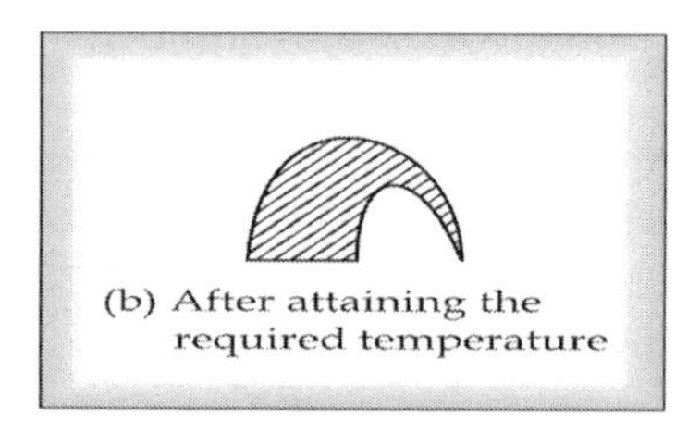

(b) After attaining the required temperature

Fig. 5.33 Seger Cones or Pyrometric Cones

The other forms of temperature-level indicators are crayon sticks, lacquer, and pill-like pellets etc.

Liquid crystals: These are perhaps the most colourful temperature indicators. The liquid crystal is a meso-phase state of certain organic compounds that shares properties of both liquids & crystals. As temperature increases, liquid crystals successively scatter reds, yellows, greens, blues and violets. These are useful from roughly – 20 °C to 100 °C with 0.1 °C as resolution. These are coated in a thin film over a blackened plastic sheet that makes the scattered light - more visible; the sheet is then affixed to the surface whose temperature is to be measured.

More Instruments for Learning Engineers

Mile 5.8

A **rotameter** measures the flow rate of liquid or gas in a closed tube. It belongs to the kind of variable area meters, which measure flow rate by allowing the cross-sectional area the fluid travels through, to vary, causing some measurable effect. The 1st variable area meter with special shaped rotating float was invented by Karl Kueppers (1908). Felix Meyer improved it with new shapes of the float and of the glass tube. The brand name Rotameter was registered by the British company GEC Rotameter Co, in Crawley, and still exists, having been passed down through the acquisition chain: KDG Instruments, Solartron Mobrey, and Emerson Process Management (Brooks Instrument) in Great Britain. In many countries such as Germany, Switzerland, Austria, Spain, Italy the brand name Rotameter is registered by Rota Yokogawa GmbH & Co. KG in Germany which is now owned by Yokogawa Electric Corp.

Rotameter

Self Assessment Questions-5.5

1. Explain what is 'Thermal radiation'? Explain its significance in instrumentation.
2. Explain about the pyrometers used as instruments for measurement of temperature. Discuss the theory behind the pyrometry.
3. Explicate the following and discuss how they are useful in measurement of temperature.
   (a) Liquid crystals (b) Seger cones
4. What is 'Pyrometer'? Explain construction and operation of optical pyrometer.
5. Discuss the transducer and the principle of transduction in the case of pyrometer.
6. What are the merits and demerits of pyrometers?
7. Narrate the construction and operation of the following instruments used for measurement of temperature.
   (a) Total Radiation Pyrometer (b) Disappearing Filament Type Pyrometer
8. Explain the operation, merits and demerits of Monochromatic Brightness Radiation Pyrometer.
9. Write short notes on
   (a) Two-colour Pyrometer (b) Infrared Pyrometer
10. Define and explain Stefan-Boltzmann law and discuss its significance in pyrometry.
11. Discuss the terms absorptivity, emissivity and transmissivity under the light of pyrometry.
12. Give the classification of pyrometers and explain any two in detail with sketches.

SUMMARY

Temperature is a surface phenomenon while heat is substantial phenomenon. The degree of hotness or coldness of a body is called temperature and the driving force or potential, causing the flow of heat energy. It is the measure of the mean K. E. of the molecules of a substance. There are many temperature scales such as Centigrade, Fahrenheit, Reaumur or Rankine and Kelvin scales related as $\frac{C-0}{5}=\frac{F-32}{9}=\frac{R-491.69}{9}=\frac{K-273}{5}$

or $\frac{\text{Reading} - \text{Lower fixed point}}{\text{Fundamental interval}} = \text{Constant}$

The relation between the temperature and the respective physical quantities are the core principles behind the operation of the instruments of temperature measurements. The physical characteristics which can be used to measure temperature are a change in dimension i.e. expansion/contraction of material (solid rod, bimetallic, liquid in glass and liquid in metal thermostats/ thermometers), electrical resistance (RTD, thermocouples, thermistors, Semi conductor - junction sensors etc.), the intensity and colour of radiation emitted by the hot body (Thermal and photon sensors, Total radiation pyrometers, Optical and two color pyrometers, Infrared pyrometers) and chemical phase (Fusible indicators, Liquid crystals). The gas laws govern the operation of Gas thermometers, Constant-volume gas thermometers and Pressure thermometer (gas, vapour & liquid filled).

The construction and operation of various temperature measurement instruments along with their merits, demerits and applications are given in this chapter. In addition various fundamentals and relevant information is also furnished wherever required.

KEY CONCEPTS

Temperature: The degree of hotness or coldness of a body or Potential causing the flow of heat energy or Measure of the mean KE of the molecules of a substance.

Heat: A form of energy caused due movement of the particles in the body.

Temperature Scale: Centigrade, Fahrenheit, Reaumur or Rankine and Kelvin.

Temperature scale principle: In any scale, $\frac{\text{Reading} - \text{lower fixed point}}{\text{Fundamental inrerval}} = \text{constant}$

Temperature scale conversion equation: $\frac{C-0}{100} = \frac{F-32}{180} = \frac{R-491.69}{180} = \frac{K-273}{100}$

Solid Rod Thermostat/Thermometer: Works on the principle of differential expansion of materials i.e., one material expands more than the other when subjected to same change in temperature.

Bimetallic Thermostat/Thermometer: Measurement by differential expansion in two dissimilar solid metals (bound together), resulting bend in bimetallic strip.

Thermometric Liquid: Liquids used in thermometers such as water, mercury and alcohol in a glass tube of uniform cross section.

Liquid-in-glass Thermometer: A relatively large bulb at the lower portion of the thermometer holds the major portion of the liquid, which expands when heated and rises in the capillary tube, and works on the property of expansion of liquids.

Liquid Filled Thermal System: Operate on the principle of volumetric expansion of the liquid caused by temperature change that provides the force required to operate the spiral and thus indicate temperature.

More Instruments for Learning Engineers

Evaporimeter (also called **Atmometer**) measures the rate of evaporation of water into the atmosphere. These are of two types. Type-I devices measure from a free water surface. The water level in a tank, often sunk into the ground and the water surface at ground level, is measured by a micrometer gauge. After determining the increase due to rain and decrease due to deliberate drain, the decrease in the water level can be attributed to evaporation. Type-II devices measure from a continuously wet porous surface according to the rate of weight loss of a wet pack of absorbent material. In Piché evaporimeter an inverted graduated cylinder of water with a filter-paper seal at the mouth is used. Evaporation from the wet filter paper depletes the water in the cylinder, so that the rate of evaporation can be read directly from the graduations marking the water level. In Livingston sphere, a wet ceramic sphere is used as the evaporating surface to simulate evaporation rates from vegetation.

Evaporimeters have limited use as their results are not exact since the rate of evaporation is too sensitive to water supply, and nature of evaporating surface.

Evaporimeter or Atmometer

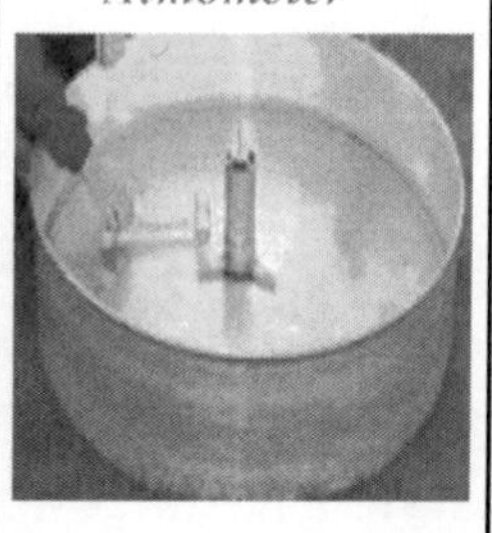

Gas-filled Thermal System: The bulb is filled with a gas and when the temperature of the gas is increased its pressure increases which is communicated to Bourdon tube and works on the principle of expansion of gas (Ideal gas law i.e., PV = nRT).

Vapour - Pressure Thermal System: The bulb is partially filled with a volatile liquid and the rest of the system contains vapour resulting in the increase of vapour pressure and is measured by pressure measuring device (principle $\log_e P = a - b/T$).

Case Compensation Method: The action of the bimetallic strip is in opposition to that of the receiving element (Bourdon tube) thereby negating the effect of changes in ambient temperature.

Full Compensation Method: The use of the dummy capillary tube, which runs along the active -capillary tube but not connected to the temperature sensing bulb thereby cancelling the effects produced by change in ambient temperature along the length of the active capillary tube.

Thermo Resistive Elements: Resistance Temperature Detectors (RTD) or resistance thermometers and thermistors.

RTD: Thermo-resistive element made of metals whose resistance increases with temperature.

Thermistor: Thermo-resistive element made of semiconductors whose resistance decreases with temperature. The short form of "***Thermal Resistors***".

Seebeck Effect: When two dissimilar metals are joined together an emf will exist between the two points which is primarily a function of the junction temperature.

Peltier Effect: The converse of Seebeck effect that when a current is passed across the junction of two dissimilar metals, heat is evolved at one junction and absorbed at the other.

Thomson Effect: The heat energy is evolved or absorbed not only at the junctions but it is absorbed or evolved all along one or both the conductors by virtue of the difference of temperatures between the two ends of the same element.

Law of Intermediate Metals or Successive Contacts: If at a given temperature, a number of metals are in successive contact so as to form a chain of elements connected in series, the electromotive force between the extreme elements, if placed in direct contact, is the sum of the electromotive forces between successive adjacent elements.

$$E_A^N = E_A^B + E_D^C + E_C^D \ldots\ldots\ldots E_M^N$$

Law of Intermediate Temperatures: The electromotive force for a couple with junction at T_1 and T_3 is the sum of the electromotive forces of two couples of the same metals one with junctions at T_1 and T_2 and the other with junction at T_2 and T_3 i.e., $E_3^1 = E_2^1 + E_3^2$.

Siemen's Three Lead Arrangement: The center lead carries no current, and the effect of the resistance of the other lead is cancelled out.

Callendar Four Lead Arrangement: Two additional lead wires inserted in the adjustable leg of the bridge so that the effect of the lead wires on the resistance thermometer is cancelled out.

Floating-Potential Arrangement: Same as Siemen's connection, with an extra lead inserted, to check the equality of lead resistance.

Thermocouples: Two unlike conducting materials used to form a thermocouple.

Base Metal Thermocouples: Combination of pure metal and alloys of Fe, Cu and Ni, used to measure lower ranges of temperature up to 1375 °C.

Rare Metal Thermocouples: Combination of pure metal and alloys of P & Rh for temperature up to 1725 °C; and W, Rh & Mo for temperature up to 2625 °C.

Non-Metallic Thermo Couples: Combination of semi con-ductors materials such as Si-Ge, used in the power packs of the Viking 1 and 2 space explorers.

Thermopile: Thermocouples connected in series

Linear Quartz Thermometer: The thermometer that works on the relationship between temperature and the resonating frequency of a quartz crystal.

Pyrometry: Study of thermal-radiation measurement of temperature.

Kirchhoff's laws: The algebraic sum of voltage drops in a closed circuit is zero; The algebraic sum of currents in a closed circuit is zero.

Stefan- Boltzmann law: The total emissive power of a black surface (ideal) is directly proportional to the forth power of the temperature of the surface expressed in °K, $E = \varepsilon \sigma T^4$

Stefan -Boltzmann constant: $5.67 \times 10^{-8} W/m^2$.

Total Radiation Pyrometer: Measures the total energy of radiation from a heated body based on Stefan-Boltzmann law.

Optical Pyrometer or Disappearing Filament Pyrometer or Monochromatic Brightness Radiation Thermometer: Temperature measurement by brightness comparison i.e., colour variation with growth in temperature is taken as an index of temperature.

Infrared Pyrometer: Uses a photo-cell and amplified & displayed.

Fusible Indicators: Among these Seger Cones (or pyrometric cones) are first to be mentioned, being used in the ceramic industry as a means of checking temperatures.

Seger (or pyrometric) Cones: Small cones made of an oxide and glass.

Liquid crystals: A meso-phase state of certain organic compounds that shares properties of both liquids & crystals.

REVIEW QUESTIONS

Short Answer Questions

1. With neat sketch explain the compensation method employed in pressure thermometer.
2. Give various configurations of bimetallic sensors.
3. Explain the principle behind the temperature measurement by radiation methods.
4. Explain the calibration procedure employed in pyrometer.
5. What is reference junction compensation? Why is it required for thermocouples?
6. Mention some advantage of thermocouples over other temperature sensors.
7. What are the different types of materials commonly used in thermistors?
8. Give the various forms of thermistors.
9. Explain the temperature - resistance relation of thermistor.

10. Explain the laws of thermocouples and their usefulness in temperature measurement.
11. What are thermocouples? How is temperature measured using thermocouples?
12. What are the different methods used to calibrate the temperature measuring systems?
13. Distinguish between thermocouples & pyrometers; give typical examples of each.
14. Discuss the following
    (a) Radiation pyrometers (b) Calibration of chromel, alumel thermocouples,
15. What are various materials used for thermocouples.
16. Explain the construction of platinum RTD.
17. With neat circuit diagram explain three wire and four wire RTD configuration
18. Give the composition and temperature range of following thermocouples.
    (i) T -type (ii) J - type (iii) K - type (iv) g - type
19. What is a thermopile? Explain tile principle of operation with neat sketch.
20. Explain the principle behind the temperature measurement by radiation methods
21. Explain the working of a total radiation in pyrometer.
22. Distinguish between RTD and thermistor.
23. Mention some advantages of thermocouples over temperatures sensors.

Long Answer Questions

1. Explain bimetallic thermometers with applications.
2. With neat sketch explain the *compensation method employed in pressure thermometer.*
3. Compare and contrast the advantages and limitations of thermocouples and resistance thermometers.
4. Explain the principles of operation and working of
   (i) Thermistor (ii) Pyrometric cones
5. With necessary diagram, explain the various methods employed in cold junction compensation.
6. What are the different types of materials commonly used in thermistors?
7. Explain the principle and working of thermistor and give their merits and demerits.
8. Give the various forms of thermistors and explain them.
9. With neat sketch explain the compensation method employed in pressure thermometer.
10. What is reference junction compensation? Why is it required for thermocouples?
11. What is pyrometery? Explain in detail total radiation pyrometer; discuss its relative merits and demerits with respect to total radiation pyrometer.
12. Explain the working of a total radiation pyrometer.
13. Explain the principle behind the temperature measurement by radiation methods. Explain the calibration procedure employed in pyrometer.

MULTIPLE CHOICE QUESTIONS

1. The units of temperature is __________.
   (a) Watt (b) Heat
   (c) °C (d) Energy
2. Thermal equilibrium of a body is attained when ______________.
   (a) There is no change in temperature between two bodies.
   (b) There is no unbalance force between two bodies.
   (c) There is no chemical unbalance between two bodies
   (d) When the resultant of moments of all forces is equal to zero.
3. 0°C is equal to __________.
   (a) 0K (b) 100K
   (c) 273.15K (d) 50K
4. The degree of hotness or coldness is called _________.
   (a) Heat (b) Temperature
   (c) Work done (d) °C
5. The upper fixed point temperature in terms of Fahrenheit (°F) is_______.
   (a) 100 (b) 212
   (c) 373 (d) 50
6. The lower fixed point temperature in terms of Rankine (°R) is_________.
   (a) 491.69 (b) 100
   (c) 273.16 (d) 0
7. In resistance thermometer the thermometric property is________.
   (a) EMF (b) Resistance
   (c) Volume (d) Pressure
8. __________ are used for non-contact temperature
   (a) Metallic plates
   (b) Radiant sensors
   (c) Bimetallic strips
   (d) Electrical sensors
9. In liquid in glass thermometer filled with alcohol, the useful range is from _____.
   (a) –75 to 129 °C (b) –65 to 430 °C
   (c) –270 to 100 °C (d) –270 to 760 °C
10. In Solid rod thermostats the transduction is from temperature to ________.
    (a) Length (b) Resistance
    (c) EMF (d) Pressure
11. Which of the following is not a desirable property for liquid thermometer.
    (a) Low adhering property
    (b) Large coefficient of expansion.
    (c) High adhering property
    (d) It should not be visible
12. Which of the following is one of the fluid-filled system.
    (a) Pressure filled
    (b) Volume filled
    (c) Temperature filled
    (d) Gas filled
13. Resistance thermometer depends on which principle _________.
    (a) Peltier effect
    (b) Seebeck effect
    (c) Wheatstone bridge
    (d) Thomson effect
14. In which of the following arrangements the centre lead carries no current.
    (a) Siemen's Three- Lead Arrangement
    (b) Calendar Four Lead Arrangement
    (c) Electrical resistance thermometer
    (d) Floating-Potential Arrangement
15. Base metal thermocouples are made of the alloys of________.
    (a) Platinum (b) Rhodium
    (c) Tungsten (d) Copper
16. The units of resistance is ________.
    (a) Coulomb (b) Ohm
    (c) Ampere (d) Tesla

17. Thermocouples connected in series is called as ________.
    (a) Thermopile
    (b) Thermostat
    (c) Equal thermocouples
    (d) Unequal thermocouples

18. Stefan-Boltzmann law is given as ________.
    (a) $E = \varepsilon\sigma T^4$ (b) $E = \varepsilon\sigma T^3$
    (c) $E = \sigma T^3$ (d) $E = \varepsilon T^3$

19. Unit of Stefan Boltzman Constant is ______.
    (a) W/m (b) W/m^2
    (c) W/m^3 (d) W/m^4

20. The thermal coefficient of expansion for Invar is ______.
    (a) 2.02×10^{-5} (b) 1.7×10^{-5}
    (c) 1.25×10^{-5} (d) 1.35×10^{-5}

ANSWERS

1.	(c)	2.	(a)	3.	(c)	4.	(b)	5.	(b)
6.	(a)	7.	(b)	8.	(b)	9.	(a)	10.	(a)
11.	(c)	12.	(d)	13.	(c)	14.	(a)	15.	(d)
16.	(b)	17.	(a)	18.	(a)	19.	(b)	20.	(b)

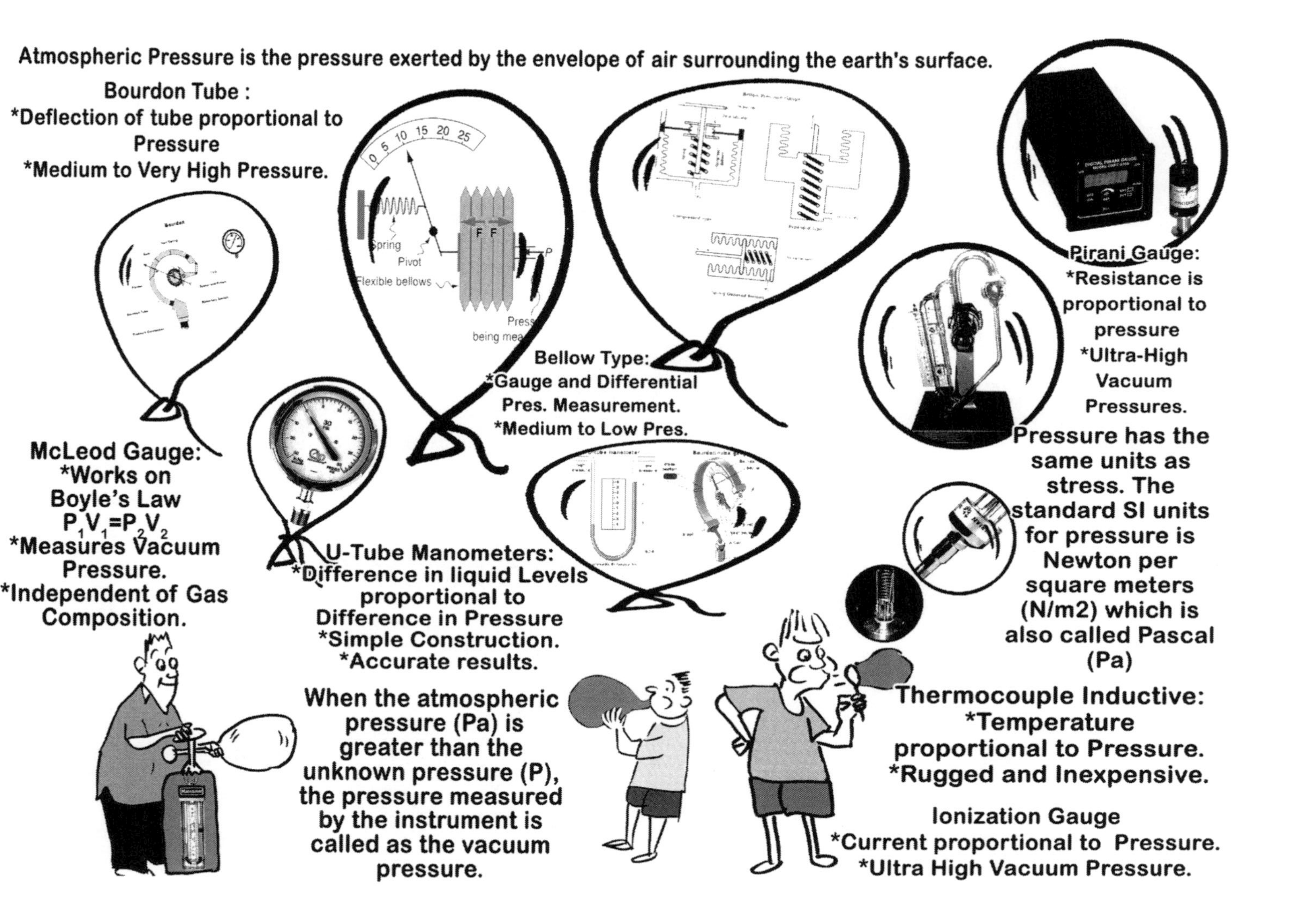
Atmospheric Pressure is the pressure exerted by the envelope of air surrounding the earth's surface.
Bourdon Tube :
*Deflection of tube proportional to Pressure
*Medium to Very High Pressure.
0 5 10 15 20 25
Spring
Pivot
Flexible bellows
F F
P
Bellow Type:
*Gauge and Differential Pres. Measurement.
*Medium to Low Pres.
Pirani Gauge:
*Resistance is proportional to pressure
*Ultra-High Vacuum Pressures.
Pressure has the same units as stress. The standard SI units for pressure is Newton per square meters (N/m2) which is also called Pascal (Pa)
McLeod Gauge:
*Works on Boyle's Law
$P_1V_1=P_2V_2$
*Measures Vacuum Pressure.
*Independent of Gas Composition.
U-Tube Manometers:
*Difference in liquid Levels proportional to Difference in Pressure
*Simple Construction.
*Accurate results.
When the atmospheric pressure (Pa) is greater than the unknown pressure (P), the pressure measured by the instrument is called as the vacuum pressure.
Thermocouple Inductive:
*Temperature proportional to Pressure.
*Rugged and Inexpensive.
Ionization Gauge
*Current proportional to Pressure.
*Ultra High Vacuum Pressure.

6

Glimpses of Measurement of Pressure

Atmospheric Pressure(Pa): Exerted by the envelop of air surrounding the earth's surface

S.No	Instrument	Transducer Principle	Construction	Operation	Application	Merits	Limitations
1.	Dead weight tester	Gravitational $P = (mg + F)/A$	A platform upon Piston and cylinder above an oil chamber fitted with plunger	Calibrate with know weight and find unknow	To Calibrate all kinds of pressure guages	Simple, wide range, P can be varied either by Wts. or piston	Accurancy is affected
2.	Simple manometer	$P = P_1 - P_2 = P = p_m gh$	Transparent tube constructed in the form of U and filled with Hg	Difference between pressure of two limbs is the function of diff in levels of fluid	Measurenment diffencial Pressure in Venturiments and flow meters	Simple in construction, gives accurate results	Chance of Breakage during transport;—— required
3.	Bourdon tube pressure gauge	Deflection of tube is proportional to pressure	Link sector-pinion pointer attached to the free end of the Bourdon tube	Due to applied Pr. cross sectiion of tube changes to move pointer on Pr. calibrated scale	Medium to very high pressure	Low cost: high accurancy at high Pressure Range	Slow response sensitive to shocks and vibrations
4.	Elastic diaphragm gauge	Deflection of tube is proportional to pressure	Diaphragm is fixed at edges and top portion is connected to link-sector-pinion-pointer	Diaphragm upward deformation causes the pointer to move on Pr. claibrated scale	Medium Pressure	Linear scale for a wide Pressure Range	For high Pr. diaphragm gets damaged
5.	Bellows gauges	Deflection of tube is proportional to pressure	Bellows contain spring inside and open end is attached to link-sector pinoin pointer	Bellows expand causing displacement of pointer to move on Pr. calibrated scale	Mediun and low pressure	measure both gauge & differential pressures	Zero shift problem; spring design is difficult
6.	McLeod gauge	DBoyle's Law $P_1 V_1 = P_2 V_2$ eflection of tube is proportional to pressure	Reference column and reference capillary tube connected to bulb	Diff. of Hts. of measuring capillary and reference capillary is measure of gas Pr.	Measure Vacum Pressure	Independent of gas composition	Applied gas should obey Boyle's law
7.	Pirani Gauge	Resistance is proportional to Pressure eflection of tube is proportional to pressure	Pirani gauge chamber encloses filament connected to wheatstone bridge ciruit	Const. current passing though filament changes in resistance is calibrated to read pressure.	Low vaccum and Ultra high vacuum pressures.	Give accurate results. Good respones to pressure changes.	Must be checked Prequently. Power is must for operation.
8.	Thermocouple conductive gauge	Temperature proportional to Pressure	Chamber contains filament welded to thermocouple	Constant current passing though filament changes temp. this measured by thermocouple calibrated to read pressure	Low vacuum and ultra high vacuum pressures	Rugged in construction and inexpen-sive ; readings can be take form distance	Filament gets burnt frequently
9.	Ionization Gauge	Current is proportional to Pressure	Cathode, grid and plate (Anode)	Due to the ionization process current flow takes place in anode which is calibrated to read Pressure	Low vacuum and ultra high vacuum pressures	Fast respone to pressure changes	Filament tempe-rature should be controlled properly

Gauge Pressure ($P_g > P_a$) : The difference above atmospheric pressure ($P_{abs} = P_a + P_g$)

Vacuum Pressure ($P_{vac} < P_a$) : The difference below atmospheric pressure ($P_{abs} = P_a - P_{vac}$)

Absolute Pressure(P_{abs}): Exerted by fluid particles due to interaction per unit area and is measured from the state of vacuum

Chapter 6

MEASUREMENT OF PRESSURE

STARTERS

To study this chapter, you should have awareness on the following concepts. For a better understanding, it is always a good idea to revise these prerequisites.

- Different types of pressures such as atmospheric, absolute, vacuum, gauge pressure etc.
- Pressure and its significance in fluids
- Concept of water head, and Bernoulli's principle
- Conversion of pressure to stress, force on a fixed area
- Relation between pressure and heat and mechanical energy
- Influence of pressure on springs.
- Relation between pressure and volume, Boyle's law ($P \propto 1/V$)
- The relation between pressure and temperature of gases i.e., Charles law ($P \propto T$)
- Ideal gas equation and Kinetic gas equation
- Effects of pressure on the physical dimensions of bodies
- Different types of elastic bodies and pressure effects on elastic bodies
- The instrument components such as U-tube, Bourdon, bellows, diaphragms etc., and their shapes/utility
- Some mechanisms to convert one type of motion to the other such as rotational to translatory or reciprocatory etc., and
- First four chapters of this book

LEARNING OBJECTIVE

After studying this chapter you should be able to

- Define and describe the terminology of pressure,
- Know the classification of pressure measuring devices,
- Understand various types of manometers and
- Understand various types of elastic transducers and their applications to measure pressure.

6.1 INTRODUCTION

You might have got filled air in tyres of your motorbike or car. Did you observe that you are maintaining the *optimum pressure*? How much is that 'optimum'?

The water from the tap in your bathroom may sometimes feebly come, when there is no *sufficient pressure*. What is that 'sufficient'?

A motor from a well cannot lift the water if the *required pressure* is not built up. What is that 'required'?

A balloon should not be fiilled with 'more' air than it can withstand. How much is 'more'?

We come across with many issues like the above. In all such cases mentioned above, we need to know and/or measure the pressure. The measurement of pressure is one of the most important measurements, as it is used in almost all industries, laboratories and in many other fields in addition to the situations in our daily life.

Some important applications of pressure measurement are:

1. Pressure measurement helps in controlling a process and to provide test data.
2. It helps in determining the liquid level in tanks and containers.
3. In many flow meters (Venturimeters, orifice meters, flow nozzles etc) pressure measurement serves as an indication of flow rate.
4. Pressure changes can indicate temperature (as used in pressure-thermometers fluid expansion type).
5. A simple case where pressure measurement plays an important role is for maintaining optimal pressure in tubes of vehicle tyres.

This chapter deals with the measurement of the pressure and the related instrumentation in industries, laboratories and applications to our daily life.

6.2 DEFINITIONS AND TERMINOLOGY

Pressure is defined as force per unit area exerted by a fluid on a containing wall.

More Instruments for Learning Engineers

Sphygmomanometer (simply **Sphygmometer**) measures blood pressure in terms of mmHg. (The blood pressure of a normal man is 120/80). It is composed of an inflatable cuff to restrict blood flow, and a mercury or mechanical manometer to measure the pressure. It is always used in conjunction with a means to determine at what pressure blood flow is just starting, and at what pressure it is unimpeded. Manual sphygmometers are used in conjunction with a stethoscope. The name comes from Greek (óõõãìüò i.e. *sphygmos* means pulse) and the scientific term manometer (pressure meter). The device was invented by Samuel Siegfried Karl Ritter von Basch (1881) and later an easy version by Scipione Riva-Rocci (1896). Harvey Cushing (1901) modernized it and popularized in medical community.

Sphygmometer

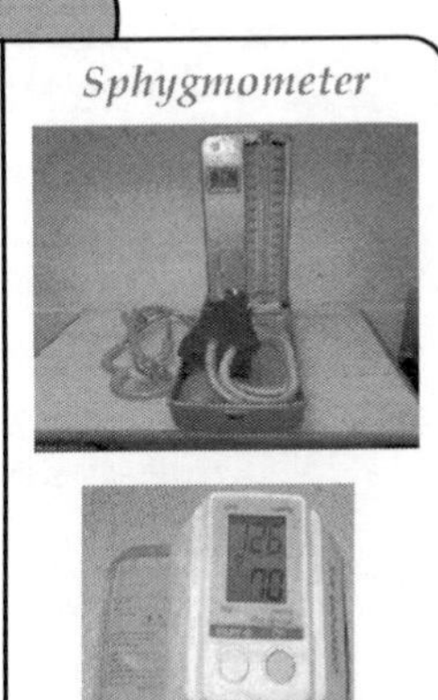

Units: Pressure has the same units as stress. The standard SI units for pressure is Newton per square meters (N/m^2) which is also called Pascal (Pa).

Pressure is also expressed in various units such as kgf/m^2, pounds per square inch (psi), atmosphere (atm), mm of Hg, cm of Hg, Bar, Torr, absolute atmosphere (ata) etc.

The Terms Related to Pressure Measurement

We often use the following terminology.

6.2.1 Atmospheric Pressure (P_A)

It is the pressure exerted by the envelope of air surrounding the earth's surface. It is usually expressed in terms of the height of a column of fluid (viz. Hg), which it will support at a temperature of 20 °C.

At standard atmospheric pressure this height is 760 mm of mercury having a density of 13.5951g/cm³ at an acceleration due to gravity of 981 cm/s²

i.e., $$\left(\frac{135951}{1000}\times(1000)^3\frac{kg}{m^3}\right)$$

$$1 \text{ atm} = 76 \times 13.5951\times 981 = 1013596.2756 \text{ dynes/cm}^2$$

$$= 1.013 \times 10^5 \text{ zn/m}^2 \text{ or pa}$$

$$1 \text{ N/m}^2 = 1 \text{ Pascal (Pa)}$$

$$1 \text{ atm} = 760 \text{ mm of Hg.} = 76 \text{ cm of Hg}$$

$$1 \text{ Bar} = 10^5 \text{ Newton/m}^2 = 100 \text{ k Pa}$$

$$1 \text{ Torr} = 1\text{mm of Hg.}$$

6.2.2 Absolute Pressure (P_{ABS})

Pressure has been defined as the force, which the fluid particles exert due to interaction amongst themselves per unit area. A zero pressure intensity will occur when molecular momentum is zero. Such a situation can occur only when there is a perfect vacuum i.e., very few gas molecules. Pressure intensity measured from this state of vacuum or zero pressure is called absolute pressure.

6.2.3 Gauge Pressure ($P > P_A$)

A pressure measuring instrument generally measures the difference between the unknown pressure (P) and the atmospheric pressure (P_a). When the unknown pressure (P) is greater than the atmospheric pressure (P_a), the pressure measured by the instrument is called as the gauge pressure.

6.2.4 Vacuum Pressure ($P_A > P$ or $P < P_A$)

A pressure measuring instrument generally measures the difference between the unknown pressure (P) and the atmospheric pressure (P_a). When the atmospheric pressure (P_a) is greater than the unknown pressure (P), the pressure measured by the instrument is called as the vacuum pressure.

6.2.5 Relation Between Absolute, Gauge, Atmospheric and Vacuum Pressure

From the Figure 6.1 we can say that absolute pressure is the algebraic sum of the gauge indication and the atmospheric pressure.

$$P_{abs} = P_a - P_g$$

$$P_{abs} = P_a - P_{vac}$$

Vacuum pressure is also called rarefaction or negative pressure.

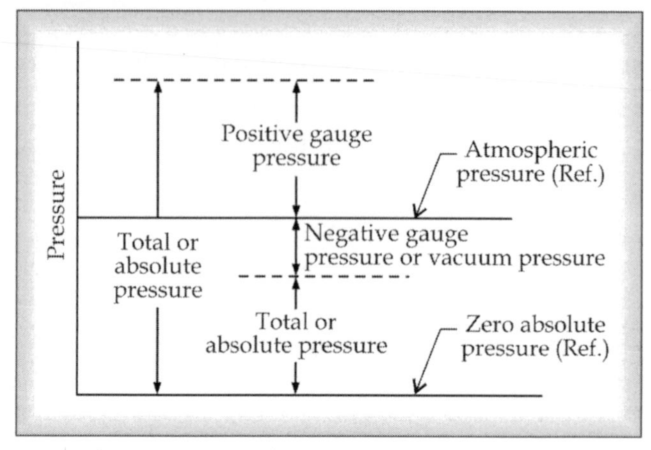

Fig. 6.1 Relation between Absolute Gauge Atmospheric and Vacuum Pressure

6.2.6 Static Pressure

The pressure exerted on the walls of a pipe due to a fluid at rest inside the pipe or due to the flow of a fluid parallel to the walls of the pipe is called static pressure. This static pressure is measured by inserting a pressure measuring tube into the pipe carrying the fluid, so that the tube is at right angle to the fluid flow path

6.2.7 Total or Stagnation Pressure

The pressure which is obtained by bringing the flowing fluid to rest isentropically is called total or stagnation pressure. Hence the pressure will be a sum of static pressure and dynamic or velocity or impact pressure.

6.2.8 Dynamic or Impact or Velocity Pressure

The pressure due to fluid velocity is called impact pressure.

Impact pressure = Total pressure – Static pressure.

Diagram showing relation between static, impact and total pressure is given in Figure 6.2.

$$h_{static} = \frac{p}{\rho g} \text{ (Static head)}$$

$$h_{static} = \frac{p}{\rho g} + \frac{v^2}{2g}$$

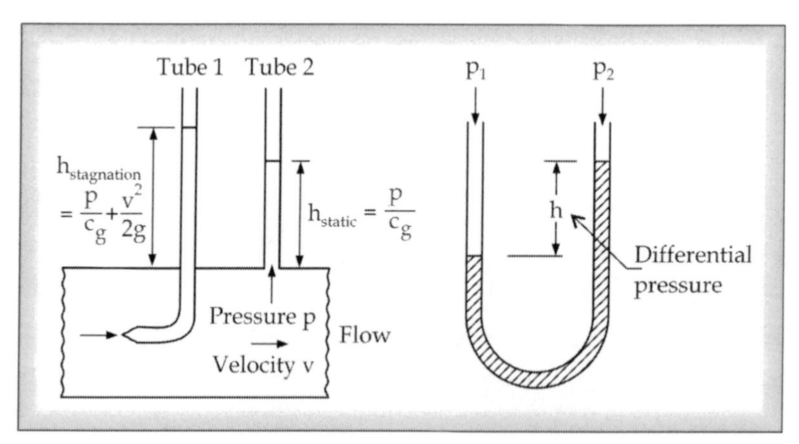

Fig. 6.2 Relation between Pressure Terms

6.3 CLASSIFICATION OF PRESSURE MEASURING DEVICES

Given below is the classification of instruments used to measure pressure under the two different kinds of transducers namely, gravitational and elastic transducers.

1. Gravitational Transducers
    1. Dead weight tester
    2. Manometers
        A. Simple manometers
            1. Piezometer
            2. U-tube Manometer
            3. Single Column Manometer
                (i) Vertical Single Column Manometer
                (ii) Inclined Single Column manometer
        B. Differential Manometers
            1. U-Tube Differential Manometer
            2. Inverted U-Tube Differential Manometer
2. Elastic Transducers
3. Bourdon Tube pressure gauge
4. Elastic diaphragm gauges
5. Bellows gauges.
    (a) Bellows gauge to measure gauge pressure
    (b) Bellows gauge to measure differential Pressure.
6. Strain gauge pressure cells.
7. Flattened tube pressure cell (Pinched tube)
8. Cylindrical type pressure cell.
9. McLeod vacuum gauge
10. Thermal conductivity gauges
11. Pirani- gauge.
12. Thermo couple type conductivity gauge
13. Ionization gauge
14. Bulk modulus or electrical resistance pressure gauge

Gravitational and Elastic Transducers

These transducers make use of the gravitational forces and the weight is the main component that plays key role in transduction process. The pressure is conveniently converted into displacement by a dead weight or the weight of the liquid used in the device. The dead weight tester and the manometers are categorized under this class. All the remaining (in the above listed) instruments use the elastic members such as springs, bourdon tube or bellows as the transducers to convert the force into the displacement. The pressure measurement instruments using gravitational transducers are described in the sections 6.4 through 6.7 and the sections from 6.8 onwards are the instruments with elastic transducers.

6.4 DEAD WEIGHT TESTER

The dead weight tester consists of the following parts:

1. Chamber filled with oil
2. Piston and Cylinder
3. Plunger
4. Platform
5. Standard weight
6. Pressure gauge

Construction

The chamber is filled with oil which is free from impurities. A piston and cylinder combination is fitted to the chamber as shown in Figure 6.3. The top portion of the piston is attached with a platform to carry weights. A plunger with a handle has been provided to vary the pressure of oil in the chamber. The pressure gauge to be tested is fitted at an appropriate place.

Operation

The dead weight tester is a device to produce and measure pressure. It is also used to calibrate pressure gauges. The following procedure is adopted for calibrating pressure gauges.

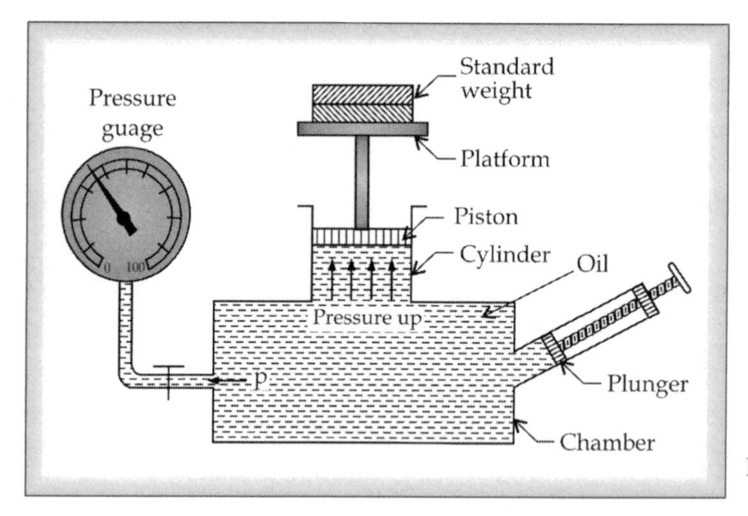

Fig. 6.3 Dead Weight Tester

More Instruments for Learning Engineers

Mile 6.2

A **barometer** is a scientific instrument used to measure atmospheric pressure, pressure tendency which can forecast short term weather changes. Though invention of barometer (1643) is credited to Evangelista Torricelli, historical documentation suggests Gasparo Berti, an Italian mathematician and astronomer, unintentionally built a water barometer between 1640 and 1643. French scientist and philosopher René Descartes described the design of an experiment to determine atmospheric pressure during 1631, but there is no evidence that he built a working barometer.

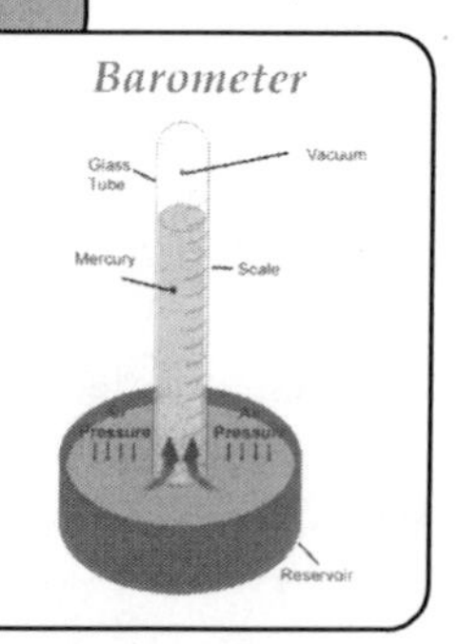

Calibration of pressure gauge means introducing an accurately known sample of pressure to the gauge under test and then observing the response of the gauge. In order to generate this accurately known pressure, the following steps are followed:

(i) Keep the valve of the apparatus closed.

(ii) Place a known weight on the platform.

(iii) Now by operating the plunger, apply fluid pressure to the other side of the piston until enough force is developed to lift the piston weight combination.

(iv) When this- happens, the piston weight combination floats freely within the cylinder between limit stops.

(v) In this condition of equilibrium, the pressure force of fluid is balanced against the gravitational force of the weight plus the friction drag.

Therefore, $PA = mg + F$

Hence, $$\text{Pressure (P)} = \left[\frac{mg + F}{A}\right]$$

Where m = Mass; kg

g = Acceleration due to gravity; m/s^2

F = Friction drag; N

A = Equivalent area of piston cylinder combination m^2.

Thus the pressure 'P', caused due to the weights placed on the platform is calculated, and then the plunger is released.

Now, the pressure gauge which is to be calibrated is fitted at an appropriate place on the dead weight tester. The same known weight which was used to calculate' P' is placed on the platform. Due to the weight, the piston moves downwards and exerts a pressure 'P' on the fluid. Now, the valve in the apparatus is opened so that the fluid pressure 'P' is transmitted to the gauge, and makes the gauge indicate a pressure value. This pressure value shown by the gauge should be equal to the known input pressure, 'P'. If the gauge indicates some other value other than 'P', the gauge is adjusted so that it reads a value equal to 'P'. Thus the gauge is calibrated.

Applications

- Used to learn what calibration is.
- Used to calibrate all kinds of pressure gauges such as industrial gauges, engine indicators and piezoelectric transducers.

Merits

- Simple to construct and easy to use
- Can be used to calibrate a wide range of pressure measuring devices.
- Fluid pressure can be easily varied either by adding weights or by changing the piston cylinder combination.

Demerits (Limitations)

- The accuracy of the dead weight tester is affected due to the friction between the piston and cylinder, and due to the uncertainty of the value of gravitational constant 'g'.

SELF ASSESSMENT QUESTIONS-6.1

1. Define pressure and give its various units
2. Define atmospheric pressure, gauge pressure and vacuum pressure and obtain the relationship among them
3. Distinguish between static and dynamic pressures.
4. Distinguish among atmospheric pressure, gauge pressure and vacuum pressure
5. Classify the pressure measurement devices and briefly describe the principles of transduction in them.
6. Describe the role of gravity transducers in pressure measurement.
7. Discuss the significance and role of elastic transducers in pressure measurement.
8. Explain the construction and operation of dead weight tester.
9. Give the applications, merits and demerits of dead weight tester.
10. Define and write a short note on the following
    (a) Vacuum pressure (b) Stagnation pressure (c) Velocity pressure

6.5 MANOMETERS

Manometers are defined as the devices used for measuring the pressure at a point in a fluid by balancing the column of fluid by a same or another column of the fluid.

Desirable Properties of Manometric Liquids

Some of the desirable characteristics of a manometric liquid are:

1. Low viscosity, i.e. capability of quick adjustment with pressure changes
2. Low coefficient of thermal expansion, i.e., minimum density changes with temperature.
3. Low vapour pressure, i.e. little or no evaporation at ambient conditions.
4. Non sticky nature negligible surface tension and capillary effects.
5. Non corrosive.
6. Non-poisonous, and stable.

Example 1: Mercury is generally used for measuring vacuums and moderate pressures of gas, vapour or water where moderate sensitivity is required, mercury does not evaporate readily, has a reasonably stable density, forms a sharp meniscus and is clearly seen, however it amalgamates or corrodes many metals, is poisonous as well as expensive.

Example 2: Water is used for measuring small vacuums and small pressure differences of gas flow with high sensitivity, Water have a fairly sharp meniscus, is cheap and readily available. However it has a tendency to evaporate and dissolve some gases in it. Further, its transparent nature renders it difficult to be seen within the manometer tube. A dye added to give it a distinctive colour, would be deposited on the tube walls when water evaporates.

For high multiplication in two liquid manometers, alcohol and kerosene are often used. Kerosene is however not satisfactory because of fractional vaporization and consequent change in density. Alcohol is also apt to change in density by taking up water.

Merits (Advantages) of Manometers

- Relatively inexpensive and easy to fabricate
- Good accuracy and sensitivity
- Requires little maintenance and are not affected by vibrations.
- Particularly suitable to low pressures and low, differential pressures
- Sensitivity can be altered easily by affecting a change in the quantity of manometric liquid in the manometer.

Demerits (Limitations) of manometers

- Generally large and bulky, fragile and gets easily broken.
- Not suitable for recording.
- Measuring medium has to be compatible with the manometric fluid used.
- Readings are affected by changes in gravity, temperature and altitude.
- Surface tension of manometric fluid creates a capillary affected and possible hysteresis.
- Meniscus height has to be determined by accurate means to ensure improved accuracy.
- Range of pressures that can be measured with manometers depends upon the manometric fluid used, the minimum displacement (which can be sensed) and the tube length. Manometers are employed to measure pressures as low as 0.35 N/m^2 and also the pressure differences in the range 1.4×10^5 to 2.1×10^5 N/m^2. Pressures higher than two or three atmospheres are invariably measured with mechanical gauges of the bourbon tube or diaphragm type.

  They are classified as:

A. Simple manometers

B. Differential manometers.

A. Simple Manometers

A simple manometer consists of a glass tube having one of its ends connected to a point t where pressure is to be measured and other end remains open to atmosphere. Common types of simple manometers are:

1. Piezometer,
2. U-tube manometer, and
3. Single column manometer.

6.5.1 Piezometer

Piezometer consists of the following parts.

1. Pipe through which liquid flows whose pressure is to be measured
2. Calibrated Tube

Construction

It is the simplest form of manometer used for measuring gauge pressures. One end of this manometer is connected to the point where pressure is to be measured and other end open to the atmosphere as shown in Figure 6.4.

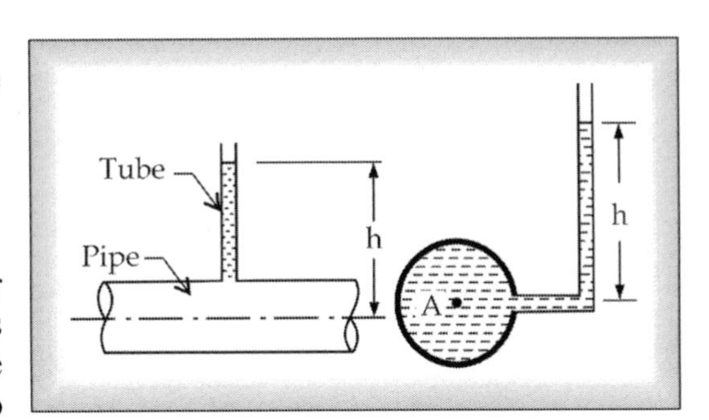

Fig. 6.4 Piezometer

Principle and Operation

The rise of liquid, gives the pressure head at that point. If at a point 'A', the height of liquid (say, water) is 'h' in piezometer tube, then pressure at A is given by

$$P = w \times h = \rho gh$$

But actually fluid pressure at the gauge point = Atmospheric pressure P_a at the free surface + pressure due to a liquid column of height h.

So, $P = P_a + \rho gh$, where ρ is the density of the liquid.

Pressures are generally described with reference to atmospheric pressure taken as zero of the pressure scale.

Obviously, then $P = \rho gh$, and the pressure thus evaluated is the gauge pressure. When using a piezometer to measure the pressure of a moving fluid, axis of the tube should be absolutely normal to the direction of flow and its bottom end must flush smoothly with the pipe surface.

Merits

- Simplest form
- Easy to operate (no manual operation is needed) and
- Cheapest to construct

Demerits

- Any burr or projection would cause obstruction resulting in changes in the pressure head.
- To reduce the tension and capillary effects, diameter of the tube must be kept at least 6 mm.
- Cannot be used to measure pressures, which are considerably excess of atmospheric pressure.
- Use of a very long glass tube would not be safe due to fragile and unmanageable.
- Gas pressures cannot be measured as gas does not form any free surface with atmosphere,
- Measurement of negative pressure is not possible due to flow of atmospheric air into the container through the tube.

However, these difficulties can be overcome by modifying the piezometer into a U-tube manometer, also called the double column manometer.

6.5.2 U-Tube Manometer

It consists of

1. A glass tube bent in U shape,
2. Mercury
3. Pipe with fluid whose pressure is to be measured

Construction

A U-tube Manometer is constructed with a glass tube bent in U-shape. One end of this tube is connected to a point at which pressure is to be measured and other end remains open to the atmosphere as shown in Figure 6.5. The tube generally contains mercury or any other liquid whose specific gravity is more than the liquid whose pressure is to be measured.

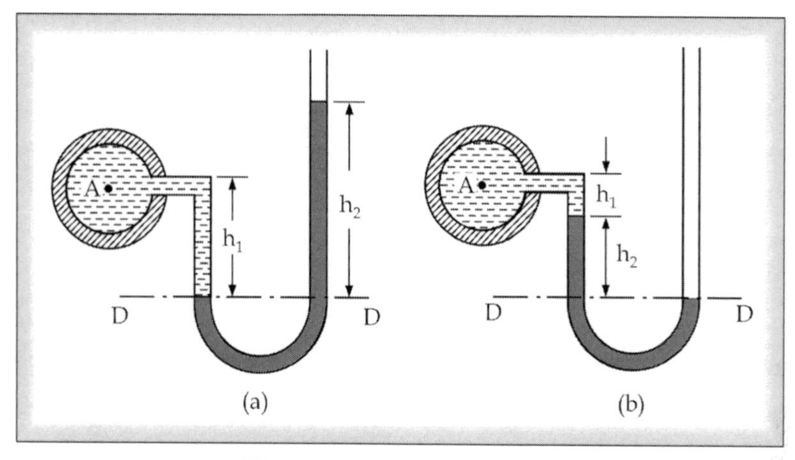

Fig. 6.5 U-Tube Manometer

Principle and Operation

(a) **For Gauge Pressure:** Let A is the point at which the pressure is to be measured, whose value in terms of pressure head is h. The datum line is D-D.

Let h_1 = Height of light liquid above the datum line

h_2 = Height of heavy liquid above the datum line.

S_1 = Sp. gr. of light liquid.

S_2 = Sp. gr. of heavy liquid.

As the pressure or pressure head is the same for the horizontal surface. Hence pressure head above the horizontal datum line D-D in the left column and in the right column of U-tube manometer should be same as shown in Figure 6.5 (a).

Pressure head above D-D is the left column = $h + h_1S_1$

Pressure head above A-A in the right column = h_2S_2

Hence equating the two pressures

$$h + h_1S_1 = h_2S_2$$

$$h = (h_2S_2 - h_1S_1)$$

(b) **For Vacuum Pressure:** For measuring vacuum pressure, the level of the heavy liquid in the manometer will be shown in Figure 6.5 (b). Then pressure head above D-D in the left column

$$h_2S_2 + h_1S_1 + h$$

Pressure head in the right column above A – A = 0.

$$\therefore \quad h_2S_2 + h_1S_1 + h = 0; \; h = -(h_2S_2 + h_1S_1)$$

ILLUSTRATION-6.1

The right limb of a simple U tube manometer containing mercury is open to the atmosphere while the left limb is connected to a pipe in which a fluid of sp.gr 0.8 is flowing. The center of the pipe is 21 cm below the level of mercury in the right limb. Find the pressure of fluid in the pipe if the difference of mercury level in two limbs is 30 cm.

Solution:

Specific Gravity of fluid, $S_1 = 0.8$

Specific Gravity of mercury, $S_2 = 13.6$

Difference of mercury level $h_2 = 30$ cm

Height of fluid from A-A, $h_1 = 30 - 21 = 9$ cm.

Let h = pressure head of fluid in pipe in terms of water head.

Equating the pressure head above A-A, we get

$$h + h_1 S_1 = h_2 S_2 \text{ or}$$

$$h + 9 \times 0.8 = 30 \times 13.6$$

$$h = 30 \times 13.6 - 9 \times 0.8$$

$$408 - 7.2 = 400.8 \text{ cm of water.}$$

(a)

The pressure P is given by P = wz or wh

where w = weight density = 1000×9.81 N/m^3 for water.

$$\frac{1000 \times 9.81}{10^6} \text{ N/cm}^3 = 0.00981 \text{ N/cm}^3$$

$\therefore$ Pressure P = $0.00981 \times 400.8 = 3.932$ N/cm^2

6.5.3 SINGLE COLUMN MANOMETER

Single column manometer is a modified form of U-tube manometer in which reservoir having a large cross sectional area (about 100 times) as compared to the area of the tube is connected to one of the limbs (say left limb) of the manometer as shown Figure 6.6. Due to large cross sectional area of the reservoir, for any variation in pressure, the change in the liquid level in the reservoir will be very small, which may be neglected and hence the pressure in other limb may be vertical or inclined. Thus there are 2 (two) types of single column manometer

(i) Vertical single column manometer

(ii) Inclined single column manometer.

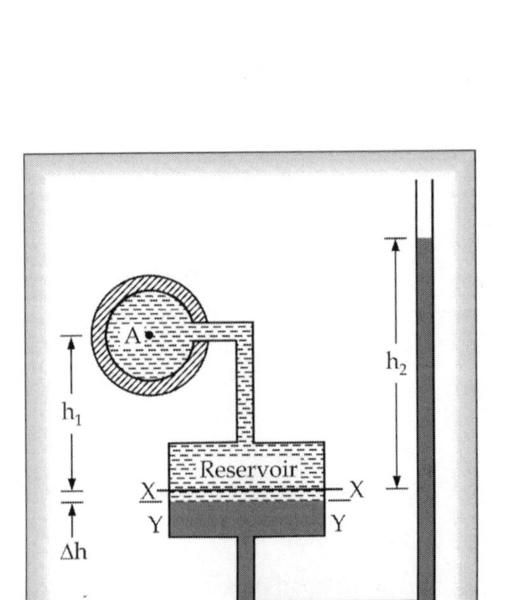

Fig. 6.6(a) Vertical Single Column Manometer

(i) **Vertical Single Column Manometer:** In vertical single column manometer let D-D be the datum line in the reservoir and in the right limb of the manometer when it is not connected to the pipe due to high pressure at A the heavy liquid in the reservoir will be pushed downward and will rise in the right limb. Refer Fig. 6.6 (a). Let Δh = Fall of heavy liquid in reservoir

h_2 = Rise of heavy liquid in right limb.

h_1 = Height of centre of pipe above D-D.

h = Pressure at A, which is to be measured

A = Cross sectional area of the reservoir.

a = Cross sectional area of the right limb.

S_1 = Sp. gr of liquid in pipe.

S_2 = Sp. gr of heavy liquid in reservoir and right limb.

Full of heavy liquid in reservoir will cause a rise of heavy liquid level in the right limb.

$$A \times \Delta h = a \times h_2$$

$$\Delta h = a \times \frac{h_2}{A}$$

Now consider the datum line y-y as shown in Figure 6.6(a), then pressure head in the right limb above Y-Y = $(\Delta h + h_2)S_2$

Pressure head in the left limb above y-y

$$= (\Delta h + h_1)S_1 + h$$

Equating these pressures, we have

$$(\Delta h + h_2)S_2 = (\Delta h + h_1)S_1 + h \text{ or}$$

$$h = (\Delta h + h_2)S_2 - (\Delta h + h_1)S_1$$

$$= \Delta h\,(S_2 - S_1) + h_2S_2 - h_1S_1$$

But $\Delta h = \dfrac{a \times h_2}{A}$

$$\therefore \quad h = \frac{a \times h_2}{A}[S_2 - S_1] + h_2S_2 - h_1S_1$$

As the A is very large as compared to 'a', hence ratio $\frac{a}{A}$ become very small and can be be neglected.

Then $h = h_2S_2 - h_1S_1$

From the above equation, it is clear that as h_1 is known and hence by knowing h_2 or rise of heavy liquid in the right limb, the pressure at A can be calculated.

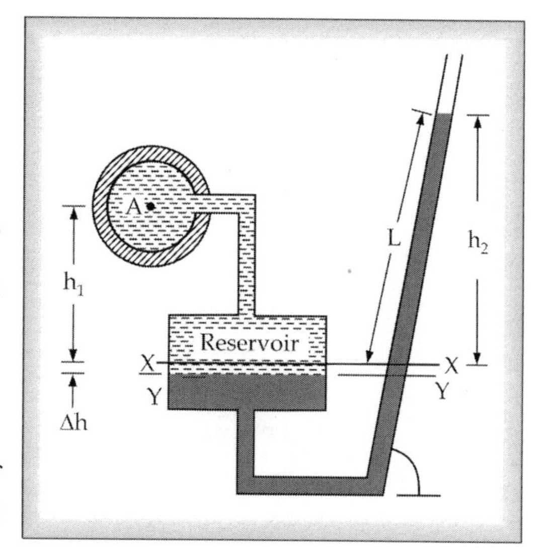

Fig. 6.6(b) Inclined Single Column Manometer

(ii) Inclined Single Column Manometer: This manometer is more sensitive. Due to inclination the distance moved by the heavy liquid in the right limb will be more. Refer Figure 6.6 (b).

Let L = Length of heavy liquid moved in right limb from X-X

θ = Inclination of right limb with horizontal

h_2 = Vertical rise of heavy liquid in right limb from X-X

$= L x \sin\theta$

From equation $h = h_2 S_2 - h_1 S_1$, the pressure head at A is

$$h = L\sin\theta \times S_2 - h_1 S_1$$

Sensitivity is defined as the ratio of change in height of the denser fluid for a given change in pressure.

ILLUSTRATION-6.2

A single column manometer is connected to a pipe containing a liquid of sp.gr 0.8 as shown in Figure 6.6(a). Find the pressure in the pipe if the area of the reservoir is 90 times the area of the tube for the manometer reading shown in Figure 6.6(a). The sp.gr of mercury is 13.6

Solution:

Sp. gr. of liquid in pipe $S_1 = 0.8$

Sp. gr of heavy liquid $S_2 = 13.6$

$$\frac{\text{Area of the Reservoir}}{\text{Area of right limb}} = \frac{A}{a} = 90 \Rightarrow \frac{a}{A} = \frac{1}{90}$$

Height of the liquid,

$$h_1 = 30 \text{ cm}$$

Rise of mercury in right limb

$$h_2 = 50 \text{ cm}$$

Let h = pressure head in pipe

Using equation

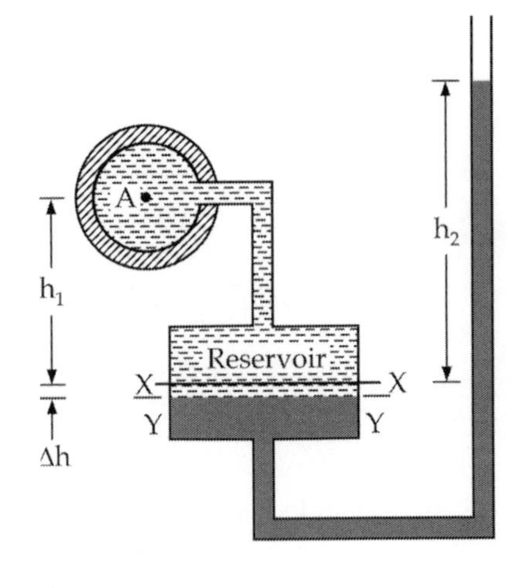

$$h = \frac{a}{A} h_2[S_2 - S_1] + h_2 S_2 - h_1 S_1$$

We get

$$h = \frac{1}{90} \times 50\,[13.6 - 0.8]$$

$$+ 50 \times 13.6 - 30 \times 0.8$$

$$= 7.1 + 680 - 24$$

$$= 663.11 \text{ cm of water}$$

$$= 6.63 \text{ m of water}$$

Pressure $p = w \times h$

$$= (1000 \times 9.81) \times 6.63 \text{ N/m}^2 \ (W = 1000 \times 9.81 \text{ N/m}^3)$$

$$= \frac{1000 \times 9.81 \times 6.63 \text{ N/cm}^2}{10^4} = 6.51 \text{ N/cm}^2.$$

ILLUSTRATION-6.3

Express the ratio of sensitivities of an inclined manometer to that of a simple manometer in terms of angle "theta". For an inclined manometer six times more sensitive than a simple manometer, what should be the angle of inclination?

Solution:

Case (i)

We have sensitivity of simple manometer = $\left(\dfrac{h}{\Delta P}\right)_{mano} = \dfrac{h_2S_2 - h_1S_1}{\Delta P}$

& that of inclined manometer = $\left(\dfrac{h}{\Delta P}\right)_{incli-mano} = \dfrac{I \sin\theta\, S_2 - h_1S_1}{\Delta P}$

$$\therefore \text{Ratio of sensitivities} = \frac{(1)}{(2)} = \frac{h_2S_2 - h_1S_1}{\Delta P} = \frac{h_2S_2 - h_1S_1}{I \sin\theta\, S_2 - h_1s_1}$$

Case (ii) $\left(\dfrac{h}{\Delta P}\right)_{incli-mano} = 6\left(\dfrac{h}{\Delta P}\right)_{mano}$

$$\frac{I \sin\theta\, S_2 - h_1s_1}{\Delta P} = 6\times\frac{h_2S_2 - h_1s_1}{\Delta P}$$

$$I \sin\theta\, S_2 - h_1S_1 = 6h_2S_2 - 6h_1S_1$$

$$I \sin\theta\, S_2 = 6h_2S_2 - 6h_1S_1 + h_1S_1$$

$$= 6h_2S_2 - 5h_1S_1$$

$$\sin\theta = \frac{6\,h_2S_2 - 5h_1S_1}{I\,S_2}$$

B. Differential Manometers

Differential manometers are the devices used for measuring the difference of pressures between two points in a pipe or in two different pipes. A differential manometer consists of a U-tube containing a heavy liquid whose two ends are connected to the points whose difference of pressure is to be measured. Most commonly used of differential manometers are

1. U-tube differential manometer and
2. Inverted U-tube differential manometer.

6.5.4 U-TUBE DIFFERENTIAL MANOMETER

Figure 6.7 shows the U-tube differential manometers.

Let the two points A and B are at different level and also contain liquids of different specific gravities. These points are connected to the U-tube differential manometer. Let the pressure at A and B are P_A and P_B

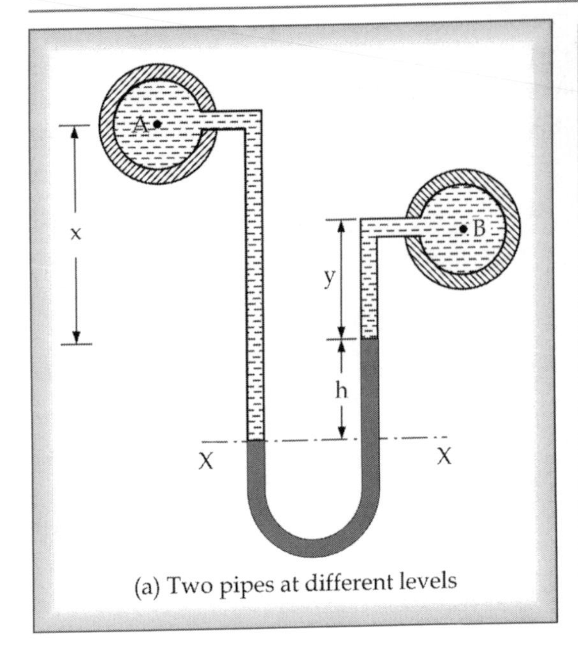

(a) Two pipes at different levels

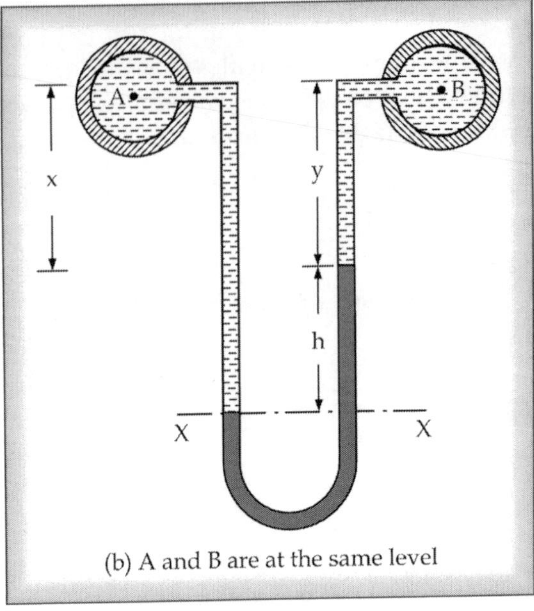

(b) A and B are at the same level

Fig. 6.7 U-Tube Differential Manometer

Let h = Difference of mercury level in the U-tube

y = Distance of the center of B, from the mercury level in the right limb.

S_1 = Sp. gravity of liquid at A.

S_2 = Sp. gravity of liquid at B.

S_3 = Sp. gravity of different liquid or mercury.

Taking datum line at X-X, Pressure head above X-X in the right limb

$$= h \times S_3 + y \times S_2 + h_B$$

Where h_B = pressure head at B

Equating the two pressure head we have

$$(h + x)\, S_1 + h_A = h \times S_3 + y \times S_2 + h_B$$

$$\therefore \quad h_A - h_B = h \times S_3 + yS_2 - (h + x)S_1$$

$$= h\,[S_3 - S_1] + y\,S_2 - x\,S_1$$ Difference of pressure head at A and B

$$= h\,[S_3 - S_1] + y\,S_2 - x\,S_1$$

Figure 6.7 (b) A & B are at same level and contains the same liquid of Sp.Gr. (S_1)

Then pressure head above X-X in right limb = $h \times S_3 + x \times S_2 + h_B$

Pressure head above X - X in left limb = $(h + x)S_1 + h_A$.

Equating the two pressure heads

$$h \times S_3 + \times S_1 + h_B = (h + x)S_1 + h_A [\therefore\ S_1 = S_2]$$

$$h_A - h_B = h \times S_3 + xS_1 - (h + x)\,S_1 = h \times S_3 - hS_1$$

$$= h\,[S_3 - S_1]$$

ILLUSTRATION-6.4

A differential manometer is connected at the two points A and B of two pipes as shown in Figure 6.8. The pipe A contains a liquid of sp.gr = 1.5 while pipe B contains a liquid of Sp gr. = 0.9. The pressures at A and B are 98.1 KN/m^2 and 176.6 KN/m^2 respectively. Find the difference in mercury level in the differential manometers.

Solution:

Specific gravity liquid at A, $S_1 = 1.5$

Specific gravity of liquid at B, $S_2 = 0.9$

Pressure at A, $P_A = 98.1$ KN/m^2 and pressure at B, $P_B = 176.6$ kN/m^2

Pressure head at A,

$$h_A = \frac{P_A}{W} = \frac{98.1 \text{ KN/m}^2}{9.81 \text{ KN/m}^3} = 10 \text{ m of water.}$$

$$\text{Pressure head at B } h_B = \frac{P_B}{W} = \frac{176.6 \text{ KN/m}^2}{9.81 \text{ KN/m}^3} = 18 \text{ m of water.}$$

Taking x-x as datum line

$$\text{Pressure head above X-X in the left limb} = h \times 13.6 + (20 + 30) \times 1.5 + \frac{P_A}{W}$$

$$= 13.6h + 7.5 + 10.0 \text{ m of water.}$$

Pressure head above to X-X in the right limb.

$$= (h + 2.0) \times 0.9 + h_B$$

$$= (h + 2.0) \times 0.9 + 18\text{m of water.}$$

Equating the two pressure heads, we get

$$13.6h + 7.5 + 10 = (h + 2.0) \times 0.9 + 18$$

$$13.6h + 17.5 = 0.9h + 1.8 + 18 = 0.9h + 19.8$$

$$(13.6 - 0.9)\, h = 19.8 - 17.$$

$$12.7\, h = 2.3$$

$$h = \frac{2.3}{12.7} = 0.181\text{m} = 18.1\text{cm}$$

6.5.5 INVERTED U-TUBE DIFFERENTIAL MANOMETER

It consists of an inverted U-tube, containing a light liquid. The two ends of the tube are connected to the points whose difference of pressure is to be measured. It is used for measuring difference of low pressures. Figure 6.8 shows an inverted U tube differential manometer connected to the two point's pressure at B.

Let h_1 = Height of liquid in left limb below the datum line X-X.

h_2 = Height of liquid in right limb.

h = Difference of light liquid

S_1 = Sp gr of liquid at A.

S_2 = Sp gr of liquid at B.

S = Sp. gr of light liquid.

h_A = pressure head at A

h_B = Pressure head at B.

Taking X-X at datum line, then pressure head in the left limb below X-X

$$= h_A - S_1 h_1$$

Pressure head in the right limb below X-X

$$= h_B - S_2 h_2 - Sh$$

Equating the two pressure heads

$$h_A - S_1 h_1 = h_B - S_2 h_2 - Sh \quad \text{or}$$

$$h_A - h_B = S_1 h_1 - S_2 h_2 - Sh$$

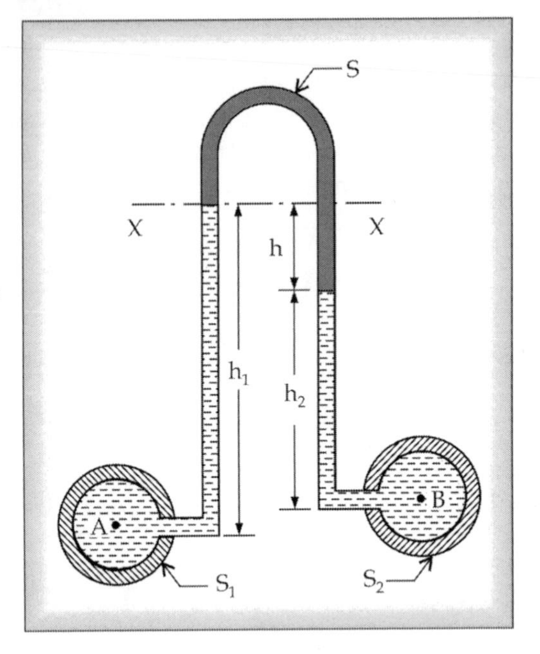

Fig. 6.8 Inverted U-tube Differential Manometer

ILLUSTRATION-6.5

In Figure 6.9 an inverted differential manometer is connected to two pipes A and B which convey water. The fluid in manometer is oil of sp. gr 0.8 for the manometer readings shown in Figure 6.9. Find the pressure difference between A and B.

Solution:

Sp. gr of oil = 0.8

Difference of oil in the two limbs = (30 + 20) – 30

= 20 cm.

Taking datum line at X-X

Pressure head in the left limb below X-X

$$= \frac{P_A}{W} - 0.30 \times 1$$

$$= \frac{P_A}{W} - 0 3$$

Pressure head in the right limb below X-X

$$\frac{P_A}{W} - 0.3 = \frac{P_B}{W} - 0.46$$

$$\frac{P_B}{W} - \frac{P_A}{W} = 0.46 - 0.30 = 0.16 \text{ m of water.}$$

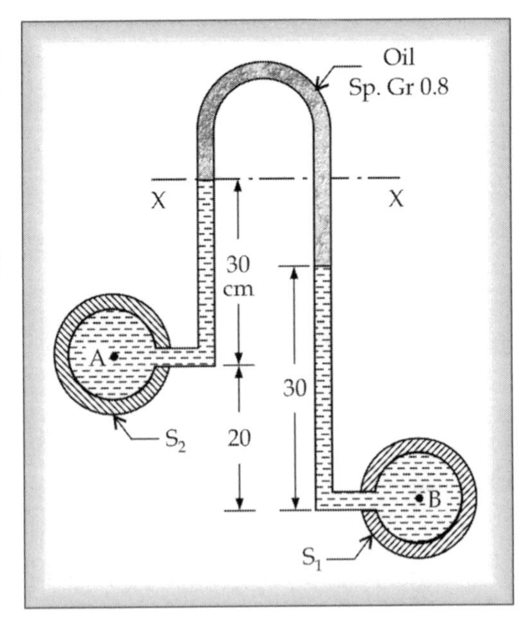

Fig. 6.9 Inverted U-Tube Differential Manometer

Self Assessment Questions-6.2

1. Classify various types of Manometers and explain any one in detail.
2. Enumerate the merits and demerits of using manometers for measurement of pressure.
3. What are the desirable characteristics of a manometric liquid?
4. Describe the construction and operation of piezometer.
5. Describe the construction and operation of simple manometer.
6. With sketches discuss the construction and operation of differential manometers. What are their merits and demerits?
7. Describe the construction and operation of single vertical column manometers. What are its merits and demerits?
8. Explicate the construction and operation of single inclined column manometers. What are its merits and demerits?
9. Distinguish between the simple and differential manometers.
10. With a neat sketch explain the construction and operation of U-Tube differential manometers. Give its merits and demerits
11. Narrate the construction and operation of Inverted U-Tube differential manometers with the help of sketch. List out its merits and demerits

6.6 MECHANICAL DISPLACEMENT TYPE GAUGES

These pressure gauges contain a liquid that merely acts as a pressure seal. The liquid density does not appear in the balance equation.

6.6.1 Ring Balance Manometer

The gauge consists essentially of

1. A hollow toroidal tube made of polythene or any other light and transparent material.
2. The tubular chamber divided into two parts by splitting and sealing
3. Suitable liquid like kerosene or paraffin.
4. Pivot
5. Ring
6. Balancing weight
7. Calibrated scale

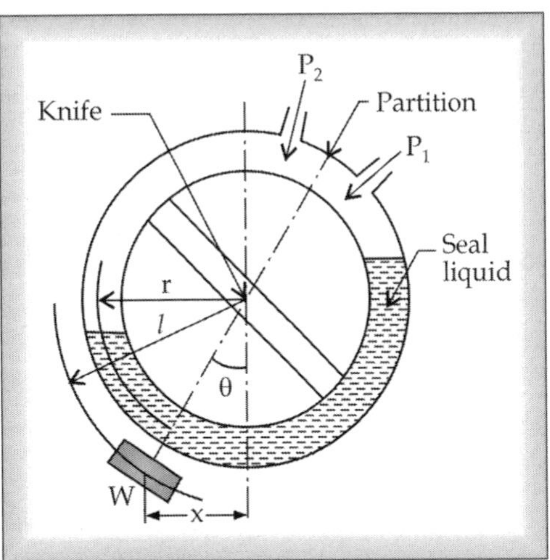

Fig. 6.10 Ring Balance Manometer

Construction

The ring balance manometer essentially uses a hollow toroidal tube made of polythene or any other light and transparent material. The tubular chamber is divided into two parts by splitting and sealing and is partially filled with a suitable liquid like kerosene or paraffin. The unit is pivoted at the center and is balanced by a weight 'w' attached to the ring. A differential gas pressure, when applied through flexible leads to the partition, tends to tilt the ring against the resisting moment exerted by the balance weight as shown in Figure 6.10.

Operation

An expression relating the tilt of the ring to the differential pressure can be deduced by taking moments about the knife edge.

Rotating moment = balancing moment.

$$(P_2 - P_1)\ ar = w\ x$$

where 'a' is the cross-sectional area of the ring; 'r' is the mean radius of the ring and 'x' is the horizontal moment of the weight.

But $\quad x/I = \sin\theta$ and $x = I \sin\theta$

$$(P_2 - P_1) = \frac{W\ I \sin\theta}{ar} = K \sin\theta \quad \text{where } k = \frac{wl}{ar} \text{ as sensitivity.}$$

Angle θ is usually small (about 30° on either side of the center) so that scale will be effectively linear. Change of range can be affected by change of weight w or its distance I from the center.

Merits/Demerits/Applications

The type of the filling liquid does not influence the calibration, yet the liquid density must be suitable for the pressure range.

The gauge has a maximum pressure range of 300 mm water gauge; 1.0% accuracy and is used for general rate flow measurements in industry.

6.6.2 Bell Type Pressure Gauges

Single element bell gauge

A single element bell gauge consists of

1. A bell
2. Sealing liquid
3. Pressure tight link lever mechanism
4. Pressure indicating assembly (calibrated)
5. A restricting spring

Principle

These gauges operate on the principle that applied pressure is balanced against a restricting spring or against a counter weight as shown in Figure 6.11.

Construction

A bell is immersed in a sealing liquid; with pressures applied on both outside and inside of the bell. Pressure differential causes movement of the bell,

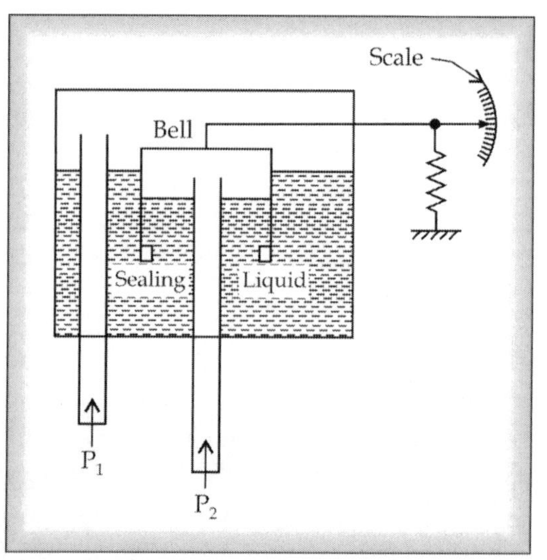

Fig. 6.11 Single Element Bell Gauge

which is taken out by a pressure tight link lever mechanism to the pressure indicating assembly. For the arrangement, the motion of the bell is opposed by a restricting spring, neglecting buoyant effect of the liquid on the bell as shown in Figure 6.11.

Operation

The result that holds good for static balance is $(P_1 - P_2) = Fh/a$, where 'F' is the spring constant of restricting spring, 'h' is the change of spring length which is directly proportional to the amount of bell movement and 'a' is the cross sectional area of the bell. Both inside and outside areas of the bell have been considered to be equal, this assumption is quite valid if the bell material is thin and the diameter is large.

Merits/Demerits/Applications

This assumption that both inside and outside areas of the bell to be equal, is valid if the bell material is thin and the diameter is large.

The gauge has a pressure range 50 to 300 mm of water gauge and 1% accuracy.

Two element bell gauge

For static balance

$$(P_{2a} - P_{1a})\, I = w\, x = w\, I \sin \theta$$

or

$$(P_2 - P_1) = \frac{wI \sin \theta}{Ia} = k \sin \theta$$

Where 'a' is the area of the bells, I is lever arm of the bell about pivot or beam x is the weight radius about pivot and θ is the angular displacement of weight from vertical.

Construction and Operation

A two element bell type differential gauge consists of two identical bells immersed in a bath of sealing liquid and connected together by a beam balance. When pressure is applied to the inside of the bells, one bell will rise higher than the other and position of balance will be achieved due to moment, of the counter weight.

Merits/Demerits/Applications

The unit is very sensitive with capability to sense pressure as small as 0.025 mm of water. Pressure range is of the order of 50 mm Water gauge. The arrangement is frequently used to control draft in a furnace.

6.7 BOURDON TUBE PRESSURE GAUGE

Elastic Transducer: Bourdon Tube

The bourdon tube essentially used in the bourdon tube pressure gauge can be considered as one of the elastic transducers. Here when it is subjected to pressure, the tube tip extends and when the pressure is withdrawn, it comes to normal state, thus acts like an elastic member.

The main parts of the bourdon tube pressure gauge are

1. Bourdon tube
2. Adjustable link
3. Sector and pinion (or Rack and Pinion)
4. Shaft
5. Pointer and calibrated scale

Principle

When any elastic transducer is subjected to a pressure, it deflects. This deflection is proportional to the applied pressure when calibrated.

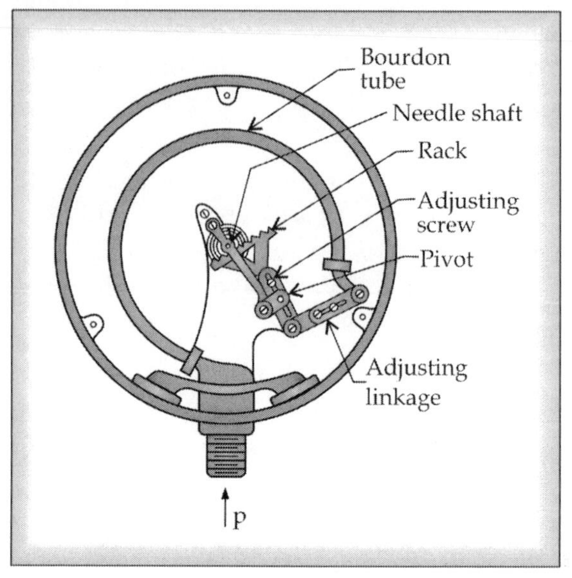

Fig. 6.12 Schematic of Bourdon - Tube Pressure Gauge

Construction

The Bourdon tube (an elastic transducer) is fixed and open at one end to receive the pressure which is to be measured. The other end of the Bourdon tube is free and closed. The cross section of the bourdon tube is elliptical. The bourdon tube is in a bent form to look like a circular arc. To the free end of the Bourdon tube is attached to an adjustable link, which in turn is connected to a sector and pinion as shown in the Figure 6.12. The shaft of the pinion is connected to a pointer, which sweeps over a pressure calibrated scale.

Operation

The pressure to be measured is connected to the fixed open end of the Bourdon tube. The applied pressure acts on the inner walls of the Bourdon tube. Due to the applied pressure, the Bourdon tube causes a displacement of the free closed end of the tube. This displacement of the free closed end of the Bourdon tube is proportional to the applied pressure. As the free end of the Bourdon tube is connected to a link-sector and pinion (or rack and pinion) arrangement, the displacement is amplified and converted to a rotatory motion of the pinion. As the pinion rotates, it makes the pointer to assume a new position on a pressure calibrated scale to indicate the applied pressure directly. As the pressure in the case containing the Bourdon tube is usually atmospheric, the pointer indicates gauge pressure.

More Instruments for Learning Engineers

Anemometer (*anemos* means wind in Greek) measures wind speed, is a common weather station instrument. and is used to describe any air speed measurement instrument used in meteorology or aerodynamics. The first description of anemometer was given by Leon Battista Alberti around 1450. Anemometers are used in two ways i.e., to measure wind's (i) speed, and (ii) pressure. However, due to the close relation between pressure and speed, an anemometer can give information about both.

Anemometer

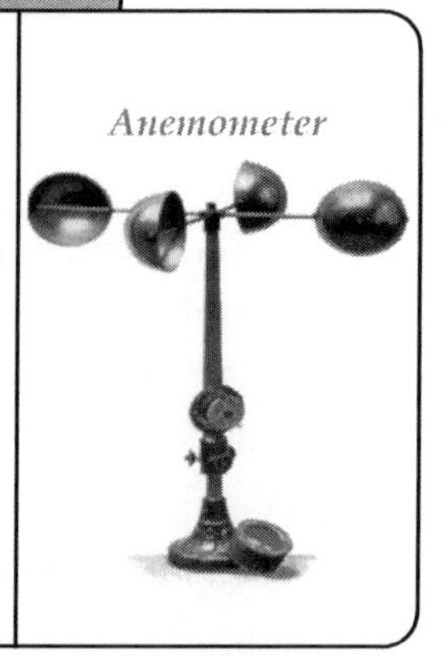

Applications

- They are used to measure medium to very high pressure.

Merits

- These are employed over a wide range of applications.
- They are less expensive.
- They are simple in construction and are available in many sizes (25.4 mm to 406.4 mm).
- They can be modified to give electrical outputs.
- They are safe even for high pressure measurement.
- Accuracy is high especially at high pressures.

Demerits

- They respond slowly to changes in pressure.
- They are subjected to hysteresis.
- They are sensitive to shocks and vibrations.
- Amplification is a must at the displacement of the free end of Bourdon tube.
- It cannot be used for precision measurement.

6.7.1 Tip Travel of Bourdon Tube

The motion of the free end, commonly called 'tip travel' is function of the tube length, wall thickness, cross sectional geometry and modulus of the tube material. For a bourdon tube, deflection, Δa of the element tip can be expressed as

$$\Delta a = 0.05\frac{ap}{E}\left(\frac{r}{t}\right)^{0.2} \times \left(\frac{x}{y}\right)^{0.33} \times \left(\frac{x}{t}\right)^{3.0}$$

where 'a' is the total angle subtended by the tube before pressurization, p is the applied pressure difference and E is the modulus of elasticity of the tube material.

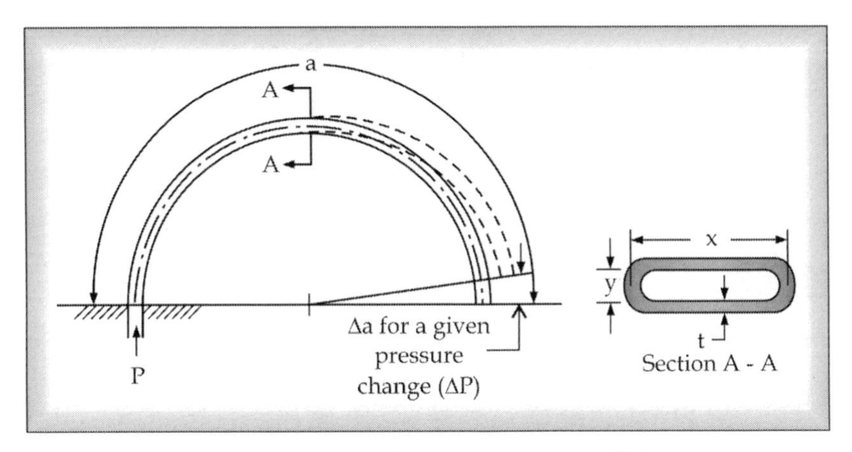

Fig. 6.13 Geometry of a C-type Bourdon Tube

The material chosen to fabricate a Bourdon tube will relate to the instrument sensitivity, accuracy and precision. For accuracy and repeatability, materials for bourdon tubes must have good elastic or spring characteristics. For a common pressure range of 100-7000 KN/m^2, tubes are made of phosphor bronze and are solid drawn. For high pressure ranges 7000-630000 KN/m^2, tubes are made of alloy steel or K-Monel. Further, the tubes are drawn, machined and heat treated to make it of enduring characteristics where corrosion is a problem, stainless steel is employed. An appropriate fluid or membrane is used to protect the gauge, if stainless steel does not meet the other gauge requirements and if protection against corrosive or viscous liquids is necessary.

6.7.2 Errors and their Rectification (in Bourdon Tube)

In general, three types of errors are found in Bourdon gauges:

(i) **Zero error or constant error:** This error remains constant over the entire pressure range. This may be due to the pointers or hands having become loose on the spindle. The error can be rectified by keeping the pointer in correct position.

(ii) **Multiplication error:** This error is possible to occur as the gauge may tend to give progressively high or low reading. The error results from a wrong setting in the multiplying mechanism between the Bourdon tube and the spindle. To rectify the error, the multiplication screw is loosened and the connecting link is moved either a little inwards (if the gauge reads low) or a little outwards (if the gauge reads high)

(iii) **Angularity error:** Quite often it is seen that a one to one correspondence does not occur when approximately linear motion of the tip is converted to the circular motion with the link-lever and pinion attachment. Because of this distortion, gauge may read correctly both at the maximum and minimum readings but may give an inaccurate reading at the midpoint. To rectify this error, angularity screw is adjusted and the pointer is set at some other point so that the gauge reads correctly at the midpoint also.

6.7.3 Bourdon Tube Shapes and Configurations

The C type Bourdon tube has a small tip travel and this needs amplification by a lever quadrant, pinion and pointer arrangement. Increased sensitivity can be obtained by using a very long length of tubing in the form of a helix, and a flat spiral indicated in Figure 6.14.

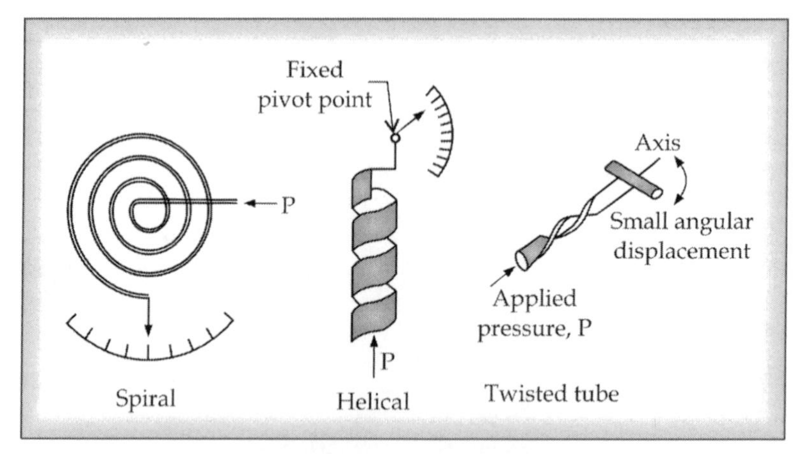

Fig. 6.14 Bourdon-Tube Configurations

The spiral tubing produces the same effect as would be given by a number of C-tubes in series. The tip travel is sufficient enough to indicate directly against a calibrated dial. Likewise, the increased numbers of turns of a helical Bourdon tube make it possible to obtain a greater angle of unwill. Spiral and helical tubes are frequently used where it is desirable to eliminate the multiplication linkages between the pressure element and the indicating or recording arm. The absence of amplification linkages makes the system more robust; wear friction and the internal effects of the linkages are eliminated.

The twisted tube has a corrosion stability, which reduces spurious output motion due to shock and vibration.

SELF ASSESSMENT QUESTIONS-6.3

1. What are the different types of mechanical displacement gauges? Explain each of them with a neat sketch.
2. Describe the principle involved in the bourdon tube pressure gauge.
3. With aid of a diagram explain the construction and operation of Ring Balance Manometer.
4. List out the application of bourdon tube pressure gauge. Explain its advantages and limitations.
5. What are the different errors found in the bourdon tube? Explain each of them.
6. Differentiate between a single element bell gauge and a two element bell gauge.
7. What do you understand by the term 'tip travel' with reference to bourdon tube?
8. What are the different shapes and configurations of bourdon tubes? Sketch them.
9. Describe the construction and operation of Bell Type Pressure Gauges.

6.8 ELASTIC DIAPHRAGM GAUGES

Transducer: Elastic type (Diaphragm)

The main parts of diaphragm pressure gauges are

1. A diaphragm, [a thin circular plate made of springy metal fixed around its edges]. The diaphragm may either be flat or Corrugated
2. Magnifying mechanism (Sector pinion arrangement or parallel plate capacitor etc.)
3. A boss of negligible weight
4. Linkage
5. A pointer over a pressure calibrated scale

Principle

When an elastic transducer (Diaphragm) is subjected to a pressure, it deflects. This deflection is proportional to the applied pressure when calibrated.

Construction

A diaphragm, which is thin circular plate made of springy metal is fixed around its edges. The diaphragm may either be flat or corrugated. A corrugated diaphragm is shown in the Figure 6.16 (a) and flat diaphragm is shown in 6.16 (b). The displacement of the diaphragm is magnified by mechanical means. [A flat diaphragm has been shown in Figure 6.16 (b), where the displacement of the diaphragm is sensed by parallel plate capacitor]. The top diaphragm is fixed with a boss of negligible weight. This boss is in turn connected to link-sector-pinion arrangement in case of a diaphragm gauge using mechanical means for displacement magnification. A pointer is connected to a pinion, which makes it to move over a pressure calibrated scale. If the displacement is sensed by a secondary transducer such as a parallel plate capacitor, its movable plate is connected to the boss as shown in Figure 6.15.

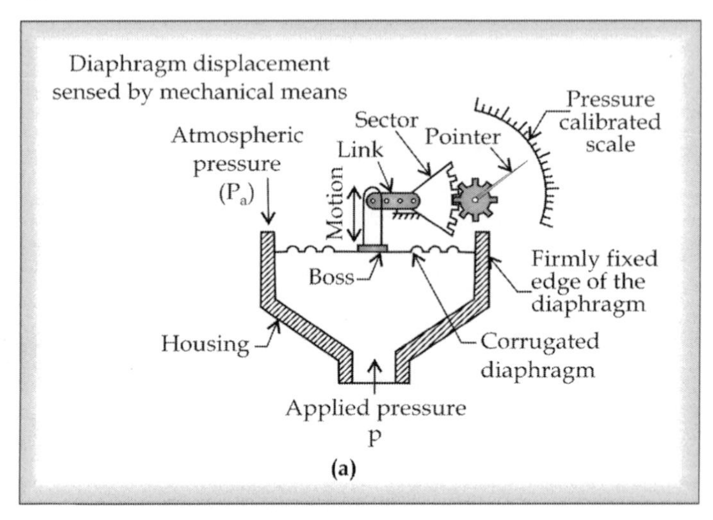

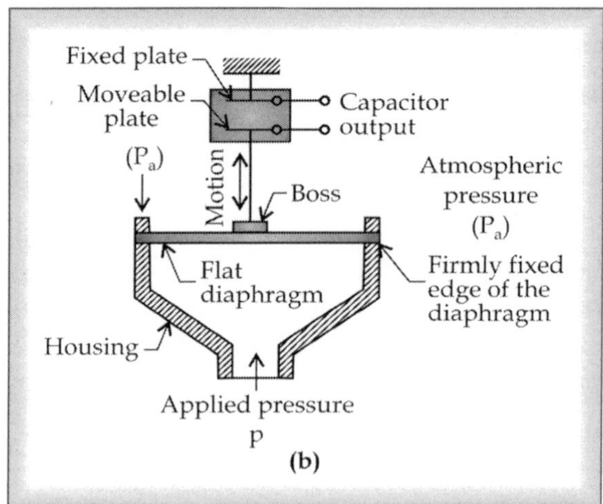

Fig. 6.15 Elastic Diaphragm Gauges (a) & (b)

Electrical resistance strain gauges may also be installed in a diaphragm. The output of these gages is a function of the local strain, which in turn, may be related to the diaphragm deflection and pressure differential. Both semi conductor and foil grade & strain gages are employed in practice, Accuracies of ± 0.5 present of full scale are typical. A Schematic diagram of LVDT - diaphragm differential pressure gauge is shown in Figure 6.16.

Operation

The bottom side of the diaphragm is exposed to the pressure, which is to be measured. The diaphragm deforms proportionally due to the applied pressure i.e., the diaphragm tends to move upwards. In a mechanical system, this deformation is magnified by the link-sector-pinion arrangement i.e., the linear displacement of the diaphragm is converted to a magnified rotary motion of the pinion. When the pinion rotates, it makes the pointer attached to it to assume a new position on the pressure calibrated scale, which becomes a measure of the applied pressure. In case of the arrangement with the parallel plate capacitor, the movable plate moves upwards, thus reducing the gap between the plates. This makes the capacitance change, which can be used as measure of the applied pressure. As the top side of the diaphragm is usually subjected to the atmospheric pressure (generally less than applied pressure), elastic diaphragm gauges usually read gauge pressure.

Merits

- They cost less
- They have a linear scale for a wide range.
- They can withstand over pressures and they are safe to be used.
- No permanent zero shift
- They can measure both absolute and gauge pressure, i.e. differential pressure.

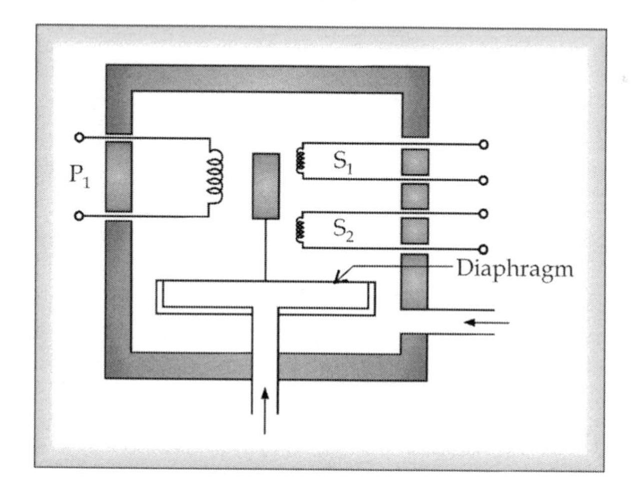

Fig. 6.16 Diaphragm LVDT Combination used as Different Pressure Gauge

Demerits

- Shocks and vibration affect their performance and hence they are to be protected.
- When used for high pressure measurement, the diaphragm gets damaged. These gauges are difficult to be repaired.

Applications

- They are used to measure medium pressures.
- But they can also be used to measure low pressures including vacuum.
- They are used to measure draft in chimneys of boilers.

Diaphragm Materials

Metals: Stainless steel, inconel, monel, nickel and beryllium copper.

Non- metals: Nylon, Teflon and Buna- N-rubber.

6.9 BELLOWS GAUGES

Principle

When elastic transducer (Bellows in this case) is subjected to a pressure, it deflects. This deflection is proportional to the applied pressure when calibrated.

Generally there are two types of bellow gauges namely

1. Bellows gauge to measure gauge pressure.
2. Bellows gauge to measure differential pressure.

The bellows element is cylindrical in shape and the wall of this cylinder is thin and corrugated. The wall of the bellow is about 0.1 mm thick and is made of some sparingly material such as stainless steel, brass or phosphor bronze. This bellows element is open at one end to receive the applied pressure and is closed at its other end. This other end is usually attached with a rod. In many cases a spring is placed inside the bellow to enable the bellows to regain its original shape when the applied pressure is relieved.

(i) **Bellow Gauge to Measure Gauge Pressure:** Bellows gauge used to measure pressure consist of the main parts -

1. Bellows whose one end is fixed and open to receive the applied pressure.
2. A rod
3. A spring
4. A link-sector-pinion arrangement
5. A pointer, over a pressure calibrated scale

Construction

The bellows are constructed such that one end is fixed and the other open to receive the applied pressure. The other end of the bellows is closed and attached to a rod externally. A

More Instruments for Learning Engineers

A **wind vane** (or weathercock) is a device that shows the direction of the wind. However, not all weather vanes have pointers. Although partly functional, weather vanes are mostly decorative, often featuring the traditional cockrel design. Other common areas where this vane is found are ships, arrows, chariots and horses etc. The name 'vane' comes from the classical English word fana means 'flag'.

Wind Vane

spring is placed inside the bellows. To this rod a link-sector-pinion arrangement is attached as shown in Figure 6.17. A pointer is attached to this arrangement, which sweeps over a pressure calibrated scale.

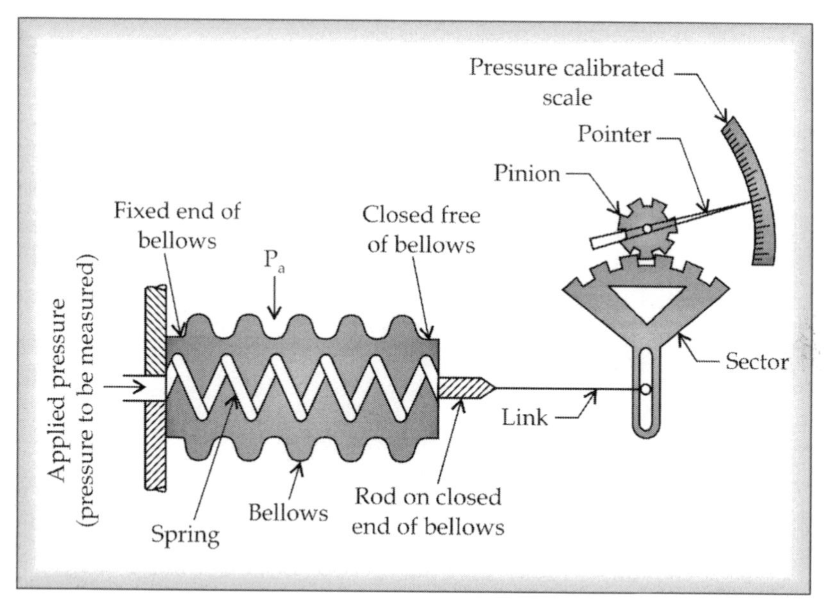

Fig. 6.17 Bellows Gauge to Measure Gauge Pressure

Operation

The pressure to be measured is applied to the fixed open side of the bellows. The applied pressure makes the bello ws expand length wise causing a linear displacement of the rod fixed at its closed free end. Obviously, this linear displacement is proportional to the applied pressure. As the rod is connected to a link-sector-pinion arrangement, the linear displacement of the rod is magnified and converted to a rotary motion of the pinion. As the pinion rotates, it makes a pointer attached to it to sweep over a scale calibrated to read pressure directly. As the pressure outside the bellows is usually atmospheric (P_a) and less than the applied pressure, the bellows gauge reads gauge pressure.

(ii) **Bellows Gauge to Measure Differential Pressure:** The main parts of the bellows gauge used to measure differential pressure are

1. Two bellows
2. Springs
3. The rods on the closed end of these two bellows
4. Equal arm level
5. A link-sector-pinion arrangement
6. A pointer over a calibrated scale.

Construction

Two bellows are fitted with springs and rods at closed ends as shown in Figure 6.18. The rods on the closed end of these two bellows are attached to an equal arm level as shown in

Figure 6.18. To this equal arm level a link-sector-pinion arrangement is attached which in turn is attachment to a pointer which sweeps over a scale.

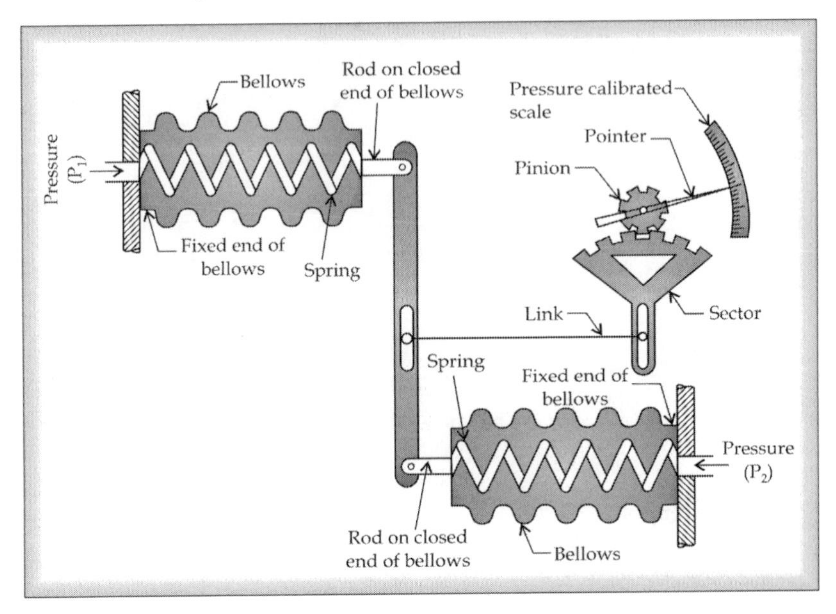

Fig. 6.18 Bellows Gauge Measure Differential Pressure

Operation

The pressure P_1 and P_2 are connected to the bellows as shown in Figure 6.19. When $P_1 = P_2$ both the bellows will expand by the same amount causing the equal arm lever to rotate. But this will not cause any movement of the sector pinion arrange-ment. The pointer will read zero on the scale. When $P_1 > P_2$ or $P_2 > P_1$ that is under differential pressure, the two bellows will expand by unequal amounts causing the equal arm lever to rotate. Now due to the movement of the equal arm lever, the link pushes the sector causing the pinion to rotate, when the pinion rotates, the pointer attached to it sweeps over the scale indicating both the direction and the magnitude of the differential pressure.

Merits (Advantages)

- They can measure both gauge and differential pressures.
- They are not very costly.
- They are simple and rugged in construction.

Demerits (Limitations)

- Zero shift problems exist
- Cannot be used for high pressure measurement.
- The springs used in the bellows are difficult to be designed.
- Temperature compensation is a must.

Applications of Bellows Gauges

- They are generally used for measuring medium and low pressures. They are widely used in low pressure measurement.

6.10 STRAIN GAUGE PRESSURE CELLS

Principle

When a closed container is subjected to pressure, it gets strained. The measurement of this strain with a secondary transducer like a strain gauge becomes a measure of the applied pressure.

More clearly, if strain gauge is attached to the container subjected to the applied pressure, the strain gauge also would change in dimension. Depending on change in dimension, the strain gauge will make its resistance to change. This change in resistance of the strain gauge becomes a measure of the pressure applied to the container (elastic container or cell).

There are two types of strain gauge pressure cells namely;

1. Flattened tube pressure cell
2. Cylindrical type pressure cell.

1. Flattened Tube Pressure Cell (Pinched Tube): The main parts of this arrangement are

1. An elastic tube, which is flat and pinched at its two ends
2. Two strain gauges
3. Closing plate
4. Resistance measurement device

Construction

Two strain gauges, one on the top and the other at the bottom are placed on an elastic tube, which is flat and pinched at its two ends as shown in Figure 6.19. One end of the elastic tube is open to receive the applied pressure and its other end is closed.

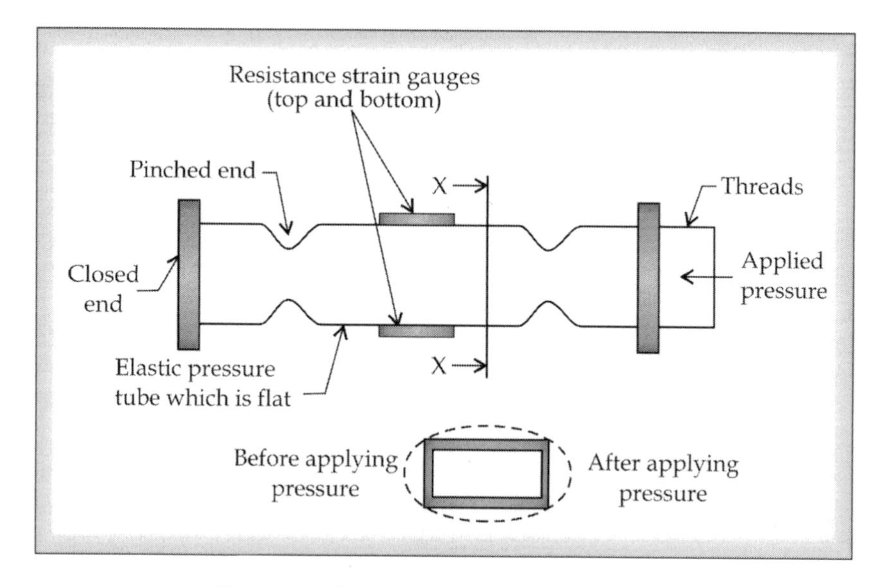

Fig. 6.19 Flattened Tube Pressure Cell

Operation

The pressure to be measured is applied to the open end of the tube. Due to the pressure, the tube tends to round off, i.e., its dimension changes (strained). As the strain gauges are mounted on the tube, the dimension of the strain gauges also changes proportional to the change in dimension of the tube, causing a resistance change of the strain gauges. The change in dimension of the tube is proportional to the applied pressure. Hence the measurement of the resistance change of the strain gauges becomes a measure of applied pressure when calibrated.

2. **Cylindrical Type Pressure Cell:** The main parts of this arrangement are
    1. A cylindrical tube with a hexagonal step at its centre.
    2. A threaded cap to close
    3. Two sensing resistance strain gauges
    4. Two temperature compensation strain gauges.
    5. Resistance measurement device

Construction

The cylindrical tube with a hexagonal step at its centre is threaded at its bottom, so as to fix this device on to the place where pressure is to be measured. The top portion of this cylindrical tube is closed and has a cap screwed on it. On the periphery of the top portion of the cylindrical tube two sensing resistance strain gauges are placed. On the cap (unstrained location) two temperature compensation strain gauges are placed.

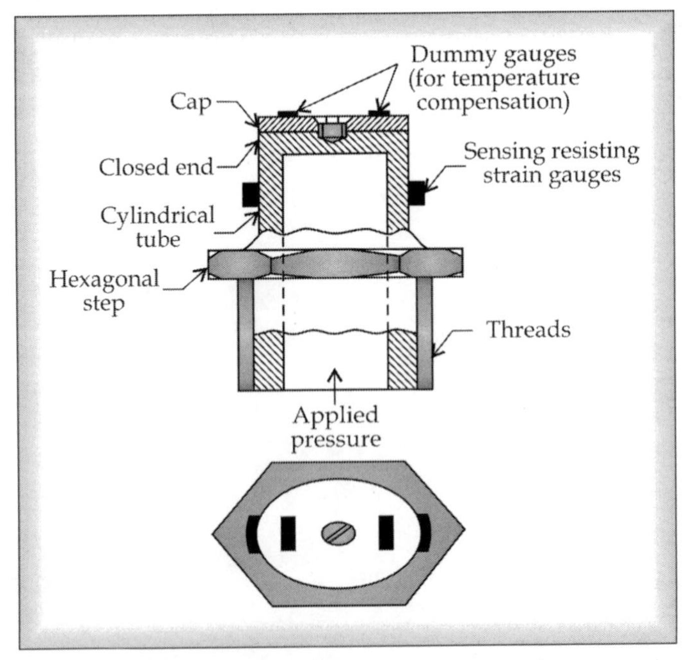

Fig. 6.20 Cylindrical Type Pressure Cell

Operation

The pressure to be measured is applied to the open end of the cylindrical tube. Due to the pressure, the cylindrical tube is strained, i.e., its dimension changes. Consequentially, the dimensions of the sensing strain gauges mounted on the cylindrical tube, also change proportional to the change in dimensions of the cylindrical tube, which in turn cause a resistance change of the strain gauges. As the change in dimension of the cylindrical tube and hence the resistance of the strain gauge is proportional to the applied pressure, the measurement of this resistance change of the sensing strain gauges becomes a measure of the applied pressure when calibrated.

Merits/Demerits/Applications

- The flattened tube pressure cells are used for low pressure measurement.
- The Cylindrical type pressure cells are used for medium and high pressure measurement.

SELF ASSESSMENT QUESTIONS-6.4

1. Describe the principle of pressure measurement using diaphragm gauges.
2. With the help of a diagram Explain the construction and operation of bellow gauge to measure gauge pressure.
3. Explain the basic principle involved in the strain gauge pressure cells.
4. What are the different types of bellow gauges? Explain each of them in detail.
5. Explain the construction and operation of Flattened Tube Pressure Cell with a sketch
6. Give the applications, advantages and limitations of elastic diaphragm gauges.
7. With the aid of a sketch explain the construction and operation of Cylindrical Type Pressure Cells
8. Enumerate the applications of strain gauge pressure cells.
9. With the help of a diagram explain the construction and operation of bellow gauge to measure differential pressure.
10. What are the different types of strain gauge pressure cells? Explain each of them.

6.11 THE McLEOD GAUGE

Operation of the McLeod gauge is based on Boyle's law, which is given by

$$P_1V_1 = P_2V_2$$

where P_1 &P_2 are initial and final pressures and V_1 & V_2 are initial and final volumes respectively.

Construction

The McLeod gauge is an integrated construction with tube made up of glass as shown in the Figure 6.21. It contains a capillary tube (C), bulb (B), vacuum space, reference column (R) as shown. It is connected to a movable reservoir with flexible pipe. The construction can be easily understood by Figure 6.21.

Operation

Initially the Hg level is made to fall below the junction of Bulb (B) and reference column (R) by releasing the pressure Hg in the Reservoir. At this condition the Bulb (B), and capillary (C) are at the same pressure as the vacuum source (P). Then by rotating the knob the pressure of Hg is made to raise until its level touches the reference mark in the reference column i.e., the Hg level in reference column is at zero point. Under this condition the volume remaining in the capillary is read directly from the scale and the difference in heights of the two capillary and reference column is the measure of the trapped pressure.

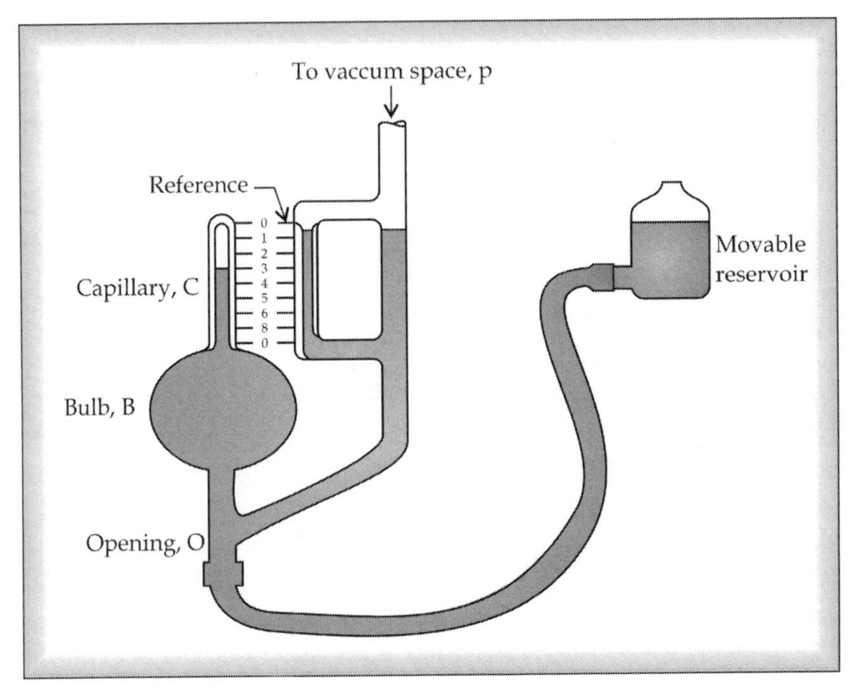

Fig. 6.21 McLeod Gauge

The initial pressure is calculated in the following way

Let

(i) The volume of the capillary tube per unit length = a

(ii) Length of the capillary tube occupied by the gas = y

(iii) Volume of the capillary tube, bulb & tube down to the opening = V_B

(iv) Pressure of the gas trapped in capillary tube = P_c

Then volume of the gas in the capillary tube occupied by the gas

$$V_c = ay$$

If we assume thermal (constant temperature) compression of the gas in the capillary tube then

$$P_c V_c = pV_B \quad \text{or} \quad P_c = \frac{pV_B}{V_C}$$

Substituting $V_c = ay$ in the above equation

We have, $$P_c = \frac{pV_B}{ay}$$

But the pressure indicated by the capillary tube is

$$p_c - p = y$$

Hence,
$$\frac{pV_B}{ay} - p = y$$

$$p\left(\frac{V_B}{ay} - 1\right) = y$$

$$p\left(\frac{V_B - ay}{ay}\right) = y$$

$$p = \frac{ay^2}{V_B - ay}$$

For most cases $ay < V_B$

Hence
$$P = \frac{ay^2}{V_B}$$

$$P = \frac{ay^2}{V_B - ay} = \frac{yV_C}{V_B - V_c}$$

6.11.1 Formula for Errors

Formula for error, fractional error and % fractional error can be derived by using

$$p = \frac{ay^2}{V_B} \text{ instead of } P = \frac{ay^2}{V_B - ay}$$

For calculating pressure in McLeod gauge.

Actually
$$P = \frac{ay^2}{V_B - ay} = \frac{y\ V_C}{V_B - V_C} \quad (\because ay = V_C)$$

But $ay \ll V_B$ or $V_B \gg V_C$

So, practically,

$$\therefore \quad p = \frac{ay^2}{V_B} = \frac{y\ V_C}{V_B}$$

Error = (Actual-practical) = Difference of above expressions i.e., $P - p$

$$\frac{yV_C}{V_B - V_C} - \frac{yV_C}{V_B} = \frac{V_B(y\ V_C) - y\ V_C(V_B - V_C)}{V_B(V_B - V_C)}$$

$$= \frac{yV_C V_B - y\ V_C V_B + y\ V_C^2}{V_B(V_B - V_C)} = \frac{yV_C^2}{V_B(V_B - V_C)}$$

$$\text{Fractional error} = \frac{\dfrac{yV_C^2}{V_B(V_B - V_C)}}{\dfrac{y\,V_C}{(V_B - V_C)}} = \frac{y\,V_C^2}{V_B(V_b - V_C)} \times \frac{(V_B - V_C)}{y\,V_C}$$

$$\text{Fractional Error} = \frac{V_C}{V_B}$$

ILLUSTRATION-6.6

A McLeod gauge has V_B=100 cm^3 and a capillary diameter of 1 mm. calculate the pressure indicated by a reading of 400 cm. What error would result if $P = \dfrac{ay^2}{V_B}$ is used ?

Solution:

$$V_B = ay = 100\text{ cm}^3 = 100\,(10\text{ mm})^3 = 100 \times 10^3 = 10^5\text{mm}^3$$

Capillary tube diameter (d) =1 mm

Length of the gas in capillary tube = 4 cm = 40 mm

$$\text{Volume of the capillary tube }(V_C) = \frac{\pi}{4} \times 1^2 \times 10 = 10p\text{ mm}^3$$

$$= 10 \times 3.14 = 31.4\text{ mm}^3$$

$$\therefore \quad P = \frac{ay^2}{V_B - ay} = \frac{y \times ay}{V_B - ay} = \frac{yV_C}{V_B - V_C} = \frac{(40.00)(31.4)}{10^5 - 31.4} = \frac{1256}{99.968.6}$$

$$= 0.01263945\text{ of Hg}$$

$$= 0.0126\text{ Torr [1 mm of Hg = 1 Torr]}$$

$$= 0.0126 \times 133.3\text{ [1 mm of Hg} = 133.3\text{ P}_a]$$

$$= 1.675\text{ Pa}$$

$$= 1.675\text{ b N/m}^2\text{[1 Pa} = 1\text{ N/m}^2\text{]}$$

The error in calculation if $\dfrac{ay^2}{V_B}$ is used instead of $p = \dfrac{ay^2}{V_B - ay}$

$$\frac{y \times V_C}{V_B} = \frac{40 \times 31.4}{10^5} = 0.01256\text{ mm of Hg}$$

$$\text{Error} = 0.01263945 - 0.01256 = 3.95 \times 10^{-6}$$

$$= 0.000004\text{ mm of Hg.}$$

$$\text{Or error} = \frac{y\, V_C^2}{V_B(V_B - V_C)} = \frac{40 \times (31.4)^2}{10^5(10^5 - 31.4)} = 3.95 \times 10^{-6}$$

$$= 0.000004 \text{ mm of Hg}$$

$$\text{Fractional error } \frac{V_C}{V_B} = \frac{31.4}{10^5} = 3.14 \times 10^4$$

$$= 0.000314 \times 100 = 0.0314$$

6.12 THERMAL CONDUCTIVITY GAUGES

Principle

A conducting wire gets heated when electric current flows through it. The rate at which heat is dissipated from this wire depends on the conductivity of the surrounding media (i.e., lower the pressure of the surrounding media, lower will be its density). If the density of the surrounding media is low, its conductivity also will be low causing the wire to become hotter for a given current flow and vice versa.

The two important thermal conductivity gauges are:

1. The pirani gauge
2. Thermocouple type conductivity gauge.

6.12.1 Pirani Gauge

At lower pressure the effecting thermal conductivity of gases decreases with pressure. The pirani gauge is a device that measures the pressure through the change in thermal conductance of the gauge.

Construction

A pirani gauge chamber encloses a platinum filament. A compensating cell is to minimize variation caused due to ambient temperature changes. The pirani gauge chamber and the compensating cell are housed on a Wheatstone Bridge circuit as shown in Figure 6.22.

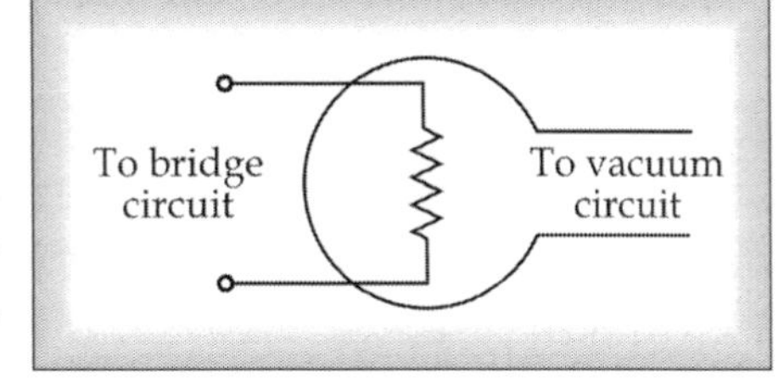

Fig. 6.22 Pirani Gauge

Operation

As shown in Figure 6.22 an electrically heated filament is placed inside the vacuum space. The heat loss from the filament is dependent on the thermal conductivity of the gas and the filament temperature.

The lower the pressure, the lower is the thermal conductivity and consequently, the higher is the filament temperature for a given electric-energy input.

The temperature of the filament could be measured by a thermocouple, but in the pirani type gauge the measurement is made by observing the variation in resistance of the filament material (tungsten, platinum etc). This may be done by an appropriate bridge.

The circuit problem with the above type of arrangement is that it does not consider the heat loss from the filament, which is also function of the ambient temperature.

To avoid the above difficulty two gauges are connected in series as shown in Figure 6.23 i.e., this arrangement compensates for possible variations in the ambient conditions.

- In this case both the measurement gauge and Scaled gauge are evacuated and exposed to the same environment conditions.
- The bridge circuit is then adjusted by moving R_2 to provide a null condition.

Now the measurement gauge is subjected to vacuum whose pressure is to be measured. The movement from the null deflection indicates the vacuum pressure compensating for changes in environment temperature.

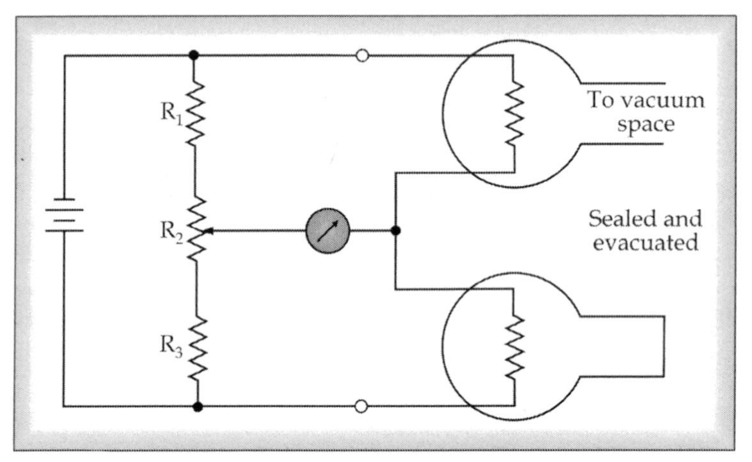

Fig. 6.23 Pirani-gauge Arrangement to Compensate for Change in Ambient Temperature

Merits

- They are rugged and inexpensive.
- Give accurate results.
- Good response to pressure changes.
- Relation between pressure and resistance is linear within the range.
- Readings can be taken from a distance.

Demerits

- The transient response is poor.
- They must be checked frequently.
- They must be calibrated for different gases.
- Electric power is a must for its operation.

Applications

- Used to measure low vacuum and ultra high vacuum pressures of the order of 0.1 to 100 Pa.

6.12.2 Thermocouple Type Conductivity Gauge

The main parts of this gauge are:

1. Chamber open one side
2. Filament
3. Battery
4. Milli-voltmeter
5. Rheostat
6. Thermocouple
7. Ammeter

Construction

A chamber has one side open to receive the applied pressure (usually vacuum). Inside the chamber, a filament is placed, which in turn is connected to a rheostat, ammeter and battery. On the filament a thermocouple is welded to measure the temperature of the filament. The thermocouple is connected to a milli volt meter. The arrangement is shown in Figure 6.24.

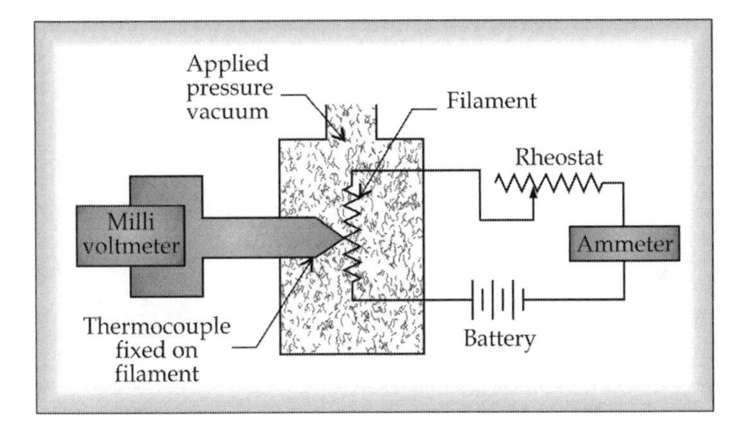

Fig. 6.24 Thermocouple Type Conductivity Gauge

Operation

A constant current is passed through the filament in the chamber. Due to this current, the filament gets heated and the filament temperature is sensed by the thermocouple welded to the filament. Now the pressure to be measured is connected to the chamber. Due to the applied pressure the density of the surrounding of the filament changes. Due to this change in density of the surrounding of the filament, its conductivity changes cause the temperature of the filament to change. This change in temperature of the filament is sensed by the thermocouple welded to the filament, which in turn becomes a measure of the applied pressure when calibrated. The milli-volt meter can be calibrated to directly read the applied pressure.

Merits

- They are less expensive and rugged.
- Relation between pressure and temperature is linear for the range of use.
- Readings can be taken from a distance.
- Easy to use.

Demerits

- Filament gets burn out frequently.
- They must be calibrated for different gases.
- Electrical Power is must for its operations.

Application

- Used to measure low vacuum and ultra high vacuum pressures.

6.12.3 Ionization Gauge

Ionization: Ionization is the process of removing an electron from an atom and thus producing a free electron and positively charged ion.

Construction

This gauge is very similar to the ordinary triode vacuum tube. A cathode grid and plate or anode placed in a chamber. The chamber is open to one end to receive the applied pressure. The grid is maintained at positive potential. The plate (anode) is negatively biased.

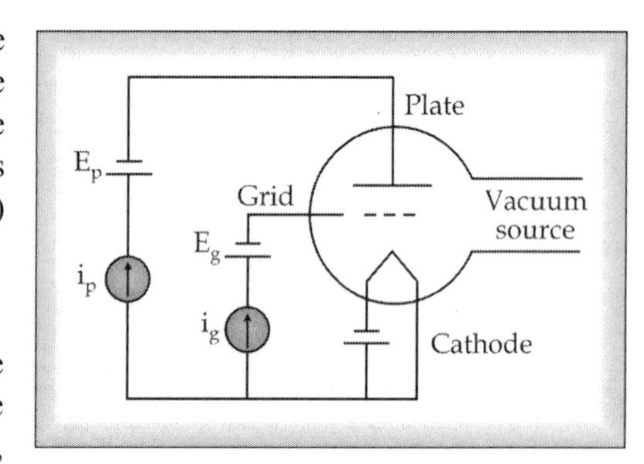

Fig. 6.25 Ionization Gauge

Operation

The pressure to be measured is connected to the chamber. The grid draws electrons from the cathode and these electrons collide with the gas molecules, thereby causing ionization of the gas molecules. The positively charged molecules are then attracted to the plate (anode), causing a current flow in the external circuit. The measurement of plate current (i_p) and grid current (i_g) becomes a measure of the applied gas pressure. The pressure of the gas is proportional to the ratio of the plate current to grid current.

It is given by $\frac{1}{S}\frac{i_p}{i_g}$ where S = sensitivity of gauge.

Merits

- Measurement can be done from a distance.
- Fast response to pressure changes.
- Have good sensitivity.

Demerits

- The filament burns out quickly.
- Filament temperature should be controlled properly.
- It has to be calibrated for different gases.
- Some gases might get decomposed due to the hot filament.

Application

- They are used to measure low vacuum and ultra high vacuum pressure, which are of the order of 0.13 to 1.3×10^{-6} Pa·

Self Assessment Questions-6.5

1. What are the different types of thermal conductivity gauges? Explain each of them
2. Explain the construction and working principle of pirani gauge.
3. List out the applications, advantages and limitations of pirani gauge.
4. Explain the construction and working principle of thermocouple type conductivity gauge.
5. Give the advantages, limitations and applications of ionization gauge.
6. With the aid of a diagram describe the operation of ionization gauge.
7. Narrate the construction and operation of McLeod gauge.
8. Derive an expression for calculating pressure using McLeod gauge.
9. Give the advantages, limitations and applications thermocouple type conductivity gauge.

SUMMARY

This chapter deals with the instrumentation of pressure measurement devices in terms of their working principles, constructions, operations, merits, demerits and applications. These devices make use of gravitational and elastic transducers. The gravitational transducers are used in a dead weight tester, manometers such as piezometer, U-tube manometer, vertical/ inclined single column manometer and U-Tube/Inverted U-tube differential manometer. The pressure is conveniently converted into displacement by the weight of the liquid used in these devices. With slight modifications, ring balance manometer and single/two element bell type pressure gauges work on balancing the gravitation.

Elastic Transducers which when are subjected to a pressure, deflects proportional to the applied pressure, are used in bourdon tube pressure gauge, elastic diaphragm gauges and bellows gauges. When a closed container is subjected to pressure, it gets strained; transducer like a strain gauge becomes a measure of the applied pressure in strain gauge pressure cells, flattened tube pressure cell (pinched tube) and cylindrical type pressure cell.

McLeod vacuum gauge operates on Boyle's law. The rate of heat dissipated from the electrically heated wire depends on the conductivity of the surrounding media pressure and hence the density in thermal conductivity gauges such as pirani- gauge, thermocouple type conductivity gauge. Ionization gauge use the principle of ionization due to pressure.

KEY CONCEPTS

Pressure: Force per unit area exerted by a fluid on a containing wall.

Units of Pressure: Newton per square meter (N/m^2) or Pascal (Pa); kgf/m^2, pounds per square inch (psi), atmosphere (atm), mm of Hg, cm of Hg, Bar, Torr, absolute atmosphere (ata) etc.

Atmospheric Pressure (P_a): The pressure exerted by the envelope of air surrounding the earth's surface, usually expressed in terms of the height of a column of fluid (Hg), at 20 °C. 1 atmosphere (atm) = 1.01325×10^5 N/m^2 (or Pa) = 760 mm of Hg; 1 Bar = 10^5 Newton/m^2 (l00 kPa); 1 Torr = 1mm of Hg.

Gauge Pressure ($P > P_a$): The pressure measured by the instrument when the unknown pressure (P) is greater than the atmospheric pressure (P_a).

Vacuum Pressure ($P_a > P$): The pressure measured by the instrument when the atmospheric pressure (P_a) is greater than the unknown pressure (P).

Absolute Pressure (P_{abs}): The force, which the fluid particles exert due to interaction amongst themselves per unit area. $P_{abs} = P_a - P_g$; $P_{abs} = P_a - P_{vac}$.

Static Pressure: The pressure exerted on the walls of a pipe due to a fluid at rest inside the pipe or due to the flow of a fluid parallel to the walls of the pipe.

Total or Stagnation Pressure: The pressure which is obtained by bringing the flowing fluid to rest isentropically.

Dynamic or Impact or Velocity Pressure: The pressure due to fluid velocity = Total pressure – Static pressure.

Static head: $h_{static} = \dfrac{p}{\rho g}$; **Total head:** $h_{total} = \dfrac{p}{\rho g} + \dfrac{v^2}{2g}$

Gravitational Transducer: The transducer that makes use of the gravitational forces and the weight as the main component that plays key role in transduction.

Elastic Transducers: The transducer that makes use of the elastic members such as springs, bourdon tube or bellows to convert the force into the displacement.

Dead Weight Tester: A pressure producing and measuring device used to calibrate pressure gauges.

Monometer: Device used for measuring the pressure at a point in a fluid by balancing the column of fluid by a same or another column of the fluid.

Simple Manometers: Glass tube having one of its ends connected to a point where pressure is to be measured and other end remains open to atmosphere.

Piezometer: One end of this manometer is connected to the point where pressure is to be measured and other end open to the atmosphere.

U-Tube Manometer: Glass tube bent in U shape, one end connected to a point at which pressure is to be measured and other remains open to the atmosphere.

Single Column Manometer: Modified form of U-tube manometer in which reservoir having a large cross sectional area (about 100 times) as compared to the area of the tube connected to one of the limbs of the manometer.

Differential Manometers: Devices used for measuring the difference of pressures between two points in a pipe or in two different pipes (consists of U-tube containing a heavy liquid in which two ends are connected to the points whose difference of pressure is to be measured).

Inverted U- Tube Differential Manometer: An inverted U-tube filled with a light liquid and the two ends connected to the points whose difference of pressure is measured.

Mechanical Displacement Type Gauges: The pressure gauges contain a liquid that merely acts as a pressure seal.

Ring Balance Manometer: The hollow toroidal tubular chamber divided into two parts by splitting and sealing, partially filled with kerosene or paraffin, and pivoted at the center, balanced by a weight attached to the ring. A differential gas pressure, when applied through flexible leads to the partition, tends to tilt the ring against the resisting moment exerted by the balance weight.

Bell Type Pressure Gauges: The applied pressure is balanced against a restricting spring or against a counter weight. (A single element bell gauge consists of a bell immersed in a sealing liquid; with pressures applied on both outside and inside of the bell).

Bourdon Tube Pressure Gauge: When Bourdon tube is subjected to a pressure, it deflects proportional to the applied pressure when calibrated.

Tip Travel of Bourdon Tube: The motion of the free end of Bourdon tube, and is function of the tube length, wall thickness, cross sectional geometry and modulus of the tube material.

Elastic Diaphragm Gauges: When Diaphragm is subjected to a pressure, it deflects proportional to the applied pressure when calibrated.

Bellows Gauges: When Bellows is subjected to a pressure, deflects proportional to the applied pressure when calibrated.

Strain Gauge Pressure Cells: When a closed container is subjected to the applied pressure, it is strained, and a secondary transducer like a strain gauge measures the applied pressure.

Cylindrical Type Pressure Cells: A cylindrical tube with a hexagonal step at its centre helps in fixing this device on to the place where pressure is to be measured.

McLeod Gauge: Works on Boyle's law, $P_1V_1 = P_2V_2$

Thermal Conductivity Gauge: Heat rate dissipated from a conducting wire gets heated when electric current flows through it depending on the conductivity of the surrounding media (i.e., lower the pressure of the surrounding media, lower will be its density).

Pirani Gauge: Measures the pressure through the change in thermal conductance of the gauge.

Thermocouple Type Conductivity Gauge: A chamber whose one side is open to receive the applied pressure (usually vacuum), inside the chamber, a filament is placed, which in turn is connected to a rheostat, ammeter and battery. On the filament a thermocouple is welded to measure the temperature of the filament.

Ionization Gauge: Device that measures pressure using ionization principle.

Ionization: The process of removing an electron from an atom and thus producing a free electron and positively charged ion.

REVIEW QUESTIONS

Short Answer Questions

1. Define pressure. What are the units of pressure in SI system?
2. What are the units of vacuum pressure?
3. If a Barometer reads 720 mm of $H_{g,}$ what is the atmospheric pressure in torr?
4. What is the conversion factor between bar and torr?
5. What is absolute pressure?
6. What is gauge pressure?

7. What is vacuum pressure?
8. Explain any one high pressure measuring device.
9. What is a bell type manometer and what is the advantage of this manometer over U tube manometer?
10. Differentiate between atmospheric pressure, gauge pressure and vacuum pressure.
11. Distinguish between static pressure and stagnation pressure.
12. Explain ring type mechanical pressure gauge.
13. Explain bell type mechanical pressure gauge.
14. Discuss the merits & demerits of elastic sensing elements
15. What are the instruments used for measurement of low pressure and low vacuum pressure?
16. With the aid of neat sketch, explain the working principle of dead -weight type tester.
17. What is a dead weight tester?
18. What are the two factors that limit the accuracies of a dead weight -tester?
19. What is the principle of conversion in a Bourdon tube pressure gauge?
20. Is it possible to measure vacuum pressure using a Bourdon tube pressure gauge? If so, how? If not, why not?
21. List out the merits and demerits of the Bourdon tube pressure sensing elements.
22. What are diaphragm and bellow gauges?
23. In a diaphragm gauge, up to what limit the pressure and deflection will have a linear relation?
24. What is the principle of conversion in a diaphragm gauge?
25. How do you increase the linear range of a diaphragm gauge?
26. What is the specialty of a diaphragm gauge?
27. List the limitations of elastic diaphragm gauge.
28. Explain the basic principle of working of bellow type pressure gauge.
29. What is a differential pressure cell?
30. List the limitations of a Mc leod vacuum gauge used to measure pressure.
31. What is a Mc Leod gauge and what is its operating range?
32. What is an Ionization gauge?
33. How pressure of the gas related to plate current and grid current?
34. What is the lowest pressure that an Ionization gauge can measure?
35. What are the disadvantages of an Ionization gauge?
36. How does an Ionization gauge differ from a pirani gauge?

Long Answer Questions

1.. Describe the construction of a U-tube manometer and explain how it can be used for measurement of absolute, gauge and differential pressures.
2. Describe different sources of errors in U-tube manometer and how corrections can be applied to minimize these errors.
3. Explain with neat sketch the principle of working of manometer.
4. Explain how sensitivity can be increased using inclined tube manometer. Describe its Construction, advantages and limitations.
5. Describe the construction and working of a Bourdon tube.
6. Discuss the working of Bourdon tube pressure gauges with relevant sketch and mention their merits and demerits.
7. What are the essential components in a Bourdon tube pressure gauge? Explain.

8. Describe the construction and working or c-type, spiral type and helical type Bourdon gauges with neat diagrams.
9. What are the errors in Bourdon tube and how can they be rectified?
10. Explain how an elastic diaphragm gauge is used to measure pressure with the help of relevant sketch.
11. Describe the application, advantages and limitations of elastic diaphragm gauges.
12. With the aid of neat sketches. Explain the principle of operation of various diaphragm gauges.
13. Explain how an elastic diaphragm gauge is used to measure pressure with the help of a relevant sketch.
14. Discuss the merits and demerits of the pressure sensing elements as Bellows.
15. Explain how pressure can be measured using Bellows gauge.
16. Illustrate the bellows arrangement used to measure differential pressure.
17. Elucidate the basic principle of operation of Mc leod vacuum gauge with necessary diagram.
18. Explain with neat sketch the principle of working of Mc leod gauge
19. What are the instruments used for measurement of low pressure and low vacuum pressure.
20. How do you compensate ambient temperature variation in a Pirani-thermal conductivity gauge?
21. Describe the working of a Pirani gauge.
22. Explain the operation of Pirani thermal conductivity gauge for pressure measurement.
23. Describe the method used for measurement of low pressure using thermocouple vacuum gauge.
24. Describe the method used of measurement of low pressure using lionization type vacuum gauge.
25. Discuss various types of elastic pressure sensing elements used in electrical transducers
26. With the help of a neat circuit diagram, explain the working principle of an Ionization gauge, specify the range & explain the characteristics.
27. Discuss the merits and demerits of electric sensing elements.
28. Explain the operation of strain gauge pressure cells.

NUMERICAL PROBLEMS

1. The right limb of a simple U tube manometer containing mercury is open to the atmosphere while the left limb is connected to a pipe in which a fluid of sp gr. 0.9 is flowing. The center of the pipe is 12 cm below the level of mercury in the right limb. Find the pressure of fluid in the pipe if the difference of mercury level in the two levels is 20 cm [2.597N/cm^2].
2. Describe the functioning of a Mcleod gauge with the help of a diagram. Derive an expression for pressure of vacuum source. State the assumptions while deriving this equation and comment on the error due to these assumptions.
3. A Mc leod gauge has V_B = l00 cm^3 and a capillary diameter of 0.75 mm. Calculate the pressure indicated by a reducing of 3.00 cm what error would result if $P = ay^2 / v_B$ is used.
4. A single column manometer is connected to a pipe containing a liquid of sp gr 0.9 as shown in Figure (A). Find the pressure in the pipe if the area of the reservoir is 100 times the area of the tube for the manometer reading. The specific gravity of mercury is 13.6

5. A differential manometer is connected at the two points A and B of two pipes as shown in Figure (B). In the pipe A contains a liquid of sp gr = 1.5, while pipe B contains a liquid of sp the differential manometer.

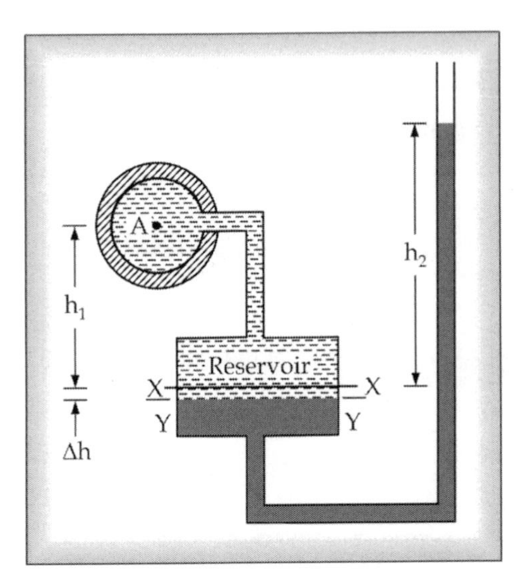

Fig. (A) Single Column Manometer

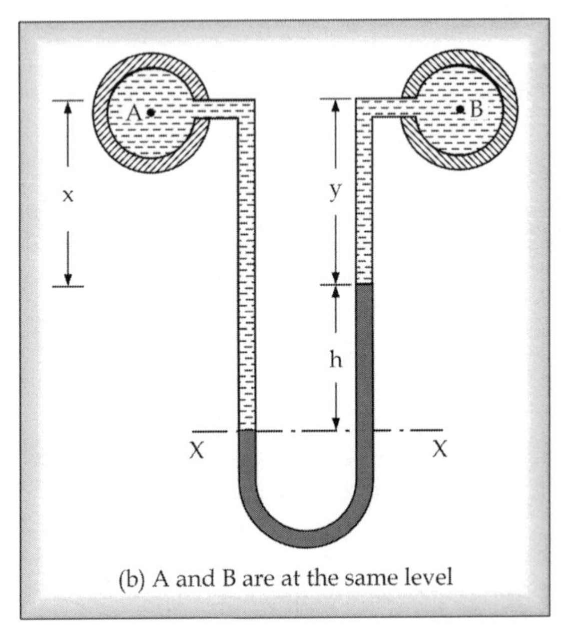

(b) A and B are at the same level

Fig. (B) Differential Manometer

MULTIPLE CHOICE QUESTIONS

1. 1 Torr = ______ mm of Hg
   (a) 1 (b) 10^2
   (c) 10^3 (d) 10^5
2. The ring balance manometer essentially uses a hollow ______ tube
   (a) Toroidal (b) spherical
   (c) Plastic (d) Glass
3. A pressure reading below the atmospheric pressure is known as _____ Pressure
   (a) Vacuum (b) Rarefaction
   (c) Negative (d) Gauge pressure
4. Manometric liquids should have
   (a) High coefficient of thermal expansion
   (b) Low coefficient of thermal expansion
   (c) Can have high or low coefficient of thermal expansion
   (d) High thermal conductivity
5. The cross section of Bourdon type is ____.
   (a) Elliptical (b) Circular
   (c) Square (d) Rectangle
6. Which of the following type of bourdon tube shape has a small tip travel & necessitates amplifications?
   (a) C-type (b) Spiral
   (c) Helical shaped (d) Straight
7. Pirani gauge produces small changes in _____, which is sensed by means of Wheatstone bridge circuit.
   (a) Resistance
   (b) Ionization current
   (c) Thermal conductivity
   (d) E.M.F
8. The _____ pressure is measured with respect to atmospheric pressure as the

datum.

(a) Vacuum (b) Gauge
(c) Absolute (d) Rarefaction

9. Which pressure gauge consists of Pressure tight link lever mechanism?
(a) Bourdon tube pressure gauge
(b) Ionization pressure gauge
(c) McLeod vacuum gauge
(d) Diaphragms

10. Differential manometers are used to determine the pressure at __ points
(a) Many (b) Two
(c) Three (d) Single

11. There cannot be negative _______ pressure.
(a) Gauge (b) Absolute
(c) Atmospheric (d) Vacuum

12. A Capsule, used in pressure measuring instruments is a pair of____ diaphragms
(a) Corrugated (b) Flat
(c) Inclined (d) Spherical Ans: a

13. In bell type pressure gauges, change of spring length varies directly with
(a) Bell movement
(b) Density
(c) Cross section area
(d) Pressure difference

14. Stagnation pressure is defined as the pressure that would be obtained if the fluid stream were brought to rest in _______ process.
(a) Isentropic (b) Isobaric
(c) Isochoric (d) Isothermal

15. The dynamic pressure is due to fluid speed and is also known as_______ pressure.
(a) Impact (b) Vacuum
(c) gauge (d) Dry bulb

16. In manometers when both limbs are open to atmosphere then left limb pressure is _____ right limb pressure.
(a) equal to (b) less than
(c) greater than (d) unequal to

17. If the level of manometric fluid in the right limb is lower than the level in the left limb. Then the apparatus is used to measure________ pressure
(a) Vacuum (b) Gauge
(c) Impact (d) Atmospheric

18. In U-tube manometer if one of the leg is replaced by a large diameter well then it is called
(a) Single column manometer
(b) Differential manometer
(c) Piezometer
(d) Inverted U-tube manometer

19. Which of the following characteristics is not desired in manometric liquid
(a) High viscosity
(b) Low viscosity
(c) Low vapour pressure
(d) Negligible surface tension

20. Ring balance manometers are divided in to two plates by splitting and _______.
(a) Dividing (b) Sealing
(c) Marking (d) Cutting

21. In bell type pressure gauge, the pressure is balanced against a ______ movement.
(a) Gear (b) Spring
(c) Piston (d) Float

22. The action of mechanical gauges is based on the deflection of________ tube.
(a) Hollow (b) Circular
(c) Spherical (d) Spiral

23. A hairspring is used in bourdon gauges for fastening the spindle to free the system from
(a) Vibration
(b) Heavy friction
(c) Backlash
(d) Power consumption

24. The reference pressure in the casing containing the bourdon tube is usually atmospheric, so that the pointer indicates the _______ pressure.
(a) Vacuum (b) Atmospheric
(c) Gauge (d) Vapour

25. The motion of the free end of the bourdon tube is called________
    (a) Displacement (b) Tip travel
    (c) Deflection (d) Error
26. A simple flat diaphragm shows a undesirable characteristics known as_______
    (a) Oil- canning
    (b) Oil-leaking
    (c) Oil-transmission
    (d) Oil-separation
27. In McLeod gauge, the gas should obey ______
    (a) Newton's law
    (b) Boyle's law
    (c) Charles' law
    (d) Gay-Lussac's Law
28. In the thermal conductivity the heat conductivity decreases with ________ pressure
    (a) Increasing (b) Decreasing
    (c) Moderate (d) Fluctuating
29. For measuring higher pressure, the bourdon tube used are having ________ cross-section
    (a) Spherical (b) Circular
    (c) Flexible (d) Moving
30. Cathode ray oscilloscope is device for measuring any mechanical quantity subjected to _
    (a) Deflection (b) Pressure
    (c) Rapid fluctuation (d) Heat
31. In cathode ray oscilloscope the voltage applied to plates deflect the electron beam in Proportion to the magnitude of____
    (a) Voltage (b) Resistance
    (c) Current (d) Pressure
32. In cathode ray oscilloscope, the horizontal field so set up that deflects the _______
    (a) Horizontal beam
    (b) Vertical beam
    (c) Beam intensity
    (d) Electrodes

ANSWERS

1.	(a)	2.	(a)	3.	(a)	4.	(b)	5.	(b)
6.	(a)	7.	(a)	8.	(b)	9.	(a)	10.	(b)
11.	(a)	12.	(a)	13.	(a)	14.	(a)	15.	(a)
16.	(a)	17.	(a)	18.	(a)	19.	(a)	20.	(b)
21.	(b)	22.	(a)	23.	(c)	24.	(c)	25.	(b)
26.	(a)	27.	(b)	28.	(b)	29.	(b)	30.	(a)
31.	(a)	32.	(a)						

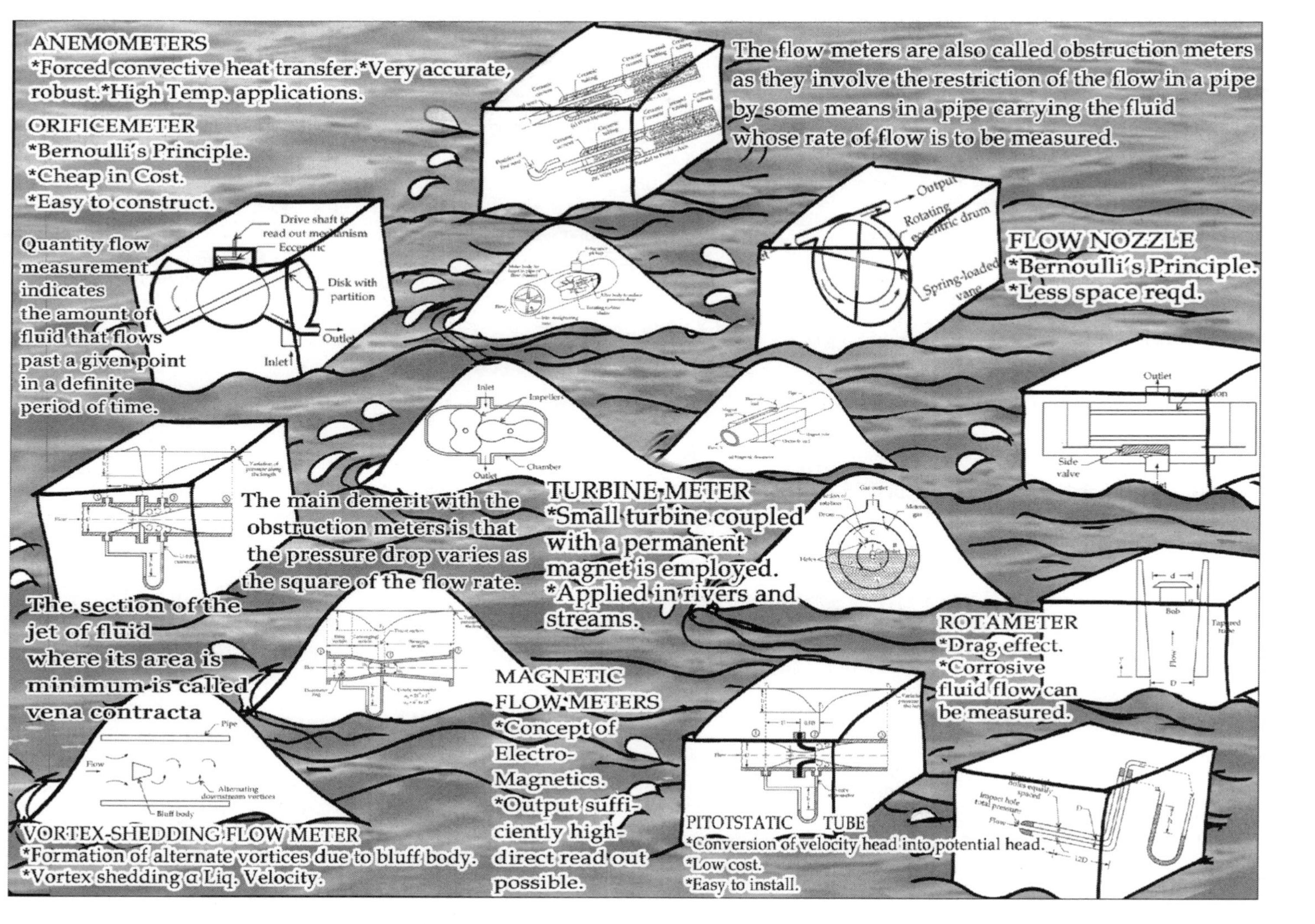

CHAPTER INFOGRAPHIC

7 Glimpses of Flow Measurement

S.No	Instrument	Principle	Construction	Operation	Application	Merits	Limitations
1. Primary or Quantity methods: In these methods the amount of fluid that flow past a given point in a definite period of time is measured. These are I(a) Weight meters, Weighters, Traps; I(b) Volumetric meters; I(c) Weight or volume tanks (These work on the weight or volume flown) I(d) Positive displacement meters							
I(d) (i)	Nutating disk meter	Nutating disk principle	Consists of disk with partition and mounting chamber	Water enters left side of meter & strikes eccentrically mounted disk must wobble/ nutate about vertical axis indicates vol.of liquid.	Used where directly volume of the liquid is required	Reliability is within 1%	Greater quantities of flow cannot be measured
I(d) (ii)	Rotary-vane flow meter	Positive displacement device	Consists of vanes which are spring loaded and maintain contact with the casing	As the eccentric drum rotates a fixed quantity of fluid is trapped in each section, which finds its way out to exit	Counter connected to shaft of eccentric drum records displaced fluid	Insensitive to viscosity & uncertainties about 0.5%	Greater quantities of flow cannot be measured
I(d) (iii)	Lobed-impeller meter		Consists impellers enclosed in a chamber	As the rotors rotate the fluid is trapped between them and is conveyed to the outlet as a result of their rotation.	Used for gas or liquid flow measurement	No. of rev. of the rotors indicates Vol. flow rate	Accurate fit between rotors and casing
I(d) (iv)	Reciprocating piston meter		Piston reciprocates in cylinder, inlet and outlet valves with open/ close controlled by slide valve and is actuated by piston itself by eccentric	By the movement of the piston, the fluid is passed from the inlet to outlet alternately through each end of cylinder.	Used for non-corrosive and low viscous liquids	Vol. of fluid flow varies as both stroke length & piston speed	Speed limited by inertia forces
I(d) (v)	Gas meter			Due to gas pressure the drum rotates sending metered quantity of gas enclosed between compartment to outlet	Used where directly volume of the gas is required	Drum revolutions directly give the amount of gas consumed	Water must be poured at reg. intervals as it evaporates
2. Flow meters: The principle of obstruction meters is that the flow is restricted by some means in a pipe carrying the fluid whose rate of flow is to be measured. The flow restriction causes a pressure drop which can be measured and when calibrated gives the flow rate							
2(i)	Venturimeter	Bernoulli's principle	Converging diverging portions joined by throat	When the fluid flows, pressure drop occurs and quantity of discharge is given by $Q=\{a_1a_2(2gh)^{1/2}\} / \{a_1^2 - a_2^2\}^{1/2}$	Horizontal and inclined orientation is used	CoD is more hence used frequently	
2(ii)	Orifice meter		A flat circular plate with a circular hole (orifice), concentric with the pipe	When the fluid flows, pressure drop occurs and quantity of discharge is given by $Q=\{a_0a_1(2gh)^{1/2}\} / \{a_1^2 - a_0^2\}^{1/2}$	Where there is space restriction to assemble Venturimeter	Cheaper than Venturimeter	Coefficient of discharge of the meter must be known
2(iii)	Flow-nozzle		Only converging part				

Contd....

S.No	Instrument	Principle	Construction	Operation	Application	Merits	Limitations
2(iv)	Variable area meter	Drag effect	A bob in a cylindrical pipe	The bob assumes a position in the tube where the forces acting on it are in equilibrium and this height gives an indication of the flow	Employed in vertical orientation and in industries	Can use on Corrosive fluids & capacity can be varied	Float not visible in opaque fluids & used only in vertical direction
3.	Turbine meter	Steam turbine	A small turbine wheel with permanent magnet is enclosed & is kept in a pipe where fluid flow is measured. Reluctance pick up is kept in pipe	As the flow takes place the turbine wheel rotates and a pulse is detected for revolution by reluctance pickup. The pulse rate is proportional to flow rate	Flow measurement of water in rivers and streams	The total pulse output is taken as an indication of total flow	Bearing maintenance low accuracy at low flow rates
4.	Magnetic flow meters	Voltage induced as magnetic field and flow direction are at 90°	A permanent magnet, pipe carrying fluid and electrodes placed perpendicular to the magnetic field	A conducting fluid when flowing through a magnetic field, a voltage is induced which gives a measure of flow rate	Where direct indication of flow rate is required	Output voltage is proportional to the flow rate	Amplification of output voltage is reqd in case of low conductivity fluids like water
5.	Vortex shedding flow meters	Vortices form when bluff body placed in flow stream	Bluff body placed in pipe where flow rate is to be measured	The liquid when flows over the bluff body the vortices are formed and the frequency of vortices is proportional to the liquid velocity which is measured by a piezoelectric sensor	Where fluid is flowing with high velocities	Where direct indication of flow rate or velocity of flow is required	Not suitable for highly viscous liquid
6. MEASUREMENT OF FLUID VELOCITIES: Flow rate is proportional to flow velocity; so measuring velocity gives flow rate. Probes are used to measure pressure at a particular point. The pressure exerted by the fluid when at rest is known as static pressure and when brought to rest is known as stagnation or total or impact pressure.							
7.	Pitot-static tube (Prandtil pitot-tube)	Pressure probe	Two concentrically placed tubes bent to form a right angled bend	The operation of pitot tube involves the introduction of tube into flow area where Pres. details are required. The P output is static & is less than that in inner tube pressure, which is total Pressure. The differential press is measured using a sensor that becomes measure of flow rate.	Pitot tubes are extensively used in laboratories to measure, velocity pressure and flow rates of fluids	Cost is very less and can be installed easily	Alignment problem, small flow rates, cannot be used for flow with suspended solids/ impurities
8. Direction sensing probes are also called as Yaw-angle probes							
9. A heated object when placed in a moving stream of fluid loses is that, which is dependent on the velocity of the fluid. If the object is electrically heated at a known power, it will reach a temperature determined by the rate of cooling. Thus its temperature will be a measure of the velocity. In either way, the heating power may be controlled by a feedback system to hold the temperature constant; In that case, heating power is a measure of velocity.							

Contd....

S.No	Instrument	Principle	Construction	Operation	Application	Merits	Limitations
10.	Hot wire anemometers	Anemometers	In these probes fine wire is heated electrically and placed in flow stream. The current reqd to maintain the temp becomes a measure of flow vel.	Fluid flowing over heated wire cools the wire and to maintain at a temperature current must be increased which becomes a measure of flow velocity	For measuring wind velocity	Very accurate	Strength is very less and flow flied must be clean
11.	Hot film anemometers	Anemometers	In probes thin film is heated electrically & placed in flow stream. The current required to maintain the temperature becomes a measure of flow	The fluid flowing over the heated film cools it; to maintain a particular temperature current must be increased which becomes a measure of flow	For measuring Wind velocity	Very accurate, robust & for high temp. applications	Strength is very less and flow flied must be clean
12.	**Scattering Measurement**: The differential Doppler system splits the laser into 2-equal intensity beams focused into an intersection point. A particle passing thru the intersection scatters light from both beams, and is collected by a photo-multiplier tube (PMT). The Doppler shift for each beam is different, by virtue of their different angles, but the intensities of the two scattered waves are now identical. The resulting beat frequency at the detector is just the difference in the Doppler shifts of the 2 beams that becomes a measure of flow of velocity.						
13.	Ultrasonic Anemometry	Doppler effect	Ultrasonic transmitter and receiver, a pair of piezoelectric or mangetostrictive materials clamped outside the pipe	As sound wave travels from source to receiver, vel.(C) in stationary fluid varies $C \pm V$ cosq, depending on wave moving up/downstream and measure flow	Measurement of flow rate in slurries and dirty liquids	Easy to use	Particles be present in the flow
14.	Flow visualization techniques are used to know the regions of separation, recirculation, pressure loss, velocity measurement and body contour.						
	Smoke Trails and Wire		Dye injection and chemical precipitates		Hydrogen bubble visualization		Laser induced fluorescence
15.	**Interferometer**: Used to obtain a direct measurement of density variations in the test section.		**Schlieren** is a device that indicates the density gradient; Shadow graph is used to view the flow phenomenon directly				
If 2-beams travel in different optical lengths, due of geometry of the system or refractive properties of any element of the optical paths, the beams will be out of phase and interfere, as a result fringes will be a function of the diff. in optical path lengths for the two beams by which we can measure of fluid velocity							

Chapter 7

Flow Measurement

STARTERS

To study this chapter, you should have awareness on the following concepts. For a better understanding, it is always a good idea to revise these prerequisites.

- Different types of flows such as laminar, transient and turbulent and the mathematical expressions connecting with them etc., and fundamentals of fluid mechanics.
- Relation between pressure and volume, Boyle's law
- Pressure and its significance in fluids
- Concept of water head, and Bernoulli's principle
- Conversion of pressure to stress, force on a fixed area
- Different types of pumps (reciprocating/centrifugal), types of turbines (impulse/reaction) and flow patterns in them.
- Awareness on impellers, vanes, traps and other fundamentals of fluid mechanics.
- The relation between pressure and fluid flow
- Effects of pressure on the volume of liquid and other parameters
- Preliminary idea on venturi, orifice, notch, nozzle, weir etc. and flow characteristics in these
- Broad idea on flow characteristics in pipes, friction losses in pipes etc.
- Concepts of nutation, trapping, and fundamentals of magnetism and electricity
- Some mechanisms to convert one type of motion to the other such as rotational to translatory or reciprocatory etc.
- Doppler effect and related topics in wave-mechanics
- Fundamentals of light, and concept of interference, ultrasonic waves
- Relation between flow and rate of cooling and basics of anemometry and
- First four chapters of this book.

LEARNING OBJECTIVE

After studying this chapter you should be able to

- Understand the nature of flow,
- Classify the flow measurements,
- Know the Measurement of Fluid Velocities,
- Explain Anemometers, and Flow-Visualization Techniques,
- Understand the interferometer, Schlieren device and
- Discuss the shadow graph and Calibration of Flow Measuring Devices.

7.1 INTRODUCTION

Measurement of fluid velocity, flow rate, and flow quantity with varying degree of accuracy are fundamental necessity in almost all the flow situations ranging from blood flow rates in a human artery to the flow of liquid oxygen in a rocket. The selection of the proper instrument for a particular application is governed by many variables, including cost. For many industrial operations the accuracy of a fluid-flow measurement is directly related to profit. A simple example is the gasoline pump at the neighborhood service station; another example is the water meter at home. A small error in flow measurement on a large natural gas or oil pipeline could make a difference of thousands of rupees over a period of time. Not only the laboratory scientists who are concerned with accurate flow measurement but an engineer in industry is also vitally interested because of the impact, the flow measurements may have on the profit and loss statement of the company.

7.2 NATURE OF FLOW

Various problems are involved in accurate measurement of flow. The flow may be an incompressible liquid, a compressible gas or a granular solid substance or a combination of any of the above. Further, the liquid flow may be laminar, turbulent, steady state or transient. A flow is described laminar when the motions of the individual particles are parallel to the pipe surface. All the particles have the same stream wise direction, but not necessarily, the same magnitude of velocity such laminar or viscous flows occur generally in smooth pipes when the flow velocity is low, and also in liquids having high viscosity. When velocity of flow increases, some particles acquire both stream wise & transverse velocity components and accordingly the laminar flow pattern gets disturbed, tends to turbulent flow. This flow is called a transition flow. As the velocity is increased further violent mixing of eddies and swirls take place, such a type of flow is called turbulent flow. This velocity of flow is called critical velocity which indicates a change from laminar to turbulent. The transition from laminar to turbulent is governed by Reynolds number R_e, a dimension less number

$$R_e = \frac{\rho v d}{\mu}$$

where v = average velocity of flow, d = diameter of pipe

ρ = density of fluid flow, μ = absolute viscosity = dynamic viscosity.

More Instruments for Learning Engineers

Solid-Particle Mass Flow Meter measures bulk solid materials continuously in a process (accuracy of 0.25% full scale and precision of 0.1%, handles bulk densities from 1-100 lb/ft^3 and flow rates of 0.35-60 ft^3/min). It can be used for mixing, blending or ratio control in the petrochemical, agricultural, aggregate, food and pharmaceutical industries. It can replace weigh belts, impact meters, loss-in-weight meters, and static weigh scales, where accuracy, low maintenance and zero drift are paramount. Solid construction, low maintenance and a compact package allow it to be installed almost anywhere.

Solid Mass Flow Meter

If $R_e < 2000$ (Laminar flow)

$R_e = 2000$ (Transition flow)

$R_e > 2000$ (Turbulent flow)

The above formulae for R_e are for pipes.

7.3 CLASSIFICATION OF FLOW MEASUREMENTS

The categorization is as follows

1. Primary or Quantity Methods
   - (a) Weight or volume tanks, burettes etc.
   - (b) Positive displacement meters
     - (i) Nutating Disk Meter
     - (ii) Rotary- Vane Flow Meter
     - (iii) Lobbed Impeller Flow Meter
     - (iv) Reciprocating Piston Meter
     - (v) Gas Meter
2. Flow Meters
   - (a) Obstruction meters
     - (i) Venturimeters
     - (ii) Flow nozzles
     - (iii) Orifices
   - (b) Variable area meters
   - (c) Turbine and propeller meters
   - (d) Magnetic flow meters (liquids only)
   - (e) Vortex shedding meters
3. Velocity Probes
   - (a) Pressure probes
     - (i) Total pressure and pilot static tubes
     - (ii) Direction sensing probes
   - (b) Hot - wire & Hot - film anemometers
   - (c) Scattering Techniques
     - (i) Laser Doppler anemometer (LDA)
     - (ii) Ultrasonic anemometer (primarily liquid).
4. Flow -Visualization Techniques
   - (a) Smoke trails and smoke wire (gauge)
   - (b) Dye injection, chemical precipitates (liquids)
   - (c) Hydrogen bubble (liquids)
   - (d) Laser induced fluorescence
   - (e) Refractive - index change
     - (i) Interferometry

(ii) Schlieren

(iii) Shadow graph.

The above lists of flow measuring methods are of primary interest for mechanical engineers but this is not the exhaust list. Obstruction meters are most often used in industrial practice.

Flow meters measure Q and/or V, while velocity probes measure V(x, y). However, the output of velocity probe can be integrated to obtain Q or V as given below

$$Q = \int_A V(x, y)dA$$

$$V = \frac{Q}{A} = \frac{1}{A}\int_A V(x, y)dA$$

7.4 PRIMARY OR QUANTITY METHODS

Quantity flow measurement indicates the amount of fluid that flows past a given point in a definite period of time. The average flow rate is then determined by dividing the total quantity of flow by the time taken to flow. Instruments of this category are generally called positive meters.

Examples

Weight meters: Weights and traps

Volumetric meters: Tank, rotating impeller, and nutating disk for liquids and bellows and liquid-sealed drums for gases.

Weight or volume tanks, burettes etc.

7.4.1 Tilting Trap Meter

Principle

As the liquid fills, the trap tilts according its center of gravity (CG). The discrete increments of liquids are weighed.

Construction & Operation

The construction and operation of this meter can be easily understood with the Figure 7.1.

When trap is filled completely, it's C.G with respect to its pivot is upset. Consequently, the trap turns and this helps in emptying. Number of tilts can be counted in a given time which gives the flow rate of liquid.

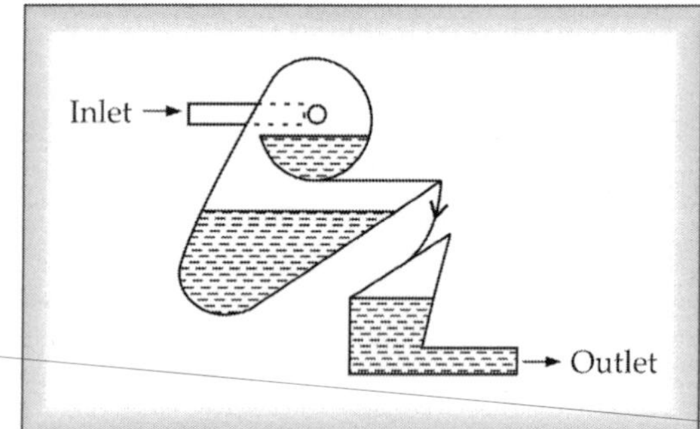

Fig. 7.1 Tilting Trap Meter

7.4.2 Positive Displacement Meters

1. Nutating Disk Meter
2. Rotary-Vane Flow Meter
3. Lobbed-Impeller Flow Meter
4. Reciprocating Piston Meter
5. Gas Meter

7.4.2.1 Nutating Disk Meter

The following are the salient components of this instrument.

1. Inlet
2. Driving shaft
3. Eccentric
4. Nutating disk with partition
5. Outlet

Principle

This meter operates on the nutating disk principle i.e., when the fluid moves through the meter, the disk wobbles or nutates about the axis.

Construction

The nutating disk is mounted eccentrically. There is an inlet and out let on either side of the disk and a meter through which the fluid has to move. In other words, a partition separates the inlet and outlet chambers of the disk and a gearing & registering arrangement is fixed to the nutating disk as shown in Figure 7.2. The meter is calibrated to measure the flow.

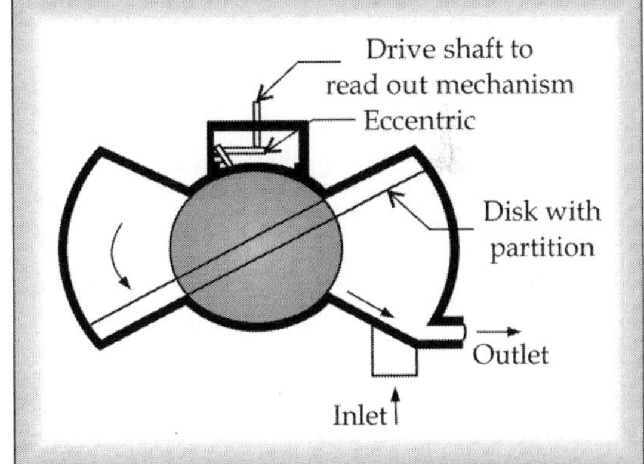

Fig. 7.2 Nutating Disk Meter

Operation

Fluid (generally water) enters the left side of the meter and strikes the disk, which is eccentrically mounted. If the fluid has to move through the meter, the disk must wobble or nutate about the vertical axis, since both the top and bottom of the disk remain in contact with the mounting chamber.

As the disk nutates, it gives direct indication of the volume of liquid, which has passed through the meter. The indication of the volumetric flow is given through a gearing and register arrangement, which is connected to the nutating disk.

Merits/Demerits/Applications

Simple and easy to construct and operate but chocks and cloggs with turbid liquids. The reliability of flow measurement of this meter is within 1%. Applicable to flows at dams, pumps and at places where there is regular flow.

7.4.2.2 Rotary-Vane Flow Meter

The main parts of this instrument are

1. Inlet
2. Spring loaded vane
3. Rotating eccentric drum
4. Outlet
5. Calibrated scale

Principle

The fluid quantity trapped by an eccentric drum is proportional to the rate of flow.

Construction

Between the inlet and outlet, an eccentric drum is made to rotate in a casing. And the vanes are spring loaded to maintain a continuous contact with the casing. A counter and calibrated scale is attached to measure the measurand. The Figure 7.3 shows the schematic of rotary-vane flow meter.

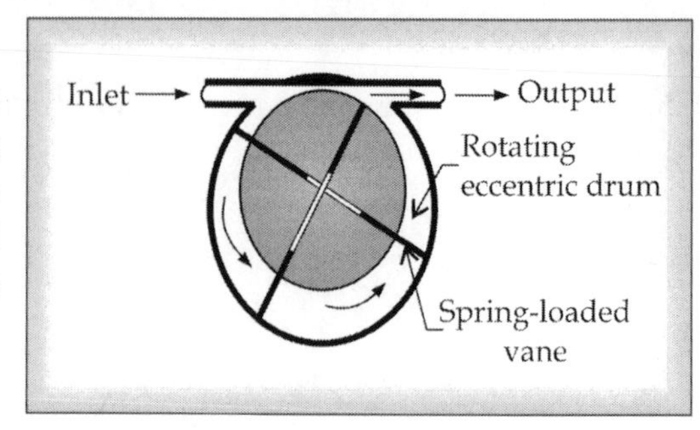

Fig. 7.3 Rotary-Vane Flow Meter

Operation

This is a positive displacement device, in which the vanes are spring loaded so that they continuously maintain contact with the casing. As the eccentric drum rotates a fixed quantity of fluid is trapped in each section, which eventually finds its way out through exit. A counter connected to the shaft of the eccentric drum records the volume of the displaced fluid.

Merits/Demerits/Applications

These are relatively insensitive to viscosity and uncertainties are of the order of 0.5%.

7.4.2.3 Lobed-Impeller Flow Meter

The following are the chief parts of this instrument.

1. Inlet
2. Chamber (casing)
3. Lobed –impellers
4. Outlet
5. Calibrated measuring device

Principle

The quantity of fluid passing through the chamber is proportional to the number of rotations of lobed-impellers as fixed amount of fluid is passed per rotation.

Construction

As shown in Figure 7.4, a pair of lobed impellers are housed exactly in a casing (chamber) so that a fixed amount of fluid can be passed per rotation. Therefore, impellers and case are carefully machined so that accurate fit is maintained. The inlet and outlet are kept one either side of the chamber.

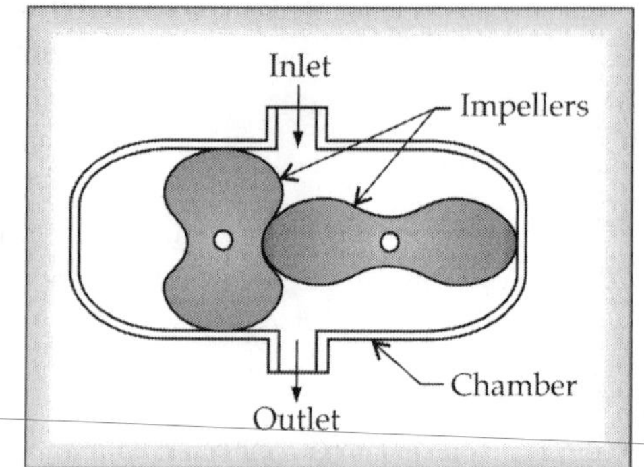

Fig. 7.4 Lobed-Impeller Flow Meter

Operation

In this meter the incoming fluid is trapped between the two rotors and is conveyed to the outlet as a result of their rotation. The number of revolutions of the rotors is an indication of the volumetric flow rate.

Application

This can be used for both gas and liquid flow measurements.

7.4.2.4 Reciprocating Piston Meter

The main parts of this meter are

1. Inlet
2. Piston & cylinder
3. Eccentric
4. Side valve
5. Outlet
6. Calibrated flow measuring device

Principle

The liquid trapped by the piston in cylinder calculated in terms of stroke length and speed of piston is proportional to the flow rate.

Construction

As shown in Figure 7.5, this consists of a piston reciprocating inside a cylinder, inlet and outlet valves with their opening and closing being controlled by a slide valve. This valve is actuated by piston itself by an eccentric. A counter calibrated for flow measurement is attached.

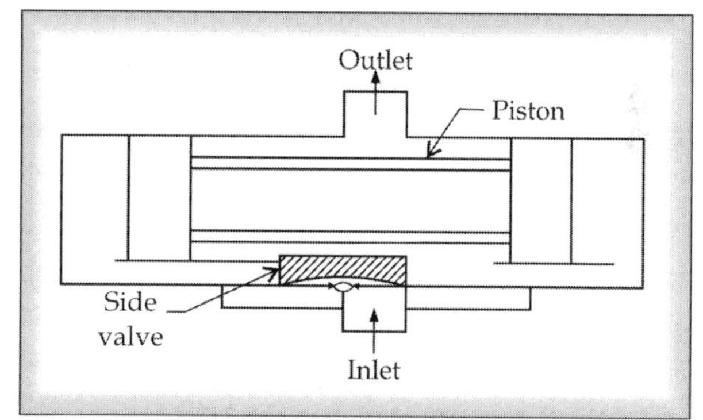

Fig. 7.5 Reciprocating Piston Meter

Operation

By the movement of the piston, the fluid is passed from the inlet to outlet alternately through each end of the cylinder. The volume of the fluid flow is directly proportional to both the stroke length and the piston speed.

Merits/Demerits/Applications

This meter is used for non-corrosive and low viscosity liquids.

7.4.2.5 Gas Meter

The main parts of this meter are

1. Gas inlet
2. 4 - Compartment Drum – with inlet/outlet holes
3. Casing
4. Water
5. Calibrated meter
6. Gas outlet

Principle

The number of rotations of drum counts the amount of gas for a given period.

Construction

The meter consists of a drum, which has four compartments with inlet and outlet holes as shown in Fig. 7.6. The drum rotates inside a cylindrical casing, which is more than half full of water, the water merely acting as the gas seal.

Operation

When the inlet hole of a compartment emerges from water, it communicates with the gas inlet. The compartment *A* of the meter is completely filled with the water, the gas is being forced out of compartment *B*, compartment *C* is full of gas and the compartment *D* whose inlet hole has just emerged from water is beginning to fill up with gas. The gas pressure causes rotation of the drum. The drum revolutions are totalized by a counter and recorded by a counting mechanism whose reading shows the amount of gas consumed.

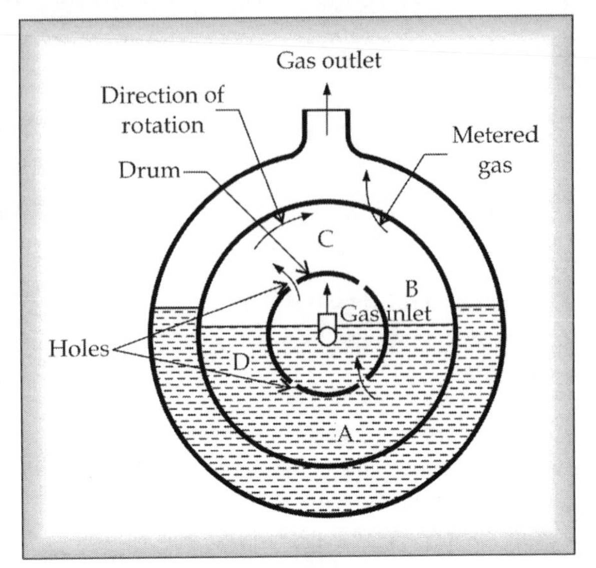

Fig. 7.6 Gas Meter

Merits/Demerits/Applications

Applicable for gas cylinders, gas pipes etc. However, much care should be taken in selection of the material for the construction of meter particularly, in the case of corrosive and/or reactive gases.

SELF ASSESSMENT QUESTIONS-7.1

1. What is transition flow?
2. What is turbulent flow?
3. Give the classification of methods of flow measurements.
4. What are the different flow meters of Primary or Quantity Methods?
5. Give the classification of Flow Meters.
6. List out various Velocity Probes.
7. What are different types of Flow -Visualization Techniques?
8. With a neat diagram explain Tilting Trap Meter.
9. Explain Nutating Disk Meter with neat sketch.
10. Discuss Rotary-vane Flow Meter with a neat sketch.
11. Discuss Lobed-Impeller Meter with a neat sketch.
12. Explain Reciprocating Piston Meter with a neat sketch.
13. What is Gas Meter? Explain it with a neat sketch.

7.5 FLOW (OBSTRUCTION) METERS

Principle

The flow meters are basically designed on basis of Bernoulli's equation. with reference to the method of measuring, these flow meters are also called obstruction meters as they involve the restriction of the flow in a pipe by some means in a pipe carrying the fluid whose rate of flow is to be measured. Due to flow restriction, the pressure in the pipe changes causing a proportional change in the flow rate. This pressure drop is measured by differential pressure sensor and when calibrated this becomes a measure of flow rate.

As discussed above, the relation for measuring flow is derived by applying Bernoulli's principle (at sections 1 and 2) as shown in Fig. 7.7. Popularly known obstruction flow meters in use are

1. Venturi meter
2. Orifice meter or orifice plate
3. Flow-nozzle meter
4. Variable area meter or rotameter

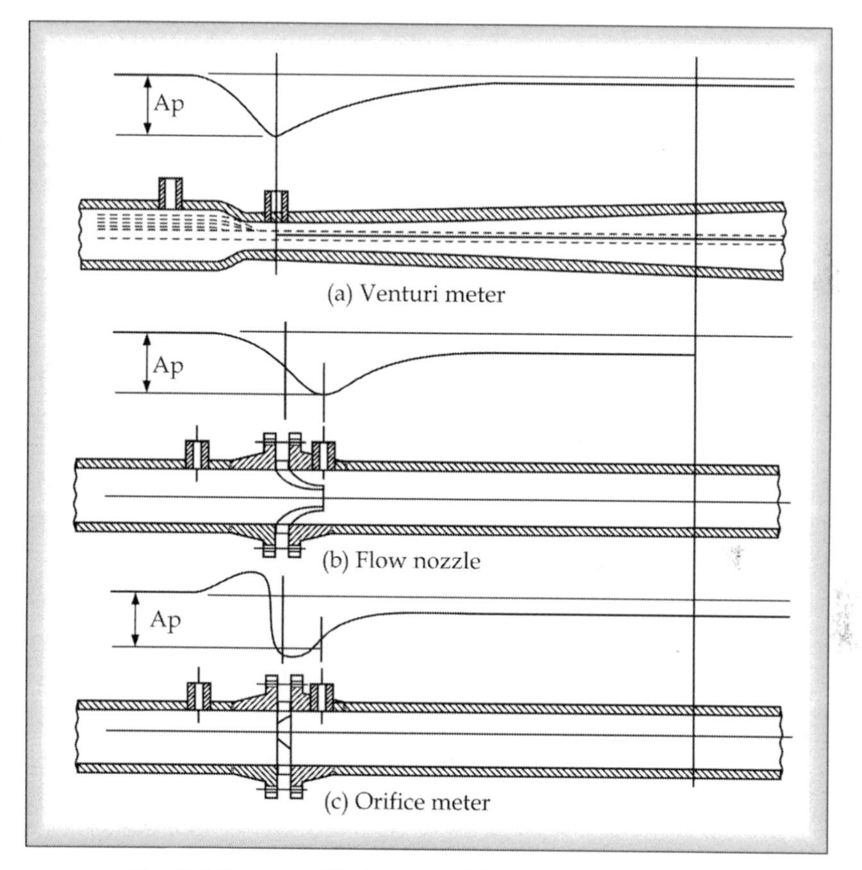

Fig. 7.7 Pressure Variation in Flow Measuring Devices

7.5.1 Venturi Meter

A venturimeter is a device used for measuring the rate of a flow of fluid flowing through a pipe. It has the following components

1. Venturi

   This consists of three parts:

   (a) A short converging part

   (b) Throat, and

   (c) Diverging part
2. U-tube manometer
3. Piezometer ring

Principle: Bernoulli's equation

Construction

The parts are arranged as shown in the Figure 7.8.

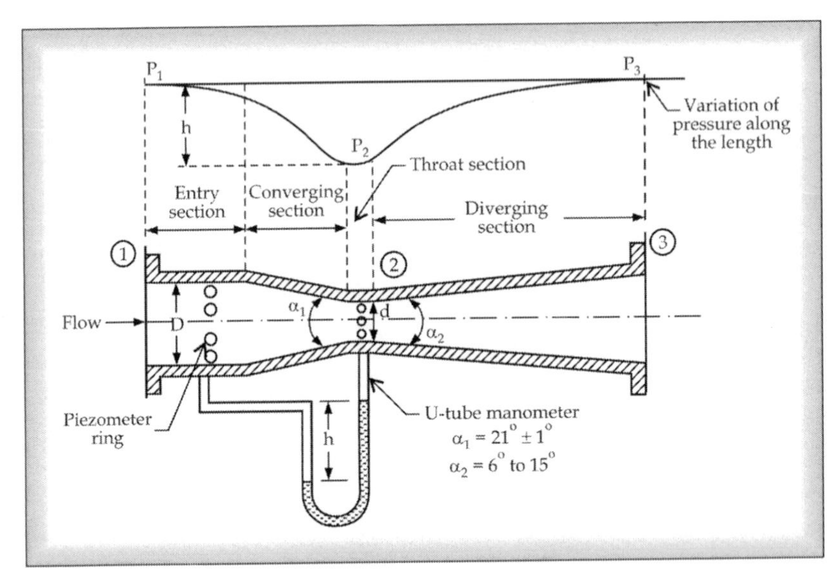

Fig. 7.8 Venturi Meter

Operation

The operation of venturimeter uses equation of discharge through venturi, derived by applying Bernoulli's equation.

Expression for the Rate of flow through venturimeter for incompressible fluid

Consider a venturimeter fitted in a horizontal pipe through which a fluid is flowing (say water), as shown in Fig. 7.8

Let d_1 = diameter at inlet or at section (1),

P_1 = pressure at section (1),

V_1 = Velocity of fluid at section (1),

$$a_1 = \text{Area at section (1)} = \frac{\pi}{4} d_1^2$$

And d_2, p_2, V_2, a_2 are corresponding values at section (2)

Applying Bernoulli's equation at section (1) and (2) we get

$$\frac{p_1}{w} + \frac{V_1^2}{2g} + Z_1 = \frac{P_2}{w} + \frac{V_2^2}{2g} + Z_2$$

As pipe is horizontal, hence $z_1 = z_2$

$$\frac{P_1}{w} + \frac{V_1^2}{2g} = \frac{P_2}{w} + \frac{V_2^2}{2g}$$

$$\frac{P_1 - P_2}{w} = \frac{V_2^2}{2g} - \frac{V_1^2}{2g}$$

But $\frac{P_1 - P_2}{w}$ is the difference of pressure heads at sections 1 and 2 and it is equal to h

or $$\frac{P_1 - P_2}{w} = h$$

Substituting this value of $\frac{P_1 - P_2}{w}$ in the above equation, we get

$$h = \frac{V_2^2}{2g} - \frac{V_1^2}{2g}$$

Now applying continuity equation at sections (1) & (2)

$$a_1 V_1 = a_2 V_2 \quad \text{or } V_1 = \frac{a_2 V_2}{a_1}$$

Substituting this value of V_1 in equation $h = \frac{V_2^2}{2g} - \frac{V_1^2}{2g}$

$$h = \frac{V_2^2}{2g} - \frac{\left(\frac{a_2 V_2}{a_1}\right)^2}{2g} = \frac{V_2^2}{2g}\left[1 - \frac{a_2^2}{a_1^2}\right] = \frac{V_2^2}{2g}\left[\frac{a_2^2 - a_1^2}{a_1^2}\right]$$

or $$V_2^2 = 2gh \frac{a_1^2}{a_2^2 - a_1^2}$$

$\therefore$ $$V_2 = \sqrt{2gh \frac{a_1^2}{a_2^2 - a_1^2}} = \frac{a_1}{\sqrt{a_1^2 - a_2^2}} \sqrt{2gh}$$

$\therefore$ Discharge, $Q = a_2 V_2$

$$= a_2 \frac{a_1}{\sqrt{a_1^2 - a_2^2}} \times \sqrt{2gh} = \frac{a_1 a_2 \sqrt{2gh}}{\sqrt{a_1^2 - a_2^2}}$$

The above equation gives the discharge under ideal conditions and is called, theoretical discharge.

Actual discharge will be less than theoretical discharge

$$\therefore \quad Q_{act} = C_d \times \frac{a_1 a_2 \sqrt{2gh}}{\sqrt{a_1^2 - a_2^2}}$$

where C_d = co-efficient of venturimeter and its value is less than 1

h = value given by differential U-tube manometer

Case (i): Let the differential manometer contain a liquid, which is heavier than the liquid flowing through the pipe.

Let S_h = Sp. gravity of the heavier liquid

S_0 = Sp. gravity of the liquid flowing through pipe.

x = difference of the heavier liquid column in *V* tube

Then
$$h = x\left[\frac{S_h}{S_0} - 1\right]$$

Case (ii): If the differential manometer contains a liquid which is lighter than the liquid flowing through the pipe, the value of h is given by

$$h = x\left[1 - \frac{S_1}{S_0}\right]$$

where S_1 = Sp. gr of lighter liquid in U tube

S_0 = Sp. gr of fluid flowing through pipe

x = difference of the lighter liquid columns in U-tube

Case (iii): Inclined venturimeter with differential U-tube manometer:

The above two cases are for a horizontal venturi meter. This case is related to inclined venturimeter having differential U-tube manometer. Let the differential manometer contain heavier liquid then *h* is given as

$$h = \left[\frac{p_1}{w} + z_1\right] - \left[\frac{p_2}{w} + z_2\right] = x\left[\frac{S_h}{S_o} - 1\right]$$

Case (iv): Similarly, for inclined venturimeter in which differential manometer contains a liquid, which is lighter than the liquid flowing through the pipe, the value of *h* is given as

$$h = \left[\frac{p_1}{w} + z_1\right] - \left[\frac{p_2}{w} + z_2\right] = x\left[1 - \frac{S_h}{S_o}\right]$$

7.5.2 Orifice Meter or Orifice Plate

An orifice meter is another device used for measuring the rate of flow of fluid through a pipe. Its working principle is same as that of venturimeter and is relatively cheaper.

It is a flat circular plate with a circular hole called orifice, which is concentric with the pipe. The hole diameter lies between 0.4 to 0.8 times the pipe diameter (d) but generally it is taken as 0.5d. A differential manometer is connected at section (1) and (2) as shown in Fig. 7.9.

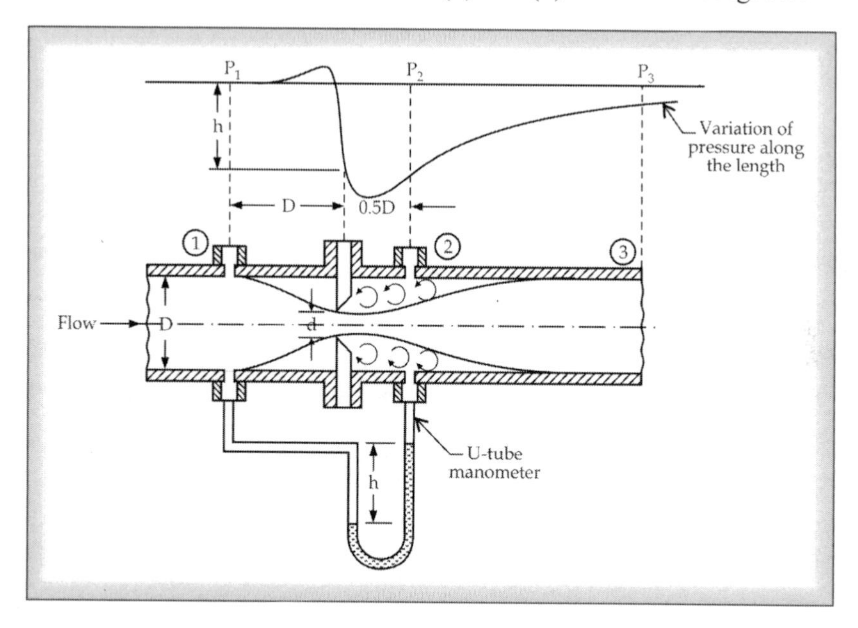

Fig. 7.9 Orifice Meter

Operation

The operation of orifice-meter uses equation of discharge through orifice, derived by applying Bernoulli's equation.

Expression for the rate of flow through orifice meter for incompressible fluid

Let P_1 = Pressure at section (1)

V_1 = Velocity at Section (1)

a_1 = area of pipe at section (1), and

P_2, V_2, a_2 are corresponding values at section (2).

Applying Bernoulli's equation at section (1) & (2)

$$\frac{P_1}{w} + \frac{V_1^2}{2g} + Z_1 = \frac{V_2}{w} + \frac{P_2^2}{2g} + Z_2$$

$$\left(\frac{P_1}{w} + Z_1\right) - \left(\frac{P_2}{w} + Z_2\right) = \frac{V_2^2}{2g} - \frac{V_1^2}{2g}$$

$$\left(\frac{P_1}{w} + Z_1\right) - \left(\frac{P_2}{w} + Z_2\right) = h = \text{differential head}$$

$$h = \frac{V_2^2}{2g} - \frac{V_1^2}{2g} \Rightarrow V_2^2 - V_1^2 = 2gh$$

$$V_2 = \sqrt{2gh + V_1^2}$$

Now section (2) is at vena contracta and a_2 represents the area at vena contracta. If a_0 is the area of orifice then

Coefficient of contraction (C_c) = Ratio of area at vena contracta to the area of the orifice

i.e., $C_c = \frac{a_2}{a_0}$

$\therefore$ $$a_2 = a_0 \times C_c$$

[**Vena Contracta:** The section of the jet of fluid where its area is minimum is called vena contracta]

$$a_2 = a_0 \times C_c$$

By continuity equation $a_1V_1 = a_2V_2$

or $$V_1 = \frac{a_2V_2}{a_1} = \frac{a_0}{a_1}C_cV_2$$

Substituting the value of V_1 in equation $V_2 = \sqrt{2gh + V_1^2}$

$$V_2 = \sqrt{2gh + \frac{a_0^2C_c^2V_2^2}{a_1^2}}$$

or $$V_2^2 = 2gh + \left(\frac{a_0}{a_1}\right)^2 C_c^2V_2^2$$

or $$V_2^2\left[1 - \left(\frac{a_0}{a_1}\right)^2 C_c^2\right] = 2gh$$

$$V_2 = \frac{\sqrt{2gh}}{\sqrt{1 - \left(\frac{a_0}{a_1}\right)^2 C_c^2}}$$

$\therefore$ Discharge $$Q = a_2V_2 = V_2 \times a_0C_c \quad \text{(since } a_2 = a_0 \times C_c\text{)}$$

$$= \frac{\sqrt{2gh}}{\sqrt{1 - \left(\frac{a_0}{a_1}\right)^2 C_c^2}} \times a_0C_c$$

$$\text{Let (coefficient of discharge) } C_d = C_c \frac{\sqrt{1-\left(\frac{a_0}{a_1}\right)^2}}{\sqrt{1-\left(\frac{a_0}{a_1}\right)^2 C_C^2}}$$

$$C_c = C_d \frac{1-\left(\frac{a_o}{a_1}\right)^2 C_C^2}{1-\left(\frac{a_o}{a_1}\right)^2}$$

Substituting $$C_c \text{ in } \frac{\sqrt{2gh}}{\sqrt{1-\left(\frac{a_0}{a_1}\right)^2 C_C^2}} \times a_0 C_C$$

$$Q = a_0 \times c_d \times \frac{\sqrt{1-\left(\frac{a_0}{a_1}\right)^2 C_C^2}}{\sqrt{1-\left(\frac{a_0}{a_1}\right)^2}} = \frac{\sqrt{2gh}}{\sqrt{1-\left(\frac{a_0}{a_1}\right)^2 C_C^2}}$$

$$= \frac{C_d a_0 \sqrt{2gh}}{\sqrt{1-\left(\frac{a_0}{a_1}\right)^2}} = \frac{C_d a_0 a_1 \sqrt{2gh}}{\sqrt{a_1^2 - a_0^2}}$$

The coefficient of discharge for orifice meter is much smaller than that of venturimeter.

7.5.3 Flow-Nozzle Meter

Flow Nozzle is a device used for measuring the ratio of flow of fluid through a pipe. This also works on the same principle as that of venturimeter. It can be treated as only the convergent part of the venturimeter. The schematic diagram is shown in Fig. 7.10.

The equation for discharge remains the same as that of venturimeter. However, the co-efficient of discharge for this meter lies between venturi-meter & orifice-meter.

(**Note:** *This is left for readers to derive the required expression or refer any book in fluid mechanics*)

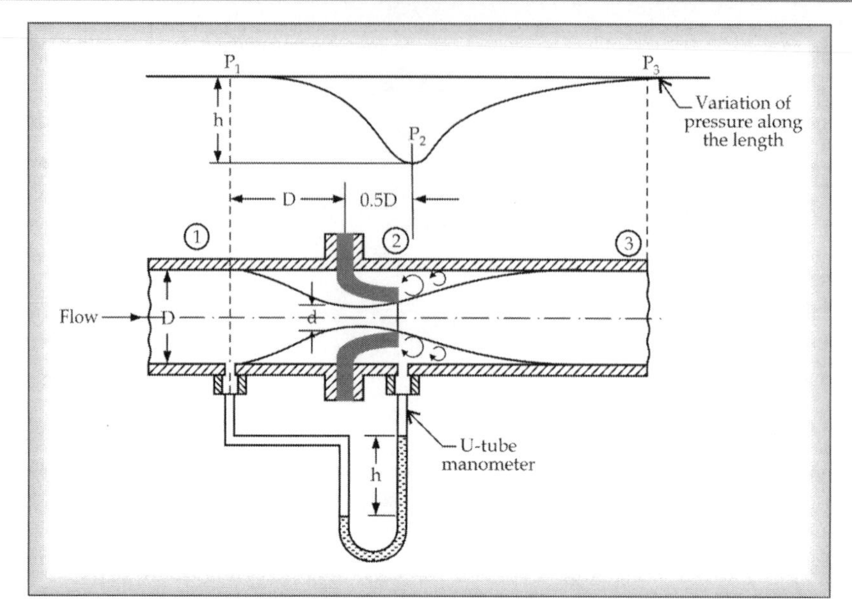

Fig. 7.10 Nozzle Meter

7.5.4 Variable Area Meter or Rotameter

The main demerit with the obstruction meters is that the pressure drop varies as the square of the flow rate. Thence if these meters are to be used over a wide range of flow rates, pressure measuring equipment of very wide range is required. Further, if this pressure range is chosen, it is obvious that the accuracy at low flow rates will be poor. The small pressure readings in that range will be limited by pressure transducer resolution. Then how do you overcome this limitation?

Well! One good idea is to use two (or more) pressure-measuring systems, one for low flow rates and another for high rates.

Exactly, this technique is employed in variable area meter. It is a device whose indication is essentially linear with flow rate selected which is also called rotameter.

This meter works with drag effect.

As shown in the Fig. 7.11 the flow enters the bottom of the tapered vertical tube and causes the bob or "float" to move upwards. The term float is somewhat a "misnomer" in that. It must be heavier than the liquid it displaces.

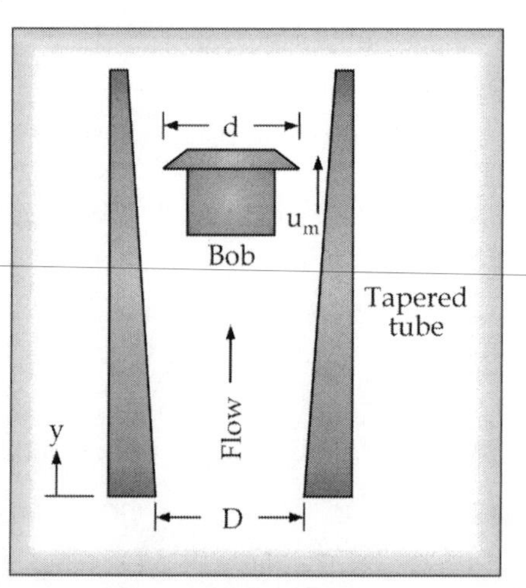

Fig. 7.11 Rotameter

Four forces act on the float. They are

1. Downward gravity force.
2. An upward buoyant force
3. Pressure force and
4. Viscous drag force.

For a given rate of flow, the float assumes a position in the tube where the forces acting on it are in equilibrium. By designing carefully the affects of changing viscosity or density can be minimized then pressure force is the only variable. As the pressure force is dependent on the flow rate and the annular area between the tube and float, the position of float will give flow rate alone.

Basic equation for volumetric rate of flow in rotameter is

$$Q = A_w C \left[\frac{2gV_f (S_f - S_w)}{A_f S_w} \right]^{1/2}$$

where

Q = The volumetric rate of flow

V_f = The volume of the float

g = Acceleration due to gravity

S_f = The float density

S_w = The liquid density

A_f = The area of the float

C = The discharge coefficient

A_w = The area of the annular orifice.

$$= \left(\frac{\pi}{4}\right)\left[(D + by)^2 - d^2\right]$$

where D = The diameter of the tube when the float is at the zero position.

b = The change in tube diameter per unit change in height

d = The maximum diameter of the float.

y = The height of the float above zero position.

As the Rotameter response is linear, its resolution is the same at both high and low flow rates.

Merits of Rotameter

1. Flow condition can be seen
2. Corrosive fluids can be handled easily
3. Very accurate (between 1 and 10% of full scale)
4. Capacity may be changed by varying float and/or tube.
5. Easy to equip with data transmission, indicating and recording devices.
6. There is a uniform flow scale over the range of instrument with the pressure loss fixed at all flow rates.

Demerits of Rotameter

1. It has to be installed in vertical position.
2. Less accurate as compared to venturi and orifice meter.

3. Float may not be visible when opaque fluids are used.
4. It is expensive where high pressure and temperatures are involved.
5. It cannot be used with liquids carrying large percentage of solids in suspension.

ILLUSTRATION-7.1

Show that there exists a linear relationship between the volume flow rate and displacement of float in the case of a rotameter. Prove that if the density of the float material is made twice that of flowing fluid, the volume flow rate becomes independent of the density of the following fluid and compression is almost complete for all flows.

Solution:

We have volume flow rate $Q = A_w C$

$$\left[\frac{2gV_f(S_f - S_w)}{A_f S_w}\right]^{1/2} \quad .....(7.1)$$

Where Q = The volumetric rate of flow

V_f = The volume of the float

G = Acceleration due to gravity

S_f = The float density

S_w = The liquid density

A_f = The area of the float

C = The discharge coefficient

$$A_w = \frac{\pi}{4}[(D + by)^2 - d^2]$$

where D = The diameter of the tube when the float is at the zero position

b = The change in the tube diameter of the float

d = The maximum diameter of the float

y = The height of the float above zero position.

For a given rate of flow, the float will be at a particular position where the forces acting on it are in equilibrium. The effects of changing viscosity or density are minimized by careful design, leaving pressure force as a variable. This again depends on flow rate and the annular area between the float and the tube. Hence the position of the float is determined by the flow rate above.

In the equation for flow rate, the values D, b and d are selected such that linear variation of A_w with y is achieved. Thus, the flow rate is a linear function of the reading.

Density of float material $(S_f) = 2 \times$ density of flowing fluid (S_w)

Substituting $S_f = 2\, S_w$ in (7.1)

$$Q = A_w C \left[\frac{2gV_f(2S_w - S_w)}{A_f S_w} \right]^{1/2}$$

$$= A_w C \left[\frac{2gV_f}{A_f} \right]^{1/2}$$

Hence volume flow rate is independent of the density of the flowing fluid.

SELF ASSESSMENT QUESTIONS-7.2

1. What is the principle of flow (obstruction) meters?
2. What is venturimeter? Discuss it with a neat diagram.
3. What is an orifice meter or orifice plate? Explain it with a neat sketch.
4. Explain Flow Nozzle meter with a neat sketch.
5. Discuss about Rotameter with a neat sketch.

7.6 TURBINE METERS

Anemometers (discussed in the section 7.10 of this chapter), used in weather stations to measure wind velocity is a simple form of free stream turbine meter. Similar rotating wheel flow-meters are used by civil engineers to measure water flow in rivers and streams.

The noteworthy parts in this meter are

1. Inlet
2. Straightening vane
3. Turbine shaft
4. Rotating turbine blades
5. Reluctance pick up
6. Casing

Principle

The rate of flow proportionally rotates turbine wheel which can be measured in terms of number of rotations or by converting into a magnetic or electrical signal/pulse (such as reluctance pick-up).

Construction

A turbine would have a shaft, and blades fitted. It will have straightening vanes at the inlet and a body set up for reducing pressure drop at the outlet. The turbine wheel body is fitted with a permanent magnet which will be connected to a reluctance pick up is attached to it at the top.

Operation

When the fluid moves through the meter, it causes a rotation of the turbine wheel and hence, the permanent magnet enclosed in turbine wheel body also rotates with the wheel as shown in the Fig. 7.12. A reluctance pick up attached at the top of the meter detects a pulse for each revolution of the turbine wheel. Since the volumetric flow is proportional to the number of wheel revolutions, the total pulse output may be taken as an indication of total flow.

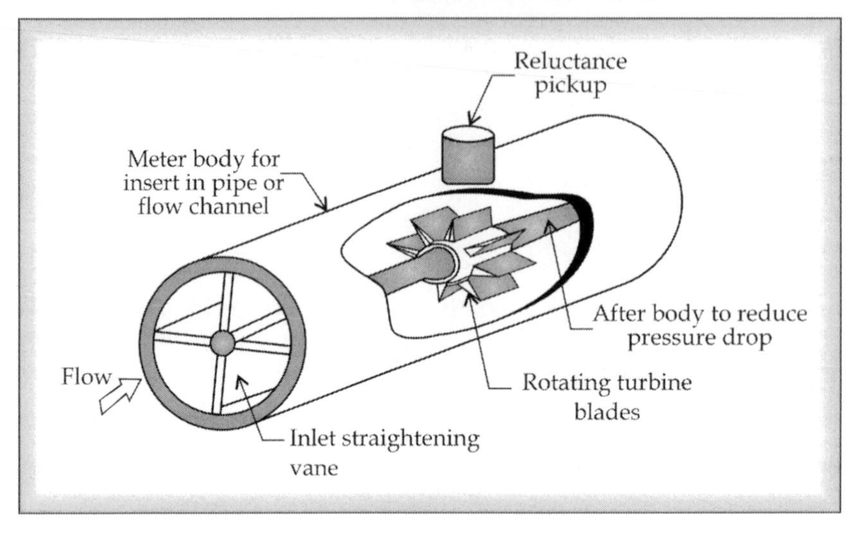

Fig. 7.12 Turbine Meter

The flow rate is given by

$$Q = \frac{f}{k}$$

where f = pulse frequency

k = flow coefficient.

The flow coefficient is dependent on flow rate and the kinematic viscosity of the fluid v.

Merits/Demerits/Applications

The pulse rate is proportional to flow rate, and the transient response of the meter is very good. The most general problems encountered in this type of meters are bearing maintenance and reduced accuracy at low flow rates.

More Instruments for Learning Engineers

Helix Meter or Woltman Meter (invented by Reinhard Woltman) considered as a type of turbine flow meter, consists a rotor with helical blades inserted axially in the flow, like a ducted fan. **Single Jet Meter** consists of a simple impeller with radial vanes, impinged upon by a single jet. **Paddle Wheel Meter** is similar to the single jet meter, except that the impeller is small with respect to the width of the pipe, and projects only partially into the flow. **Multi Jet Meter** is a velocity type meter with an impeller that rotates horizontally on a vertical shaft. The impeller element is housed in which multiple inlet ports direct the fluid flow at the impeller causing rotation in a specific direction in proportion to the flow velocity. This meter works like a single jet meter except that the ports direct the flow at the impeller equally from several points around the circumference of the element, thus minimizes uneven wear on the impeller and its shaft.

Mechanical Fluid Flow Meters

7.7 MAGNETIC FLOW METERS

The major parts of these meters are

1. Inlet
2. Pipe though which fluid flows
3. Magnetic field around pipe
4. Outlet
5. Voltmeter calibrated to measure flow

Principle

The flow of the conducting fluid is directly proportional to the voltage induced.

Construction

Flow meter arrangement is shown in Fig. 7.13. The flowing medium is passed through a pipe of which a short section is subjected to a transverse magnetic flux. The fluid itself acts as the conductor having dimension 'L' equal to pipe diameter and velocity v roughly equal to the average fluid velocity. The emf is detected by electrodes placed in the conduct walls. Either an alternating or direct magnetic flux may be used. However, if amplification of the output is required, the advantage lies with the alternating field.

Operation

These meters are useful for conducting fluids. Since the fluid represents a conductor moving in the magnetic field, as shown in the Figure 7.14, there will be an induced voltage according to the following relation.

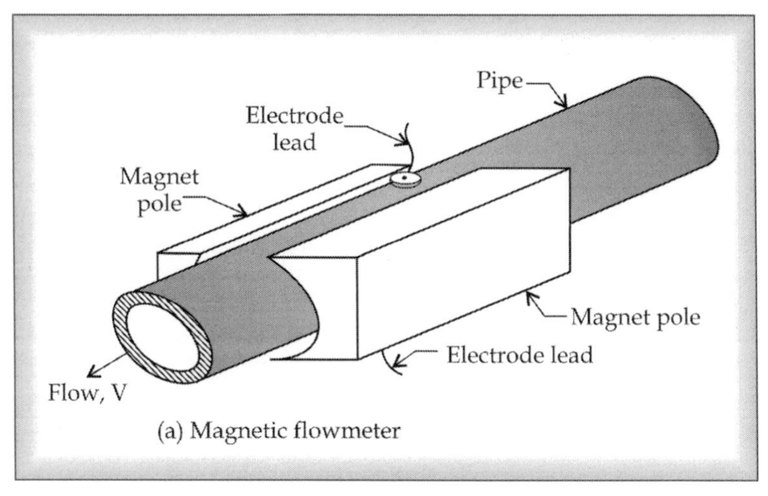

(a) Magnetic flowmeter

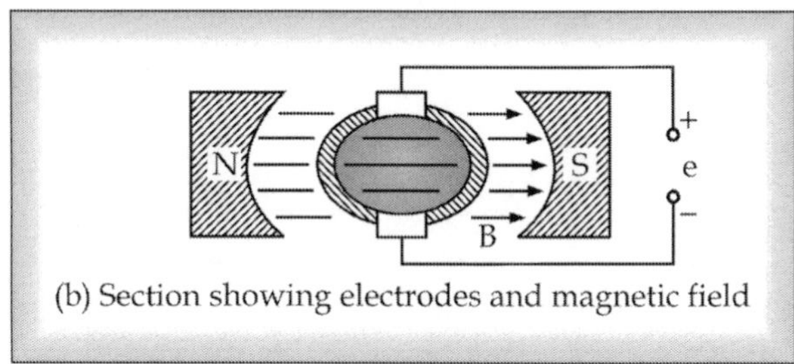

(b) Section showing electrodes and magnetic field

Fig. 7.13 Magnetic Flow Meter Arrangement

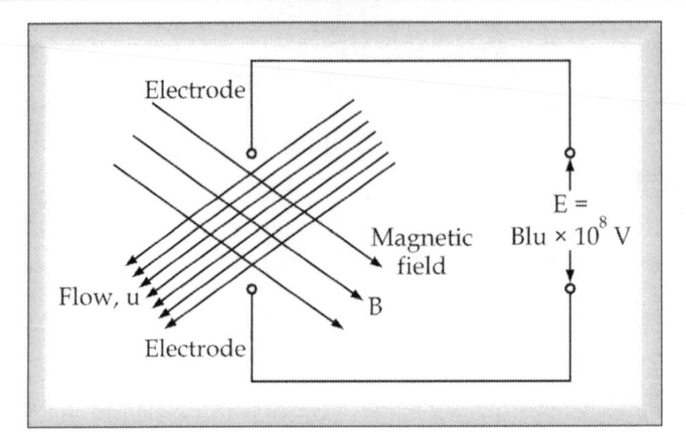

Fig. 7.14 Flow of Conducting Fluid through a Magnetic Field

$$E = BLu \times 10^{-4} \text{ Volts}$$

where

E = induced voltage in volts

B = magnetic flux density in gauses

L = length of the conductor in m

u = velocity of the conductor in *m/s.*

Two types of magnetic flow meters are used commercially.

The first type is used for fluids with low conductivities, like water which has a non-conducting pipe-liner. The electrodes are mounted such that they are flush with the non-conducting liner and make contact with the fluid. Alternating magnetic fields are normally used with these meters since the output is low and requires amplification.

The second type of magnetic flow meters is used for high conductivity fluids, like liquid metals. A stainless steel pipe is used in this case with the electrodes attached directly to the outside of the pipe diametrically opposite to each other. The output of this type of meter is sufficiently high that it may be used for direct read-out purpose.

7.8 VORTEX SHEDDING FLOW METERS

Principle

Vortex flow meters are based on the fact that when a bluff body is placed in a stream, vortices are alternately formed, first to one side of the obstruction and then to the other as shown in Figure 7.15.

Construction & Operation

We know that the frequency of vortex shedding is directly proportional to the liquid velocity. So, a piezoelectric sensor arranged inside the vortex shedder detects the vortices, and subsequent amplifications circuits can be used to indicate either the instantaneous flow rate or a totalized flow over a selected time interval. The meter is pre-calibrated by the manufacturer for a specific pipe size. It is generally not suitable for use with high viscous liquids.

The frequency of formation (f) is a function of flow rate, given by:

$$f = \left(\frac{S_t}{D}\right) v \text{ (for compressible fluids)}$$

where

S_t = A calibration constant called Strouhal number

v = flow velocity

D = dimension of the obstruction transverse to the flow direction.

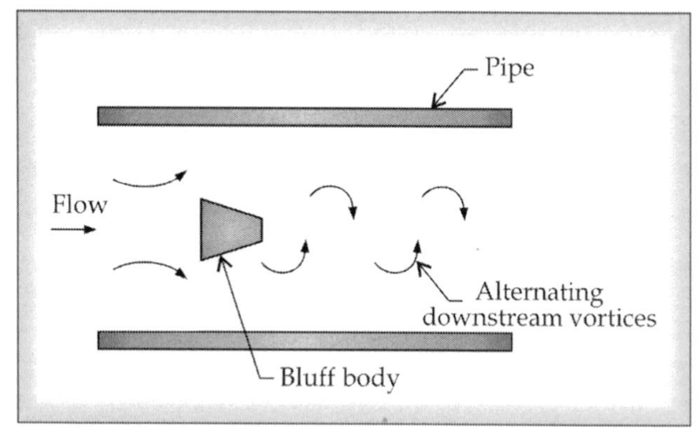

Fig. 7.15 Vortex-Shedding Flow Meter

7.9 MEASUREMENT OF FLUID VELOCITIES

Flow velocity and flow rate often confuse us as they seem to be same but are little different.

Flow Rate & Flow Velocity

The flow velocity is defined as the displacement made by the fluid in unit time, whereas the flow rate is the mass of fluid passed per unit time. Flow rate is proportional to flow velocity, therefore by measuring the velocity we can get an idea of flow rate also. However, these instruments generally measure local velocity rather than average velocity.

Pressure Probes

Pressure probes measure pressure at a particular point. There are many types of pressure probes whose selection for a given purpose depends on the availability of space, pressure gradients, flow magnitude and direction. They measure either of the two pressures i.e. total pressure or static pressure. Hence, it is necessary to know about static & total pressures. (These are discussed in detail in the chapter-6).

Static Pressure

The pressure exerted by the fluid when at rest is known as static pressure.

Total or Stagnation or Impact Pressure

The pressure exerted by the flowing fluid when brought to rest is known as stagnation or total pressure.

Total Pressure = Static Pressure + Dynamic Pressure

= Static Pressure + Velocity Pressure

7.9.1 Total - Pressure Probe (Pitot Static Tube)

Simple Pitot-Tube as shown in Fig. 7.16 will measure total or impact pressure.

Construction

Usually, the Pitot tube is combined with static openings, constructed as shown in Fig. 7.17. This is known as a Pitot static tube, or sometimes as a Prandtl-Pitot tube. The unit consists of two concentrically placed tubes bent to form a right-angled bend as shown in Fig. 7.17. The outer tube is sealed and has a stream-lined shape.

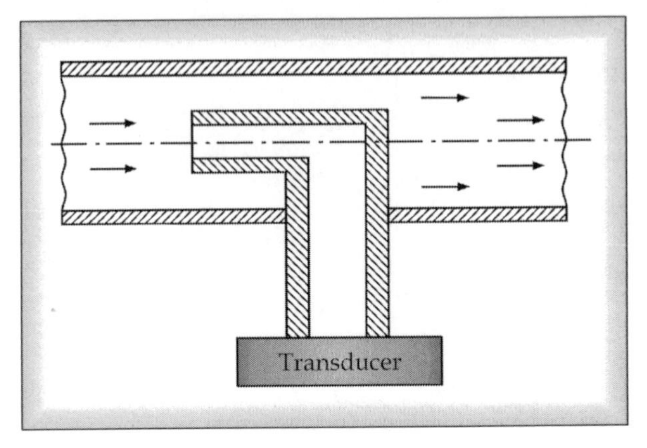

Fig. 7.16 Measurement of Total Pressure

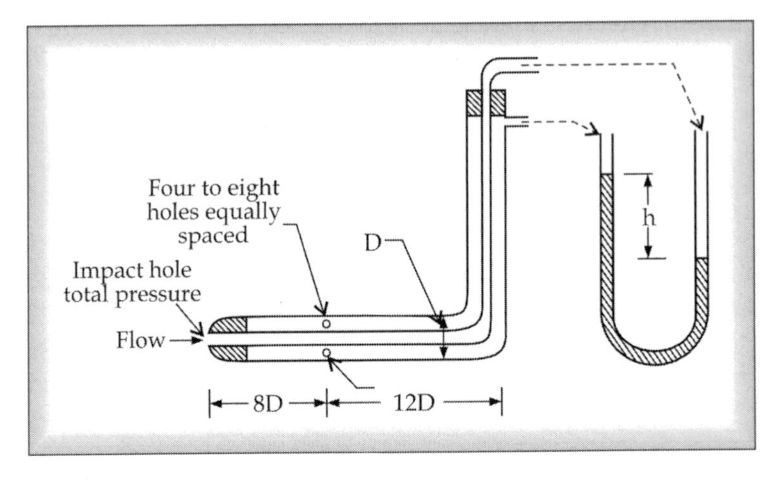

Fig. 7.17 Pitot-Static Tube

Between the nose and the stem, at a distance of *3d* (where d is diameter) from the tip of the tube, a number of holes (6 to 8) are drilled through the surface of the outer tube. The right-angled bend is at a further distance of *8d* to *10d.* Thus, it has one impact opening and six (6) to eight (8) static openings.

Operation

The Pitot tube operation involves the introduction of the tube into the flow area where pressure details are required. The pressure in the out tube is static pressure and pressure in inner tube is greater than the static pressure which is total pressure. The differential pressure (p_t–p_s) is measured using a differential pressure sensor. This pressure becomes a measure of flow rate at that point where the Pitot tube is present in the flowing fluid.

Merits

- Their cost is very less and easy to construct and install.

Demerits

- The proper alignment problem,
- Cannot be used in fluids with suspended solids & impurities and
- Small flow rates cannot be measured.

Applications

- Pitot tubes are extensively used in laboratories to measure, velocity pressure and flow rates of fluids.

7.9.2 Direction - Sensing Probes or Yaw-Angle Probe

Figure 7.18 shows two forms of direction sensing or Yaw-angle probes. Each of these probes uses two impact tubes. In each case the probe is placed transferable to flow and is rotatable around its axis. The angular position of the probe is then adjusted until the pressures sensed by the openings are equal. In such case, the flow direction will correspond to the bisector of the angle between the openings. Probes with third openings are also available which when properly aligned senses maximum impact pressure.

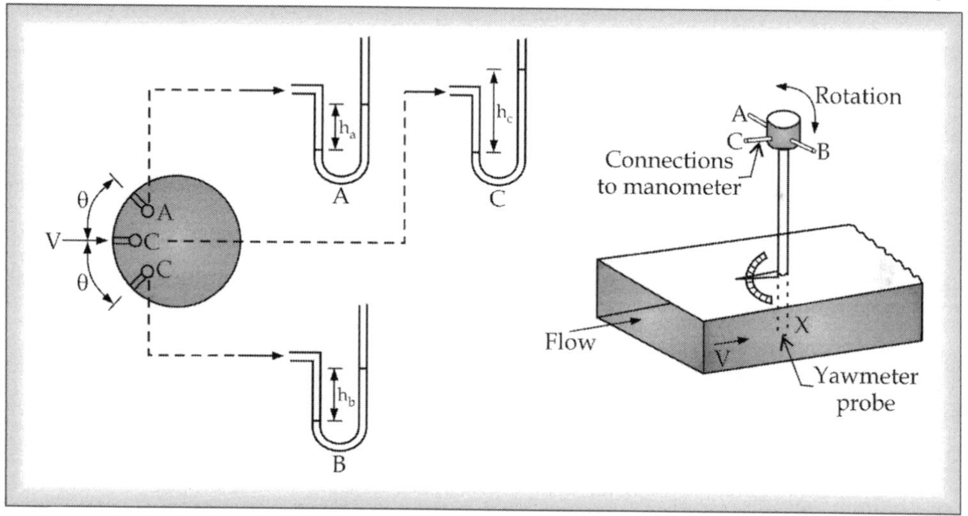

Fig. 7.18 Yaw-angle Probe

Self Assessment Questions-7.3

1. Explain Turbine Meter with a neat sketch.
2. With a neat sketch discuss magnetic flow meter.
3. Discuss Vortex shedding flow meter With a neat sketch.
4. What are different methods of the measurements of fluid velocities?
5. Briefly explain the following
   (a) Flow Rate (b) Flow Velocity
   (c) Pressure Probes (d) Static Pressure
   (e) Total or Stagnation or Impact Pressure
6. Discuss about the Pitot tube with a neat diagram.
7. Explain direction- sensing probe with a neat diagram.

7.10 ANEMOMETERS

Principle

A heated object placed in a moving stream of fluid loses its heat, which is proportional to the velocity of the fluid. If the object is electrically heated at a known power, it will reach a temperature determined by the rate of cooling. Thus its temperature will be a measure of the velocity.

Further, the heat power can be controlled by a feedback system to hold the temperature constant. In such case, the heat power is a measure of velocity. The above relations are the basis for thermal anemometry. Of course, the rate of cooling the wire depends upon the

(i) Dimension and physical properties of the wire,

(ii) Difference of the temperature between the wire and the fluid.

(iii) Physical properties of the fluid and stream velocity under measurement.

For a simple hot wire anemometer the first three conditions are effectively constant and the instrument response is then a direct measure of the velocity. The most important thermal velocity probes are

1. Hot wire anemometers
2. Hot film anemometers

We shall now discuss these in detail in the sections to follow.

7.10.1 Hot Wire Anemometers

These anemometers consist of

1. A fine wire supported by two larger diameter prongs
2. Device to provide electric current
3. Ceramic tubing
4. Ceramic cement
5. Calibrated scale

Principle

The temperature drop is proportional to the fluid velocity.

Construction

A fine film is supported by two large diameter prongs. The prongs are made in such a way that the film can be mounted either parallel or normal to the flow. Typically, the wire is 4 to 10 mm in diameter and l mm in length made of platinum or tungsten. A device is attached to this to provide electric current. The prongs are covered by ceramic cement and cement tubing. This set up is housed in inconel tubing. A calibrated scale is arranged to read the output at outlet.

Operation

An electric current heats the wire to a temperature well above the fluid temperature. The wires are extremely fragile, so hot wire probes are used only in clean gas flows. There are two types of hot wire anemometers as shown in Figure 7.19 (a) and (b).

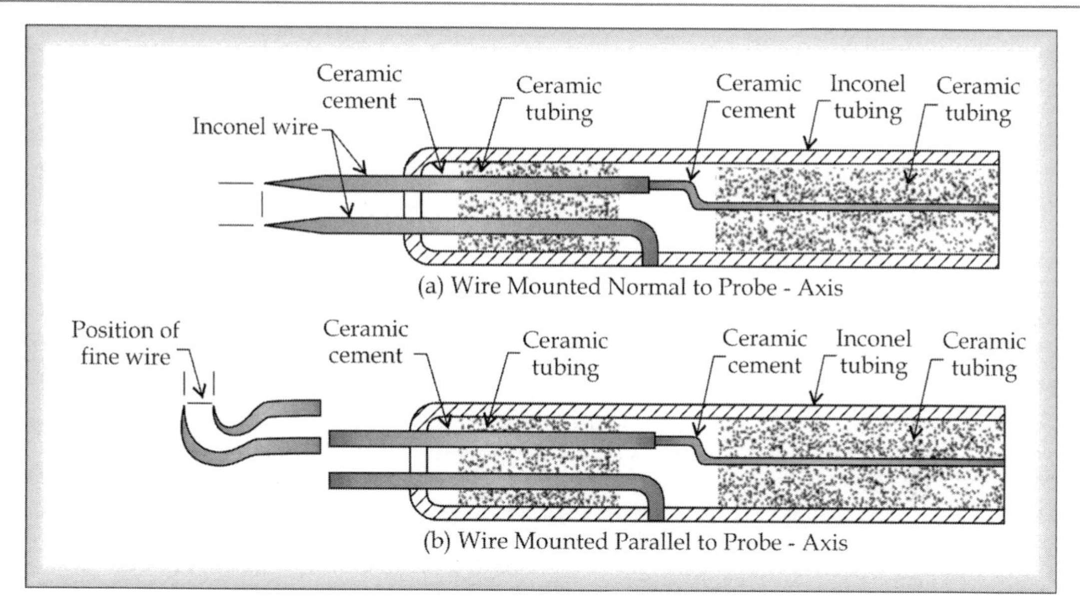

Fig. 7.19 Two Forms of Hot-wire Anemometer Probes

As shown in Fig. 7.19 (a & b) types of anemometer probe, fine wire is heated electrically and placed in the flow stream. The heat transfer rate from the wire can be expressed as

$$q = A\,(T_w - T_\alpha)$$

$$q = (a + b\sqrt{u}\,)\,(T_w - T_\alpha)$$

where
A = heat transfer area
T_w = wire temperature
T_α = free stream temperature of fluid
u = fluid velocity
a, b = constants obtained from calibration of the device.

The heat transfer rate must also be given by

$$q = i^2R_w = i^2R_o[1 + \alpha(T_w - T_\alpha)]$$

where
i = electric current
R_o = Resistance of the wire at the reference temperature T_o
α = temperature coefficient of resistance.

Equilibrium temperature is attained when the joule heating in the wire is just balanced by the convective heat loss

$$i^2R_w = K_c\, hA\,(T_w - T_\alpha)$$

where R_w is the resistance of the probe, and K, is a conversion factor from thermal to electrical quantities

Substituting $h = a + b\sqrt{h}$

$$
\begin{aligned}
i^2R_w &= K_c(a+b).A.(T_w - T_\alpha)\\
&= KA\,(a+b)(T_w - T_\alpha)\\
&= K_c Aa\,(T_w - T_\alpha) + K_c Ab\,(T_w - T_\alpha)\\
&= [K_c A(T_w - T_\alpha)a] + [K_c A(T_w - T_\alpha)b]\\
&= C_o + C_1
\end{aligned}
$$

where $C_o = K_c\,A\,(T_w - T_\alpha).a$

and $C_1 = K_c\,A\,(T_w - T_\alpha).b$

∴ The relation is linear between i^2 & R_w. This functional relationship is obtained by the process of calibration.

For measurement purposes the hot wire is connected to a bridge circuit as shown in Fig. 7.20.

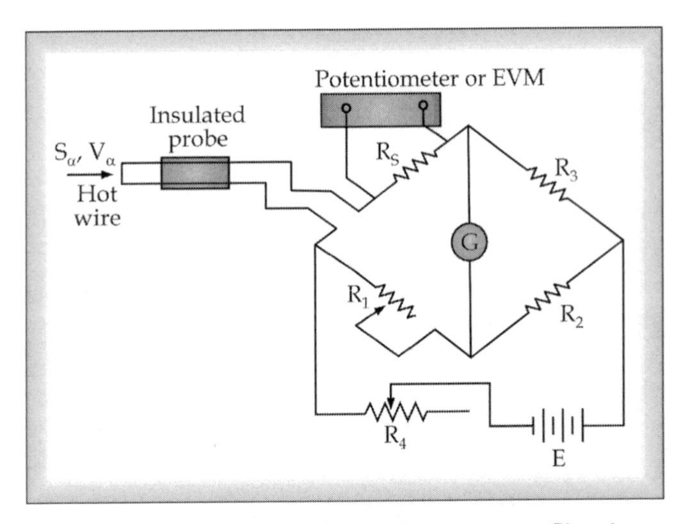

Fig. 7.20 Hot Wire Flow - Measurement Circuit

The current is determined by measuring the voltage drop across the standard resistance R_s and the wire resistance is determined from the bridge circuit. Null condition may be used for steady-state measurements while for transient measurement, oscilloscope output is used.

Knowing i & R_w, u can be calculated by using $i^2R_w = C_o + C_1$.

When the hot wire is employed for rapidly changing flow patterns, full amount of transient response of both the thermal and electrical resistance characteristics of the wire, two types of electrical compensation are employed in practice. They are

(a) **A constant current arrangement:** In constant current arrangement a large resistance is connected in series with the hot wire and a thermal compensating circuit is then applied to the output ac voltage. Refer figure 7.21.

(b) **A constant temperature arrangement:** In constant temperature arrangement a feedback control circuit is added to vary the current so that wire temperature remains nearly constant. Refer Figure 7.22.

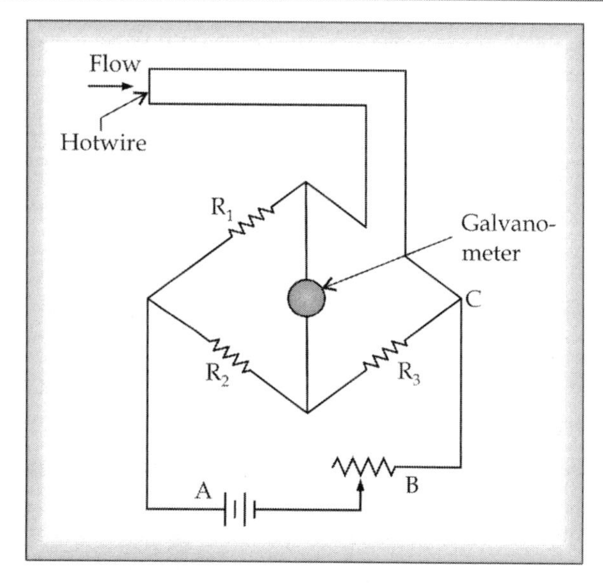

Fig. 7.21 Constant Current Hot Wire Bridge Circuit

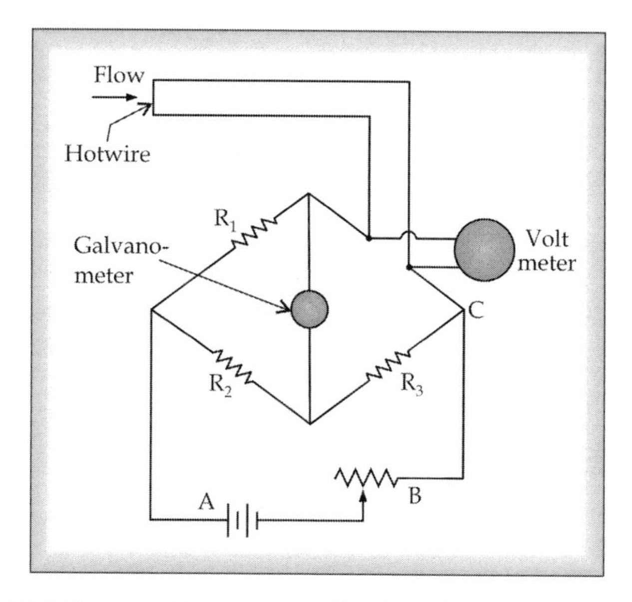

Fig. 7.22 Constant Temperature Hot Wire Bridge Circuit

The response of a hot wire placed normal to the flow velocity is given by

$$E^2 = E_0^2 + BV^n$$

where E = instantaneous value of this bridge voltage. E_o & B are constants which depend on cold resistance of the wire and on the properties of fluid medium.

The graph between E^2 and $\sqrt{V}$ is shown in Fig. 7.23 calibrates the relation.

Demerits of Hot Wire Anemometer

1. Limited Strength
2. Calibration changes caused by the accumulation of impurities on the wire.
3. Vibration of wire resulting in quick damage and flutter effects.
4. Unless the flow field is clean, its use is not recommended.

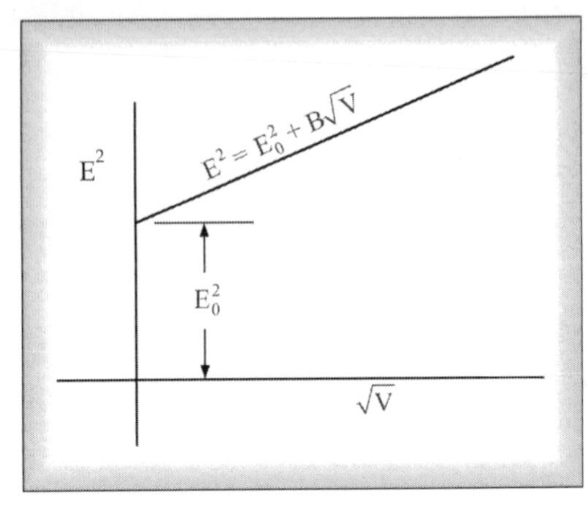

Fig. 7.23 Hot Wire Response

7.10.2 Hot Film Anemometers

This is similar to the above type of hot wire anemometers. In liquids, a variation of hot wire anemometer often called, thin film anemometer, is used. A thin film of platinum is coated on pyrex glass wedge and the connections are taken through heavy silver plates. It is very robust, easy to clean and can be used at higher temperatures. This device is approximately called a hot film probe, some probes are water cooled, which may be employed in high temperature applications.

7.11 MEASUREMENT BY SCATTERING

Principle

The principle of the measurements of flow by scattering measurement is based on the fact that light or sound waves undergo a Doppler shift when they are scattered off by the particles in a moving fluid. The important aspect of this shift is its proportionality to the particle velocity.

We know that the Doppler frequency shift in the scattered waves is proportional to the speed of the scattering particle. Therefore, by measuring the frequency difference between the scattered and unscattered waves, the speed of the particle or flow can be calculated.

The laser light and ultrasound wave sources are the most often used in fluid velocimetry. The Doppler shift in a scattered wave from a moving particle depends on the particle direction relative to the incident wave, and the observer's position as shown in Figure 7.24.

According to this theory, the frequency shift observed is given by

$$\Delta f = \left(\frac{2V}{\lambda}\right)\cos\beta\sin\left(\frac{\alpha}{2}\right)$$

where Δf = The Doppler frequency shift

V = The particle velocity

λ = The wave length of the original wave before scattering.

β = The angle between the velocity vector and the bisector of the angle SPQ.

α = The angle between the observer and the axis of the incoming wave.

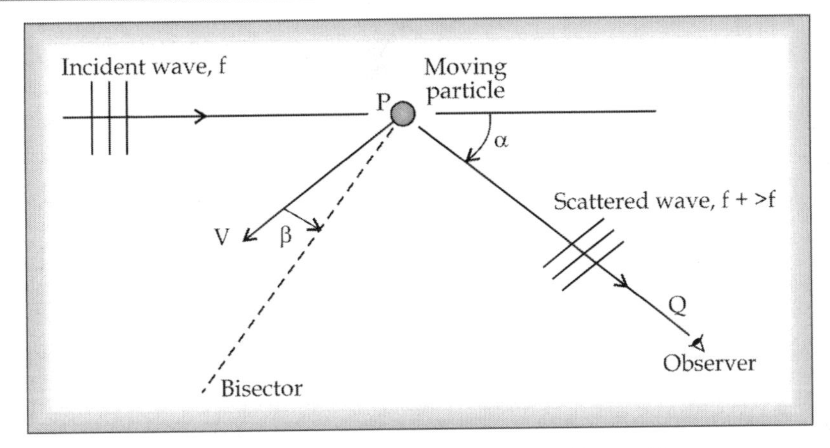

Fig. 7.24 Doppler Frequency Shift of Incident Wave of Frequency (*f*) on Moving Particle with Velocity (*v*)

When an incident wave of frequency (f) is scattered from a particle moving at speed V, the observer can notice a scattered wave of frequency f +Δf, where Δf is the Doppler frequency shift.

If we can measure this shift, we can find the particle speed. As the particle moves with the flow, this speed should be equal to the fluid speed.

Since the frequencies of laser light and ultrasonic waves are relatively high, the Doppler shift is only a small fraction of the original wave's frequency. For instance, the fractional frequency change in scattered laser light may be only $1{:}10^8$. The Doppler frequency is usually resolved by heterodyning the scattered wave with an unshifted reference wave to produce a measurable beat frequency.

7.11.1 Laser Doppler Anemometry

In the early days, Laser Doppler Anemometer (LDA) used separate scattering and reference beams to create on optical heterodyne at a photo detector as shown in Figure 7.25 (a). The light scattered by particles in the flow interfere with the light from the reference beam to produce beats at a frequency of one half the Doppler shift. Unfortunately, these systems are difficult to align because the intensity of the reference beam must be nearly equal to that of the scattered light in order to achieve an acceptable heterodyne. Consequently, reference beam systems have largely given way to the differential Doppler approach shown in Fig. 7.25 (b).

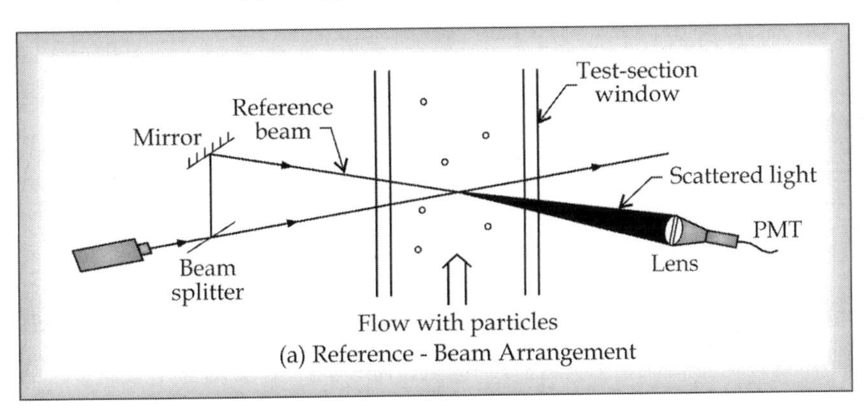

(a) Reference - Beam Arrangement

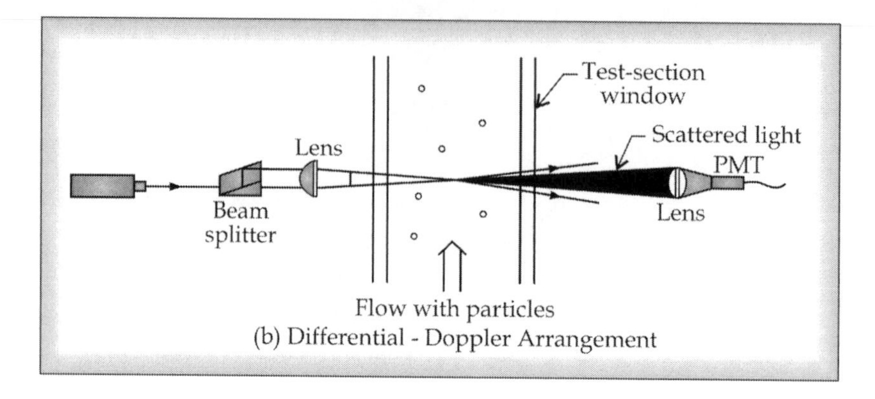

(b) Differential - Doppler Arrangement

Fig. 7.25 Laser-Doppler Optical Systems

The main components of this arrangement are:

1. Light source (Laser beam)
2. Beam splitter
3. Lens (or Mirror)
4. Photo-Multiplier Tube (PMT)
5. Flow through pipe
6. Calibrated scale

Construction

The laser is arranged to produce a beam through beam splitter and focusing lens, which is made to fall on the particles flowing in the test pipe as shown in the Figure 7.26. On the other side of this test pipe (through which the fluid flows), a photo-mltiplier tube (PMT) is arranged. The PMT consists of a receiver lens, photo detector and output signal wire.

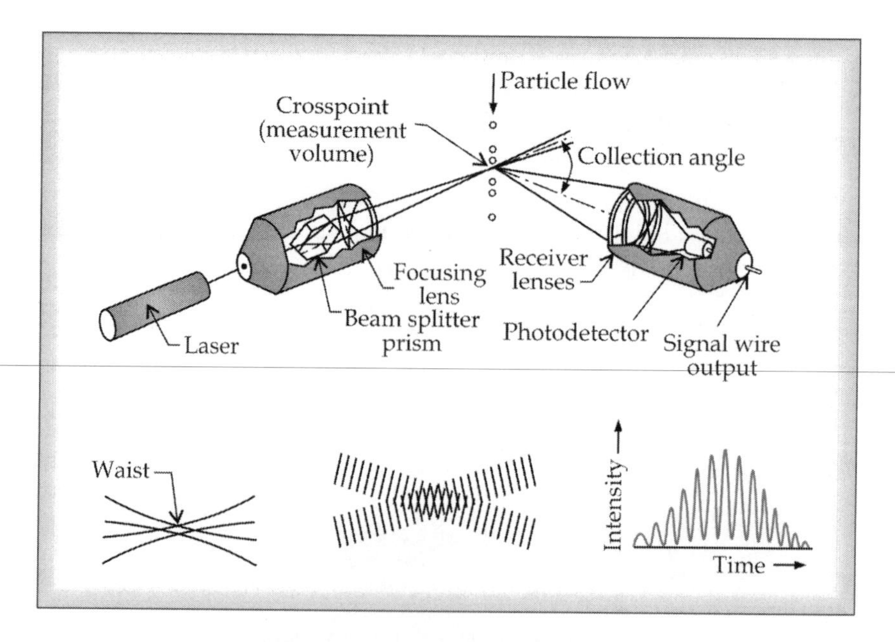

Fig. 7.26 LDA Transmitter and Receiver

Operation

The differential Doppler system splits the laser into two equal intensity beams, which are focused into an intersection point. Now, a particle (of the fluid) passing through the intersection scatters light from both beams, and this light is collected by a photo-multiplier tube (PMT). Obviously, the Doppler shift for each beam is different, by virtue of their different angles. However, the intensities of the two scattered waves are identical. Thence, resulting beat frequency at the detector is just the difference in the Doppler shifts of the two beams.

The signal of the differential LDA can be interpreted in terms of the light and dark interference fringes produced at the beam crossover point. The distance between these fringes is

$$\delta = \frac{\lambda}{2\sin\left(\theta/2\right)}$$

where δ = fringe spacing
λ = laser wavelength
θ = angle between the two beams.

A small particle crossing the fringe pattern produces a burst of scattered light whose intensity varies as the particle crosses each fringe. The frequency of this Doppler burst is just the particle velocity divided by the fringe spacing.

$$f_D = \frac{V_X}{\delta} = \left(\frac{2V_X}{\lambda}\right)\sin\left(\frac{\theta}{2}\right)$$

where f_D = The Doppler shift frequency
V_x = particle velocity in the direction normal to the fringes.

The burst frequency depends only on the velocity component normal to the plane of the fringes. Also note that the Doppler frequency is now independent of the position of the photo multiplier tube.

The PMT signal is processed electronically to find a frequency f_D and thus the particle velocity, V_x.

Merits/Demerits/Applications

The particles for scattering are usually seeded into the flow. For liquids, natural impurities may provide acceptable seed particles. If not, adding small polystyrene spheres or even a little milk will do.

In gases, an aerosol of non-volatile oil can be used. For good signal quality, the scattering particles should generally be of diameter smaller than the fringe spacing. Particle diameter of about 1ìm is common for gas flows.

Commercial LDA systems are expensive and their use is justified only when local and non intrusive measurements are absolutely imperative. If intrusion can be accepted and if flow reversal or flow borne particulates are not a problem, thermal anemometers are more economical alternative.

7.12 ULTRASONIC ANEMOMETRY

Ultrasonic waves can be conveniently applied in the range of hundreds of kilohertz (kHz) to several megahertz (MHz) to Doppler flow measurements in liquids. The ultrasonic transmitter and receiver may be piezoelectric and are generally designed to clamp outside of a pipe. Ultrasonic Doppler Anemometer (UDA) has been widely employed to flow metering. Further, the portable models are available at moderate cost.

Just like LDA, Ultrasonic Doppler Anemometer (UDA) also needs the particles to be present in the flow. Since ultrasonic flow meters are often used in industrial settings where deliberate seeding is inconvenient, they are largely applied to the measurement of slurries and dirty liquids which already contain particulates. UDA flow meters show accuracy between 1 and 5% of the flow rate.

Ultrasound can also be used for 'time of travel' measurements of mean flow velocity. A pair of piezoelectric or magnetostrictive transducers is placed on the outside of a conduit a few inches apart, one to serve as a 100 kHz source and the other as the pickup. When the sound wave travels from the transmitter (source) to the receiver, its ordinary velocity in the stationery fluid, C, either increases or decreases due to the Doppler Effect resulting from the liquid velocity V. For example, if the sound wave crosses the pipe at an angle 'θ' relative to the flow direction, then the effective velocity of the wave is $C \pm V\cos\theta$ depending on the direction of wave i.e. upstream (–) or downstream (+), since a wave travels slowly in the upstream than in the downstream. The flow velocity can be determined from the difference in travel time or the relative phase shift between upstream and downstream. Two ultrasonic transducers are reversed ten times per second.

SELF ASSESSMENT QUESTIONS-7.4

1. What is the principle of anemometers?
2. Explain Hot Wire Anemometers with a neat sketch.
3. What is the hot film anemometer?
4. What is the principle of the measurements of flow by scattering measurement?
5. Discuss Laser Doppler Anemometry with a neat sketch.
6. Explain ultrasonic anemometry.

7.13 THE INTERFEROMETER

The interferometer is used to obtain a direct measurement of density variations in the test section. This instrument consists of the components listed below

1. Light source (or transmitter of input)
2. Two Collimator Lenses
3. Two Splitter plates
4. Two Mirrors
5. Flow through pipe (to be measured)
6. Screen (or calibrated scale of output)

Principle

The number of fringes will be a function of the difference in optical path lengths for the two beams, which will become the measure of flow characteristics (speed or density of the flowing fluid.

Construction

A light source (input transmitter), a lens, a splitter plate and a mirror are fixed on one side and another splitter plate, mirror, lens and a screen (output receiver) are placed on the other side of the flow field test section as shown in the Figure 7.27.

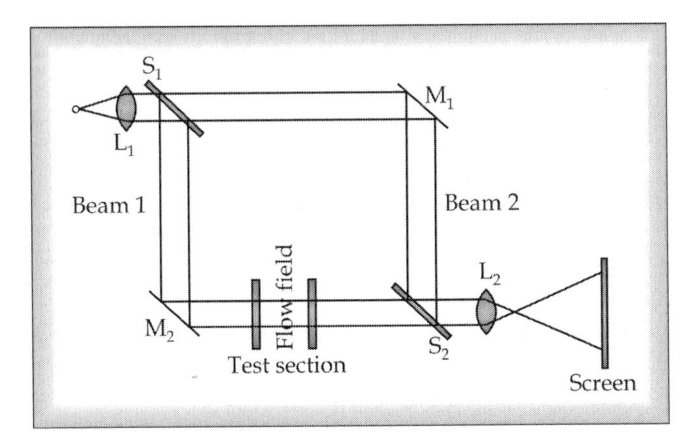

Fig. 7.27 Mach - Zehnder Interferometer

Operation

The light source (an input from transmitter) is collimated through lens L_1 onto the splitter plate S_1. This plate permits half of the light to be transmitted to mirror M_2 while reflecting the other half toward mirror M_1. Thence, *beam-1* passes through the test section, while *beam-2* travels an alternative path of equal length. The two beams are brought together again by means of the splitter plate S_2 and finally focused on the screen (or output receiver). If these two beams travel different optical lengths, because of geometry of the system or refractive properties of any element of the optical paths, the beams will be out of phase and would interfere. This action results in formation of alternate dark & bright regions called fringes. The number of fringes will be a function of the difference in optical path lengths for the two beams which are measured by a calibrated scale.

7.13.1 The Schlieren

The Schlieren is a device that indicates the density gradient. As shown in Figure 7.28, light from a slit source as is collimated by the lens L_1 and focused at plane-1 in the test section. After the light passes through lens L_2, an inverted image of the source at the focal plane-2 is produced. Lens L_3 then focuses the image of the test section on the screen at plane-3.

Now, let us consider the imaging process in more detail. The pencil of light emerging at point 'a' occupy a different portion of the various lenses from those emerging from point 'b' or any other point in the slit source. The regions in which these pencils overlap are shown in Fig. 7.28. Pencils of light pass through the image plane cd in the test section and the source image plane b'a'. An image of test section at d'c' is then uniformly illuminated since the image at b'a' is uniformly illuminated. This means that all points in the plane b'a' are affected in the same manner by whatever fluid effects may take place in the test section.

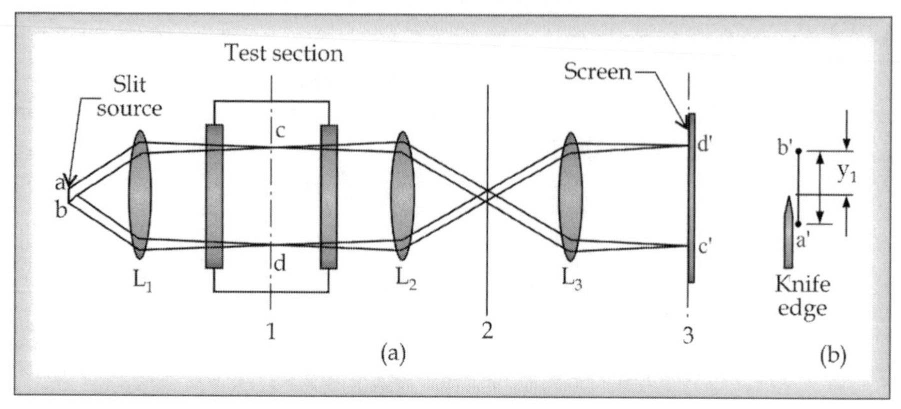

Fig. 7.28 Schlieren Flow Visualization

If the test section is completely uniform in density, the pencils of light appear as shown in Fig. 7.28, a pencil originating at point c is deflected by the same amount is a pencil originating at point d. This is consistent with the observation that all pencils of emerging in plane cd completely fill the image plane b'a'.

7.14 THE SHADOW GRAPH

In this method, direct viewing of the flow phenomenon is possible as shown in the Figure 7.29. Let the flow take place in the *x*-direction & having density gradients in *y*-directions. In the regions where there is no density gradient, the light rays will pass straight through the test section with no deflection. In other regions where velocity gradient exists deflection occurs. The result of this is the formation of bright & dark spots. The illumination will depend on the relative deflection of the light rays

$\frac{d\in}{dy}$ and hence $\frac{d^2\rho}{dy^2}$.

It is a very simple optical tool, and its effect may be viewed in several everyday phenomenon using naked eye & local room lighting.

Example: The free-connective boundary layer on a horizontal electric hot plate is clearly visible when viewed from the edge.

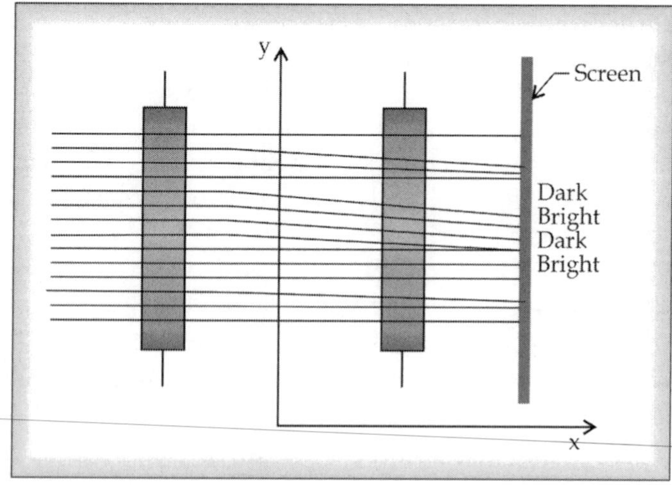

Fig. 7.29 Shadow Graph

7.15 CALIBRATION OF FLOW MEASURING DEVICES

For calibration of a flow meter, standardized flow producing facilities are required. Fluid at known rate of flow is passed through the meter and the rate compared with the meter is read out. The calibration can be done in two ways

1. Primary calibration and
2. Secondary calibration.

In primary calibration, the basic flow input is determined through measurement of time and either volumetric flow or mass flow.

In secondary calibration, the meter for which primary calibration has been done is used as a secondary standard for standardizing other meters.

Primary calibration in terms of volumetric flow is done by collecting fluid at constant flow rate for a given period of time. Volumetric displacement of a liquid may be measured in terms of the liquid level in a carefully measured tank or container. For a gas, volume may be determined through use of an inverted bell type gasometer. In terms of mass, it is accomplished by means of weight tank. Figure 7.30 illustrates a method of primary calibration for direct weight measurement in which a tank of known capacity is used as a collector and the time of collection is noted.

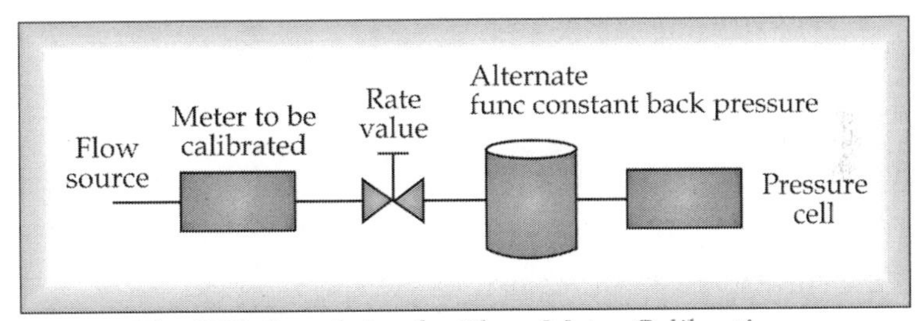

Fig. 7.30 Standpipe for Flow-Meter Calibration

Secondary calibration may be either direct or indirect. Direct secondary calibration is accomplished by placing a secondary standard in series with the meter to be calibrated and their respective readings are compared over the desired range of flow rate.

Turbine type meters are particularly useful as secondary standards for calibration of orifice or venturimeters.

Indirect calibration is based on maintaining equal Reynolds number in two different meters.

i.e.,
$$\frac{\rho_1 D_1 V_1}{\mu_1} = \frac{\rho_2 D_2 V_2}{\mu_2}$$

[Subscript 1 for standard meter; Subscript 2 for meter to be calibrated].

The practical significance lies in the fact that provided the similarity is maintained discharged coefficients of the two meters will be directly comparable.

7.16 FLOW-VISUALIZATION TECHNIQUES

Sometimes, we may encounter difficulties in predicting complex flow form. In such cases, an experimental visualization of the flow field becomes is necessary and forms an integral part of the design process.

The merit of this method is that it can illustrate an entire flow field, where as velocity probes yield information only at a single point. Regions of separation, recirculation, and pressure loss can also be identified without detailed velocity measurements or calculations and body contour may be adjusted to reduce the drag effects etc.

The following are some of common flow visualization techniques, which either use some visible material such as dye into the flow or density variations in the fluid itself.

7.16.1 Dye Injection

A colored dye is mixed through a small hole or holes. The dye track shows the path taken by the liquid as it passes the object.

7.16.2 Chemical Precipitates

A chemical change is produced in electrolytic process, to cause a solution to change colour or formation of a fine colloidal precipitate.

7.16.3 Smoke Trails and Smoke Wire

A thin steel wire of 0.1 mm (approximately) coated with oil is placed in an air flow. Now, as the electric pulse resistively heats the wire, the oil coating burns to form smoke. The thin line of smoke is carried along with the flow showing the fluid path lines.

7.16.4 Hydrogen Bubble Visualization

In this method, a very fine wire, often platinum of 25 to 100 ìm diameter, is placed in water flow. A second, flat electrode is placed nearby in the flow. When dc power of about 100 *V* is applied to the wire, electrolysis of water takes place at the wire surface creating tiny hydrogen bubbles. These are very small to experience buoyancy, and they instead flow with the water and become visible markers.

7.16.5 Laser Induced Fluorescence

Fluorescence is the tendency of some molecules to absorb light of one colour/frequency and re-emit light of a different colour/frequency. Fluorescent dyes are added to water, as in regular dye injection, but a thin sheet of lesser light is used to excite the dye in a specific plane of the flow. The resulting fluorescence can be used for gas flows as well.

Self Assessment Questions-7.5

1. What is the principle of the interferometer? Explain its construction and operation with a neat sketch.
2. What is the Schlieren? Explain it with a neat sketch.
3. Explain the shadow graph used for flow measurement.
4. In how many ways the calibration can be done? What are they in flow measurement.
5. Explain Primary calibration in flow measuring devices.
6. Explain Secondary calibration in flow measuring devices.
7. List out some Flow-Visualization techniques. Briefly explain them for measurement of flow.

SUMMARY

For the flow measurement, we can use some primary (or quantity) methods like weight/ volume tanks, burettes etc. The tilting trap meter uses principle of center of gravity (CG). Positive displacement meters such as Nutating Disk Meter (operates on the nutating disk principle i.e. the disk wobbling about the axis), Rotary-vane Flow Meter (fluid trapped by an eccentric drum), Lobed-Impeller Meter (the number of rotations of lobed-impellers), Reciprocating Piston Meter (liquid trapped in terms of stroke length and speed of piston), Gas Meter (number of rotations of drum) are narrated. The flow obstruction meters work on Bernoulli's equation which include Venturimeter, Orifice meter and Flow nozzle meter. Yet another flow meter is Rotameter which makes use of Drag effect. The turbine meters measure flow by number of rotations or by converting into a magnetic or electrical signal/ pulse such as reluctance pick-up. The Magnetic Flow Meters measure the flow by the voltage induced. The Vortex flow meters are based on vortices of a buff body placed in a stream. Pressure probes such as Pitot-Tube measure pressure at a particular point can also indicate the flow. Another most widely used physical relation, flow and heat/temperature is used in anemometers. The Doppler shift principles can also be used for measurements of flow for which Laser Doppler Anemometer (LDA) is good example. Even ultrasonic waves can be conveniently applied to Doppler flow measurements in liquids. The interferometers obtain a direct measurement of density variations in the test section using the concepts of light interference. The Schlieren is a device that indicates the density gradient. The construction, operations, merits/demerits and applications of all these instruments are described at sufficient lengths.

Some common flow visualization techniques, such as Dye Injection, chemical change, Smoke Trails and Smoke Wire, Hydrogen Bubble and Laser Induced Fluorescence are discussed.

This chapter explains the fundamentals for measuring fluid velocity, flow rate, flow quantity with varying degree of accuracy at all the situations such as studying ocean or air currents, blood flow rates and the flow of liquid oxygen in a rocket, oil pipelines, pumps, compressors and turbines etc, which have engineering importance because it is necessary to measure or detect a fluid flow rate for controlling processes and operations in any engineering applications.

KEY CONCEPTS

Laminar Flow: A flow when the motions of the individual particles are parallel to the pipe surface. $R_e < 2000$.

Transition flow: When velocity of flow increases, some particles acquire both stream wise & transverse velocity components and accordingly the laminar flow pattern gets disturbed, tends to turbulent flow. This flow is called a transition flow. $R_e = 2000$.

Turbulent flow: As the velocity is increased further violent mixing of eddies and swirls take place, such a type of flow is called turbulent flow. $R_e > 2000$.

Reynolds number R_e: A dimension less number $R_e = \frac{\rho v d}{\mu}$ that governs the transition from laminar to turbulent i.e., if $R_e < 2000$ (Laminar flow); $R_e = 2000$ (Transition flow); $R_e > 2000$ (Turbulent flow).

Primary or Quantity Meter: Indicates the amount of fluid that flows past a given point in a definite period of time, e.g., Weight meters, Traps, Volumetric meters like tank, Rotating impeller, and Nutating disk.

Tilting Trap Meter: Weighs discrete increments of liquids.

Nutating Disk Meter: Operates on the nutating disk principle. Water strikes the eccentrically mounted disk both the top and bottom of the disk remain in contact with the mounting chamber, and if the fluid has to move through the meter the disk must wobble or nutate about the vertical axis.

Rotary-vane Flow Meter: The vanes are spring loaded to maintain continuous contact with the casing and a fixed quantity of fluid is trapped in each section when the eccentric drum rotates.

Lobed-Impeller Meter: Measures the number of revolutions of the rotors when the incoming fluid is trapped between the two rotors.

Reciprocating Piston Meter: Meter for non-corrosive and low viscous liquids in which a piston reciprocates inside a cylinder with controlled inlet/outlet valves actuated by piston itself by an eccentric, and fluid is passed from the inlet to outlet alternately through each end of the cylinder. The volume of the fluid flow is directly proportional to both the stroke length and the piston speed.

Gas Meter: Consists of a drum of four compartments with inlet and outlet holes rotates inside a cylindrical casing, having more than half full of water, to act as gas seal. The gas pressure causes rotation of the drum and drum revolutions are totalized by a counter.

Obstruction Flow Meters: The flow is restricted causing a pressure drip by some means in a pipe carrying the fluid whose rate of flow is to be measured.

Venturi Meter: A device works on Bernoulli's equation, with short converging part, throat, and diverging part used to measure the rate of a flow of fluid.

Orifice Meter or Orifice Plate: A flat circular plate with a circular hole (orifice) concentric with the pipe measures the pressure drop.

Vena Contracta: The section of the jet of fluid where its area is minimum.

Flow-Nozzle Meter: A device to measure the ratio of flow in a pipe and can be treated as only the convergent part of the venturimeter.

Rotameter: A device works with drag effect using two (or more) pressure-measuring systems, one for low flow rates and another for high rates.

Turbine Meters: The fluid moving through the meter causes a rotation of the small turbine wheel, where a pulsating permanent magnet is enclosed so as to rotate with the wheel which is detected by a reluctance pick up attached to the top of the meter.

Magnetic Flow Meters: The flow of conducting fluid through a magnetic field produces induced voltage, which can be measured.

Vortex Shedding Flow Meters: A buff body placed in a stream produces vortices alternately, first to one side of the obstruction and then to the other which are measured.

Pressure Probes: Measures either of the two pressures i.e., total pressure or static pressure.

Static Pressure: The pressure exerted by the fluid when at rest.

Total or Stagnation or Impact Pressure: The pressure exerted by the flowing fluid when brought to rest. Total pressure = static pressure + dynamic pressure = static pressure + velocity pressure.

Total - Pressure Probe: Simple Pitot-tube measures total or impact pressure.

Pitot - static tube or Prandtl - Pitot tube: The Pitot tube combined with static openings.

Direction - Sensing Probes or Yaw-angle probes: Two impact tubes placed transferable to flow, rotatable around its axis and the angular position of the probe is then adjusted until the pressures sensed by the openings are equal.

Anemometers: A heated object placed in a moving stream loses its heat depending on fluid velocity and the rate of cooling is the measure of velocity.

Hot Wire Anemometers: Consist of a fine wire supported by two larger diameter prongs and an electric current heats the wire to a temperature well above the fluid temperature.

Hot Film Anemometers: A thin film of platinum is coated on pyrex glass wedge and the connections are taken through heavy silver plates.

Principle of Scattering Measurements: The light or sound waves undergo a Doppler shift when they are scattered off by the particles in a moving fluid and the particle or flow speed is the frequency difference between the scattered and unscattered waves.

Laser Doppler Anemometry (LDA): The light scattered by particles in the flow interfere with the light from the reference beam to produce beats at a frequency of one half the Doppler shift.

Ultrasonic Anemometry: The ultrasonic transmitter and receiver may be piezoelectric and are Ultrasonic Doppler techniques are adopted to flow metering.

Interferometer: The light source is collimated through lens onto the splitter plate which permits half of the light to be transmitted to mirror while reflects the other half toward mirror. Two such beams are brought together again by means of the splitter plate and at last focused on the screen. The number of fringes will be a function of the difference in optical path lengths for the two beams.

Schlieren: The flow measurement device that indicates the density gradient using interferometry.

Shadow Graph: Direct viewing of the flow in the x-direction & having density gradients in y-directions.

REVIEW QUESTIONS

Short Answer Questions

1. Why flow measurement is important?
2. On what basic principle, the obstruction meters work?
3. How is a venturi flow meter used to measure flow?
4. List the applications, advantages and limitations of a flow nozzle and venturi flow meter?
5. Describe the working of a rotameter.
6. How rate of flow is measured using a Pitot tube?
7. Where are magnetic flow meters used?
8. What is an ultrasonic flow meter?
9. Explain the turbine type anemometer.

10. Enumerate and explain the various methods for the measurement of velocity of flow at a point.
11. How can you classify the fluid flow? What is Reynolds number?
12. Discuss the types of vortex flow meter.
13. Write a short note on flow visualization methods.
14. List the various methods of flow measurement.
15. What are the advantages of rotameter?
16. List out the importance of calibration of flow measuring instruments.
17. Explain the principle of operation of Laser-Doppler' flow meter?
18. Explain the principle of operations of Turbine meter for the measurement of fluid velocity.

Long Answer Questions

1. Explain the construction and working of an orifice meter.
2. Explain the construction and working of a flow nozzle.
3. Explain the working of a magnetic flow meter.
4. Explain the ultrasonic flow meter using the travel time difference method.
5. Explain the ultrasonic flow meter using the oscillating loop system.
6. How does the measurement of compressible flow differ from that of incompressible flow? What devices are used for compressible flows?
7. Compare the relative merits and demerits of venturimeter, nozzle meter and orifice meter together with their sketches.
8. Sketch and explain the principle of working of

   (a) Turbine flow meter (b) Electromagnetic flow meter
9. Sketch and explain the principle of working of a hot wire anemometer. Discuss the constant current and constant temperature mode of operation. How the anemometer measuring the flow of liquids differs from that used for gases?
10. Explain the principle of working of positive displacement meters. Describe any are with neat sketch.
11. Enumerate the devices for the measurement of velocity of flow at a point, average velocity of flow, direction of flow, rate of flow and total flow.
12. Explain how “a wire mounted normal to probe axis”, type hot wire anemometer is used in flow measurement. Enumerate the principle of operation and its limitations.
13. With neat sketches explain in detail the functioning of any two flow measuring instruments working on Faraday’s law of induced voltage.
14. With neat sketch describe the principle of operation, construction, advantages and limitations of rotameter.
15. Explain the ultrasonic flow measurement. Why can’t it be used for measuring gas flows?
16. Write short notes (a) Magnetic flows meter (b) Ultrasonic flow meter
17. Explain the principle of hot- wire anemometer. What are its limitations and merits

18. List the various quantity flow meters and explain the working of a nutating disk flow meter.
19. Certain meters are known as variable head meters. Explain clearly what is meant by the designation variable head.
20. Explain the construction and working of a hot wire anemometer.

MULTIPLE CHOICE QUESTIONS

1. Flow rate of any fluid can be measured by____
   (a) Speedometer (b) Vibro meter
   (c) Quantity meter (d) Anemometer
2. The instruments used for flow, measurement must be ____
   (a) Large (b) Colored
   (c) Exact (d) Small
3. The overall accuracy of the instrument depends on the accuracy of measurement of pressure and____
   (a) Velocity
   (b) Temperature
   (c) Flow rate
   (d) Area of cross section
4. Volume tank is a ____ type of
   (a) Level meter
   (b) Venturimeter
   (c) Quantity meter
   (d) Flow meter
5. Orifice, nozzle, venturi and variable area meters come under____ category.
   (a) Obstruction
   (b) Velocity probes
   (c) Special methods
   (d) All the above
6. Velocity probes can be classified as probes.
   (a) Obstruction pressure
   (b) Static pressure
   (c) Total pressure
   (d) Both static and total pressure
7. Vortex shielding phenomenon comes under ____ methods.
   (a) Obstruction
   (b) Velocity probes
   (c) Special
   (d) Flow visualization
8. Flow visualization method consists of
   (a) Venturi
   (b) Total pressure probes
   (c) Turbine flow meters
   (d) Interferometry
9. ___ is the most extensively used form of the variable area flow meter.
   (a) Orifice meter (b) Venturimeter
   (c) Rota meter (d) Pressure meter
10. In Rota meter the pressure differential across the annular orifice is paranormal to___ of its flow area.
    (a) Twice (b) Thrice
    (c) Equal (d) Half
11. The tube materials of Rota meter may be of____
    (a) Wood, Fiber
    (b) Wood, Metal
    (c) Glass, Ceramics
    (d) Glass, Metal
12. The accuracy of Rota meter is ___ than venture and orifice meters.
    (a) Greater (b) Lower
    (c) Equal (d) Cannot say
13. The permanent magnet attached to the body of rotor is polarized at 90 degrees to the axis of Rotation, this is the principle of___ Meter.
    (a) Turbine flow
    (b) Magnetic flow
    (c) Hot wire anemometer
    (d) Ultrasonic

14. In hot wire anemometer, the sensor is a 5 micron diameter with ___wire.
    (a) Aluminum (b) Copper
    (c) Tungsten (d) Silicon

15. _____ meter depends upon the Faraday's law of electromagnetic induction.
    (a) Magnetic flow
    (b) Turbine flow
    (c) Hot wire anemometer
    (d) Ultrasonic

16. The word Ultrasonic is related to ___ energy.
    (a) Light (b) Sound
    (c) Chemical (d) Mechanical

17. With reference to flow measurement, LDA stands for _____
    (a) Length, Diameter, Area
    (b) Light Detection Anemometer
    (c) Laser Doppler Anemometer
    (d) Line, Diameter, Arm

18. Which of the following is not a flow measuring instrument
    (a) Venturimeter (b) Tachometer
    (c) Anemometer (d) Rotameter

19. Which of the following is a flow measuring instrument
    (a) Viscometer (b) Anemometer
    (c) Dynamometer (d) Tachometer

20. Venturimeter is a rate meter whereas _________ is the quantity meter
    (a) Nutating disk (b) Orifice meter
    (c) Nozzle meter (d) Rota meter

21. The Pitot-static tube measures the _____ velocity.
    (a) Mean (b) Median
    (c) Local (d) Low

22. Of the various methods of measuring discharge through a pipe line, the one with the least loss of energy and direct readings is by means of:
    (a) Transversing a Pitot static tube
    (b) Orifice meter
    (c) Notch
    (d) Venturimeter

23. In flow measurement, the expansion factor 'Y' depends upon:
    (a) g, P_2/P_1 and A_2/A_1
    (b) g , R and A_2/A_1
    (c) R, P_2/P_1 and A_2/A_1
    (d) g, R and P_2/P_1

24. The fluid flow between the electrodes of an electromagnetic flow meter generates e.m.f, which is a function of
    (a) Dynamic pressure
    (b) discharge
    (c) Flow velocity
    (d) Temperature

25. A quantitative estimate of density changes can be made by
    (a) Interferometer
    (b) Shadow graph
    (c) Schlieren system
    (d) Gas meter

26. An electromagnetic flow meter must be
    (a) Mounted horizontally
    (b) Rotated at constant speed to develop proper emf
    (c) Mounted vertically
    (d) Can be mounted in any position

27. Which of the following are used for clean fluids only
    (a) Turbine flow meter
    (b) Hot Wire Anemometer
    (c) Ultrasonic flow meter
    (d) Laser Doppler Anemometer

28. Which of the following is a positive displacement meter
    (a) Turbine flow meter
    (b) Rota meter
    (c) Nutating disc meter
    (d) Magnetic flow meter

29. Electromagnetic flow meters can be used for measuring the flow of
    (a) Non-conducting fluids
    (b) Fluids having some minimum electrical conductivity
    (c) Gases
    (d) Fluids which contain no solid matter

30. Performance of most flow measuring devices generally are independent of
    (a) The viscosity of the fluid
    (b) The compressibility of the fluid
    (c) Whether the flow is steady or not
    (d) Color of the fluid

ANSWERS

1.	(d)	2.	(c)	3.	(b)	4.	(a)	5.	(a)
6.	(d)	7.	(c)	8.	(d)	9.	(c)	10.	(a)
11.	(d)	12.	(b)	13.	(a)	14.	(c)	15.	(a)
16.	(b)	17.	(c)	18.	(b)	19.	(b)	20.	(a)
21.	(c)	22.	(d)	23.	(a)	24.	(c)	25.	(c)
26.	(d)	27.	(b)	28.	(c)	29.	(b)	30.	(d)

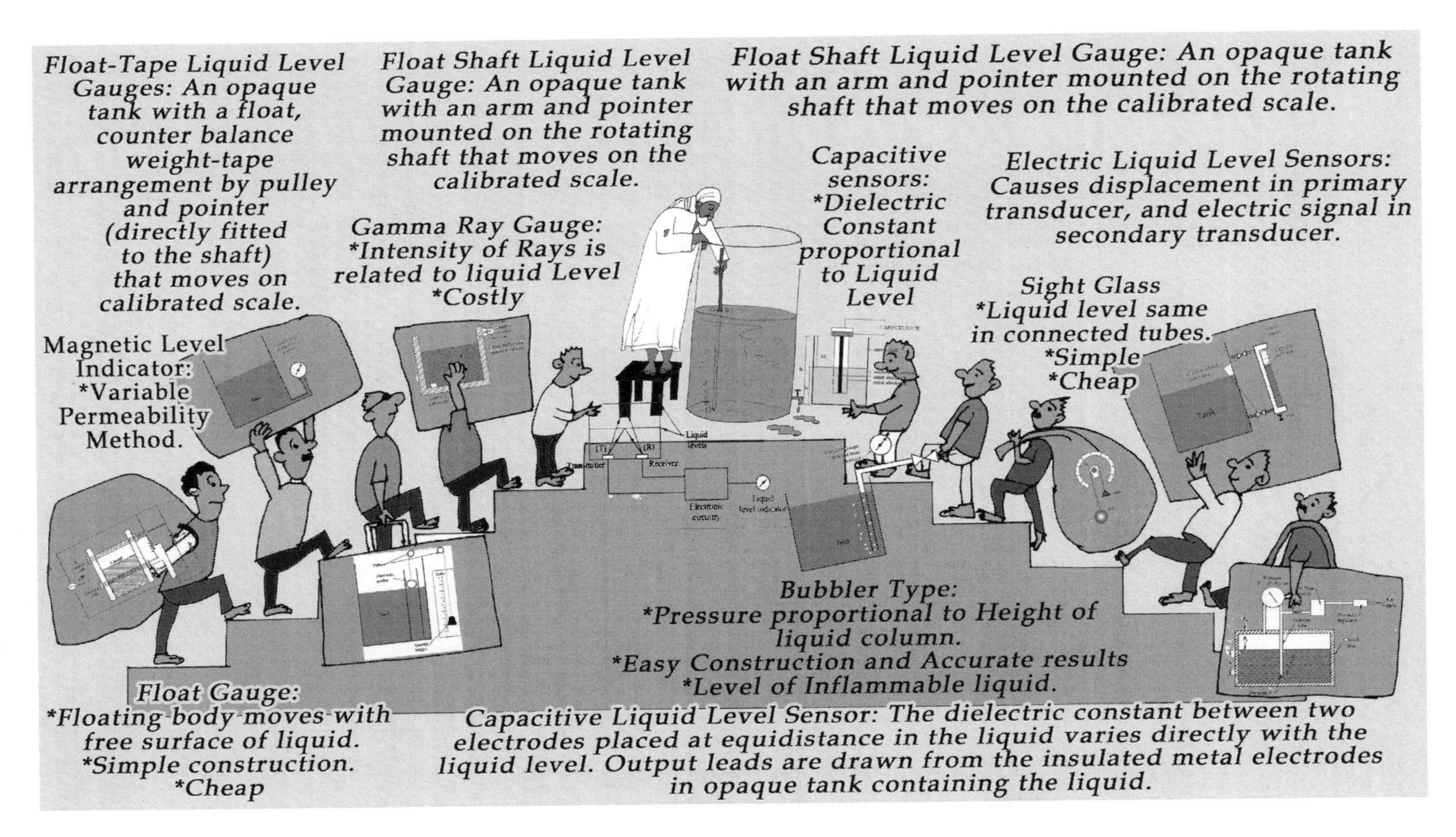

CHAPTER INFOGRAPHIC

Glimpses of Measurement of Liquid Level

S.No	Instrument	Transducer Type and Principle	Construction	Operation	Application	Merits	Limitations
1. Direct Methods							
1a.	Sight glass	Same liquid level in all connected tubes	Glass tube attached to the side of tank	Level of liquid in glass tube indicates level in tank	For both clean & colored	Simple and cheap	Not suitable for corrosive
1. (b) Buoyancy method							
1(b) (i)	Float Tape liquid level gauges	Freely floating body on liquid surface always follows liquid level	Float, counter balance weight-tape arrange-ment, runs over pulley with pointer attached	As liquid level rises/falls float moves and rotates pulley and hence the pointer	Used to measure levels of both clean, colored liquids and semi liquids	Simple and cheap	The density of the float material should be lower than the density of the material
1(b) (ii)	Float shaft liquid level gauges		A float attached to one end of arm and pointer to other end	When the liquid level/falls float also moves up or down			
2. Indirect Methods							
2.(a) Hydrostatic pressure devices (bubbler or purge systems)							
2.(a) (i) & (ii)	Hydrostatic pressure devices for open/closed tanks	The pressure of liquid is directly proportional to liquid level $P \propto h$ (or $P = h\rho g$)	A bubbler tube is immersed in the liquid contained in the open tank whose level is to be measured	When air pressure sent from the bubbler tube = pressure of liquid at bottom of tank, then bubbles come out. This (differential) air pressure indicates liquid level.	Used for measuring level of inflammable liquids	Easy to use and accurate indication	Compressed air is required
2.(b) Electric liquid level sensors							
2(b) (i)	Rheostat operated by a float	The resistance varies with liquid level and hence output voltage	Float moves on liquid level attached to arm, other end moves over resistance	As the end of arm moves over resistance, voltage changes & indicates level	If liquid level is controlled/ monitored from control room	Automatic control is possible	Effectiveness depends on the primary transducer
2(b) (ii)	Capacitive liquid level sensor	Dielectric const. b/n two equidistant electrodes immersed in liquid varies with liquid level	Insulated metal electrodes immersed in a liquid in an opaque tank	As liquid level rises/falls capacitance changes which indicates level		Can be used for both conductive/ non-conductive	The gap between the electrodes must be same

S.No	Instrument	Transducer Type and Principle	Construction	Operation	Application	Merits	Limitations
2(c)	Gamma ray liquid level gauge	γ rays intensity reaching source is governed by height of the liquid level	γ-ray source is placed at bottom of tank and g ray sensor at the top of tank	Higher liquid level, more is g ray absorption of and so is lower output of sensor and vice versa			Costly
2(d)	Ultrasonic liquid level gauge	Reflection of an acoustic wave from a liquid surface	Ultrasonic receiver and transmitter are arranged at bottom of tank	Measured either by time of reflect-wave reaches receiver or on basis of phase change	Can be used in submarines	Quick response	Involves complex circuitry
2(e)	Magnetic level indicator	Employ variable permeability method	Consists two coils L_1 and L_2 wound around a steel tube containing liquid	Inductance of search coil L_1 is used to measure of liquid level changes as conducting liquid moves into plane of coil L_2	For measuring level of condu-cting liquids	Can be used for good conducting materials like mercury	Only conducting liquid levels can be measured
2(f)	Cryogenic fuel level indicator	Wheatstone bridge circuit	Hot wire resistance elements are fixed to tank at known height intervals and are connected in series to form Wheatstone bridge	HT coeff. changes when resistance is immersed in rising liquid; brings down temp. & resistance; imbalances bridge ckt, hence measures level	For filling fuel tanks in rocker engines with cryogenic liquid fuels	Very accurate	Costly

Chapter 8

MEASUREMENT OF LIQUID LEVEL

STARTERS

To study this chapter, you should have awareness on the following concepts. For a better understanding, it is always a good idea to revise these prerequisites.

- Different types of liquids and their characteristics such as cohesiveness, adhesiveness, meniscus, density, solubility, miscibility, chemical reactivity, corrosiveness etc.
- Fundamentals of fluid mechanics and hydraulics
- Relation between pressure and volume, Boyle's law
- Pressure and its significance in liquids
- Concept of water head, and Bernoulli's principle
- Concept of float, bubblers, purges, buoyancy and Archimedes principle
- Conversion of pressure to stress, force on a fixed area
- Different types of pumps (reciprocating/centrifugal), types of turbines (impulse/reaction) and flow patterns in them.
- Awareness on impellers, vanes, traps and other fundamentals of fluid mechanics.
- The relation between pressure, fluid flow and level
- Effects of pressure on the volume of liquid and other parameters
- Preliminary idea on venturi, orifice, notch, nozzle, weir etc. and flow characteristics in these
- Broad idea on flow characteristics in pipes, friction losses in pipes etc.
- Concepts and fundamentals of magnetism, electricity, ultrasonic and gamma rays and
- First four chapters of this book

LEARNING OBJECTIVE

After studying this chapter you should be able to

- Understand the liquid level and classify the methods for liquid level measurement,
- Explain the direct and indirect liquid level measurement devices,
- Classify and discuss the float gauges,
- Describe the hydrostatic pressure devices for open and closed tanks,
- Discuss liquid level sensors, gauges and
- Understand magnetic level indicator and cryogenic fuel level indicator.

8.1 INTRODUCTION

Definition

Liquid Level: Liquid level refers to the position or height of a liquid surface above a datum line.

8.2 MEASUREMENT OF LIQUID LEVEL

The study of liquid level measurement is important as this is to be very precisely controlled in many places, some of which have been listed below.

1. Liquid level is an important process variable in industrial plants that use large quantities of liquids such as water, chemicals and solvents.
2. The level of water in the feed tank attached to the boiler is to be monitored continuously so that it is fed with water continuously.
3. The pressure is to be determined in the design of containers or tanks, which is a function of liquid level.
4. The level of water in reservoirs and distribution tanks must be measured, so that proper distribution of water is achieved and also the quantity available for future use can be known.

8.3 CLASSIFICATION OF LIQUID-LEVEL MEASUREMENT METHODS

The measurement of liquid depends on the characteristics of the liquids and properties of the liquids. The main properties that can influence the method of measuring the liquid level are:

- Color of liquid
- Viscosity
- Corrosiveness
- Volatility
- Reactivity
- Density

The liquid level measurement can be made directly by using the principles of buoyancy, Bernoulli's equation etc. while there are some indirect methods by using electrical sensors / transducers and magnetic, hydrostatic and ultrasonic methods. The Figure 8.1 clearly gives the classification of level measurement.

More Instruments for Learning Engineers

Nephelometer measures concentration of suspended particulates in a liquid or gas colloid by using a light beam (source) and a light detector generally set at 90°. It uses the principle that the particle density is dependent on properties of the particles such as their shape, color, and reflectivity and hence is a function of the light reflected into the detector from the particles.

Nephelometer

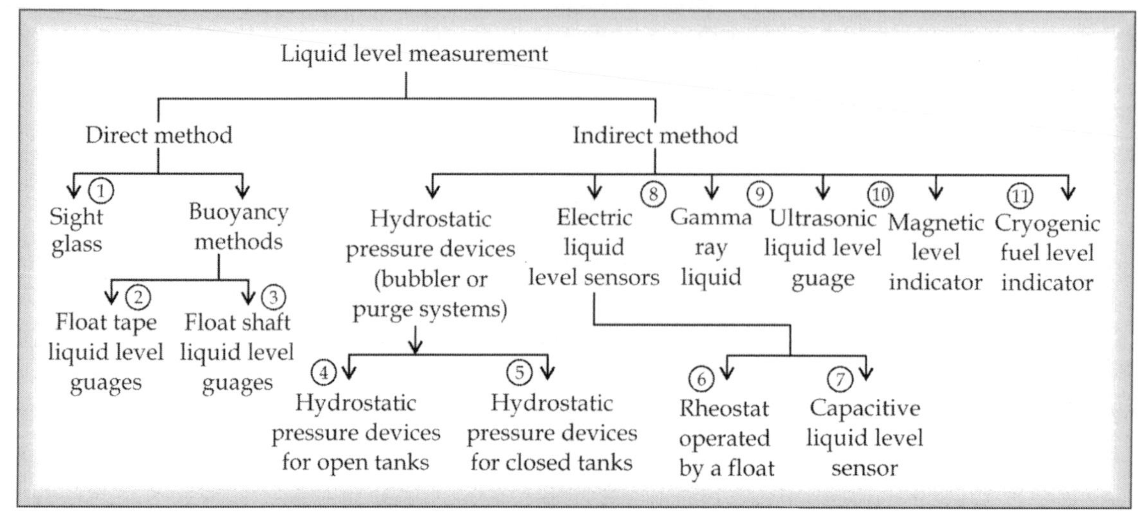

Fig. 8.1 Methods of Liquid Level Measurement

We shall discuss the above methods in detail in the following sections. The liquid level can be measured in number of ways as given as Figure 8.1.

8.4 DIRECT LIQUID LEVEL MEASUREMENT DEVICES

As described above, the direct methods use the Bernoulli's levels concept and/or buoyancy concept. These methods are quite simple to construct and operate. Under direct methods, we often come across the following three types of arrangements

1. Sight glass tube
2. Float-tape arrangement (Float-Tape Gauge)
3. Float-shaft arrangement (Float-Shaft Gauge)

8.4.1 Sight Glass Tube

Perhaps this is the simplest and cheapest direct-measuring instrument for liquid level and is most widely employed which can be often found at the municipal water tanks of colonies/towns, reservoirs tanks etc. The chief parts of this arrangement are:

1. Opaque tank containing liquid level
2. A transparent glass tube
3. Calibrated scale

Construction

An opaque tank with an opening at its bottom contains the liquid whose level is to be measured. A transparent glass tube is fitted on the outside of the tank and connected with a small tube through the opening at the bottom of the tank. Adjacent to this glass tube, a calibrated scale is fitted to read the liquid level as shown in the Figure 8.2. The tubes are generally made of toughened glass provided with a metal protective cover around it.

More precisely, the glass tube itself may be graduated and calibrated to read the water level directly.

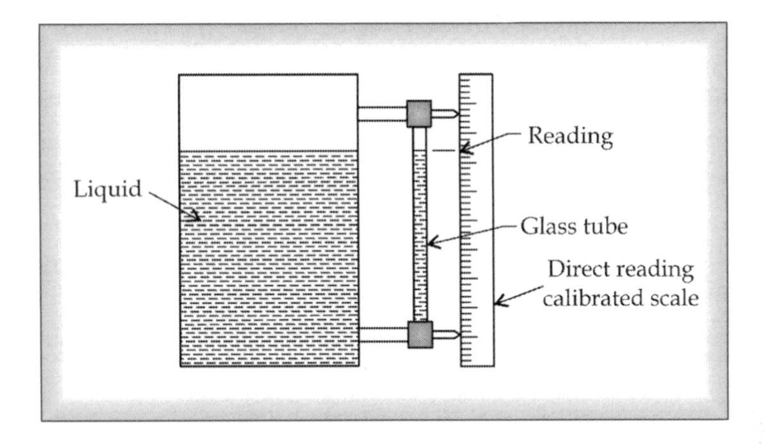

Fig. 8.2 Sight Glass Liquid Tube

Operation

When the liquid flows into the tank, it also flows into the transparent glass tube through the connected tube at the bottom.

The level in the glass tube exactly matches the level of the liquid in the tank.

The level of liquid in the transparent glass tube can be directly read from the scale near the transparent glass tube.

Merits

- It is simplest and cheapest
- Easy to construct
- No operation is required (directly shows the level)
- It is used to measure levels of both clean and colored liquids.

Demerits

- It is not convenient for corrosive liquids and those forms a scaling
- It is not suitable for viscous liquids as they do not raise or fall easily in the transparent glass tube.
- When the liquid is not clean (turbid), the foreign particle may choke the connecting tube and hence cannot read the level.

Applications

Most suitable for:

- Clean liquids
- Coloured liquids
- Non-corrosive liquids
- Non-reactive liquids

Precaution

The diameter of the glass tube must neither be too large to change the container liquid level by an appreciable amount, nor too small to cause capillary action on the tube. Further, the glass tube should be tough.

8.5 FLOAT GAUGES

A freely floating body on the surface of the liquid would always follow the varying liquid level and can be used for continuous liquid level measurement. .

8.5.1 Float-Tape Liquid Level Gauges

The float-tape liquid level gauge yet another simplest instrument to measure the level and can be constructed easily. The parts of this arrangement are:

1. Tank with liquid
2. Float
3. Cable running on a Pulley(s)
4. Compensatory weight
5. Pointer connected to pulley
6. Calibrated scale

Construction

An opaque tank contains the liquid whose liquid level is to be measured. A float tied to a cable and cable is run over a pulley and a compensatory (balancing) weight is tied to the other end of the cable. A pointer may be directly fitted to the shaft of the pulley (or another pulley fitted in parallel). A calibrated scale is fitted so that the point moves on it. The arrangement is shown in Fig. 8.3. The float is usually light weight metallic or ceramic or synthetic or wood and is in the shape of a sphere or cylinder (hallow).

Operation

When the liquid level rises or falls, the float also moves by the same distance and consequentially cable moves and hence rotates the pulley. The shaft also rotates along with the pulley and thence the pointer connected to the shaft moves on the calibrated scale. Thus the pointer attains a suitable position on the liquid level calibrated scale. The reading on the scale corresponding to the pointer will be the indication of the liquid level in the tank.

More Instruments for Learning Engineers

A **pH meter** measures acidity or alkalinity (pH) of a liquid. A typical pH meter consists of a special measuring probe (a glass electrode) connected to an electronic meter that measures and displays the pH reading. The probe of a pH meter is a rod like structure made up of glass with a bulb, containing a sensor at the bottom. To measure the pH of a solution, the probe is dipped into the solution. The probe is fitted in an arm known as the probe arm.

pH meter

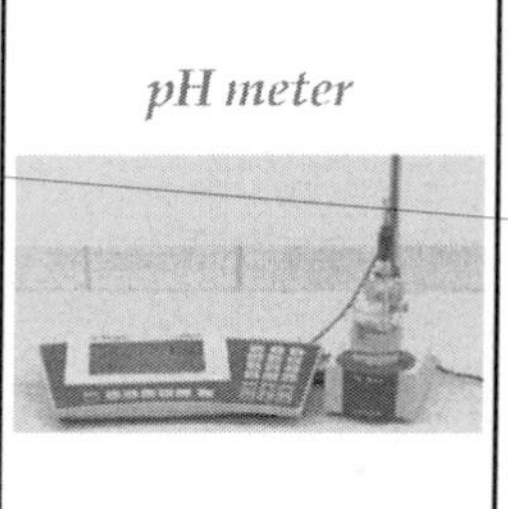

Merits

- Simple and cheap
- Easy to construct
- Easy to operate (automatically operates, no manual operation is required)
- Can be used for colorless liquids and viscous liquids also.

Demerits

- Difficult to use for reactive or volatile liquids

Applications

- Household overhead tanks
- Water tanks
- Boiler tanks

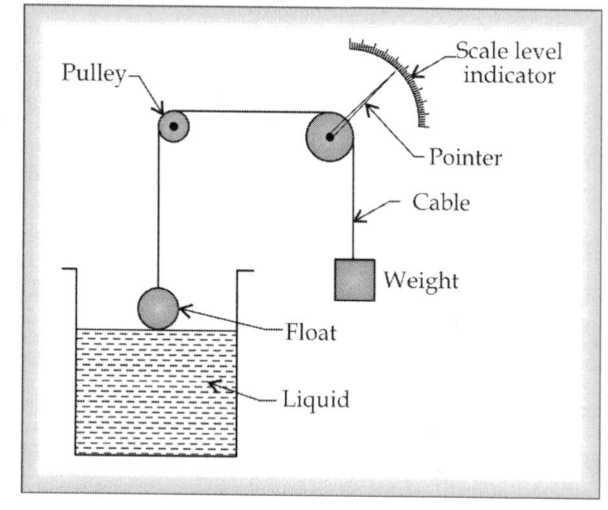

Fig. 8.3 Float-Tape Liquid Level Gauge

Precautions

- The material of the float should be so selected that it should not be corroded by the liquid whose level is measured.
- The density of the material of float should be less than the density of the liquid and the supporting material.

8.5.2 Float Shaft Liquid Level Gauge

Yet another direct measuring arrangement is that the float may be connected to pointer on rotating shaft. The important parts of this gauge are:

1. Tank with liquid
2. Float
3. Pointer arm
4. Rotating shaft
5. Calibrated scale

Construction

An opaque tank contains the liquid whose level is to be measured. An arm with pointer is mounted on a rotating shaft. At the bottom side of the arm, a float is attached and the top side of the arm (with pointed end) itself acts as a pointer, which moves on the calibrated scale as shown in Fig. 8.4.

More Instruments for Learning Engineers

Mile 8.3

A **fuel gauge** (or **gas gauge**) indicates the level of fuel contained in a tank of automobiles such as cars, lorries etc. including underground storage tanks in bunks. It consists of two parts: sensing unit and indicator. The sensing unit uses a float connected to a potentiometer. As fuel drains from tank, the float drops and slides a moving contact along the resistor, increasing its resistance. Further, it can also indicate 'low fuel' at a certain point. The indicator unit mounted on the dashboard displays the electrical current flowing through the sensing unit. Maximum current flows if the tank is full, at which the needle points 'F' and if empty, the least current flows to show 'E' on the indicator. Though the system is fail-safe; the corrosion or wear problems of the potentiometer are not ruled out.

Fuel/Gas Gauge/ Meter

Operation

When the liquid level in the tank increases or decreases the float moves up or down. As the float changes its position, the arm attached to the float rotates the shaft by a certain angle proportional to the liquid level. The arm acting as pointer turns by same angle. Thus relative position of the pointer measures liquid level on a calibrated scale.

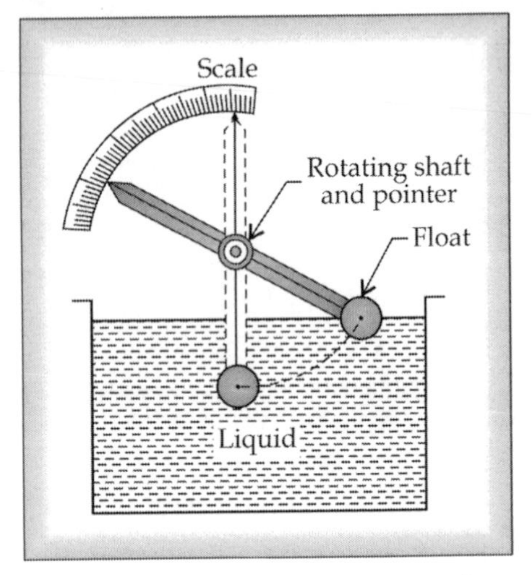

Fig. 8.4 Float-Shaft Liquid Level Gauge

Merits

- Direct measurement and no need of manual operation
- Faster response
- Simple to construct and easy to operate
- Comparatively cheaper to indirect methods.

Demerits

- Can be used for smaller range only, since it depends on the length of the arm
- Difficult to operate in corrosive liquids.

Applications

- This gauge can be used both in closed and open tanks
- Also it can be used to measure levels of semi liquids.
- This arrangement is widely used as control system of flush tanks.
- Can be used as fuel gauge.

Precautions

- The density of the float material should be lower than the density of the material and
- Float material should be so selected that it should not be corroded by the liquid.

Self Assessment Questions-8.1

1. Classify various methods for liquid level measurement.
2. Describe the operation of sight glass liquid tube. Explain its advantages and limitations.
3. What are the different types of direct liquid level measurement devices? Explain each of them.
4. What is the difference between float-tape and float shaft liquid level gauges?
5. With the aid of diagram explain the operation of float shaft liquid level gauge.
6. Give the applications and precautions of float shaft liquid level gauge.

8.6 INDIRECT LIQUID LEVEL MEASUREMENT DEVICES

The indirect methods use the principles of hydraulics, electrical, magnetic and ultrasonic mechanisms. The hydrostatic pressure devices take an advantage of bubbler or purge system. In the electrical devices, the principles involved in rheostat, capacitance can be integrated with the operation of the float. The principles of magnetic induction, gamma ray and ultrasonic rays can be used to measure the liquid levels. The most common indirect methods used to measure liquid level are listed below.

1. Hydrostatic pressure devices
   (a) For open tanks
   (b) For closed tanks
2. Electric liquid level sensors
   (a) Rheostat operated by a float
   (b) Capacitive liquid level sensor
3. Gamma ray liquid level sensor
4. Ultrasonic liquid level gauge
5. Inductive methods or magnetic level indicator
6. Cryogenic fuel level indicator

We shall now discuss the construction and operation of these measuring systems in detail.

8.7 HYDROSTATIC PRESSURE DEVICES (BUBBLER OR PURGE SYSTEM)

These devices are the indirect methods of measuring liquid level and are further classified into two types with reference to the type of tank used, namely,

(i) Hydrostatic pressure devices for open tanks.
(ii) Hydrostatic pressure devices for closed tanks

Principle

The principle for both the devices is as follows:

The pressure ($P = \rho gh$) of the liquid is directly proportional $(P \propto h)$ to the liquid level (h).

As shown in Fig. 8.5, a pressure gauge, fitted at the bottom of the tank is used to measure liquid level. The pressure gauge scale is calibrated to read liquid level directly based on the expression:

$$P = \rho gh$$

where

P = Pressure in N/mm^2

ρ = Density in kg/mm^3

g = Acceleration due to gravity in mm/sec^2

h = Height of liquid column in mm.

As the liquid level in the tank raises or falls, the pressure gauge reading also raises or falls respectively. Thus the reading on pressure gauge which is directly proportional to the liquid level indicates the measurement when calibrated accordingly.

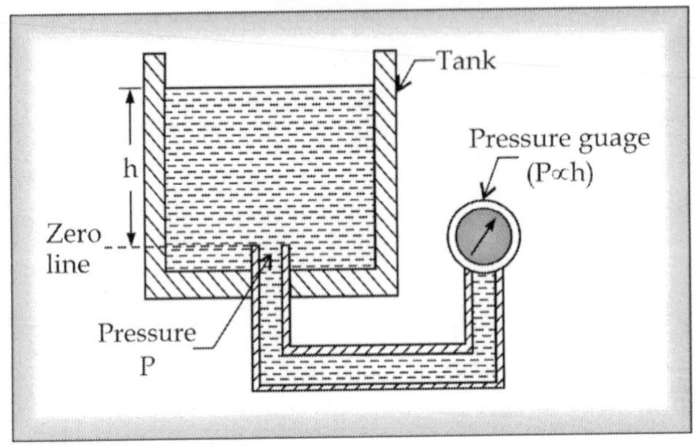

Fig. 8.5 Hydrostatic Pressure Devices

8.7.1 Hydrostatic Pressure Devices for Open Tanks

This arrangement consists of the following main parts:

1. An opaque open tank which contains the liquid
2. A bubbler tube
3. A pressure gauge
4. Air flow meter and
5. A pressure regulator

Construction

The opaque open tank contains the liquid whose level is to be measured. The bubbler tube is immersed in the liquid in the tank deep up to a point of reference (the minimum level, just above the bottom of the container) from which the level is measured. The bubbler tube is attached to an air flow meter along with the pressure regulator. A pressure gauge is fitted on the tube between the level and the air flow meter, as shown in Fig. 8.6.

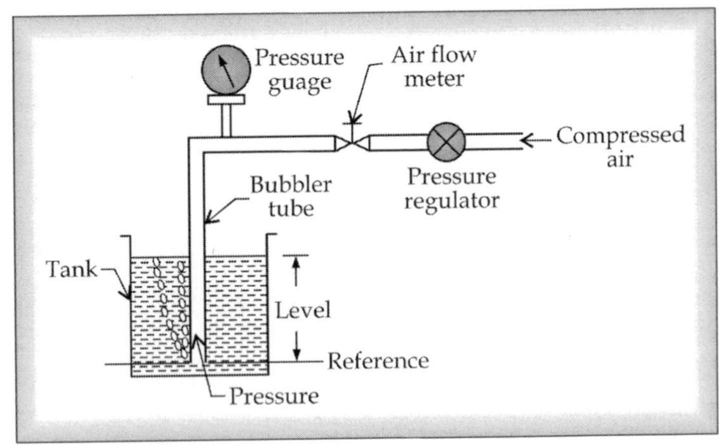

Fig. 8.6 Bubbler or Purge Type Hydrostatic

Operation

Air is sent through the pressure regulator into the bubbler tube whose opening is at the bottom of tank (tube is immersed in the liquid). Now, the air flowing through this bubbler tube is adjusted by means of an air flow meter until air bubbles start coming out of the open end of the bubbler tube i.e., at the bottom of the tank.

At this moment, the pressure of the air is equal to the pressure of the liquid at the bottom of the tank, which is often known as the hydrostatic pressure of the liquid. This pressure reading obtained from the pressure gauge (the hydrostatic pressure of the liquid), is proportional to the liquid level "h", which means higher the liquid level, higher will be the pressure read by the pressure gauge and vice - versa. The scale of the pressure gauge can be calibrated so that it directly refers reading of the liquid level in the tank. Generally, air and preferably, nitrogen is used for bubbling, particularly when used for measuring the level of inflammable liquids.

8.7.2 Hydrostatic Pressure Devices for Closed Tanks: Bubbler or Purge System

This is also same as the above described device i.e. hydrostatic pressure gauge for open tank (refer section 8.7.1) but for the difference that closed tank has differential pressure gauge instead of simple pressure gauge of open tank. The hydrostatic pressure device for closed tank consists of the main parts as listed below:

1. Closed opaque tank containing liquid
2. Bubbler tube
3. Air flow meter
4. Pressure regulator
5. Differential pressure gauge

Construction

The closed opaque tank contains the liquid level whose level is to be measured. A bubbler tube is immersed in the liquid deep up to a reference point (the minimum level, just above the bottom of the container) from which the level is measured. An air flow meter along with pressure regulator is connected to the bubbler tube as shown in Fig. 8.7. A differential pressure gauge connected between the top portion of the tank to sense the pressure of air in the tank above the liquid, P_i and the pressure of air flowing through the bubbler tube, P. The differential pressure gauge reads the pressure p is read as $p = P - P_i$ which is proportional to the height of the liquid in the tank.

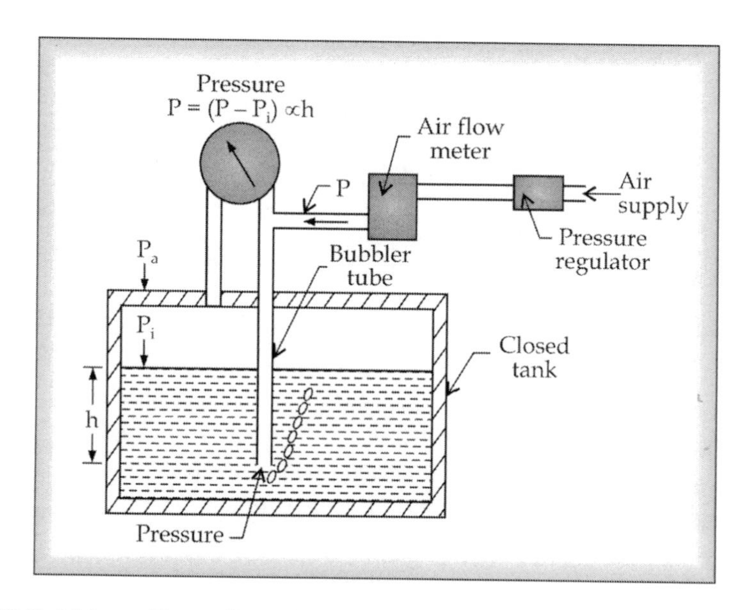

Fig. 8.7 Bubbler or Purge Type Hydrostatic Pressure Device for Closed Tanks

Operation

The air is sent through a pressure regulator into a bubbler tube whose opening is at the bottom of the tank. The flow of air through the bubbler tube is adjusted by means of an air flow meter until air bubbles start coming out of the open end of the bubbler tube. At this instance, the differential pressure indicated by the gauge caused to the pressure of air (P) passing through the bubbler tube and the pressure of air in the tank above the liquid (P_i) will be proportional to the level of liquid in the tank. [Since the pressure gauge is connected between the tank and the bubbler tube, the pressure indicated by the gauge is a differential pressure]. As the differential pressure $P-P_i$ is proportional to the liquid level "h" in the tank, a suitable scale of the differential pressure gauge can be calibrated to give a direct reading of the liquid level in the tank.

Contrast and Comparison between Hydrostatic Pressure Devices of Liquid Level Measurement in Open and Closed Tanks

1. The hydrostatic pressure devices for both the tanks are almost similar. Both the methods are based on the governing equation, Pressure (P) = ρgh where ρ is Density, g is Acceleration due to gravity and h is Height of liquid. Thus pressure is directly proportional to the height of the liquid. However, for open tank we take gauge pressure, $p \propto h$, while for the closed tank we take the gauge pressure (p) as differential pressure i.e., $(P - P_i) \propto h$, where P is the air pressure in bubbler tube and P_i is the air pressure in the tank.
2. The main parts are same in both the methods except for the difference that the simple pressure gauge is used for open tanks while differential pressure gauge for the closed tanks.
3. Construction and operation are same in both except for the difference that the closed tanks, as the name so suggests, have closed cover. Obviously, the cost of closed tanks is a little high. (Cost of construction increases due to differential pressure gauge also).

Merits

- Easy to construct
- Easy to operate (No manual operations are required)
- Can be used for corrosive, volatile (closed tank) liquids

More Instruments for Learning Engineers

Mile 8.4

A **stadimeter** is an optical device for estimating the range to an object of known height by measuring the angle between the top and bottom of the object as observed at the device. It is similar to a sextant, in that the device is using mirrors to measure an angle between two objects but differs in that one *dials in* the height of the object. It is one of several types of optical rangefinders, and does not require a large instrument, and so was ideal for hand-held implementations or installation in a submarine's periscope. A stadimeter is a type of analog computer

Stadimeter

Demerits

- Measurement is made with the reference point which is not exactly bottom of the tank but just a little above so that the bubbles can escape out easily. This gap creates an error and it is to be considered while calibrating the scale.
- Not comfortable to for viscous fluids

Applications

- Applicable for liquid even reactive, corrosive or volatile.
- Applicable for clean and coluorless liquids also

8.8 ELECTRIC LIQUID LEVEL SENSORS

Principle

Liquid level produces displacement through a primary transducer, which is used to give an electric signal using a secondary transducer. Thus, this electric signal can become a measure of liquid level. Based on the type of electric signal required, the electric liquid level sensors can be categorized into the following two types

1. Rheostat operated by a float
2. Capacitive liquid level sensors

8.8.1 Rheostat Operated by a Float

The electrical components can be conveniently used in measurement of liquid level for accuracy and quick response. The main components of this instrument are:

1. An opaque tank containing the liquid
2. Float
3. Wiper/Slider
4. Resistance
5. Voltmeter to measure Output Voltage

Principle

This instrument operates on the following principle.

More Instruments for Learning Engineers

A **dumpy level**, **builder's auto level**, **leveling instrument**, or **automatic level** is used to establish or check points in the same horizontal planes, applicable surveying and building to measure or set horizontal levels. In 1832, English civil engineer William Gravatt, who worked with Marc Isambard Brunel and his son Isambard on the Thames Tunnel, it was commissioned by Mr. H.R. Palmer to examine a scheme for the South Eastern Railway's route from London to Dover. Forced to use the then conventional Y level, during the works, Gravatt devised the more transportable and easier to use, called dumpy level.

The liquid level varies circuit resistance, which in turn changes the output voltage and becomes the measure of the liquid level.

Construction

An opaque tank, which contains the liquid whose level is to be measured. A float is connected to a wiper/slider which can move over a potential divider or a rheostat or a resistance as shown in Fig. 8.8. The potential dividers or rheostats output terminal is connected to a voltmeter to read the output voltage.

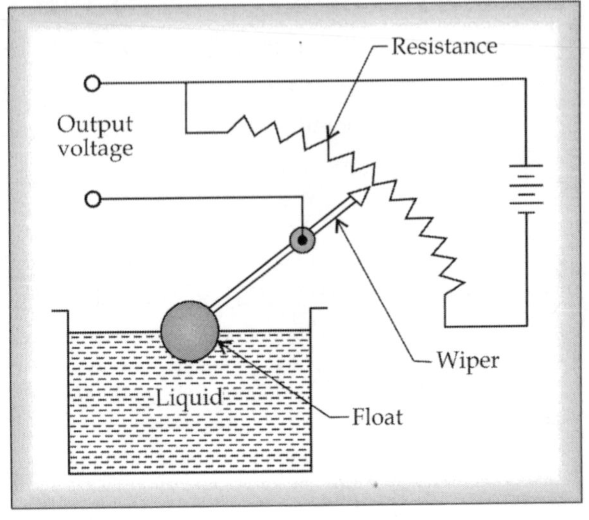

Fig. 8.8 Float Operated Potentiometer

Operation

When the liquid level rises or falls, the float floating on the liquid surface in the tank is lifted up or lowered down and then the wiper attached to it moves over the resistance or potential divider or rheostat. The movement of wiper alters the resistance accordingly. Thence the output voltage will change as the circuit resistance is changed. Thus, reading of the voltmeter becomes a measure of liquid level in the tank.

Merits

- More accurate
- Good speed of response
- No manual operation is required and can be operated from a remote place

Demerits

- Probabilities of short circuit, therefore care should be taken for proper insulation wherever required.

Application

- When liquid level is to be controlled or monitored from a control room, which is situated at a distance, this method is most suitable.

8.8.2 Capacitive Liquid Level Sensor

The main parts of these devices are:

1. An opaque tank containing the liquid
2. Metal electrode(s)
3. Output leads
4. Calibrated scale or measuring device of capacitance

Principle

The dielectric constant between two electrodes placed at equidistance in the liquid varies directly with the liquid level, which means, greater the height of liquid level, greater will be capacitance, and vice-versa.

Construction

An opaque tank contains the liquid whose level is to be measured. The insulated metal electrodes are immersed in the liquid. The output leads insulated of these electrodes are drawn out of the metallic tanks. The metal tank itself will act as an electrode while in non-metallic tanks, two electrodes have to be used as shown in Figure 8.9.

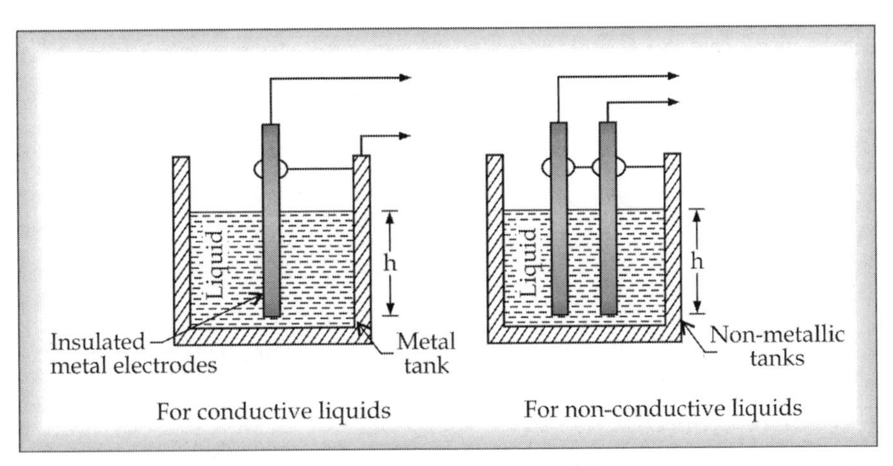

Fig. 8.9 Capacitive Liquid Level Sensor

Operation

These devices are used for measuring liquid levels in metal and non metal tanks. The operation slightly differs, as given below:

(i) **In Metallic Tanks:** The metallic tank contains liquid. This device can be used for the level measurement both non conductive as well as conductive liquids. Here, only one insulated metal electrode is sufficient. However, the gap between the electrodes and the wall of the tank should be even. For the non conductive liquids, the electrode and the tank wall act as the plates of a parallel plate capacitor and the liquid between them acts as dielectric, where as for conductive liquids, the electrode and the liquid act as the plates of the parallel plate capacitor, and the insulation between them acts as dielectric.

(ii) **In Non-Metallic Tanks:** In the case of non-metallic tanks, two insulated metal electrodes are immersed in the liquid whose level is to be measured. The gap between the two insulated electrodes should be constant. These electrodes will act as two plates of the parallel plate capacitor.

An amount of capacitance corresponding to the level of the liquid is generated in the system. That is to say, greater the liquid level, larger is the capacitance and vice versa. Thus the output in the form of capacitance when calibrated becomes a measure of liquid level in the tank.

Merits

- Can be used for any type of liquid conductive or non-conductive, corrosive on non-corrosive etc.
- Good input-output response.
- Accurate method.

Demerits

- Problem of scaling cannot be avoided, and measurement is difficult or inaccurate in such situations
- Extra care is to be taken to avoid electrical shocks and/or other electricity related problems

Applications

- Applicable for conductive liquids as well as non-conductive liquids.
- It is difficult to measure the level of turbid and very high dens liquid.

Self Assessment Questions-8.2

1. What are the different types of indirect liquid level measurement devices? Explain each of them.
2. Explain the construction and operation of bubbler type hydrostatic pressure device for open tanks
3. Elucidate the construction and operation of bubbler type hydrostatic pressure device for closed tanks
4. Compare and contrast the hydrostatic pressure devices used to measure the levels of an open tank and that of a closed tank.
5. Illustrate the differences in construction of metal tank and non-metallic tank of capacitive liquid level sensor.
6. How the electric liquid level sensors are classified. Describe each of them.
7. Explain the basic principle of capacitive liquid level sensor with a sketch.
8. Which device works on the principle of air pressure? Describe it in detail.

8.9 GAMMA RAY LIQUID LEVEL SENSOR

The main parts of this device are:

1. An opaque tank with the liquid
2. The gamma ray source
3. The gamma ray sensor (Geiger Muller Tube)

More Instruments for Learning Engineers

Pedometer is a portable electronic/electromechanical instrument that counts each step a person takes by detecting the motion of the person's hips. These are originally used in the fields of sports and physical fitness, but today these are very popular as daily exercise measurer and motivator. By wearing on the belt and kept on all the day, it can record the number of steps the wearer has walked that day and thence the distance (number of steps ×step length). Pedometers help getting fit and losing weight. A total of 10,000 steps per day, equivalent to 8 km (5 miles), are often considered as the benchmark for an active lifestyle. Presently, these are integrated into portable consumer electronic devices such as music players, smartphones, and mobile phones

Pedometer

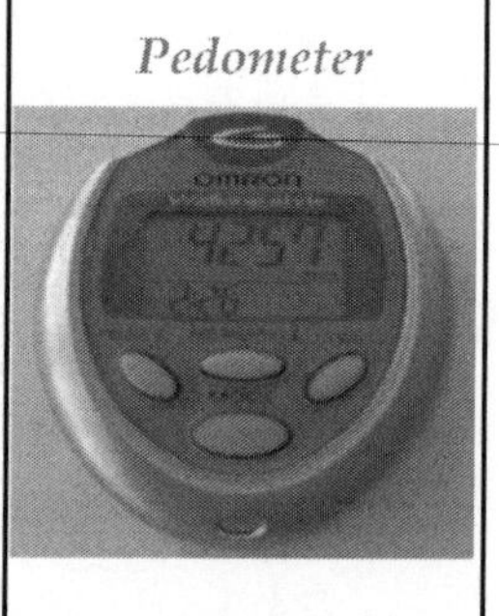

Principle

The intensity of gamma rays reach the sensor from a source is governed by the height of the liquid level in between the sensor and the source.

Construction

An opaque tank contains the liquid whose level is to be measured. A gamma ray source is arranged at the bottom of the tank. A gamma ray sensor (Geiger Muller Tube) is fitted at the top of the tank as shown in Fig. 8.10.

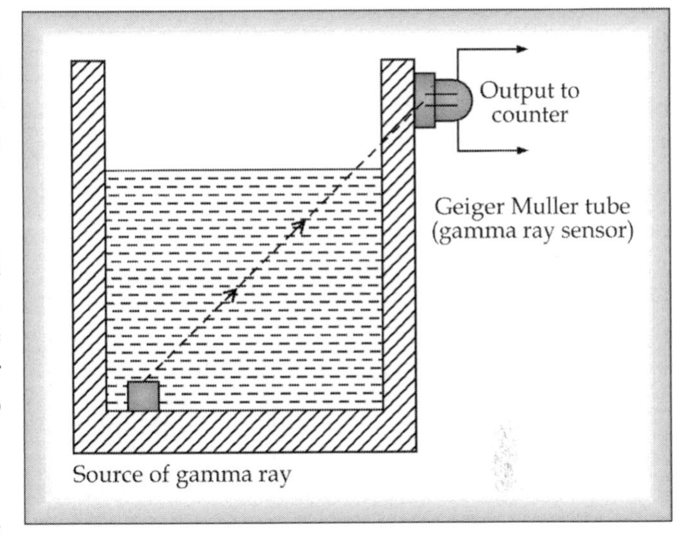

Fig. 8.10 Liquid Level Measurement by using Gamma Rays

Operation

Whenever the level of liquid in the tank is to be measured, the gamma rays from the source are impinged and these rays would pass through the liquid in the tank and reach the gamma ray sensor, which would sense the intensity of gamma rays which is proportional to the level of liquid in the tank. i.e., higher the liquid level, greater is the absorption of gamma rays by the liquid and hence lower will be the output of the gamma ray sensor and vice versa. Thus the output from the gamma ray sensor would become a measure of liquid level in the tank when calibrated.

Merits

- Easy to construct and simple to use
- Can be operated only when required
- Can be operated from a remote place

Demerits

- Gamma rays are somewhat harmful if impinged excess
- Gamma ray sources and sensors are relatively costly

Applications

- Useful for sophisticated equipment
- Useful when it is required to operate from a longer distance

8.10 ULTRASONIC LIQUID LEVEL GAUGE

The main parts include:

1. An opaque tank with the liquid
2. The ultrasonic ray transmitter (source)
3. Ultrasonic ray sensor (receiver)

Principle

It is based on the principle of reflection of an acoustic wave from a liquid surface.

Construction

An opaque tank contains the liquid whose level is to be measured. The ultrasonic transmitter (T) and receiver (R) set are fitted at the bottom of the tank as shown in the Fig. 8.11.

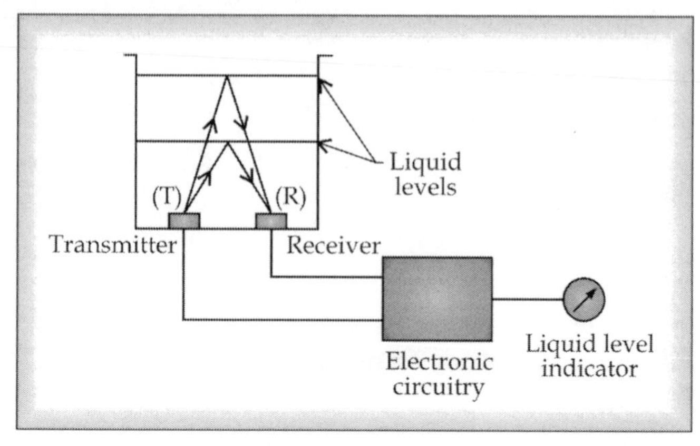

Fig. 8.11 Ultrasonic Liquid Level Sensor

Operation

The ultrasonic wave sent by a transmitter *T* is reflected at a surface of liquid, which is received by a receiver *R*. As the liquid level changes the transmission time of the wave also changes. Measurement of level can be made either by the time taken by the reflected wave to reach the receiver or on the basis of change of phase of the wave. Piezoelectric crystal such as quartz or barium titanate can be used as transmitter and receiver.

Merits

- Easy to construct and simple to use
- Can be operated only when required
- Can be operated from a remote place

Demerits

- Ultrasonic wave sources and sensors are relatively costly
- The principle although is simple, the actual instrument involves complex circuitry.

Applications

- Useful for sophisticated equipment
- Useful when it is required to operate from a longer distance

8.11 INDUCTIVE METHODS OR MAGNETIC LEVEL INDICATOR

The inductive level transducers are mostly used for measurement of level of conductive liquids employing variable permeability method. The arrangement is shown in Fig. 8.12. It uses two coils L_1 and L_2 wound around a steel tube containing the liquid. The coils are connected in series through a resistance and the circuit is energized by an alternating current (AC) source.

The inductance of each coil is initially equal say about 250 μH. Here, one coil say, L_1 acts as the search coil. It can be set to a predetermined level, and the inductance of the search coil changes rapidly as the conducting liquid moves into the plane of the coil. This method works well since the tape material of the coil is weakly magnetic and the liquid metal is a conductor that allows eddy currents to flow through it. Note that the relationship between the output voltage and the liquid level is non-linear.

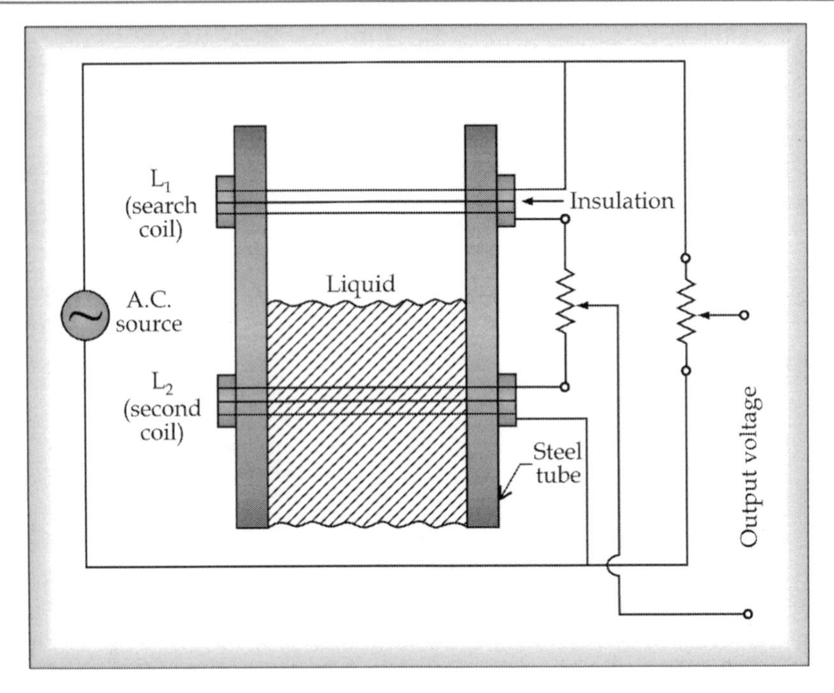

Fig. 8.12 Measurement of Liquid Level with Variable Permeability Method

Another method uses the loading of secondary winding of a transformer. This method is applicable to good electrical conducting materials like mercury. This set up is shown in Fig. 8.13.

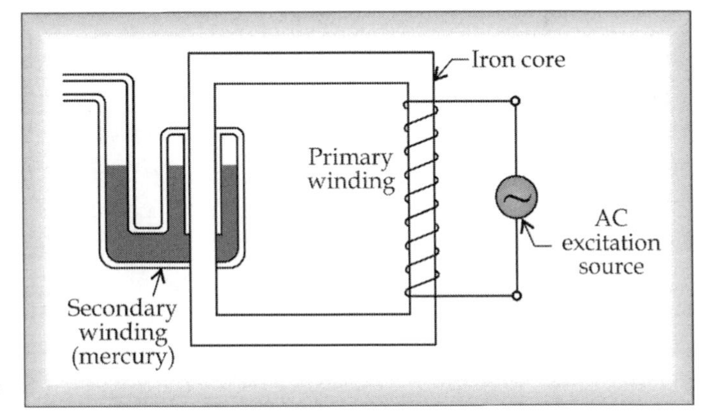

Fig. 8.13 Liquid Level Determination by Loading of Secondary Winding

A coil is wound around one core of a two-limbed transformer. The mercury column surrounding the iron core forms the secondary of the transformer. The resistance of the secondary winding depends upon the height of the mercury column and therefore the power consumption as monitored on the primary side is indicative of the liquid level.

8.12 CRYOGENIC FUEL LEVEL INDICATOR

In this case, hot wire resistance elements are located along the height of the tank as shown in Fig. 8.14. The level is measured by using a Wheatstone bridge. A number of hot wire resistance elements are fixed to the tank at known intervals of height and are connected in series to form one arm of the Wheatstone bridge. When a resistant element is immersed in the rising liquid, its heat transfer coefficient drastically changes, bringing down the temperature and the resistance values of the element. Thus the balance of the bridge circuit is affected and the unbalanced voltage, V_a indicates the change in the level. The voltage output V_a changes in steps as per the arrangement of hot wire resistance elements in the tank.

Applications

This method of level measurement is most suitably adopted in filling up fuel tanks in rocket engines with cryogenic liquid fuels.

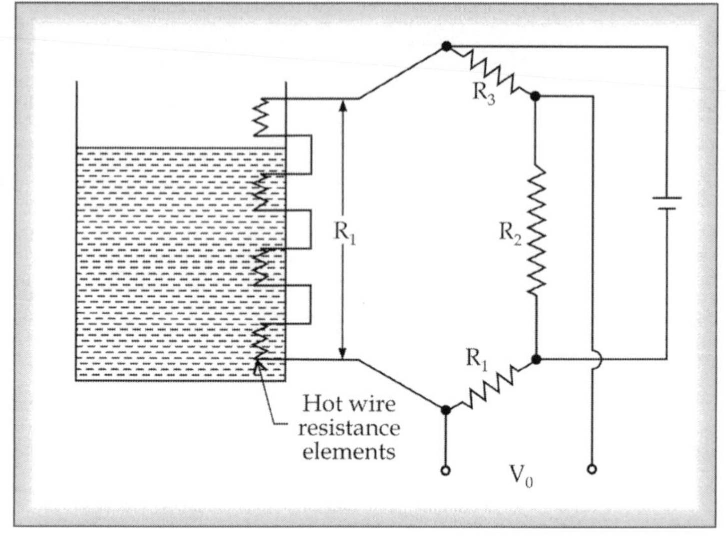

Fig. 8.14 Resistor Level Indicator (Cryogenic Fuel Level Indicator)

Self Assessment Questions-8.3

1. Describe the working principle of cryogenic fuel level indicator.
2. What is the basic principle of ultrasonic liquid level gauge? Explain its operation.
3. Elucidate the operation of gamma ray liquid level sensor.
4. Draw the circuit diagram of cryogenic fuel level indicator.
5. How the liquid level measurement is done by using magnetic level indicator?
6. Differentiate between direct and indirect liquid level measuring devices.

More Instruments for Learning Engineers

Mile 8.7

A **bevameter** measures the mechanical properties of soil to study vehicle mobility. The test consists of penetration test to measure normal loads and shear test to determine shear loads exerted by vehicle.

Bevameter

SUMMARY

Liquid level refers to the position or height of a liquid surface above a datum line. The measurement of liquid depends on the characteristics of the liquids and properties of the liquids such as Color, Viscosity, Corrosiveness, Volatility, Reactivity, Density etc. The direct methods used in Sight glass tube, Float-Tape Gauge, Float-Shaft Gauge work on Bernoulli's levels and/or buoyancy concept. The indirect methods use the principles of hydraulics, electrical, magnetic and ultrasonic mechanisms. The hydrostatic pressure devices take an advantage of bubbler or purge system. In the electrical devices, the principles involved in rheostat, capacitance can be integrated with the operation of the float. The principles of magnetic induction, gamma ray and ultrasonic rays can be used to measure the liquid levels. Inductive or Magnetic level indicators are mostly used for measurement of level of conductive liquids employing variable permeability method. Hot wire resistance elements are located along the height of the tank in cryogenic fuel indicator. The construction and operation of these instruments are narrated and merits, demerits and applications are enumerated at appropriate places in this chapter.

KEY CONCEPTS

Liquid Level: The position or height of a liquid surface above a datum line.

Sight Glass: A transparent glass tube with calibrated scale is fitted on the outside of the tank.

Float Gauges: A freely floating body on the surface of the liquid follows the varying liquid level.

Float-Tape Liquid Level Gauges: An opaque tank with a float, counter balance weight-tape arrangement by pulley and pointer (directly fitted to the shaft) moves on calibrated scale.

Float Shaft Liquid Level Gauge: An opaque tank with an arm and pointer mounted on the rotating shaft moves on the calibrated scale works on the principle $p = \rho gh$.

Hydrostatic Pressure Devices for Open Tanks: A pressure regulator connected to bubbler tube, which is immersed in the liquid contained in the tank.

Hydrostatic Pressure Devices for Closed Tanks: Bubbler or Purge System: A bubbler tube connected with air flow meter and pressure regulator, immersed in the liquid in a closed opaque tank containing the liquid level. A differential pressure gauge connected at the top of the tank senses the pressure of air in the tank above the liquid.

Electric Liquid Level Sensors: Causes displacement in primary transducer, and electric signal in secondary transducer.

Rheostat Operated by a Float: The circuit resistance change, depending on the liquid level and output voltage is proportional to liquid level. A float in opaque tank is connected to a wiper/slider moves over a potential divider or rheostat.

Capacitive Liquid Level Sensor: The dielectric constant between two electrodes placed at equidistance and liquid varies directly with the liquid level. Output leads are drawn from the insulated metal electrodes in opaque tank containing the liquid.

Gamma Ray Liquid Level Sensor: The intensity of gamma rays reach the sensor (Geiger Muller Tube) from a source is governed by the height of the liquid level in between the sensor and source.

Ultrasonic Liquid Level Gauge: Works on the principle of reflection of an acoustic wave from a liquid surface. The ultrasonic wave sent by a Piezoelectric crystals such as quartz or barium titanate transmitter is reflected at surface of liquid and receiver measures the transit time.

Magnetic Level Indicator: Measures level of conductive liquids using variable permeability method. It uses two coils of equal inductance wound around a steel tube containing the liquid, are connected in series through a resistance and the circuit is energized by an alternating current source to measures power consumption.

Cryogenic Fuel Level Indicator: Hot wire resistance elements are located along the height of the tank and the level is measured by using a Wheatstone bridge.

REVIEW QUESTIONS

Short Answer Questions

1. What is liquid level measurement and what is its importance?
2. Give classification of various liquid level measurements.
3. What do you understand by direct method of level measurement?
4. How do you measure liquid level using a direct method? Give examples.
5. What is an indirect method of level of measurement?
6. When do you use an indirect method of level measurement?
7. What are the advantages and disadvantages associated with direct methods of liquid level measurement?
8. What are the advantages and disadvantages associated with indirect methods of liquid level measurement?
9. List the principle of conversion used in the indirect methods of level measurement
10. How do you use a capacitance method to detect level?
11. Among capacitance and ultrasonic level measurement identify which is contact type and which is of non contact type?
12. What is round trip transit time in ultrasonic level measurement?
13. What is the basic measurement that is done in an ultrasonic level transmitter?
14. What is ultrasonic cone in the context of ultrasonic transmitter?
15. What is the principle of conversion in a magnetic level indicator?
16. What is a cryogenic temperature range?
17. Give some cryogenic liquids and their boiling points?
18. What is a cryogenic fuel? Where is it used? Identify some of the applications?
19. What is a bubbler level indicator?
20. Where is the bubbler indicator used extensively and why?
21. What is the primary measuring device used in bubbler level indicator?

Long Answer Questions

1. What is the importance of liquid level measurement? Explain in detail, with few examples in daily life and industrial application.
2. Explain the relative features direct and indirect methods of liquid level measurement. Suggest one type of each method for measuring the water level in a boiler shell.
3. Explain the direct level measurement of liquid level by sight glass. Give examples.
4. Explain how float gauges are used for measuring liquid level. Give examples.
5. How do you use a bubbler level indicator of level measurement in an industrial situation? What are the salient features of a bubbler level indicator? Describe the advantages and disadvantages?
6. Write about the advantages of liquid level measurement by using electrical methods.
7. Explain the liquid level measurement by using Rheostat. Discuss its merits and demerits.
8. Explain how capacitance change can be utilized to build a liquid measuring system.
9. Explain how gamma ray (Radio active) liquid level indicator is used fur measuring liquid level. Give its merits and demerits.
10. With the help of a neat diagram, explain the functioning of an ultrasonic liquid level sensor. Describe the salient features & applications
11. (a) Write the principle of operation of a bubbler purge and float raised level indicator.
    (b) With neat sketch explain anyone type of ultrasonic fluid level indicator
12. Explain in detail with neat sketches.
    (a) Liquid level measurement using capacitive transducer.
    (b) Cryogenic fuel level indicator
13. Enumerate the principle of operation of the following and explain briefly
    (a) Capacitive level indicator (b) Ultrasonic level measuring instrument
    (c) Magnetic level indicator (d) Cryogenic fuel level indicators
14. (a) List out the advantages and limitations of direct method level measurements.
    (b) Describe with neat sketch the functioning of any two types of displacement type liquid level measuring instruments.
15. List out various methods for measurement of level. Briefly indicate the features of each.

MULTIPLE CHOICE QUESTIONS

1. The measurement of liquid depends on __________ of the liquids.
   (a) Properties (b) Quantity
   (c) Purity (d) Density
2. In the following, which one of the method is the direct method?
   (a) Gamma ray liquid method
   (b) Capacitive liquid level sensor method
   (c) Buoyancy method
   (d) Electric liquid level sensor method
3. Which one of the following liquid measuring device is mostly used for domestic purposes?
   (a) Sight glass tube
   (b) Float-tape gauge
   (c) Float-shaft gauge
   (d) Electric liquid level sensor
4. In sight glass tube, if the diameter of the tube is low then which of the following error

may occur?
(a) Capillary
(b) Viscous
(c) Surface Tension
(d) Frictional losses

5. In float-Tape liquid gauges the density of the float should be_______ the density of the liquid to be measured.
(a) Greater (b) Less
(c) Same as (d) we can't judge

6. Float-shaft liquid level gauge is difficult to operate in ________ liquids.
(a) High viscosity (b) High density
(c) Corrosive (d) Low density

7. Hydrostatic pressure devices are also called as ____________.
(a) Bubbler or purge system
(b) Bernoulli's Device
(c) Universal Equipment
(d) Ideal Device

8. In Hydrostatic pressure devices the liquid level "*h*" is directly proportional to _____.
(a) Temperature of the liquid
(b) Volume of liquid
(c) Hydrostatic pressure of liquid
(d) Density of the liquid

9. In Hydrostatic device for closed tanks the Pressure gauge is connected between ________ and Bubbler tube.
(a) Air meter
(b) Tank
(c) Pressure regulator
(d) free surface of the liquid

10. In capacitive sensor, in which case the tank itself acts as an electrode?
(a) If it is an Insulated tank
(b) If it is a metallic tank
(c) If it is a non-metallic tank
(d) If the tank contains negative pressure

11. Which of the following is Geiger Muller Tube?
(a) Gamma ray sensor
(b) Source of gamma ray
(c) The tube in which contains the liquid
(d) Tube calibrated with the scale

12. ________ can be used as Ultrasonic Transmitter.
(a) Metal (b) Quartz
(c) Glass (d) Ceramics

13. In Cryogenic fuel level indicator which principle is used?
(a) Wheat stone bridge
(b) Di-electric method
(c) Free surface measurement method
(d) Pascal's Principle

14. Which of the following instruments measure the liquid level by the principle of buoyancy
(a) Hydrostatic Pressure devices
(b) Ultrasonic liquid level guage
(c) Magnetic level indicator
(d) Cyogenic fuel level indicator

15. Which of the following is instruments uses direct method to measure liquid level
(a) Float-tape guage
(b) Cryogenic fuel level indicator
(c) Magnetic level indicator
(d) Ultrasonic fuel level indicator

16. Which of the following instrument uses indirect method to measure liquid level (secondary transduction)
(a) Float - type guage
(b) Float - shaft gauge
(c) Sight glass
(d) Hydrostatic pressure system

17. Float - shaft device used for measurement of liquid level works on the principle of
(a) Boyle's law
(b) Buoyancy
(c) Wheatstone bridge
(d) Capillary principle

18. The flush - tank in toilets will have level indicators are usually
    (a) Magnetic level indicators
    (b) Ultrasonic level indicators
    (c) Cryogenic level indicators
    (d) Float type guages
19. Most suitable device to measure the level of viscous fluids is
    (a) Bubbler/purge system
    (b) Sight glass
    (c) Gamma ray liquid level indicator
    (d) Hydrostatic pressure devices
20. Which of the following instrument/device can be operated from remote place.
    (a) Sight glass
    (b) Float - tape gauge
    (c) Float - shaft gauge
    (d) Gamma ray level indicator
21. The most suitable device to measure the level of corrosive/volatile liquids is ________.
    (a) Gamma ray level indicator
    (b) Ultrasonic level indicator
    (c) Rheostat operated float
    (d) Magnetic level indicator
22. Which of the following liquid level measuring device uses Bernouli's Principle.
    (a) γ-ray level indicator
    (b) Ultrasonic level indicator
    (c) Magnetic liquid level indicator
    (d) Sight glass
23. The Principle of reflection of an accoustic wave from a liquid surface is used in________.
    (a) γ-ray level indicator
    (b) Ultrasonic level indicator
    (c) Magnetic liquid level indicator
    (d) Sight glass
24. Variable permeability is core concept of ________ instrument used for measuring liquid level.
    (a) γ-ray level indicator
    (b) Ultrasonic level indicator
    (c) Magnetic liquid level indicator
    (d) Sight glass
25. Sight - glass tube ued to indicate liquid level works on principle of
    (a) Bernouli (b) Buoyancy
    (c) Boyle's law (d) Newton's law

ANSWERS

1.	(a)	2.	(c)	3.	(a)	4.	(a)	5.	(b)
6.	(c)	7.	(a)	8.	(c)	9.	(b)	10.	(b)
11.	(a)	12.	(b)	13.	(a)	14.	(a)	15.	(a)
16.	(d)	17.	(b)	18.	(d)	19.	(c)	20.	(d)
21.	(c)	22.	(d)	23.	(b)	24.	(c)	25.	(a)

CHAPTER INFOGRAPHIC

9

Glimpses of Measurement of Speed

Speed is defined as the distance per unit time. To measure it we need two mechanisms; one for measuring distance and the other for measuring time. A device that directly indicates the angular speed is called tachometer

S.No	Instrument	Principle	Construction	Operation	Application	Merits	Limitations
Tachometers							
I. MECHANICAL (Use mechanical parts)							
I(a).	Revolution counter and timer or speed counter	Worm gear arrangement	The worn gear attached to the rotating part actually a pointer on a calibrated dial	When the m/c member rotates; worm, in turn gear rotates and thus will be rotating the pointer on the dial	2000-3000 rpm	Average rotational speed is measured	Time is to be measured by separate timer and low speed measurement
I(b)	Tachoscope	Worm gear arrangement	Separate resolution counter and timer and are started simultaneously the shaft	As contact point is pressed against the rotating shaft both counter & timer are started simultaneously & rotational speed is obtained	Up to 5000 rpm	Average rotational speed is measured and separate time measurement is not required	Separate calculation is required i.e., we have to divide the reading obtained in counter and timer
I(c)	Hand speed indicator	Worm gear arrangement	Separate counter and timer will be there but these two are started simultaneously by the additional start button	When the contact point is pressed against the rotating shaft both resolution counter & the timer are started simultaneously and rotational speed is obtained	20000-30000 rpm		
I(d)	Slipping clutch tachometer	Slipping clutch tachometer	Consists two shafts, input soft an indicating shaft both are conducted via a slipping clutch. The pointer is attached to indicating shaft	When the input shaft is pressed against the rotating member it rotates and in turn rotates the indicating shaft via slipping clutch against the torque of spring. A pointer attached to the indicating shaft moves on a calibrated dial indicating speed	For medium speed	Separate timer is not required	After some time the slipping clutch may not be effected
I(e)	Centrifugal force tachometer	Centrifugal force	Two fly balls are arrange about a central spindle	When two shaft pointers due to centrifugal force. Fly balls try to rotate at large radius, by doing to the sleeve moves, which is communicated to the pointer moving on the calibrated dial	Up to 40000 rpm	By using attachments linear speed also can be measured	Range of instrument must be selected
I(f).	Vibrating reed tachometer	Speed/vibration are interrelated	Consists of set of vibrating reed, each having its own natural frequency placed on base plate	When the plate is in contact with the frame of a routing machine vibration responds most frequently. This when calibrated gives speed	Hermatically sealed refrig-erator comp-ressor 600 to 1800 rpm	Used where shafts are inaccessible an concealed	Calibration must be one carefully
II(a) (i)	Drag cup tachometer	Eddy current	A permanent magnet attached to the input shaft and a fixed steel cup. In between these two a non magnetic conducting cup is placed conducted to the pointer moving on a dial	When the shaft rotates magnet rotates which produces eddy current in a drag cup and produces a torque on the cup by which the cup moves which in turn moves pointer over calibrated dial	Automobile speedometer up to 13000 rpm	Easy to use	Calibration must be one carefully

Contd...

S.No	Instrument	Principle	Construction	Operation	Application	Merits	Limitations
II. ELECTRICAL TACHOMETER: II (a) contact type (contact between the instrument an test shaft is essential)							
II(a) (ii)	Commutated capacitor tachometer	Alternately charging/is charging capacitors	Consists of two units tachometer head in which a reversing switch is locate an a indicating unit containing voltage source, capacitors, milli-ammeter & calibration ckt	When the spindle is attached to the rotating unit, it rotates charging and discharging a capacitor ends in each revolution. The average valve of charging and discharging is reduced by electricity and gives in terms of speed	700 to 10000 rpm	Easy to use	Calibration must be one carefully
II(a)(iii) Tachogenerators							
(a)	D.C tachoge-nerators	Permanent magnet D.C electrical gen	A shaft attached to the coil that rotates in a permanent magnet field	When the shaft rotates in permanent magnetic fields producing pulsating D.C voltage measuring this voltage, speed can be measured when calibrated	Automation speed measurement is possible	For measurement direction rotation	Commutation and brush maintenance is required
(b)	A.C Tachometer	Permanent magnet A.C electrical generator	Rotating permanent magnet coupled to the shaft whose speed is to be measured	When the shaft rotates magnet also rotates inducing an A.C voltage in the stator coil which is rectified and measured, which in turn gives speed when calibrated		For measuring diff in speed of 2 sources by differently connecting stator	Calibration must be done carefully
II(b) Electrical Cantacless Tachometers (Contact between the instrument and test is not required)							
II (b) (i)	Inductive pick-up tachometer	Induction	A permanent magnet with a coil wound around is placed near a metallic toothed wheel fixed on shaft whose speed is measured	As wheel rotates magnetic flux linking magnet and coil changes which induces voltage in coil. Knowing frequency of pulse and number of teeth on wheel, speed of shaft can be measured	Automatic measurement of speed is possible	Contact is not required	Inductive circuit must be nearer to the shaft whose speed is to measured
II(b) (ii)	Capacity type pick-up tachometer	Capacitance change	A vane attached to the shaft whose speed is to be measured. This vane while rotating moves between parallel plate capacitor	The movement of the vane between parallel plane capacitor when the shaft rotates, changes the capaci-tance of the circuit. The frequency gives the speed of the shaft			The arrangement must be nearer to the shaft whose speed is to measured
II (b) (ii)	Photo electric tachometer	Photo electric transducer	An opaque disk with evenly spaced holes on its periphery is attached to the shaft whose speed is to be measured. A light detector placed on either side disc	When the shaft rotates based on the number of holes the photo cell produces voltage pulses whose frequency is a measure of the shaft	Remote application up to three million rpm	Very accurate	Calibration circuit must be designed carefully
II (b) (iii)	Stroboscope	Freezing the motion	It consists of a flashing light source, which provides reputed short duration light flashes	The light flashes are directed on the rotating or oscillating objects whose speed it to be determined. A separate mask is placed on the disk attached to the shaft and frequency of light flashes I adjust such that is appears stationery and this frequency is a measure of speed	Remote application	Very accurate	Aliasing and ambient light will affect the reading

Chapter 9

MEASUREMENT OF SPEED

STARTERS

To study this chapter, you should have awareness on the following concepts. For a better understanding, it is always a good idea to revise these prerequisites.

- Definitions of terms and basics of linear motion, and mechanics
- Equations of motion such as $v=u+at$, $s=ut+ \frac{1}{2} at^2$, $v^2 - u^2 = 2as$ etc.
- Definitions of terms and equations of angular motion, angular velocity, angular acceleration
- Relation between the terms of linear and angular motions such as v=rù
- Some mechanisms of motion conversion (translatory to rotatory and vice-versa)
- Concepts of speed of light, sound rays.
- Simple gear mechanisms for calibration into suitable scale
- Relation between speed and electrical parameters such as resistance, capacitance etc.
- Conversion of mechanical to light, electrical, magnetic systems
- Fundamentals of mechanics, Newton's laws, fundamentals of rotatory motion, harmonic motion, vibrations and wave motion and
- First four chapters of this book.

LEARNING OBJECTIVE

After studying this chapter you should be able to

- Define speed and its related terms
- Classify the Tachometers,
- List out and explain Mechanical Tachometers,
- List out and explain the Electrical Tachometers and
- Explain contact type and non-contact type electrical tachometers.

9.1 INTRODUCTION

Speed is defined as the distance per unit time. To measure this we need two mechanisms – viz. mechanism of counting number of revolution (angular displacement) and mechanism of measuring.

A device that directly indicates the angular speed is generally called *tachometer.*

9.2 CLASSIFICATION OF TACHOMETERS

Tachometers are classified as shown in Fig. 9.1.

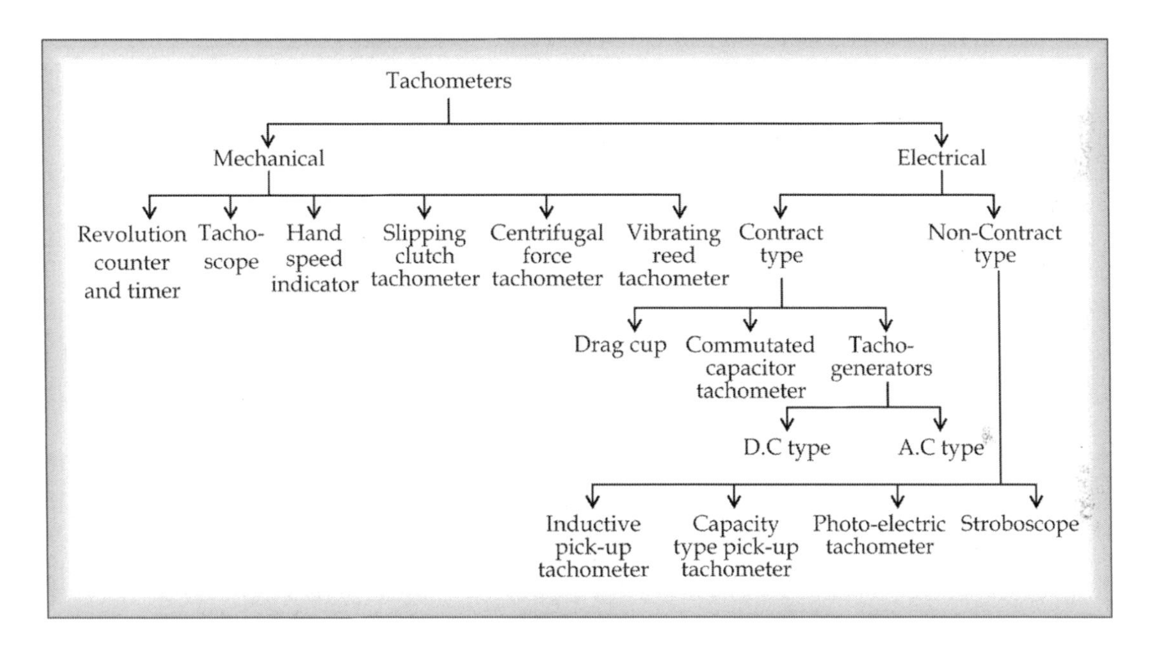

Fig. 9.1 Classification of Tachometers

More Instruments for Learning Engineers

Mile 9.1

A **disdrometer** is an instrument measures the drop size distribution and velocity of falling hydrometeors. Some disdrometers can distinguish between rain, graupel, and hail. The disdrometers can be used for traffic control, scientific examination, airport observation systems and hydrology. The latest instruments are equipped with microwave or laser technologies and 2D video disdrometers can be used to analyze individual snowflakes.

Disdrometer

9.3 MECHANICAL TACHOMETERS

Principle

The mechanical tachometers operate on the principle that linear/rotational speed of the source can be converted to angular/linear displacement by suitable transduction and measured on a calibrated dial.

Classification

The tachometers, which use mechanical parts & mechanical movements, are called mechanical tachometers. They are classified as follows

1. Revolution Counter and Timer
2. Tachoscope
3. Hand speed indicator
4. Slipping clutch tachometer
5. Centrifugal force tachometer and
6. Vibrating reed tachometer.

9.3.1 Revolution Counter and Timer

It is also called *speed counter.* These are used to measure the rotational speeds. The salient parts are:

1. Shaft (or spindle) with tip
2. Worm gear
3. Spur gear
4. Frame
5. Calibrated dial
6. Locking mechanism
7. Handle

Principle

The rotation of the source is transmitted to the shaft by a worm shaft which is transferred to a spur gear whose rotations are measured.

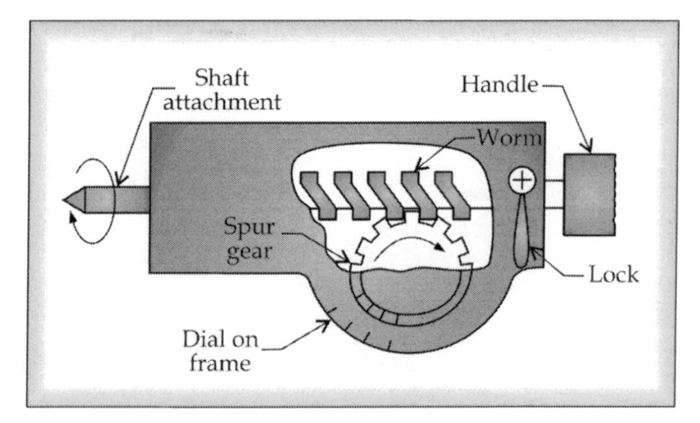

Fig. 9.2 Revolution Counter or Speed Counter

Construction

It consists of a worm gear, attached or integrated to the shaft which is driven by the speed source whose speed is to be measured as shown in the Fig. 9.2. A spur gear is aligned to the worm gear. A locking mechanism may be designed to lock the gears. This set up is housed in a suitable frame. A pointer is connected to this spur gear which moves on a calibrated dial fitted on the frame. A handle is fitted to the shaft for holding.

Operation

When a speed is to be measured, the tip of the shaft is kept in contact so that the speed source rotates the shaft. Thence the shaft rotates the worm which drives the spur gear. This spur gear in turn actuates the pointer on a calibrated dial. The pointer indicates the number of revolutions turned by the input shaft in a certain time. The unit requires a separate timer to measure the time interval usually a stop watch. The revolution counter thus gives an average rotational speed rather than an instantaneous rotational speed.

Merits/Demerits/Applications

These speed counters are easy to operate but limited to low speed engines which permit reading the counter at definite intervals. Another limitation is that the timer (stop watch) which is separately operated, is to be started simultaneously and exactly at the same time. This is difficult and often proned to some human errors.

These counters are applied to measure the speeds of rotating parts such as wheels, shafts, discs, winders and pulleys etc. This can be used satisfactorily in the range of 2000-3000 rpm.

9.3.2 Tachoscope

(Modified Revolution Counter and Timer)

Principle: Same as above.

Construction & Operation

This is the modified form of the above described device in section 9.3.1 (Revolution counter & timer). If the revolution counter and the timer are mounted integrally and started simultaneously when the contact point is pressed against the rotating shaft then it is called *Tachoscope.* By this arrangement the difficulty in starting a counter and a watch at exactly the same time is eliminated.

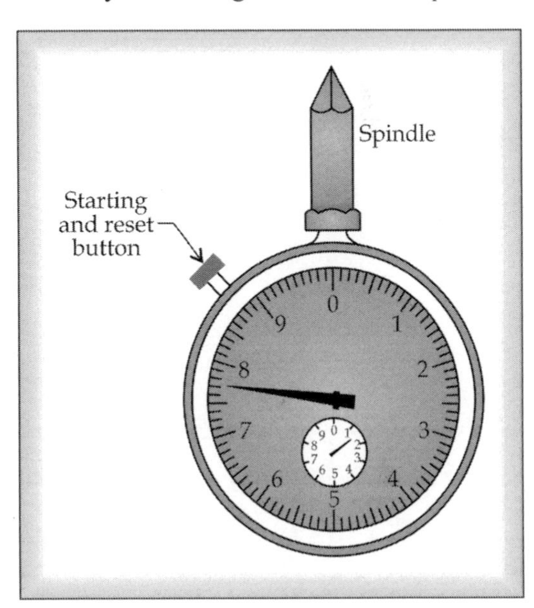

Fig. 9.3 Tachoscope

The instrument as shown in Figure 9.3 runs until the contact point is removed from the shaft. The rotational speed is computed by dividing the readings of the counter with that of the timer. These are used for measuring up to 5000 rpm.

9.3.3 Hand Speed Indicator (Modified Tachoscope)

Principle: Same as above.

Construction & Operation

This is similar to Tachoscope discussed above (section 9.3.2) but only difference is that the counter starts when start button is pressed. The engaging with the counter is possible by the automatic clutch and after fixed time interval (usually 3 or 6 seconds) the revolution counter automatically gets disengaged (see Figure 9.4). These are used for speeds within the range of 20,000 to 30,000 rpm.

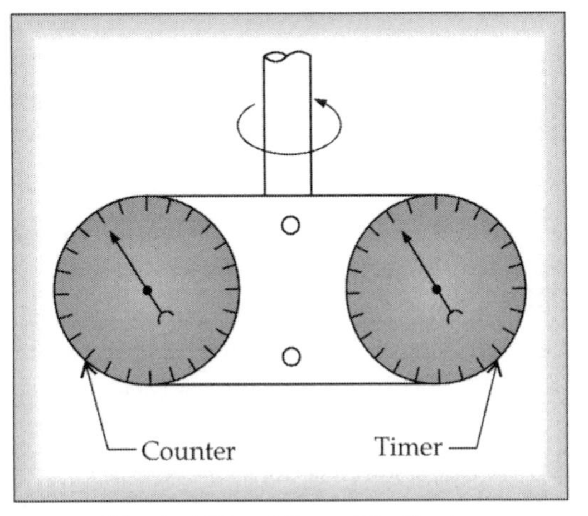

Fig. 9.4 Hand Speed Indicator

9.3.4 Slipping Clutch Tachometer

The main parts are:

1. Input shaft
2. Indicator shaft
3. Friction material
4. Slipping clutch
5. Spiral spring
6. Calibrated scale

Construction

This consists of two shafts, input shaft and indicator shaft separated by slipping clutch as shown in Fig. 9.5. The pointer is attached to indicating shaft. A spiral spring is mounted on indicator shaft. A pointer is also attached to the indicator shaft so as to move on a calibrated scale.

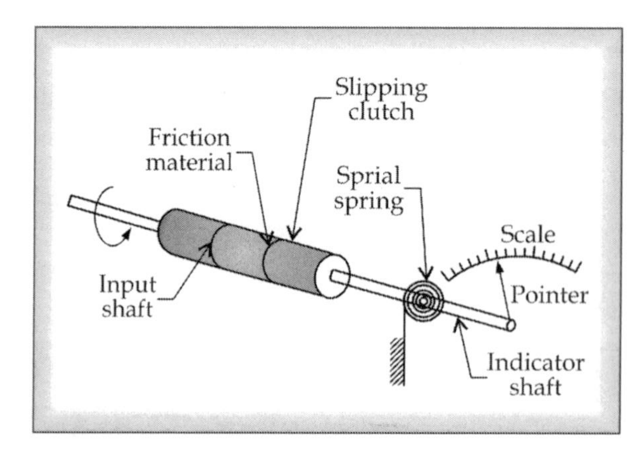

Fig. 9.5 Slipping Clutch Tachometer

Operation

The rotating shaft whose speed is to be measure drives the input shaft which in turn transmits the rotation to the indicating shaft through a slipping clutch. The pointer which is attached to indicating shaft moves over a calibrated scale against the torque of a spring. The position of pointer over the scale gives a measure of the shaft speed.

More Instruments for Learning Engineers

A **light meter** measures the amount of light and is often used in photography to check the sufficient exposure for a photograph. Generally a light meter will include a computer, either digital or analog to enable the photographer to choose shutter speed and f-number for an optimum exposure. Light meters are also used in cinematography and scenic design to decide the optimum light level for a scene. They may be used in the general lighting so as to reduce the wastage of light, light pollution outdoors, and plant growing to ensure proper light levels. It has a special use in cricket to decide if the light is sufficient to continue the play.

Light meter

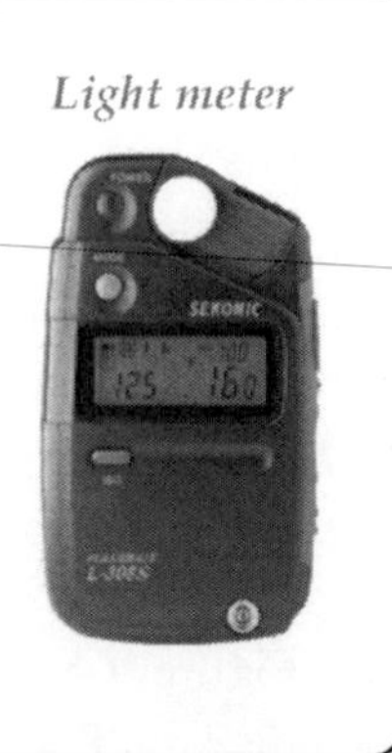

9.3.5 Centrifugal Force Tachometer

Main parts of this device are:

1. Central spindle
2. Helical spring
3. Fly balls
4. Sleeve
5. Pivot
6. Sector and pinion mechanism
7. Pointer and calibrated scale

Principle

This device operates on the principle that centrifugal force is proportional to the square of the angular speed (ω).

Construction

Two fly balls (small weights) are arranged about a central spindle upon which a helical spring is wound. The spring and the fly balls are fixed to the sleeve that moves on the central spindle as shown in Figure 9.6. A pivot is connected to the sleeve at one end and to the sector & pinion mechanism on the other end. A pointer movable on the calibrated scale is mounted on the sector & pinion mechanism.

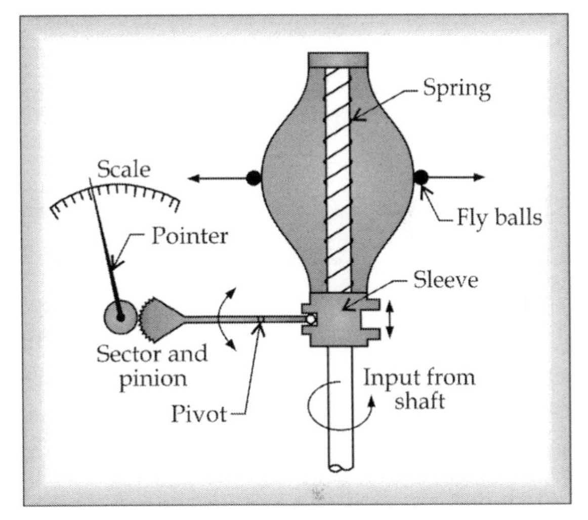

Fig. 9.6 Centrifugal Tachometers

Operation

The central spindle is kept in contact with the rotating shaft whose speed is to be measured. As the central spindle rotates, due to centrifugal force, fly balls try to rotate at larger radius. By doing so, they compress the spring as a function of rotation speed. A grooved collar or sleeve on the central spindle connected to the spring is pulled up which in turn will move the pivot and the motion of this pivot is magnified and communicated to the pointer of the sector & pinion mechanism to indicate the speed on a calibrated scale. By using suitable attachments, linear speed can also be measured.

Merits/Demerits/Applications

Centrifugal tachometers are generally made in multiple range units. The change from one range to another is accomplished by a gear train between the fly ball shaft and the spindle that communicates with the rotating shaft. While using this instrument, generally, high speed range is not selected since over speeding will seriously damage the instrument. Further, a change from one range to another is not to be made while the instrument is in use. These tachometers are frequently used to measure rotational speed up to 40,000 rpm.

9.3.6 Vibrating Reed Tachometer

The main components of this type of tachometer are:

1. Set of vertical reeds
2. Base plate or/and a mounting block

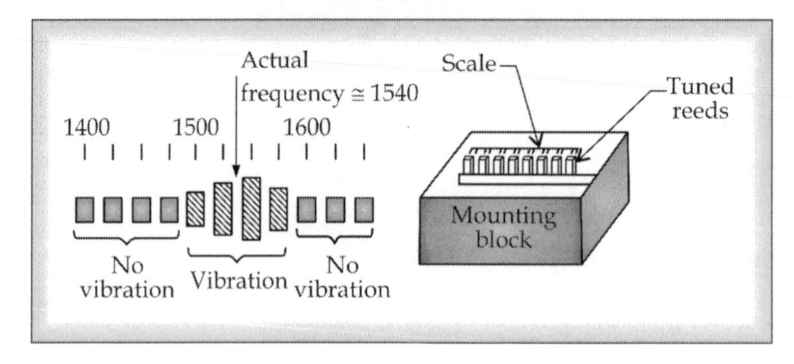

Fig. 9.7 Vibrating Reed Tachometers

Principle

These tachometers are based on the fact that speed and vibration in a body are interrelated.

Construction

This consists of a set of vertical reeds each having its own natural frequency of vibration. These reeds are fastened to a base plate and fixed on mounting block. The reeds are lined up in increasing order of their natural frequency, fastened to a base plate at one end, and are free to vibrate at the other end as shown in the Figure 9.7.

Operation

When the base plate is in contact with the frame of a rotating machine, a reed tuned to resonance with the machine vibration responds most frequently. This reed can be calibrated to indicate speed of rotating machine.

Merits/Demerits/Applications

These tachometers are widely used where the shafts are inaccessible & concealed such as those of a hermetically sealed refrigerator compressor. These are effectively used in the speed range of 60 to 10,000 rpm.

Self Assessment Questions-9.1

1. Give the classification of tachometers.
2. What are the various types of mechanical tachometers?
3. List the various types of electrical tachometers.
4. How the tachoscope differ from revolution counter and timer? Explain it.
5. What is the difference between tachoscope and Hand Speed Indicator?
6. Describe the principle of Slipping Clutch Tachometer with a neat sketch.
7. With the help of a diagram explain the working of Centrifugal Force Tachometer
8. Explain the construction and operation of vibration reed tachometer.
9. What is centrifugal force? Explain the tachometer which works based on centrifugal force.
10. Name the components involved in slipping clutch tachometer. State the purpose of each component.

9.4 ELECTRICAL TACHOMETERS

Principle

Electrical tachometers operate on the principle that an electrical signal can be generated in proportion to the speed of shaft.

Classification

Based on the type of the transducer, electrical tachometers can be constructed in a variety of ways. They can be broadly classified as contact and noncontact type.

I. Under contact type they are as follows:
   1. Drag Cup tachometer
   2. Commutated capacitor tachometer
   3. Tachogenerator
      (i) DC type
      (ii) AC type

II. Under non-contact type they are as follows:
   1. Inductive pick-up tachometer
   2. Capacity type pick-up tachometer
   3. Photo-electric tachometer.
   4. Stroboscope

I. Contact Type Electrical Tachometers

In these tachometers physical contact between the instrument & test-shaft is essential. The most commonly used contact type electrical tachometers are drag cup, commutated capacitor tachometers and tacho-generators (AC type of DC type). These are described in the sections to follow.

9.4.1 Drag Cup Tachometers (Automobile Speedometers)

The salient components in this device are listed below.

1. Input shaft
2. Permanent magnet
3. Fixed steel cup
4. Non-magnetic conducting cup
5. Spring
6. Pointer
7. Calibrated scale

More Instruments for Learning Engineers

Speedometer or speed meter (fitted universally to motor vehicles from 1910 onwards) measures the instantaneous speed of a vehicle. Speedometers for other vehicles have specific names such as pit log for a boat, airspeed indicator for an aircraft. Charles Babbage is credited with creating an early type of a speedometer that was fitted to locomotives. The electric speedometer invented by the Croatian Josip Belušiæ (1888) was originally called a velocimeter

Speed meter- Speedometer - Velocimeter

Principle

The proportionally induced eddy current in permanent magnet due to rotating test shaft can be a measure of speed.

Construction

The rotating shaft which contacts with the speed source, whose speed is to be measured, is connected to a permanent magnet. A non-magnetic conducting drag cup or disc is kept against this magnet in which the eddy currents are induced. The magnet and cup are housed in fixed steel cup. The non-magnetic conducting drag cup is connected to a spring to which a pointer is attached. A calibrated scale is arranged so as the pointer moves on it as shown in Fig. 9.8.

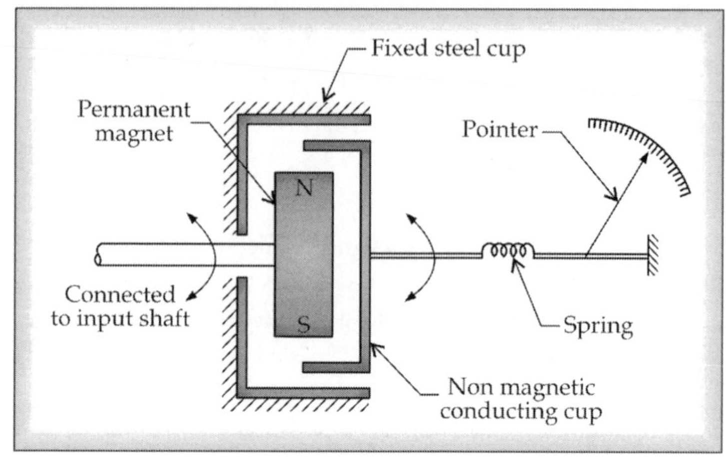

Fig. 9.8 Drag Cup Tachometer

Operation

The test shaft rotates a permanent magnet which induces eddy current in a drag cup or disc held close to the magnet. This eddy current generated interacts with the magnetic field and produces a torque on the cup. The torque so produced is proportional to the relative angular velocity between the magnet and the cup. This causes the cup to rotate till balanced by a spring torque. A pointer attached shows the rotational speed on a scale (refer Fig. 9.8).

Merits/Demerits/Applications

These devices are easy to construct and operate and hence are widely employed. The automobile speedometers operate on this principle and measure the angular speed of the wheels. Locomotive tachometers incorporate a stationary magnet and a revolving magnetic field is produced by a soft iron rotor. These tachometers are used for measuring rotational speed up to 12,000 rpm.

9.4.2 Commutated Capacitor Tachometer

This device consists of the following parts:

1. Spindle
2. Tachometer head containing reverse switch
3. RPM indicator
4. Capacitor
5. Milli-Ammeter
6. Battery

Principle

The operation of this tachometer is based on alternately charging and discharging a capacitor, which is controlled by the speed of the shaft.

Construction

The construction of this instrument can be understood as a combination of the following two units as shown in the Fig. 9.9.

Operation

When the switch is kept closed in one direction, the capacitor gets charged from D.C. supply and current flows through the ammeter. Again, when the spindle operates the reversing switch to close it in opposite direction, capacitor discharges through the ammeter with the current flow direction remaining the same. This device is so designed that the indicator responds to the average current, proportional to the rate of reversal of contacts. This in turn is proportional to the speed of the shaft. The meter scale is graduated to read in rpm instead of milli amperes if calibrated.

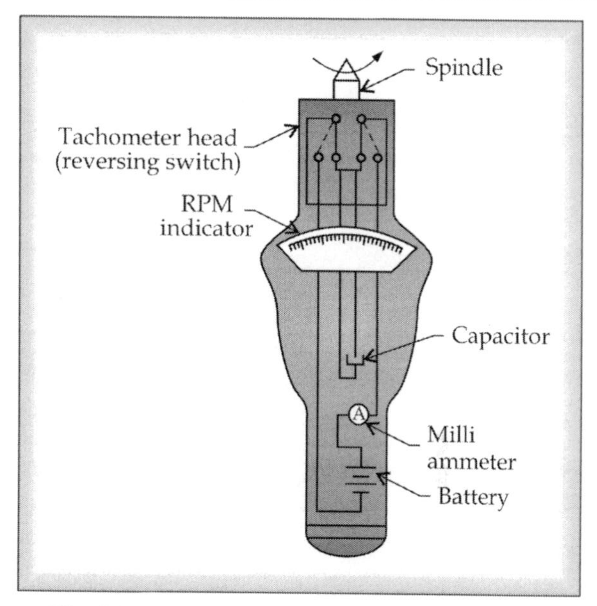

Fig. 9.9 Commutated Capacitor Tachometer

Merits/Demerits/Applications

This device is relatively more accurate, and gives fast response. Though the principle looks simple, the circuit is complex. The tachometer is used in wide range, within the range 200 – 10,000 rpm.

9.4.3 Tacho Generators

Principle

These are generally of permanent magnet type electrical generators, AC/DC which translate the rotational speeds into DC/AC voltage signal.

1. **DC Tacho Generator:** This instrument is composed of the following main components

   1. Shaft
   2. Coils
   3. Horse-Shoe magnet
   4. Moving coil DC Voltmeter
   5. Calibrated scale and pointer

   ### Construction

   A shaft wound with a coil is arranged so as to rotate in the magnetic field of a horse shoe magnet as shown in the Fig. 9.10(a). The output leads of the coil are connected to voltmeter which is connected to a pointer moving on a calibrated scale with zero (0) at center of the scale as shown in Fig. 9.10(b).

   (i) A tachometer unit, in which a reversing switch is located and is connected to a spindle, which reverses twice with each revolution.

   (ii) Indicating unit containing a voltage source, a capacitor, a milli ammeter and a calibrating circuit.

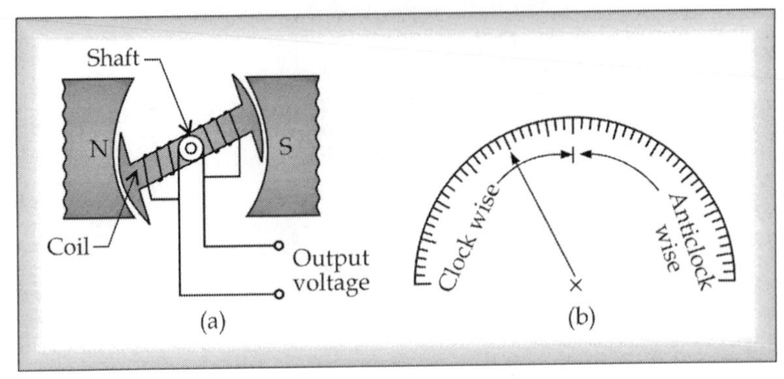

Fig. 9.10 DC Tacho Generator

Operation

The shaft enables the coils to rotate in a permanent magnetic field of a horse - shoe magnet. A pulsating dc voltage proportional to the shaft speed is produced, which can be measured with the help of a moving coil d.c voltmeter having uniform scale and calibrated directly it terms of speed. The important thing to consider here is that the output depends on direction of rotation of shaft. This can be utilized for knowing direction of rotation just by making the scale with its zero point at mid-scale.

Merits/Demerits/Applications

These devices require some form of commutation and present the problem of brush maintenance.

2. **AC Tacho Generator:** The AC tacho-generator also consists of the similar parts as that of the DC tacho-generator.

    1. Shaft
    2. Coils
    3. Magnet (Rotating)
    4. Moving coil AC Voltmeter
    5. Calibrated scale and pointer

Construction

This tacho-generator consists of a rotating permanent magnet, which is coupled to the shaft whose speed is to be measured. This set up is shown in Fig. 9.11. The output leads of the coil are connected to voltmeter which is connected to a pointer moving on a calibrated scale with zero (0) at one end of the scale.

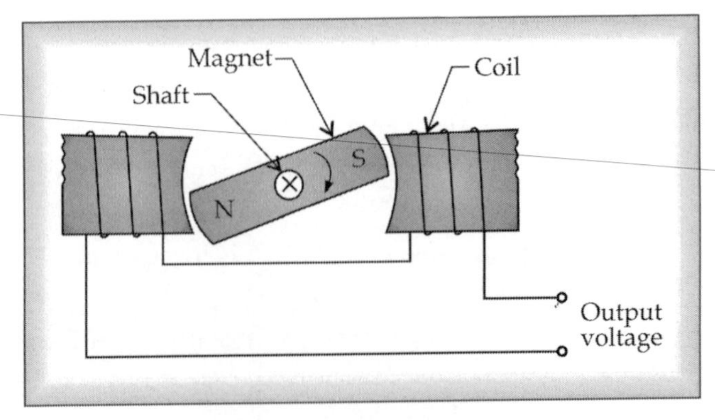

Fig. 9.11 AC Tacho Generator

Operation

The shaft of this tachogenerator when rotates, the connected magnet rotates in a stator surrounding it (usually a multiple pole piece). Due to the rotation of magnet, an ac voltage is induced in the stator coil, which is rectified and measured with a permanent magnet moving coil instrument.

Merits/Demerits/Applications

The most meritorious feature of this tachogenerator (tachometer) is that it can be used for measuring a difference in speed of two sources by differentially connecting the stator coil. These can be successfully employed for measuring speeds up to 500 rpm.

Distinction between AC and DC Tachogenerators

The two tacho-generators work on the same principle but have few distinct differences as tabulated below.

No.	AC Tacho-generators	DC Tacho-generators
1.	The magnet is attached to shaft and acts as rotor	The coils rotate in a horse-shoe magnetic field
2.	AC voltage is produced	DC voltage is produced
3.	The scale is calibrated from one end (full scale)	The scale is calibrated from center of the scale in both directions (two half scales)
4.	The direction of the speed cannot be measured	The direction of the speed can also be measured.

Self Assessment Questions-9.2

1. What is the main difference between DC tacho-generator AC tacho generator? Explain the principle involved in it.
2. What is the function of a magnet used to read the speed of a shaft in the drag cup tachometer?
3. State the need of capacitor in commutated capacitor tachometer.
4. What is the principle involved in the commutated capacitor tachometer? How the capacitor is useful for reading the speed?
5. Name the two types of tachogenerators. Explain each of them with a diagram.
6. List the various contact type and non-contact type electrical tachometers.
7. Draw the circuit diagram for commutated capacitor tachometer.

II. Electrical Contactless Tachometers

Principle

These tachometers produce a pulse by some arrangement on the rotating shaft without any contact between the speed transducer and the shaft.

The Important advantage is that the transducer is not subjected to any load.

These are of four (4) types

1. Inductive pick-up tachometer
2. Capacity type pick-up tachometer
3. Photoelectric tachometer
4. Stroboscope.

9.4.4 Inductive Pick-up Tachometer

The chief parts of this tachometer include:

1. Shaft whose speed is to be measured
2. Toothed wheel
3. Permanent magnet wound with a coil
4. Pulse shaper/Amplifier
5. Counter

Principle

The induced magnetic flux due to rotating shaft is proportional to the speed.

Construction

A shaft whose speed is to be measured is connected to a toothed wheel. This is made to rotate in a permanent magnet wound with a coil. An amplifier or pulse shaper is attached to it, which will be connected to a counter as shown in Fig. 9.12.

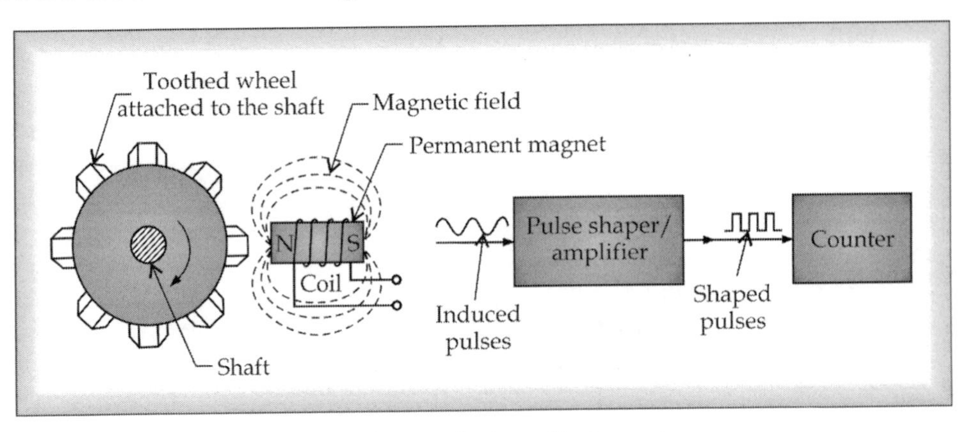

Fig. 9.12 Inductive Pick-up Tachometer

More Instruments for Learning Engineers

Mile 9.4

A **photometer** measures light intensity or optical properties of solutions or surfaces. Before electronic light sensitive elements were developed, photometry was done by the eye by comparing the luminous flux of a standard. By 1861, three most precise types were developed: Rumford's photometer, Ritchie's photometer, and photometers that used the extinction of shadows

Photometer

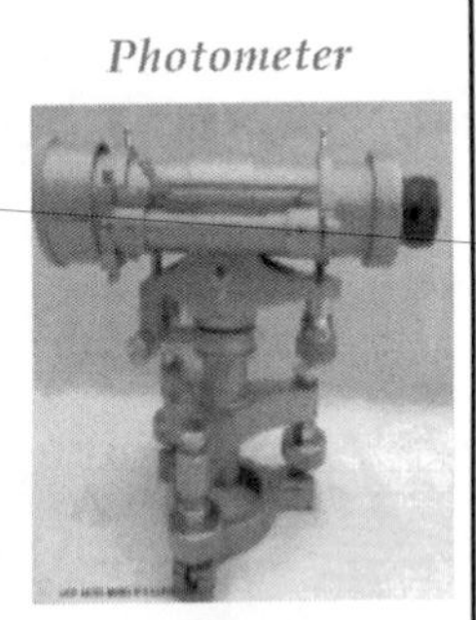

Operation

Metallic toothed wheel is fixed on the shaft whose speed is to be measured. When this rotates, the permanent magnet with a coil wound around placed near to it gets a change in the magnetic flux between the magnet and the coil. As result of this, the field expands or collapses and a voltage is induced in the coil. The frequency of the pulse depends on the number of teeth on the wheel and speed of the shaft.

Mathematically,

Speed of the shaft = Pulse per second /Number of teeth on the wheel.

If number of teeth on the rotor = T

Speed of the shaft in rps = N and

Number of pulses per sec = f_p

Then $N = f_p/T$ rps

ILLUSTRATION-9.1

A toothed gear having 80 teeth is coupled to a shaft whose speed is to be measured. The pulses induced in the inductive-type magnetic pick-up are registered as 4000 per sec. Estimate the shaft speed and also if the meter gives the counts within ± 6 Hz, calculate the range within which the shift speed can lie

Solution:

Given Number of teeth (T) = 80

Frequency of pulse If$_p$ pulses /sec) = 4000

Speed of the shaft (N) =?

$f_p = 4000$

We have $N = f_p /T = 4000/80 = 50$ rps

Accuracy of measurement = ± 6 Hz = ± 6/50 rps = ± 0.12 rps

Hence range of shaft speed = 50 ± 0.12 rps

= 50.12 to 49.88 rps

= 50.12 × 60 to 49.88 × 60 rpm = 3007.2 to 2992.8 rpm

9.4.5 CAPACITY TYPE PICK-UP TACHOMETER

This instrument consists of the main parts as listed below.

1. Rotating shaft
2. Rotating vane
3. Capacitor plates
4. Pulse shaper/Amplifier
5. Counter

Principle

This works on the principle that the speed of rotation is directly proportional to the capacity.

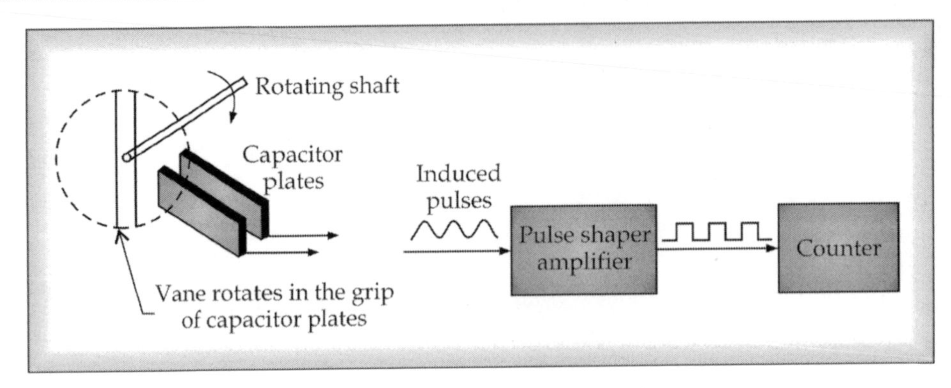

Fig. 9.13 Capacity Type Pick-up Tachometer

Construction

A vane attached to one end of the rotating shaft whose speed is to be measured as shown in Fig. 9.13. The capacitor is attached to an amplifier or pulse shaper which is connected to the counter.

Operation

When the shaft rotates, the vane moves between the fixed capacitive plates, which causes a change in capacitance. The capacitor forms a part of an oscillator tank so that number of frequency changes per unit of time is a measure of the shaft speed. The pulses produced are amplified, squared and are fed to frequency measuring unit or to a digital counter so as to provide a digital output of the shaft rotation.

9.4.6 Photo Electric Tachometer

In this Tachometer, the main parts, we can observe are:

1. Rotating shaft
2. Light source
3. Disk with holes
4. Light sensitive photocells
5. Display device

Principle

The light passing through the rotating disk is proportional to the speed of rotating shaft.

Construction

An opaque disk with evenly spaced holes on its periphery is attached to the shaft whose speed is to be measured as shown in the Fig. 9.14. A light source is placed on one side and a light sensitive transducer on the other side of the disk. Both are placed exactly in opposite directions in alignment with the holes. A display unit is connected to this set up.

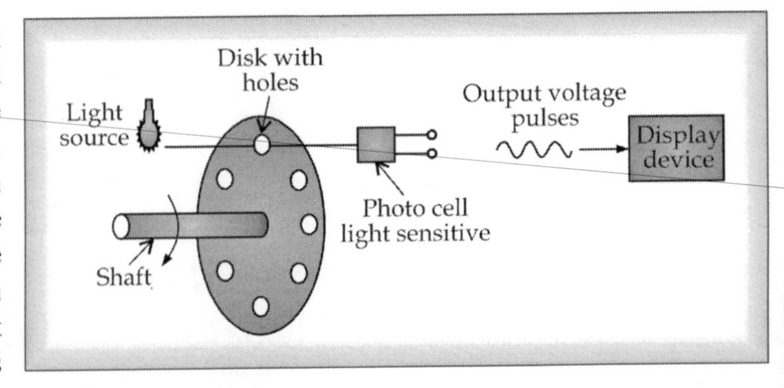

Fig. 9.14 Photo-Electric Tachometer

Operation

As the disk rotates the intermittent light falling on the photo cell produces voltage pulses whose frequency is a measure of the speed of the shaft. This is displayed on the display unit.

Another similar method consists of placing an intermittent reflections (white) and non reflecting (black) surface on the shaft, as shown in the Fig. 9.15.

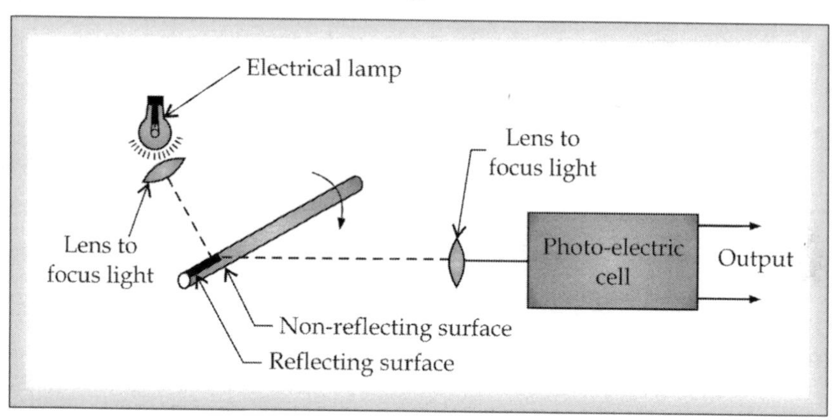

Fig. 9.15 Photoelectric Pick-up Tachometer

A beam of light is directed to fall on these reflecting and non reflecting surfaces. When light falls on reflecting surface, the light is reflected and is focused onto the photo electric cell. The frequency of light pulses is proportional to the shaft speed, and so will be the frequency of electrical output pulses from the photo electric cell.

By suitable circuitry, the pulse rate can be measured by an electronic counter whose scale can be calibrated to indicate speed directly in units of revolutions per sec or rev/min. The optical pick ups have been designed to measure speeds up to 3 million rpm.

9.4.7 Stroboscope

It is a device used for measuring the speed of objects having repeating or cyclic motion e.g. rotating shaft and reciprocating mechanism. The term stroboscope is derived from two Greek words *'strobe'* meaning "whirling" and *'scope'* meaning "to watch". Thus this device is a measuring instrument by watching a flash light through whirling disk.

The following are the main parts of this instrument.

1. Test shaft
2. Moving machinery
3. Mask/plate with hole
4. Light source
5. Calibrated scale

Principle

This uses the concept of making the object appear motionless (freezing the motion) by adjusting the timing of a flashing light illuminating the object, relative to the timing of periodic motion.

The above concept can be easily understood with the function of early stroboscopes, which were using a whirling disk as shown in Figure 9.16.

Construction

A flash of light is arranged on disk attached to test shaft. A mask or a plate with an opening is kept in the line of the sight of light. A screen is arranged to capture the image along with a device to measure the frequency or a calibrated scale.

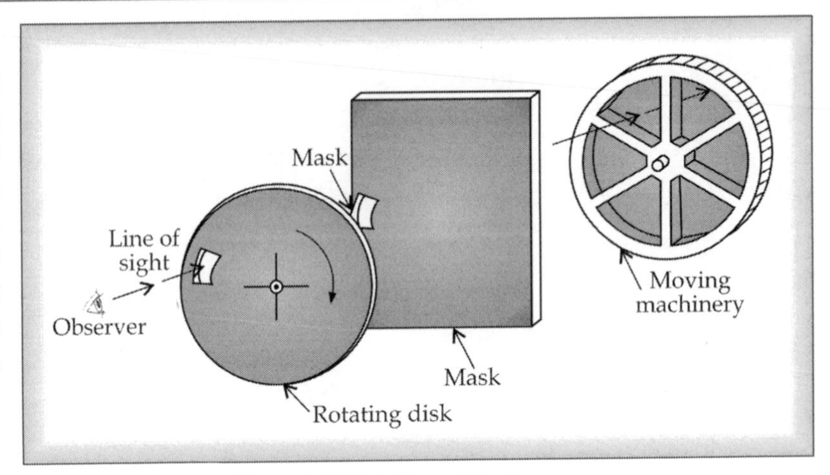

Fig. 9.16 Disk-Type Stroboscope

Operation

When opening of the disk and the stationary mark coincide, the observer can catch glimpses of an object behind the disk. If the speed of the disk is made equal to the object, the object would appear as motionless. It can be understood as 'just the inverse of the illusion' produced by the motion picture projector. Also if disk is made to rotate with slightly less or greater speed than the object, then it appears to creep either forward or backward. The speed of the rotating disk can be known when the object appears to be stationery. By this it is possible to observe directly the rotating gears, shaft whip, helical spring surge when the devices were in operation.

Modern stroboscopes use a flashing light source, which provides repeated short duration light flashes. The frequency of these flashes can be controlled by a variable frequency electronic oscillator operating the flashing bulb as shown in Fig. 9.17.

A separate mark can be placed on the disk or the key connecting the disk and shaft can itself be used. The frequency can be adjusted by a knob and the value read off on a scale. The light flashes can be directed on the rotating or oscillating object whose speed is to be determined. The flashing frequency is normally variable from 1- 250 Hz.

More Instruments for Learning Engineers

Mile 9.5

A **Potometer** sometimes known as a transpirometer— is a device used for measuring the rate of water uptake of a leafy plant shoot. The causes of water uptake are photosynthesis and transpiration. Everything must be completely water tight so that no leakage of water occurs. There are two main types of potometers used - the bubble potometer, and the mass potometer. The mass potometer consists of a plant with its root submerged in a beaker. This beaker is then placed on a digital balance; readings can be made to determine the amount of water lost by the plant. The mass potometer measures the water lost through transpiration of the plant and not the water taken up by the plant.

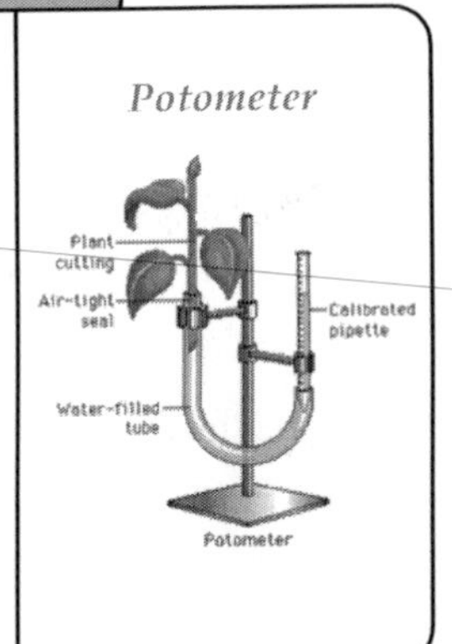

The precautions, to be taken while using this instrument are:

1. It requires that the ambient light should not be bright.
2. One or more distinguishing marks are required on the object i.e. a key way or spokes-working with the multiple mark (e.g. 3 blades of a fan) is more difficult than with a single mark.
3. The object appears to be stationary even in multiple ratios of flashing rate to the objects true cycling rate. This is an example of *aliasing*.

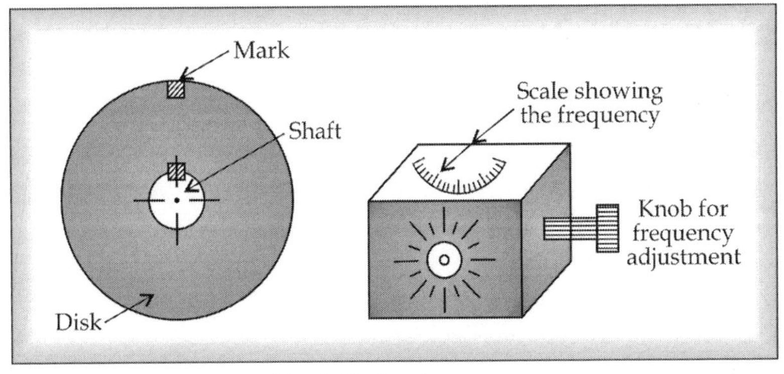

Fig. 9.17 Stroboscope

Actual cycling rate or frequency can be determined as explained below.

- Determine a flashing rate N_1 that freezes the motion.
- Now, reduce the rates, slowly till the motion is frozen once again. Note this rate N_2
- Then, the actual cycling rate of the object $N_o = N_1 N_2 / (N_1 - N_2)$

Merits

1. Measurement can be made from remote area.
2. Measurement can be made even when devices are in operation.

Demerits

1. One or more distinguishing marks are required on the object whose speed is to be measured.
2. Aliasing must be taken care.
3. Ambient light affects the reading to certain extent.

Self Assessment Questions-9.3

1. How the magnetic flux change takes place in inductive pick up tachometer ? Explain the principle behind it.
2. Explain the principle of capacity type pick up tachometer with a diagram.
3. With the aid of diagram explain photo electric tachometer.
4. Describe the working of stroboscope with a neat sketch.
5. What is the instrument used for measuring the speed of objects having repeating or cyclic motion?
6. List the precautions, merits and demerits of stroboscope.
7. Illustrate the range of speed measured by the following tachometers
   (i) Inductive pick up tachometer (ii) Photo electric tachometer
   (iii) Capacity type pick up tachometer

SUMMARY

Speed is defined as the distance per unit time. Speed measured using two mechanisms viz. (i) counting number of revolution and (ii) measuring. A device that directly indicates the angular speed is generally called *tachometer.* The mechanical tachometers operate on the principle of conversion of linear/rotational speed to angular/linear displacement by suitable transduction and measured on a calibrated dial. In speed counters and tachoscopes, hand speed indicators, the rotation of the source is transmitted to the shaft by a worm shaft which is transferred to a spur gear whose rotations are measured. In Slipping Clutch Tachometer, the input shaft transmits the rotation to the indicating shaft through a slipping clutch. Centrifugal Force Tachometer uses centrifugal force that varies as the square of the angular speed (ω). Vibrating Reed Tachometer makes use of the relation between speed and vibration. Electrical tachometers operate by an electrical signal generated in proportion to the speed of shaft. We have two types of Electrical Tachometers viz. contact and non-contact type. Under contact type, Drag Cup Tachometers (Automobile Speedometers) work on the proportionally induced eddy current in permanent magnet due to rotating test shaft. A Commutated Capacitor Tachometer uses alternately charging and discharging capacitor controlled by the speed of the shaft. In DC Tacho Generator, the shaft enables the coils to rotate in a permanent magnetic field of a horse-shoe magnet and a pulsating dc voltage proportional to the shaft speed is measured by moving coil d.c voltmeter. Whilst, in AC tachogenerators the shaft rotates, the connected rotating magnet induces ac voltage in the stator coil, which is rectified and measured with a permanent magnet moving coil instrument.

In the Electrical Contactless Tachometers, a pulse is produced by some arrangement on the rotating shaft without any contact between the speed transducer and the shaft. In Inductive Pick up Tachometer the induced magnetic flux due to rotating shaft is proportional to the speed ($N = f_p/T$ rps). Capacity Type Pick-Up Tachometer works on the principle that the speed of rotation is directly proportional to the capacity. In Photo Electric Tachometer the light passing through the rotating disk is proportional to the speed of rotating shaft. Stroboscope uses repeating or cyclic motion e.g. rotating shaft and reciprocating mechanism i.e., the concept of making the object appear motionless (freezing the motion) by adjusting the timing of a flashing light illuminating the object, relative to the timing of periodic motion.

The details of constructions, working principle, operation, calibration methods, merits/demerits, and applications of these instruments and accordingly different problems are illustrated in this chapter.

KEY CONCEPTS

Tachometer: A device that directly indicates the angular speed.

Revolution Counter and Timer: Also called *speed counter,* consists of a worm gear in the shaft attachment (driven by speed source) drives the spur gear which in turn actuates the pointer on a calibration dial to measure the revolutions, (Range: 2000-3000 rpm).

Tachoscope: The revolution counter and the timer are mounted integrally and started simultaneously when the contact point is pressed against the rotating shaft.

Hand Speed Indicator: Similar to Tachoscope but for the difference that counter starts when start button is pressed. (Range: 20,000 to 30,000 rpm).

Slipping Clutch Tachometer: Rotating shaft drives an indicating shaft through a slipping clutch to move attached pointer over a calibrated scale.

Centrifugal Force Tachometer: Works on principle, $F_{centrifugal} \propto (\omega^2)$, Two fly balls (small weights) rotated due to centrifugal force, compress the spring as a function of rotation speed (Range 40,000 rpm).

Vibrating Reed Tachometer: Works on speed-vibration relationship, consists of a set of vertical reeds fastened to a base plate at one end in increasing order natural frequency of vibration, and the other end is free to vibrate and as base plate in contact with rotating part, a reed resonates which is calibrated (speed range: 60 to 10,000 rpm).

Electrical Tachometers: Operate on the principle that electrical signal can be generated in proportion to the speed of shaft.

Drag Cup Tachometers: The test shaft rotates a permanent magnet, induces eddy current in a drag cup or disc, in turn produces a torque on the cup, proportional to the relative angular velocity between the magnet and the cup, read by a pointer (Range: 12,000 rpm).

Commutated Capacitor Tachometer: Works on alternately charging and discharging a capacitor, which in turn is controlled by the speed of the shaft (Range 200 - 10,000 rpm).

Tacho-Generators: Permanent magnet type ac or dc electrical generators which translate the rotational speeds into dc or ac voltage signal.

DC Tacho-Generator: The shaft makes the coils to rotate in a permanent horseshoe magnetic field and a pulsating dc voltage proportional to the shaft speed is measured with a moving coil dc voltmeter having uniform calibrated scale.

AC Tacho-Generator: Consists of a rotating permanent magnet, coupled to the test-shaft that rotates in a stator induces ac voltage, which is rectified and measured with a permanent magnet moving coil instrument (Range: 500 rpm).

Inductive Pick-Up Tachometer: A permanent magnet with a coil wound around is placed near a metallic toothed wheel, fixed on the test-shaft and as the wheel rotates the magnetic flux linking the magnet and the coil and thence induced voltage is measured.

Capacity Pick-Up Tachometer: Consists of a vane attached to one end of the rotating shaft causes a change in capacitance, the capacitor, a part of an oscillator tank and number of frequency changes per unit time is a measure of the shaft speed.

Photo Electric Tachometer: Light source on one side and a light sensitive transducer on the other side of the disk exactly aligned are in opposite directions of an opaque disk with evenly spaced holes on periphery, attached to the test-shaft. As the disk rotates the intermittent light falling on the photo cell produces voltage pulses whose frequency is a measure of the speed of the shaft. (Range: up to 3 million rpm).

Stroboscope: A device used to measure the repetitive (reciprocating or rotating) uses the concept of making the object appear motionless (freezing the motion) by adjusting the timing of a flashing light illuminating the object, relative to the timing of periodic motion. When openings in the disk and the stationary mark coincided, the observer would catch feeling glimpses of an object behind the disk and adjusts till it appears as motionless. (Range: 1-250 Hz).

Aliasing: The state of the object to appear to be stationary even in multiple ratios of flashing rate to the objects true cycling rate.

REVIEW QUESTIONS

Short Answer Questions

1. Define speed.
2. Define Tachometer.
3. Mention various ways in which speed can be measured.
4. What is the principle involved in measuring speed by stroboscope?
5. What is tachoscope?
6. Classify Tacho generators.
7. What is the difference between Inductive and capacitative type pick up tachometers?
8. What are the three essential components in a mechanical tachometer?
9. What is a typical output from an electrical tachometer?
10. Mention the disadvantages of mechanical tachometer.
11. Mention the advantages of electrical tachometer over mechanical tachometer.
12. What is a non contact type of tachometer?
13. Write about the advantages and disadvantages of stroboscope.
14. What is aliasing?

Long Answer Questions

1. Describe the different methods used for measurement of speed, describe their advantages and disadvantages
2. Explain briefly all mechanical tachometers
3. Mention all the disadvantages associated with mechanical tachometers
4. Explain the following mechanical tachometers
   (i) Revolution counters timer
   (ii) Tachoscope
   (iii) Hand speed Indicator
   (iv) Slipping clutch tachometer
   (v) Centrifugal force tachometer.
   (vi) Vibrating reed Tachometer.
5. Give a brief account of electrical contact type of tachometer.
6. Give a brief account of electrical non-contact type of tachometers with special emphasis on stroboscope.
7. Discuss the relative features of a stroboscope and a tachometer for speed measurement.
8. Explain the construction, working and theory of a drag cup type tacho generator.
9. Explain the following Electrical Tachometers:
   (i) Drag up tachometer
   (ii) Commutated capacitor tachometer.
   (iii) Tachogenerators (both *AC* & *DC* type)
   (iv) Inductive pick up tachometer
   (v) Capacitive type pick up tachometer
   (vi) Photo electric tachometer
   (vii) Stroboscope.
10. Compare and contrast contact and non-contact type of electrical tachometer.
11. A 50 teeth ferrous gear wheel was used with an electromagnetic pick-up tachometer to measure the angular velocity of a hydraulic turbine. What reading will be indicated by the frequency meter if the turbine shaft is turning at 10 p radians per second?

12. An assignment was given to students for using stroboscope to measure the speed of a shaft turning at 3600 rpm. For a given stroboscope, the available flashing rates are within the range of 30 to 90 per second. At what frequency settings, the students expect to get a single, double & triple steady image of a dark spot marked on the rotating shaft.
13. A bike has 36 spokes. When a stroboscope is directed on to the spokes, a true pattern is observed at the highest flash frequency equal to 96 flashes per minute. What will be the other flash frequencies which produce (i) 36 spokes & (ii) 72 spokes?
14. When a stroboscope is directed onto the ceiling fan, a true pattern of it is observed at the highest flash frequency equal to 300 flashes per minute. What will be the other flash frequencies which produces (i) same wings (ii) double the number of wings.
15. Explain the working of non-contact type of tachometer?
16. Explain the principle of any one non-contact type of tachometer.
17. What are the various types available in non-contact type of tachometers? Explain briefly with necessary sketches.
18. Give the construction and working of a vibrating reed tachometer for measuring speed.
19. Write short notes on stroboscope.
20. Describe the different methods used for measurement of speed. Describe their advantages and disadvantages.
21. Give detailed classification of speed measuring devices with examples.
22. What are mechanical tachometers? Explain with examples. Describe the disadvantage- tags of mechanical tachometers.
23. Describe the construction and working of a D.C. tachometer generator. Explain its advantages and disadvantage
24. Explain the construction and working of an A.C. tachometer generator describes its limitations.
25. Explain working of non-contact type-tachometer.
26. Write short note on induction sensors for speed measurement.
27. Explain the construction and working of a photoelectric tachometer. Explain its advantages and disadvantages.
28. Describe with a neat sketch the principle and working of stroboscope.

MULTIPLE CHOICE QUESTIONS

1. Speed is a variable defined as the _____ of motion.
   (a) Mass (b) Acceleration
   (c) Rate (d) Deflection
2. Angular speed measurements are made with a device called
   (a) Speedometer (b) Vibrato meter
   (c) Sextant (d) Tachometer
3. Tachometers may be broadly classified in to two categories as
   (a) Mechanical, Electrical
   (b) Electronic, Pneumatic
   (c) Hydraulic, Electronic
   (d) All of the above
4. Revolution counter, is also called a _____ counter.
   (a) Speed (b) Angular
   (c) Mechanical (d) Circular
5. Revolution counter consists of ____ gear.
   (a) Herring (b) Helical
   (c) Bevel (d) Worm

6. The measuring range of revolution counter is ______ rpm
   (a) 2000-3000 (b) 500-2000
   (c) 1500-3000 (d) More than 3000
7. The difficulty of starting a counter and a watch at exactly the same time led to the development of
   (a) Stop watch (b) Dial gauge
   (c) Tachoscope (d) Tachometer
8. Hand speed indicator has an integral ___ and counter with automatic disconnect.
   (a) Spindle (b) Stop watch
   (c) Dial gauge (d) Pointer
9. The accuracy of hand speed indicator is
   (a) 1% (b) 2.5%
   (c) 1.5% (d) 25%
10. A spring is used in
    (a) Slipping clutch tachometer
    (b) Tachoscope
    (c) Hand speed indicator
    (d) Revolution counter
11. Eddy current tachometer is also known as ___ type tachometer.
    (a) Commutated capacitor
    (b) Tacho generator
    (c) Drag
    (d) Contactless electrical
12. The _____ type tachometer utilizes the phenomenon of vision when an object is viewed intermittently.
    (a) Photo-electric
    (b) capacitive pick-up
    (c) Inductive pick-up
    (d) Stroboscope
13. _____ refers to the repeated cyclic oscillations of a system.
    (a) Angular momentum
    (b) Whirling speed
    (c) Impulse
    (d) Vibration
14. The oscillations are caused when _____ is applied to the machine alternately in two directions.
    (a) Pressure (b) Acceleration
    (c) Force (d) Momentum
15. Units of vibration is
    (a) Hertz (b) Decibels
    (c) Microns (d) Henry
16. F = ma is derived from Newton's
    (a) first law of motion
    (b) Second law of motion
    (c) third law of motion
    (d) Newton's law of gravitation
17. Displacement sensing accelerometer is also known as______ accelerometer.
    (a) Mechanical (b) Vibrato meter
    (c) Seismic (d) Non-seismic
18. In a ______ accelerometer, the sensing mass is mounted on a cantilever beam.
    (a) Strain gauge (b) Stress gauge
    (c) Seismic type (d) Piezo-electric
19. Units of linear speed are
    (a) Rad/sec (b) m/sec
    (c) N-m/sec (d) No units
20. Units of angular speed are
    (a) Rad/sec (b) m/sec
    (c) N-m/sec (d) No units
21. RPM refers to
    (a) Rotations per meter
    (b) Revolutions per meter
    (c) Revolutions per minute
    (d) Rounds per meter
22. The instrument used to measure angular velocity of shaft, either by registering the number of rotations during the period of contact, or by indicating directly the number of rotations per minute is
    (a) Vibrato meter (b) Speedometer
    (c) Tachometer (d) Accelerometer
23. In revolution counter, worm gear drives gear.
    (a) Bevel (b) Worm
    (c) Helical (d) Spur

24. The speed counter gives _____ rotational speed.
(a) Exact (b) Approximate
(c) Average (d) Instantaneous

25. ______ are limited to low speed engines.
(a) Tacho-generators
(b) Tachometers
(c) Revolution counters
(d) Tacho-scopes

26. Tachoscopes have been used to measure speeds up to_______ rpm.
(a) 3000 (b) 4000
(c) 5000 (d) 8000

27. In hand speed indicator, the fixed time interval ranges from
(a) 1-2Sec (b) 5-11Sec
(c) 3-6Sec (d) 2-4Sec

28. Hand speed indicator (HSI) is used for ______ speed engines.
(a) Low (b) High
(c) Medium (d) For all

29. SCT Stands for _______ tachometer
(a) Slipping clutch tachometer
(b) Speed control tachometer
(c) Super centrifugal tachometer
(d) Speed capacitive tachometer

30. Centrifugal force is___ speed of rotation.
(a) Inversely proportional to
(b) Directly proportional to
(c) Equal to
(d) Twice of

31. Vibrating reed tachometer works on
(a) Velocity (b) Acceleration
(c) Vibration (d) Speed

32. In vibrating reed tachometer (VRT), the reeds are lined up in order of their frequency.
(a) Forced (b) Natural
(c) Transverse (d) generated

33. An electrical tachometer depends for its indications upon
(a) Computer output
(b) Mechanical response
(c) Electrical signal
(d) Pneumatic pressure

34. The magnet used in eddy current tachometer is
(a) Temporary
(b) Horse-shoe
(c) Electro-magnet
(d) Permanent

35. Eddy current tachometers are used for measuring rotational speeds up to _____ rpm with an accuracy of
(a) 6000, 3% (b) 8000, 2%
(c) 12000, 3% (d) 14000, 2%

36. In commutated capacitor tachometer, spindle reverses _____ times that of one revolution.
(a) Thrice (b) Equal
(c) Twice (d) Five

37. In commutated capacitor tachometer (CCT), the indicating unit comprises of voltage source, capacitor, milli ammeter and
(a) Inductor
(b) Voltmeter unit
(c) Insulator
(d) Calibrating circuit

38. The rate of reversal of contacts is proportional to
(a) Speed (b) Voltage
(c) Current (d) Resistance

39. Contactless electrical tachometers produce pulse from a rotating shaft without any physical contact between _____ and the shaft.
(a) Speed transducer
(b) Speed reducer
(c) Sensor
(d) Electromagnet

40. Inductive pick-up tachometer has a rotor of_______ shape.
(a) Gear type (b) Plane circular
(c) Toothed (d) Helix toothed

41. Vanes are used in
    (a) Speed counter
    (b) Tacho generators
    (c) Capacitive tachometer
    (d) Stroboscope
42. Photo electric tachometer is a_____ type of tachometer.
    (a) Disk (b) Contactless
    (c) Capacitor (d) Vibrating
43. Catastrophic failure is a result of stress caused by
    (a) Strain (b) Small force
    (c) Point load (d) Fatigue
44. The value of acceleration due to gravity 'g' is_______ m/sec^2
    (a) 9.81 (b) 10
    (c) 5.2 (d) 9
45. Displacement sensing accelerometer is also known as
    (a) Piezo-electric (b) Photo-electric
    (c) Mechanical (d) Seismic
46. __________ transducer is used in displacement sensing acclerometer.
    (a) Electrical-displacement
    (b) Mechanical
    (c) Pneumatic
    (d) Hydraulic
47. The output of strain gauge accelerometer is input for
    (a) Seismic accelerometer
    (b) Meter bridge
    (c) Tachometer
    (d) Wheatstone bridge
48. Reed vibrometer uses
    (a) Accelerometer (b) Stroboscope
    (c) Vibrato meter (d) Tachometer
49. Reed vibrometer can measure up to______ hertz.
    (a) 1000 (b) 10000
    (c) 5000 (d) 100
50. The device which converts given input into readable output is
    (a) Transducer (b) Indicator
    (c) Manipulator (d) Translator

ANSWERS

1.	(c)	2.	(d)	3.	(a)	4.	(a)	5.	(d)
6.	(c)	7.	(c)	8.	(b)	9.	(a)	10.	(a)
11.	(c)	12.	(d)	13.	(d)	14.	(b)	15.	(a)
16.	(b)	17.	(c)	18.	(a)	19.	(b)	20.	(a)
21.	(c)	22.	(c)	23.	(d)	24.	(c)	25.	(c)
26.	(c)	27.	(a)	28.	(a)	29.	(a)	30.	(a)
31.	(c)	32.	(b)	33.	(c)	34.	( )	35.	(a)
36.	(c)	37.	(d)	38.	(a)	39.	(a)	40.	(c)
41.	(c)	42.	(b)	43.	(d)	44.	(a)	45.	(d)
46.	(a)	47.	(d)	48.	(b)	49.	(b)	50.	(a)

CHAPTER INFOGRAPHIC

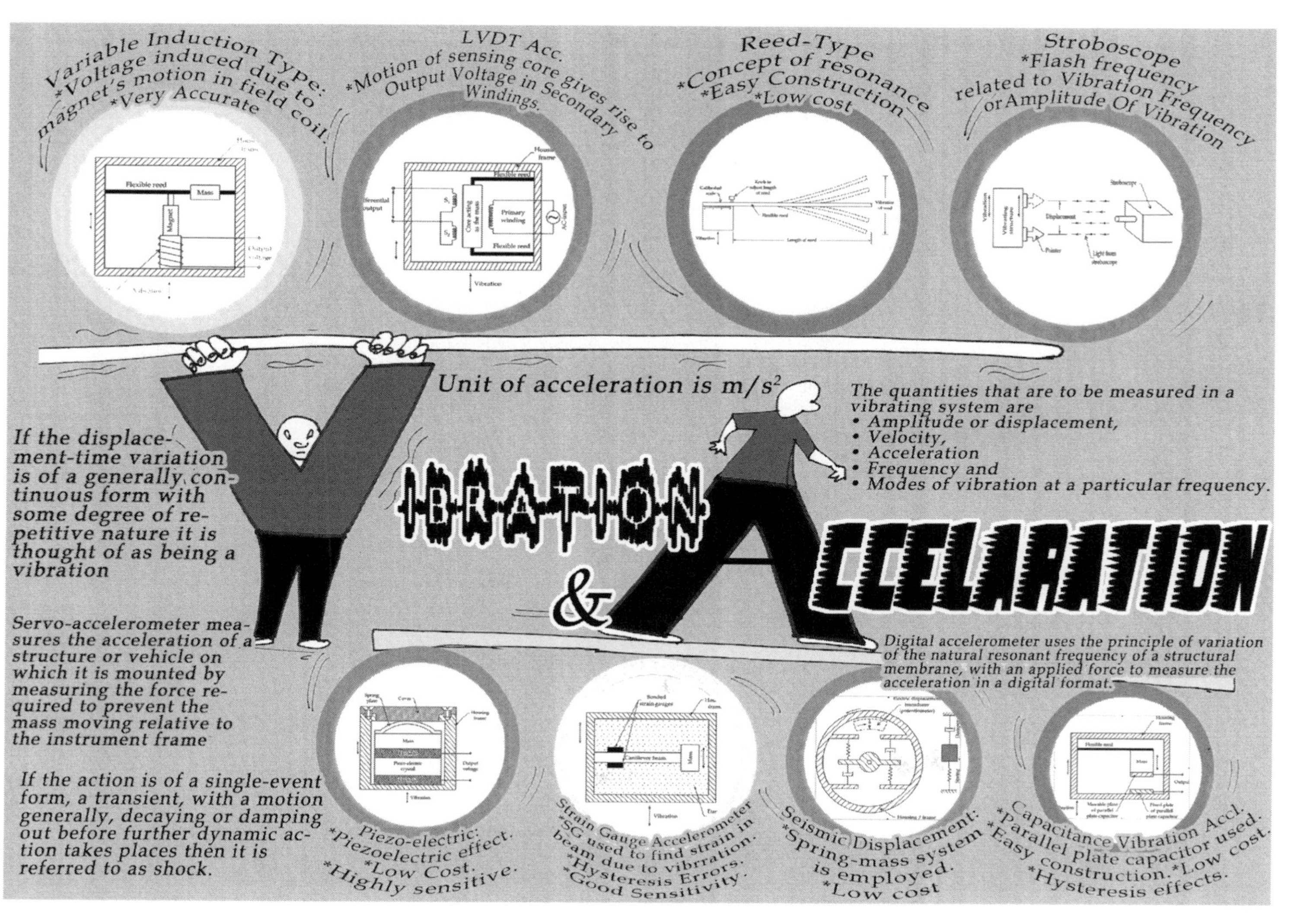

10 Glimpses of Measurement of Vibration and Acceleration

S.No	Instrument	Type of Transducer	Construction	Operation	Application	Merits	Limitations
1.	Piezo-Electric Accelerometer	Piezo-Electric Effect	Piezo-Electric crystal placed between two electrodes	A force is exerted on the Piezo-Electric crystal due to which a voltage is generated which is picked by electrodes	To measure acceleration of the structures	Small size, rugged, less cost and highly sensitive	Hysteresis errors, sensitive to temperature changes
2.	Strain-Gauge Accelerometer	Change in resistance of strain gauge when subjected to strain	Two strain gauges mounted on a cantilever beam loaded at its free end in a damping fluid	Instrument is mounted structure strains cantilever, in turn strains the gauges, resulting in change of resistance measures acceleration		Can be Digitized; less cost, highly sensitive	Hysteresis errors, sensitive to temperature changes
3.	Seismic Displacement sensing Accelerometer	Displacement of the spring mass damper system by acceleration	A seismic mass is suspended from housing of accelerometer with spring	Due to the acceleration, the seismic mass experienced a displacement which varies with acceleration		Less cost, rugged, and good sensitivity	Inertia efforts to be minimized
4.	Capacitance vibration sensor/ Accelerometer	Variation of gap between parallel plate capacitor subjected to vibration	A flexible reed one end fixed to housing and other end carries a mass which is a movable plate of capacitor	Under vibration, gap between fixed & movable plate of capacitor varies which is calibrated to measure vibration / acceleration		Less cost, easy construction	Hysteresis effects
5.	Vibration induction / reluctance type accelerometers	Output voltage from a field coil which is proportional to the amount by which a magnet is dipped into the field coil	A magnet attached at center of flexible reed is made to move in the field coil whose output varies with vibration	Due to vibration when mass attached to flexible reed displaces the magnet attached to it also displaces and moves in the field coils. The output from this coils varies with vibration		Very accurate, automation can be achieved	Mass to be kept on the flexible reed must be found
6.	LVDT Accelerometer	The differential output voltage of the secondary winding varies with vibration displacement by sensing mass	Central core attached to flexible reed moves b/n primary& secondary windings change output,	When the instrument is subjected to vibration the central core moves in between primary and		Very accurate, automation can be	
7.	Reed type vibrometer	Adjust reed length so that when freq. of vibration resonate with frequency of reed, and maximum displacement occurs	Consists of a flexible reed	A flexible reed connected to structure and reed coincide resonance occurs and reed will vibrate with maximum amplitude, hence when calibrated measures amplitude of vibration	To measure acceleration of the structures	Very easy in construction, less cost	Calibration is different
8.	Stroboscope	Matching the frequency of light flashes to that of the vibrating structure	A pointer to be attached to vibrating structure and a stroboscope for varying frequency of light flashes			Where other accelerometers cannot be fixed to vibrating body	High cost, calibration to be done carefully

Chapter 10

MEASUREMENT OF VIBRATION AND ACCELERATION

STARTERS

To study this chapter, you should have awareness on the following concepts. For a better understanding, it is always a good idea to revise these prerequisites.

- Definitions of terms and basics of linear motion and acceleration due to gravity
- Equations of motion such as $v = u + at$, $s = ut + \frac{1}{2} at^2$, $v^2 - u^2 = 2as$ etc.
- A study on relation between Force, mass and acceleration i.e., $F = ma$ or $a = F/m$; also the relation between acceleration and stress.
- Definitions of terms and equations of angular motion, angular velocity, angular acceleration
- Relation between the terms of linear and angular motions such as $v = r\omega$, $a = r\alpha$
- Some mechanisms of motion conversion (translatory to rotatory and vice-versa)
- Simple gear mechanisms for calibration into suitable scale
- Relation between speed and electrical parameters such as resistance, capacitance etc.
- Conversion of mechanical to light, electrical, magnetic and piezo-electric systems
- Fundamentals of mechanics, Newton's laws, fundamentals of rotatory motion, harmonic motion, vibrations and wave motion.
- Concept and applications of simple differentiation, integration and Laplace Transforms and
- First four chapters of this book.

LEARNING OBJECTIVE

After studying this chapter you should be able to

- Define vibration,
- Understand the principle of the seismic instrument,
- List out and discuss the transducers/instruments used to measure vibration/ acceleration and
- Explain servo accelerometer and digital accelerometer.

10.1 INTRODUCTION

Measurement of vibration is done in many situations, some of which are as follows:

- The vibration of rotating and reciprocating machines are measured, which in turn gives information about probable problems that may occur.
- Vibration monitoring is done on turbines of power station to detect early problem.
- Due to vibration, a lot of noise will be created which becomes a disturbance. Hence vibration is to be controlled.
- The intesity of earth quakes id estimated by measuring vibrations.

10.2 VIBRATION MEASUREMENT

If the displacement-time variation is of a generally continuous form with some degree of repetitive nature it is thought of as being a vibration. On the other hand, if the action is of a single-event form, a transient, with a motion generally, decaying or damping out before further dynamic action takes places then it may be referred to as shock.

The quantities that are to be measured in a vibrating system are

- amplitude or displacement,
- velocity,
- acceleration,
- frequency, and
- modes of vibration at a particular frequency.

Further, another important point to be noted here is that the above listed quantities are inter-related, for instance, displacement, velocity and acceleration may be transformed one to another by using differentiation or an integration. [Velocity is the rate of change of displacement with time and acceleration is the rate of change of velocity with time].

Thus, by using a suitable transducer circuit, one of the quantities may be measured and converted to other or calibrated to measure the vibration. The following are the three essential elements in such transducer circuit for measuring vibration.

More Instruments for Learning Engineers

Mile 10.1

An **audiometer** evaluates hearing loss, used at ENT (ear, nose, throat) clinics and in audiology centers. They consist of an embedded hardware unit with a pair of headphones and a test subject feedback button controlled by a PC. Alternatively software audiometers are also available in different configurations. As these can be calibrated to fractions of a decibel, these are more accurate than hardware audiometers. They can also be used with bone vibrators, to test conductive hearing mechanisms. Audiometers are also used to conduct Industrial Audiometric Testing. [The specifications, requirements and the test procedure are given in IEC 60645, ISO 8253, and ANSI S3.6 standards].

Audiometer

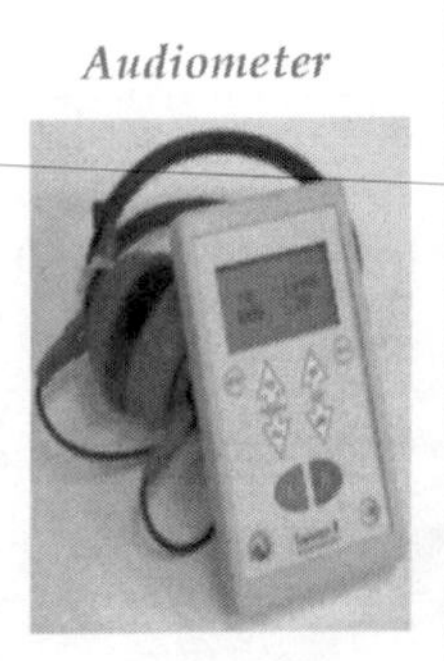

1. **Transducer element:** Transducer element converts the measured variable into a usable form e.g., vibration signal is converted to an electrical signal, by an electro-mechanical transducer. (These are discussed in the next section i.e., 10.3).
2. **Signal conditioning element:** Signal conditioning element such as amplifiers or filters or differentiators, integrators etc., convert the signal to a recordable or displayable form according to the requirement.
3. **Display or recording element:** Display or recording elements are used for displaying or recording the output obtained from signal conditioning unit.

10.3 TRANSDUCERS USED IN VIBRATION MEASUREMENT

There are basically two types of transducers in use for measurement of vibration, namely-

1. Fixed reference type transducer
2. Seismic type transducer.

We shall discuss about these now.

10.3.1 Fixed Reference Type Transducer

The two popular kinds of fixed reference type transducers are shown in the Figure 10.1(a) and (b), which are self explanatory.

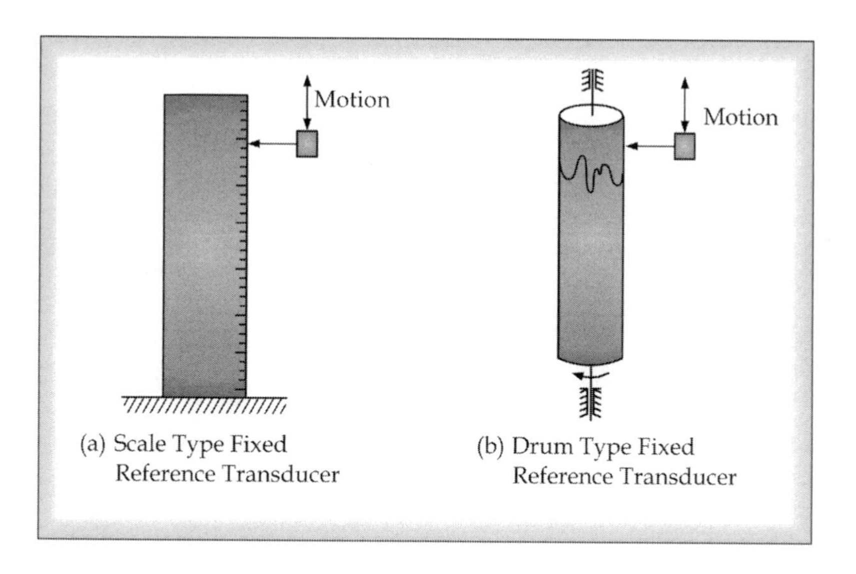

Fig. 10.1 Fixed Reference Type Transducers

In these transducers the motion of the moving object is measured relative to the fixed datum (earth) as shown in Fig. 10.2 and 10.3.

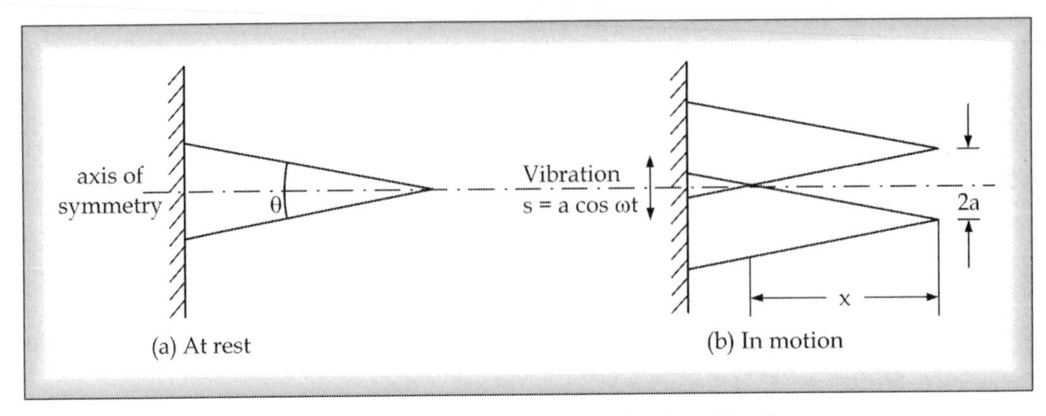

Fig. 10.2 Simple Wedge as a Device for Amplitude/ Displacement Measurements through Vibration

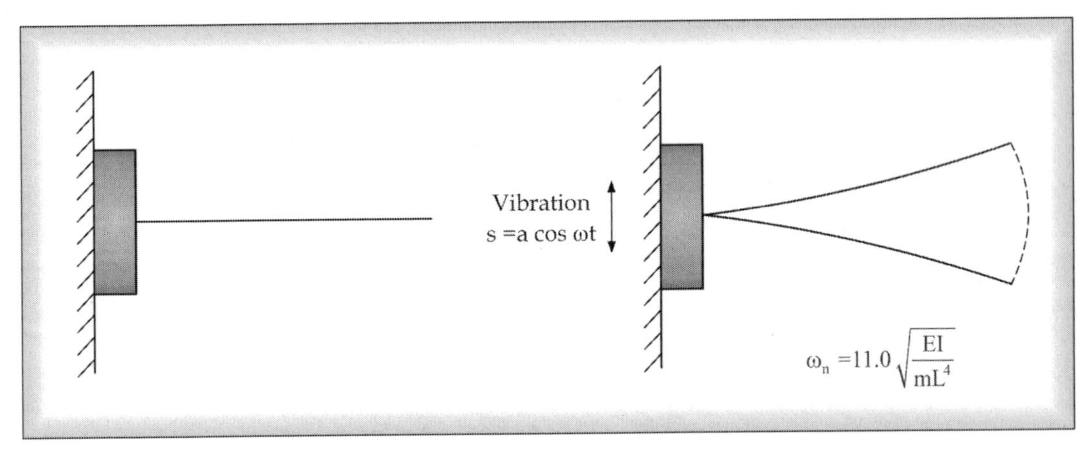

Fig. 10.3 Cantilever Beam used as a Frequency Measurement Device

More Instruments for Learning Engineers

Mile 10.2

A **tachometer (revolution-counter**, **tach**, **rev-counter**, **RPM gauge**) is an instrument measuring the rotation speed of a shaft or disk, as in a motor or other machine. The device usually displays the revolutions per minute (RPM) on a calibrated analogue dial, but digital displays are increasingly common. The word comes from Greek ταχος (*tachos* "speed") and *metron* ("measure").

A **tachymeter** or **tacheometer** is a type of theodolite used for rapid measurements (electronic/electro-optic) such as the distance to target. Such tachymeters are often used in surveying. **Tachymetry** or **tacheometry** is the process of measuring distance indirectly. This can be done by measuring time and speed in a moving vehicle or by sighting through small angle a distant scale transverse to the line of sight.

Tachometer - Tacheometer (or Tachymeter)

If the amplitude of motion is greater than 0.8 mm, a simple device for measuring amplitude of vibration is vibrating wedge of paper or of thin material, attached to the surface of vibrating members shown in Fig. 10.2 (a). The wedge is kept in such as way that the axis of symmetry of wedge is at right angles to the motion. When the member vibrates, the wedge also vibrates and assumes two extreme positions as shown in Fig. 10.2 (b) location of the point *P*, where images overlap and measuring the distance 'x', the amplitude of motion can be determined. At this distance, the width of the wedge is equal to the double amplitude 'a' of the motion.

$$a = x \tan \theta/2$$

where θ is the included angle of the wedge, 'a' is the amplitude of motion and x is the distance as shown in Fig. 10.2 (b).

Yet another simple device for measuring frequency can be constructed with a small cantilever beam mounted on a block which is placed against the vibrating surface as shown in Fig. 10.3. The beam length can be varied by appropriate arrangement. The beam length is slowly adjusted to the length of beam at which resonance occurs and its natural frequency exactly matches (equals) the frequency of vibrating surface. The length of beam can be calibrated in terms of frequency.

10.3.2 Seismic Type Transducer

These transducers are normally employed in practical situations where a fixed datum is not available such as for measurement of vibrations of a moving vehicle, a bridge or a machine located in an industrial environment where the disturbance due to surroundings results in non-availability of the fixed datum.

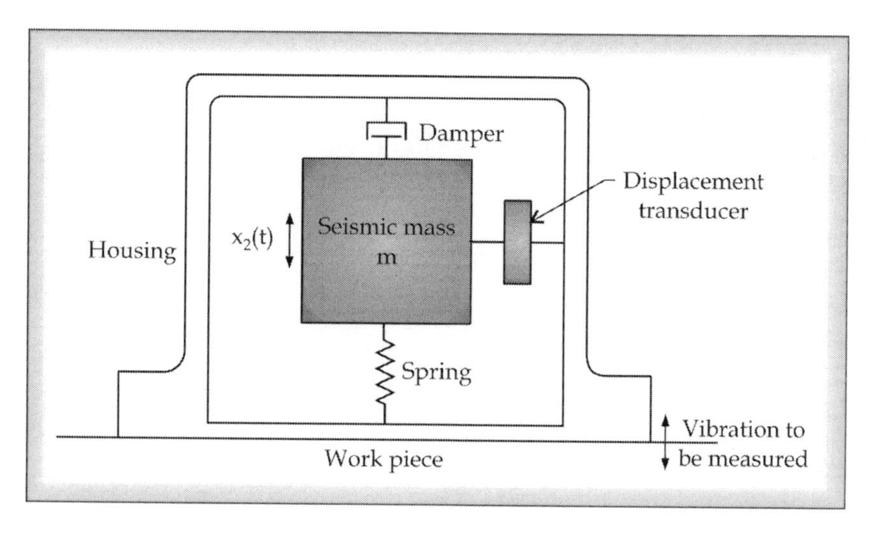

Fig 10.4 Typical Seismic Instrument

Seismic transducer consists a mass mounted on a spring and damper. The relative motion with respect to the frame of the instrument is measured by displacement transducer, which is a measure of the unknown vibration to be measured as shown in Fig. 10.4. This frame is connected to the vibration source whose characteristics are to be measured. The mass tends to remain fixed in its spatial position so that the vibration of motion is sensed by an appropriate transducer and recorded as a relative displacement between the mass and the housing frame.

By proper selection of mass, spring, and damper combinations, the seismic instrument may be used for either displacement or acceleration measurements. In general, large mass and soft springs are used for vibration, displacement measurements, while a relatively small mass and stiff spring are employed for acceleration measurements.

Self Assessment Questions-10.1

1. Name the basic three elements for measuring vibration.
2. Describe the Fixed reference type transducers.
3. Explain the principle of the seismic instrument (seismic type transducer)
4. What are the functions of following elements in vibration measurement devices?
   (i) Transducer element (ii) Signal conditioning element
   (iii) Display or recording element.
5. What are the different types of transducers available for the measurement of vibrations? Explain each of them.

10.4 LIST OF TRANSDUCERS/ INSTRUMENTS USED TO MEASURE VIBRATION/ACCELERATION

1. Piezo Electric Accelerometer.
2. Seismic Displacement Sensing Accelerometer.
3. Strain Gauge Accelerometer.
4. Capacitance Vibration Sensor/Accelerometer.
5. Variable Induction Type Accelerometer
   (Inductive Vibration Sensor) or Variable Reluctance Accelerometer.
6. LVDT - Accelerometer.
7. Reed Type Vibrometer.
8. Vibration Measurement using a Stroboscope.

More Instruments for Learning Engineers

Mile
10.3

A **Bettsometer** tests/ measures fabric degradation or the integrity of fabric coverings (and associated stitching) on aircraft and their wings. It is a pen-like instrument (which works like a spring balance) and a smooth round needle or pin. The needle is inserted into the fabric and then the instrument is pulled to exert a specific force on the fabric in order to test. A visual inspection is made to check for any rips or tears at the needle insertion point. The Bettsometer test (i.e. the areas of sail and stitching to be tested and the force to be exerted) is required for the annual 'permit' renewal for an aircraft and ships.

Bettsometer

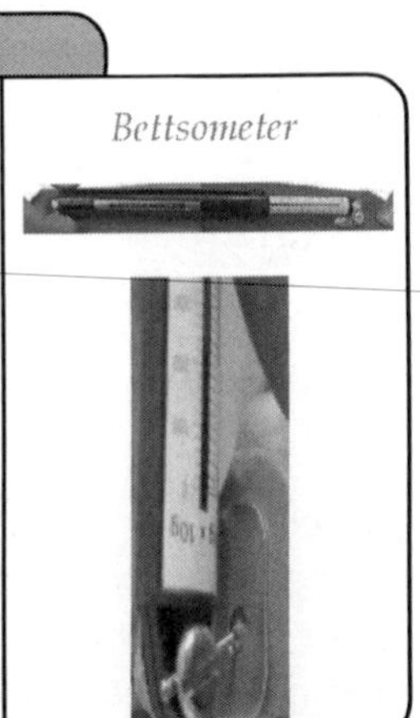

10.4.1 Piezo Electric Accelerometer

The piezoelectric accelerometers consist of the following main parts

1. Piezoelectric crystal
2. Electrodes
3. Mass
4. Spring plate
5. Housing frame
6. Output leads
7. Calibrated display unit

Principle

This accelerometer is based on the piezo-electric effect. When a piezoelectric crystal is subjected to a mechanical forces or stresses due to acceleration or vibration along specific planes, a voltage is generated across the crystal, which becomes a measure of the acceleration when calibrated.

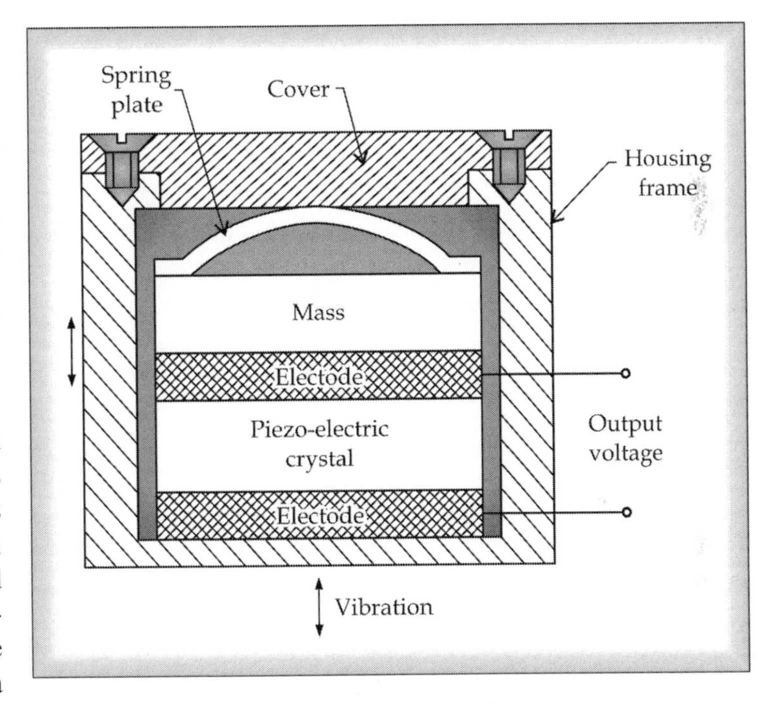

Fig 10.5 Piezo Electric Accelerometer

Construction

A piezoelectric crystal is placed between two electrodes. A mass is kept on the crystal-electrodes combination as shown in Fig. 10.5. The crystal is loaded using a spring plate. This crystal-electrode-mass-spring plate arrangement is housed in a suitable frame. The output leads from the electrodes are connected to a calibrated display unit. The accelerometer is fixed on the setup so as to measure the acceleration.

Operation

Due to the acceleration, the spring presses the mass and a force is exerted on the piezoelectric crystal.

We know that, Force = Mass × Acceleration ($F = ma$)

Keeping mass constant, the force is proportional to acceleration ($F \propto a$)

Since, the mass is constant here; the generated force is proportional to the acceleration. Due to this force, a voltage is generated across the crystal, which is picked up by the electrodes. Thus voltage becomes a measure of acceleration when calibrated.

Merits

- The size of this accelerometer is small.
- The instrument is rugged and relatively cheaper.

- It can be employed on a wide range of applications, since its natural frequency is high.
- The crystal has high output impedance.
- The

of the instrument is high.

Demerits

- The instrument is sensitive to changes in temperature
- To avoid loading effect, a voltage monitoring source should be used.
- The instrument is often subjected to hysteresis errors.

10.4.2 Seismic Displacement Sensing Accelerometer

There are two types of seismic displacement sensing accelerometers namely

(a) Linear seismic accelerometer. (b) Rotational seismic accelerometer.

Both of the above mentioned seismic accelerometers work on the same principle and as shown in Fig. 10.6.

The seismic displacement sensing accelerometer consists of the following main parts

1. Electric displacement transducer (Potentiometer)
2. Springs
3. Dampers
4. Housing frame
5. Display unit (accelerometer)

Principle

The mass is displaced proportionally when a spring-mass-damper system is subjected to acceleration. Hence a measure of displacement of the mass becomes a measure of acceleration.

Construction

A seismic mass is suspended from the housing of the accelerometer through a spring. A damper is connected between the seismic mass and the housing of the accelerometer. The seismic mass is connected to an electric displacement transducer as shown in Figure 10.6. The display unit (accelerometer) is fixed on to this structure.

More Instruments for Learning Engineers

Seismometer or Seismograph measures motions of the ground, including those of seismic waves generated by earthquakes, volcanic eruptions and other seismic sources. **Seismometry** is a branch of seismology. Seismometer (in Greek óåéóìüò, *seismós*, means shaking or quake) and was coined by David Milne-Home (1841), to describe an instrument designed by Scottish physicist James David Forbes.A simple seismometer sensitive to up/down motions of the earth like a weight hanging on a spring. The spring & weight are suspended from a frame that moves along with the earth's surface. As the earth moves, the relative motion between the weight and the earth provides a measure of the vertical ground motion. If a recording system, such as a rotating drum attached to the frame, and a pen attached to the mass, records the history of ground motion, called a seismogram or seismograph

Seismometer

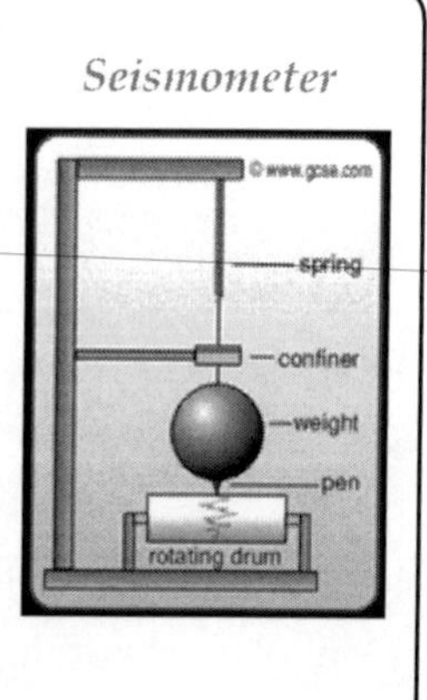

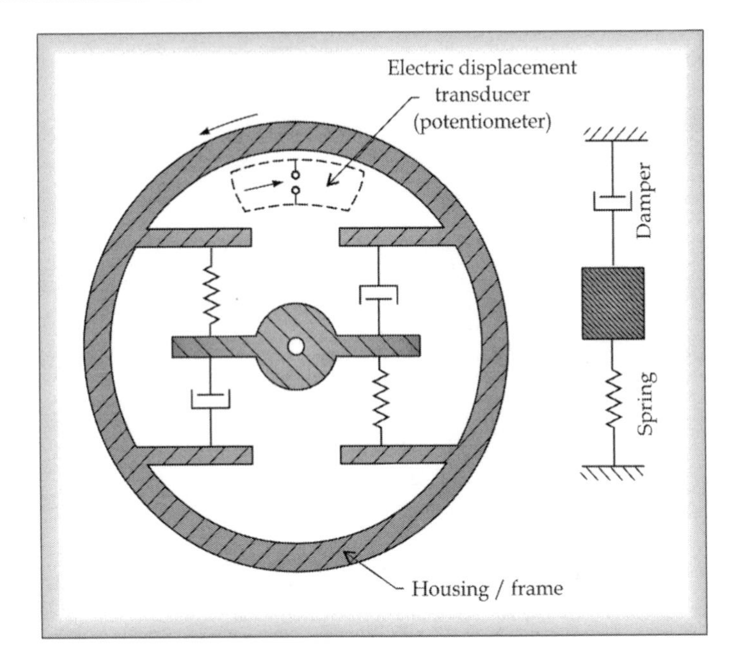

Fig. 10.6 Rotational Seismic Transducer

Operation

Because of the acceleration, the seismic mass is subjected to a displacement and obviously, this displacement of the mass is proportional to the acceleration. As the mass is connected to an electric displacement transducer, the output of the transducer depends on the extent to which the mass is displaced. Hence, the output of the transducer is calibrated to give a direct-indication of the acceleration characteristics of the structure.

10.4.3 Strain-Gauge Accelerometer

The salient components of strain gauge accelerometer are:

1. Bonded Strain gauges
2. Cantilever beam
3. Mass
4. Damping fluid
5. Housing frame
6. Display unit (accelerometer)

Principle

A cantilever beam, attached with a mass at its free end when subjected to vibration, displacement of the mass takes place that deflects the beam causing strain. The resulting strain is proportional to the displacement of the mass and hence the vibration/acceleration.

Construction

One end of a cantilever beam is fixed to the housing frame and a mass is fixed to the free end. Two bonded strain gauges are mounted on the cantilever beam as shown in the Fig. 10.7. Damping is provided by a viscous fluid filled inside the housing. The display unit (accelerometer) is fixed on to the structure whose acceleration is to be measured.

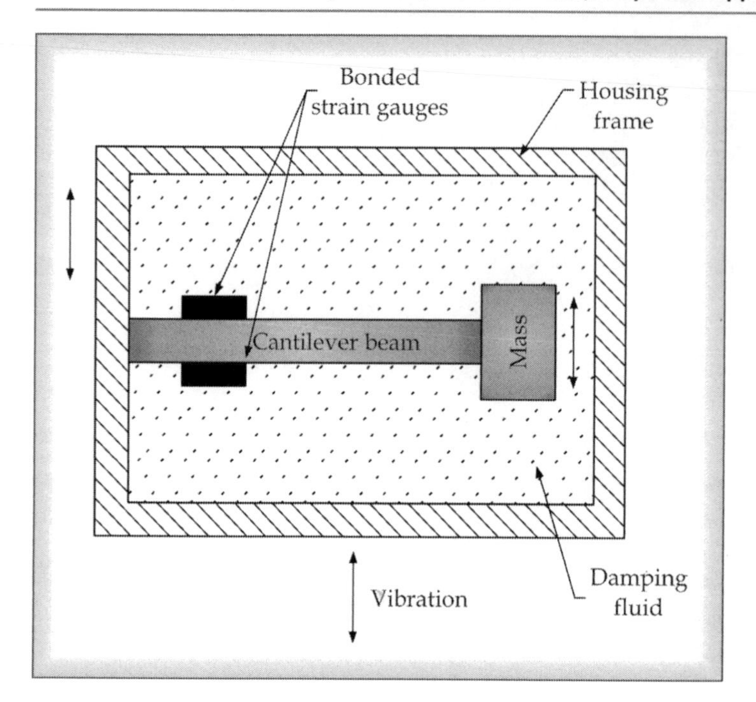

Fig. 10.7 Strain Gauge Accelerometer

10.4.4 Capacitance Vibration Sensor/Accelerometer

The major parts of this instrument are:

1. Flexible Reed
2. Mass
3. Movable plate of parallel plate capacitor
4. Fixed plate of parallel plate capacitor
5. Housing frame
6. Display unit (Accelerometer)

Principle

The output of a parallel plate capacitor depends on the gap between its movable and fixed plates.

Due to vibration, the change in gap between the plates of parallel plate capacitor changes its capacitance which becomes a measure of vibration/acceleration.

Construction

One end of a flexible reed is fixed to the housing of the instrument and other end carries a mass. The mass carries the movable plate of a parallel plate capacitor. The fixed plate of the parallel plate capacitor is placed in the housing as shown in Fig. 10.8. The accelerometer is fixed on to the structure whose acceleration is to be measured.

Operation

The vibration causes displacement of the mass, which in turn affects the cantilever to be strained. So, the strain gauges mounted on the cantilever beam are also strained and thence their resistance changes. A measure of this change in resistance of the strain gauges becomes a measure of the acceleration.

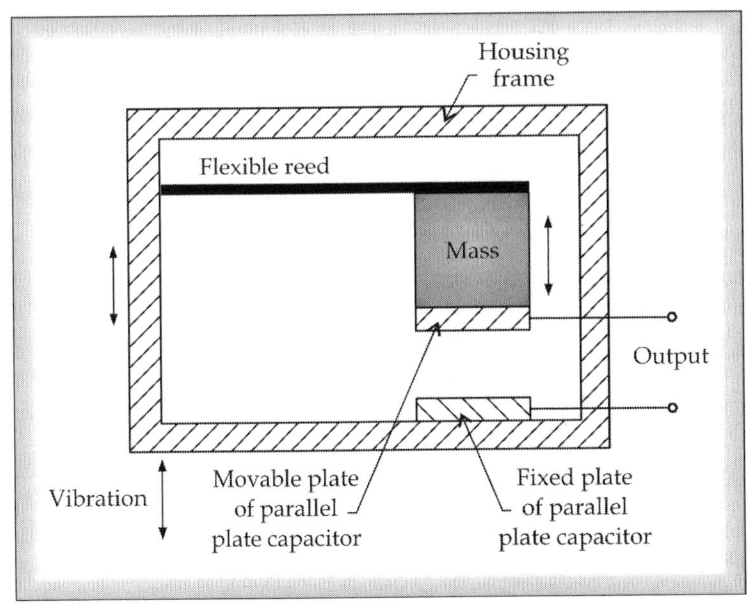

Fig. 10.8 Capacitance Vibration Sensor

Self Assessment Questions-10.2

1. List the instruments used to measure vibration/ acceleration.
2. Describe the basic principle of piezo electric accelerometer.
3. Explain the operation of piezo electric accelerometer with a neat sketch.
4. Give the advantages and limitations of piezo electric accelerometer
5. Draw and describe the main parts of a seismic displacement sensing accelerometer
6. Explain the basic principle involved in the seismic displacement sensing accelerometer
7. Discuss the operation of seismic displacement sensing accelerometer
8. What is the function of bonded strain gauges in the strain-gauge accelerometer?
9. Describe the main parts of strain gauge accelerometer with the help of a diagram. Explain its operation.
10. On what principle does capacitance vibration sensor/accelerometer work? Explain it.

More Instruments for Learning Engineers

Durometer is one of several measures of the hardness of a material. Hardness may be defined as a material's resistance to permanent indentation. The durometer scale was defined by Albert Ferdinand Shore, who developed a measurement device to measure Shore hardness in the 1920s. The term durometer is often used to refer to the measurement as well as the instrument itself. Durometer is typically used as a measure of hardness in polymers, elastomers, and rubbers. The two most common scales, using slightly different measurement systems, are the ASTM D2240 type A and type D scales. The A scale is for softer plastics, while the D scale is for harder ones.

Durometer

10.4.5 Variable Induction Type Accelerometer

This instrument consists of the following components

1. Flexible reed
2. Mass
3. Magnet
4. Induction Coil
5. Housing frame
6. Voltmeter
7. Display unit (Accelerometer)

Principle

The output voltage (magnitude of inductance) from a field coil is proportional to the amount by which a magnet is dipped (displacement due to vibration) into the field coil, which in turn is proportional to the vibration/ acceleration.

Construction

A mass is suspended on a flexible reed as shown in Fig. 10.9. A magnet is connected to the center of this flexible reed. A field coil is placed just below the magnet. Suitable devices are connected to the field coil to measure the induced voltage. The accelerometer is fixed to this setup to measure the acceleration.

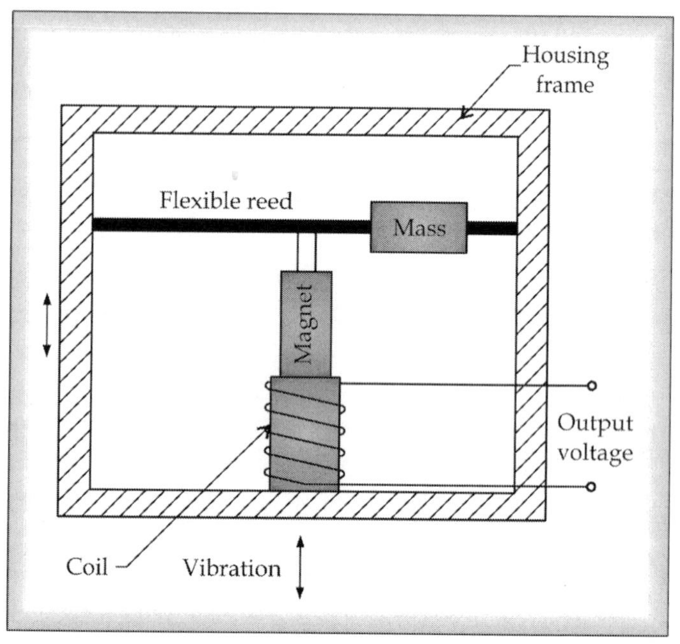

Fig. 10.9 Variable Induction Type Accelerometer

Operation

Due to vibration, displacement of the mass occurs proportionally to the vibration/ acceleration. When the mass is displaced, the magnet attached to the reed is also gets same displacement

More Instruments for Learning Engineers

A densitometer measures the degree of darkness (the optical density) of a photographic or semitransparent material or of a reflecting surface. The densitometer is basically a light source aimed at a photoelectric cell. It determines the density of a sample placed between the light source and the photoelectric cell from differences in the readings. Modern densitometers have the same components, but also have electronic integrated circuitry for better reading. There are two types: **Transmission densitometers** (for transparent materials) and **Reflection densitometers** to measure reflection.

Densitometer

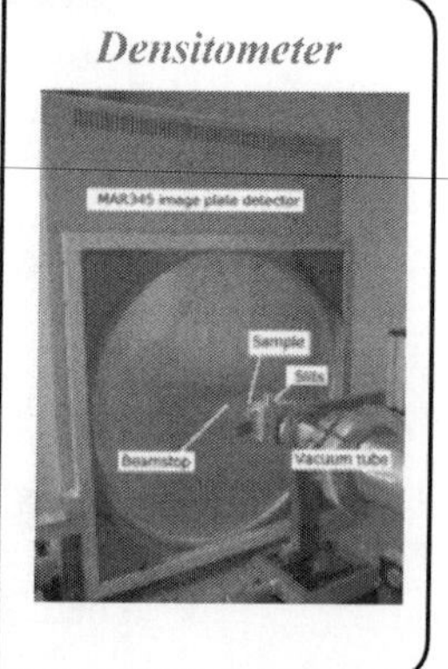

causing it to dip itself in the field coil. If this happens, the magnetic field is cut by the field coil and this induces voltage in the field coil. Thus this induced voltage is proportional to the displacement of the mass and thence the vibration/acceleration. The induced voltage is measured and when calibrated becomes a measure of vibration/acceleration.

10.4.6 LVDT - Accelerometer

The major parts of this instrument are listed below.

1. Linear Variable Differential Transformer
2. Flexible reeds
3. AC input
4. Device to measure differential output
5. Housing frame

Principle

The differential output voltage of the secondary winding of the LVDT is proportional to the displacement experienced by the sensing mass (core) caused due to vibration/acceleration.

Construction

A sensing mass (core) is attached between two flexible reeds mounted to the housing of the instrument. Primary & secondary windings are placed as shown in Fig. 10.10. The secondary windings have equal number of turns and are identically placed on either side of rim. The accelerometer is fixed on to this structure to measure acceleration on a calibrated scale.

Operation

Displacement takes place proportionally in the sensing mass (core) due to the vibration. As the core moves up and down due to vibration, the secondary windings give a differential output voltage. This

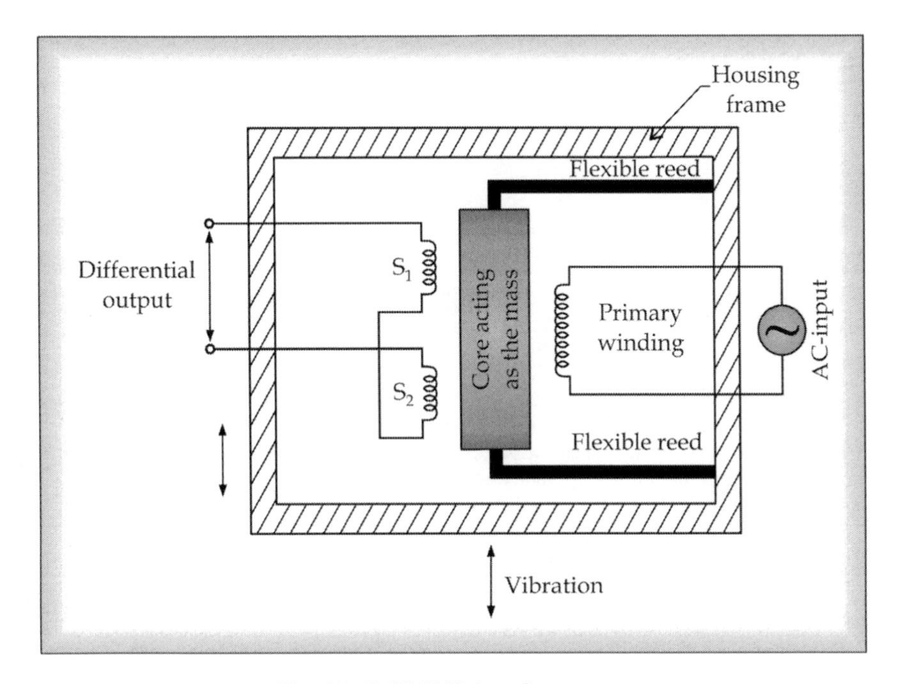

Fig. 10.10 LVDT Accelerometer

differential output is proportional to the displacement of the mass (core) and thence the vibration/ acceleration. The differential output of the secondary windings is measured and when calibrated becomes a measure of vibration/acceleration.

10.4.7 Reed Type Vibrometer

The main parts of this vibrometer are

1. A flexible reed,
2. A knob to adjust the reed length
3. A calibrated scale

Principle

The vibrations in the reed are matched with those of vibrating body.

Construction

The flexible reed is mounted on a frame and knob is fixed on the reed to adjust the length. A calibrated scale is attached to this set up as shown in Fig. 10.11.

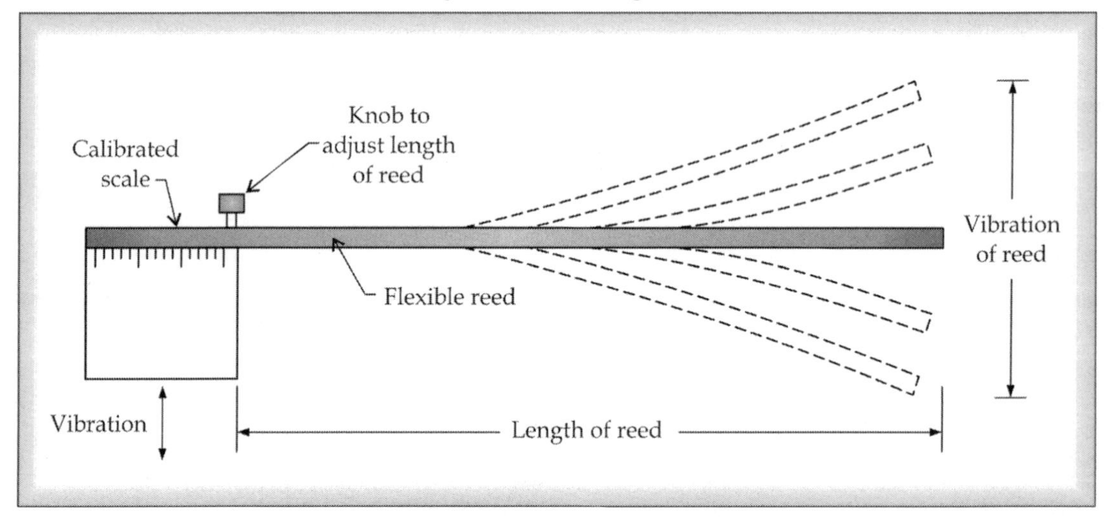

Fig. 10.11 Reed Type Vibrometer

More Instruments for Learning Engineers

Mile 10.7

A **Snickometer**, invented by Allan Plaskett (UK) in mid-1990s is used in televising cricket to graphically analyse sound/video if a fine noise (snick) occurs as ball passes bat. It was launched by Channel 4 in UK, who also introduced **Hawk-Eye** & **Red Zone**. The 3rd umpire views in a slow motion tele-replay to decide if ball touched bat. If there is sound of leather on willow, usually a short sharp sound in synchrony with the ball passing bat, then the ball touched the bat. Other sounds such as ball hitting batsman's pads, or bat hitting pitch, and so on, tend to have a fatter shape. The umpire has no benefit of Snicko, and has to rely on his sight & hearing, and his judgement. When **DRS (Decision Review System)** was introduced to Test Cricket, Snicko was not considered accurate enough, and so another detecting tool, **Hot Spot** was used.

Snicko Meter

Operation

The reed vibrometer is fixed to the structure whose vibration characteristics are to be measured. Now, the length of the reed is adjusted until the frequency of the reed equals the frequency of the vibrating structure. At this condition, the reed vibrates with maximum amplitude.

The reed length is calibrated to give a direct indication of its frequency and hence the frequency of the vibrating structure.

10.4.8 Vibration Measurement using Stroboscope

The chief parts are

1. Vibrating structure
2. Pointer
3. Stroboscope
4. Light source

Principle

The reflection of the flash light by the vibrating pointer is measured by the stroboscope.

Construction

In this method of measuring vibration, a pointer is attached to the vibrating structure. A stroboscope with a flash light source is aligned to it in line as shown in Fig. 10.12.

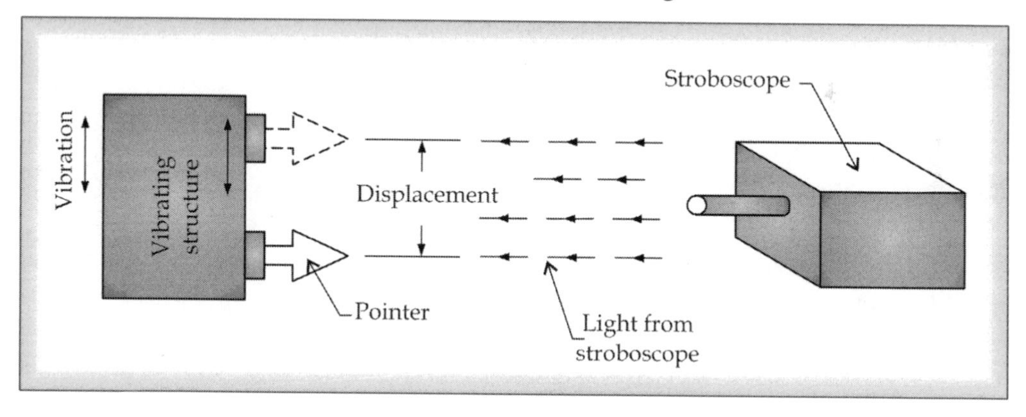

Fig. 10.12 Vibration Measurement using Stroboscope

Operation

The flashing light from a stroboscope is made to fall on the pointer as shown in Fig. 10.12. The frequency of light flashes from the stroboscope is adjusted until a stationary image of the pointer is obtained. Now the flash frequency of the stroboscope becomes a measure of frequency or amplitude of vibration.

Sometimes vibration measurement is done using a reed or a pointer-stroboscope combination, in situations where the various accelerometers cannot be mounted on the structure whose vibration characteristics are to be measured.

Merit

- Finds application where other accelerometers cannot be fixed to the vibrating structure.

Demerits

- High cost
- Calibration must be done carefully.

SELF ASSESSMENT QUESTIONS-10.3

1. Describe the construction and principle of variable induction type accelerometer.
2. With the aid of a diagram explain the operation of LVDT – accelerometer
3. Sketch and explain the vibration measurement using following instruments
   (i) Stroboscope (ii) Reed type vibrometer
4. How variable induction type accelerometer and LVDT accelerometer differs in the working principle?
5. What are the advantages and disadvantages of stroboscope?

10.5 FACTORS AFFECTING THE CHOICE OF TRANSDUCERS

The following points are to be considered while selecting a vibration transducer.

1. The type of transducer required: fixed reference type or seismic type.
2. The nature of the transducer: active or passive.
3. The magnitude of motion: very small (μm), medium or large (mm)
4. The frequency range over which vibrations are likely to be encountered and over which the transducer is having linear frequency response.
5. The input-output relation i.e. to which output physical quantity (such as displacement, velocity, acceleration, force, strain, voltage, capacity etc.) must be proportional to the input physical quantity i.e., vibration/acceleration.
6. The contact or proximity/contactless type required.
7. The type of related circuit and its complexity.
8. The cost involved in constructing and operating the instrument.

More Instruments for Learning Engineers

Mile 10.8

Diffracto Meter uses electron or neutron (as ë<10^{-9}) to study crystal structure similar to X-ray diffraction. Electrons do not penetrate as deep as X-rays, so, electron diffraction can identify the structure of the surface; further, neutrons are advantageous as they have an intrinsic magnetic moment causing interaction differently with atoms having different alignments. The device consists of a radiation source, a monochromator (to select ë), slits , a sample and a detector. A goniometer can also be used for fine adjustment of sample and detector positions.

Diffractometer

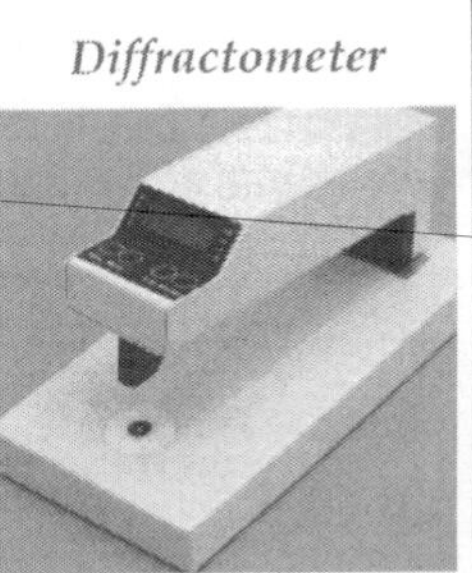

10.6 CALIBRATION OF VIBRATION TRANSDUCERS

There are different calibration methods for different types of vibration transducers. For instance, standard calibration equipment is avai l able for piezoelectric accelerometers, as shown in Fig. 10.13.

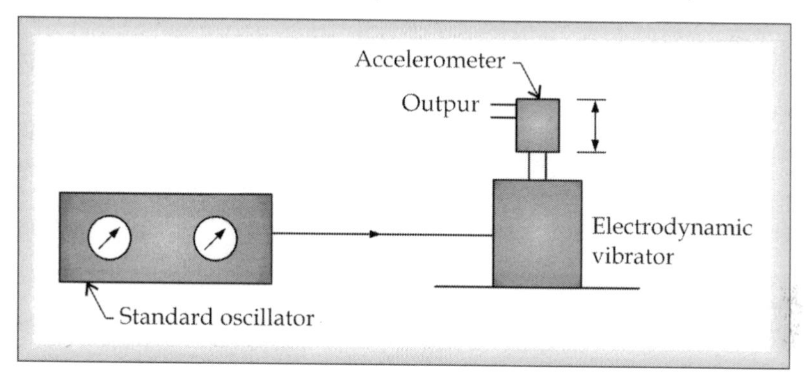

Fig. 10.13 Standards Calibration Equipment

It consists of an electro-dynamics vibrator driven by an oscillator. On this electro-dynamics vibrator, the given accelerometer is mounted and is subjected to '1g' of acceleration. The output of the accelerometer is noted and gives the sensitivity of the accelerometer which can be compared with manufacturer's rating.

Another method of calibration, mostly found in laboratories uses variable frequency oscillator, power amplifier and vibrator as shown in Fig. 10.14. The frequency 'ω' of vibrations of the vibrator, to which the accelerometer is subjected, can be changed by the oscillator while the amplitude of vibration 'X_o' is changed by the setting of the power amplifier. The displacement amplitude 'X_o' may be read by a microscope or a standard proximity transducer. The amplitude of acceleration is $\omega^2 X_o$ and the output voltage can be measured on an oscilloscope or a recorder.

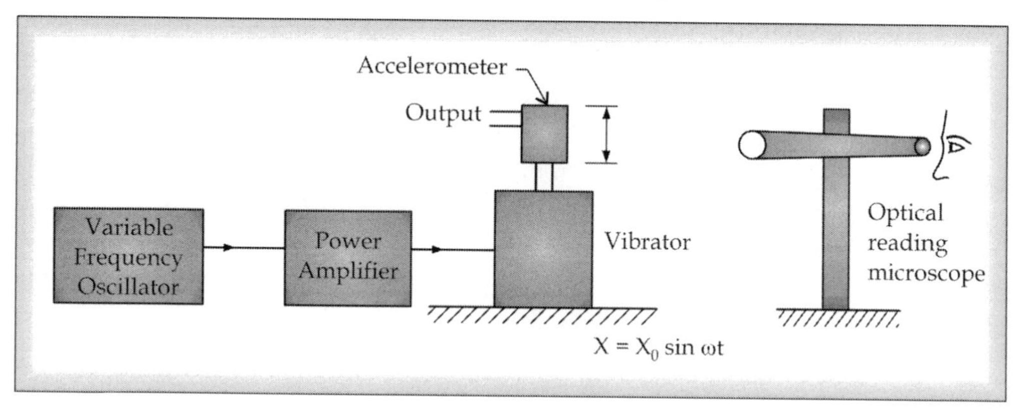

Fig. 10.14 Calibration of Vibration Transducer by using Variable Frequency Oscillators

10.7 SERVO ACCELEROMETER

Special problem occurs in the vibration measurement of large structures. The natural modes of large massive structures tend to occur at low frequencies and the expected acceleration associated that many conditions of measurement are small. Hence the accelerometer should be very sensitive and suitable for measuring such small values of acceleration right from the static conditions. Unlike many other applications, the motion of large structure is relatively unaffected by the loading of the accelerometer mass during measurement. This allows the use of heavy accelerometers. For such applications, a servo (Null-Balance or Force Balance) accelerometer is ideally suitable which operates on the measuring forces required to keep the seismic mass at rest with respect to the instrument frame under acceleration. This device is originally developed for inertial guidance systems in aircraft and missiles.

The major components of this instrument are

1. Displacement transducer
2. Spring
3. Restoring force coil suspended in Magnetic field
4. Amplifier
5. Damping device
6. Frame
7. Calibrated scale

Principle

The force required to prevent the mass moving relative to the instrument frame is proportional to the acceleration/vibration.

Construction

The displacement transducer connected between the housing frame and a damping device. In parallel, a spring attached to a mass is also connected between the frame and a restoring coil suspended in a magnetic field which is connected to the displacement transducer and an amplifier 'as a' or 'in a' circuit as shown in Fig. 10.15.

Operation

A displacement transducer gives a signal proportional to the relative movement of the mass with respect to the instrument frame and this signal is amplified and fed back as direct current to the force coil suspended in a magnetic field. The effect of this current is to generate the required restoring force for equilibrium. Therefore, the current needed to constrain the mass is a measure of the input acceleration on the frame along the direction in which the mass is free to move. Thus servo-accelerometer measures the acceleration of a structure or vehicle on which it is mounted by measuring the force required to prevent the mass moving relative to the instrument frame. The schematic arrangement of the servo accelerometer is shown in Fig. 10.15.

More Instruments for Learning Engineers

Mile 10.9

Dosimeter or Radiation Dosimeter measures exposure to ionizing radiation. The two main uses are: protect humans from radiation and measure the dose of radiation in both medical and industrial processes. **Electronic personal dosimeter (EPD)** useful in high dose areas, has merits like continual monitoring, alarm & live readout. It can be reset and re-used. **Film badge dosimeter (FBD)** is for one-time use only. The radiation absorption is indicated by a change to film emulsion, shown when film is developed. **Quartz fiber dosimeter (QFD)** is charged to a high voltage and the gas gets ionized by radiation & leaks away which is calibrated. **Thermoluminescent dosimeter (TLD)** measures intensity of visible light emitted by radiation from a crystal in detector when heated. Both QFD/FBD are superseded by TLD/EPD.

Dosimeter

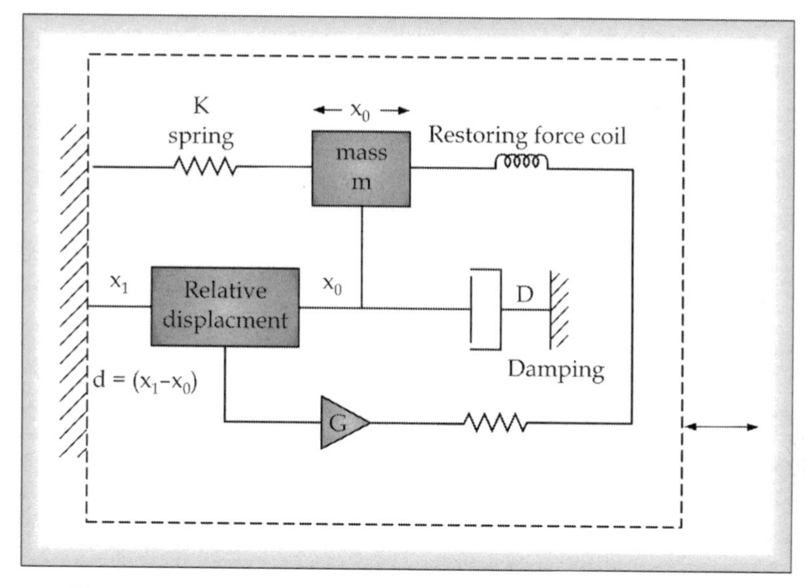

Fig. 10.15 Schematic Arrangement of the Servo Accelerometer

The theoretical/analytical expressions and physical relations can be deduced as follows.

Referring to the Figure 10.15,

We have $m\ddot{x}_0 = Kd + FI + Cd$

Therefore $m\ddot{x}_i = Kd + FI + Cd + \dot{m}d$

Then, under the condition I = *Gd*, Where G is the gain of the amplifier.

$$\text{Therefore } \ddot{x}_i = \left[\frac{m}{FG}S^2 + \frac{c}{FG}S + \frac{k}{FG} + 1\right]\frac{FI(s)}{m}$$

Where S is the Laplace operator.

The static sensitivity is then expressed as

$$\frac{1}{\ddot{x}_i} = \frac{G}{S^2 + \frac{c}{m}s + \frac{k+FG}{m}}$$

This is the closed loop dynamic response equation giving a value for the current 'I' required to make the seismic mass stationary for an input acceleration $\ddot{x}_1$. The undamped natural frequency is given by

$$w_n = \sqrt{\frac{k+FG}{m}}$$

For large values of *G*, *K*<< *FG*, $w_n = \sqrt{\frac{FG}{m}}$

In steady state $\frac{1}{\ddot{x}_i} = \frac{Gm}{K + FG}$

And for large values of *G*, $K << FG$, $\frac{1}{\ddot{x}_i} = \frac{m}{F}$

The open loop characterizes is given by (assuming $F = 0$)

And the undamped natural frequency in this case is

$$w_n = \sqrt{\frac{k}{m}}$$

A frequency response of 2 to about 500 *Hz* can be obtained in these devices with the high sensitivities.

10.8 DIGITAL ACCELEROMETER

The major components of digital accelerometer are

1. Masses
2. Resonators (double-ended quartz tuning fork)
3. Sensing element
4. Housing frame
5. Input (vibrating/accelerating body)
6. Display device with accelerometer

Principle

Digital accelerometer uses the principle of variation of the natural resonant frequency of a structural membrane, with an applied force to measure the acceleration in a digital format.

The output (frequency) of the instrument is directly proportional to the acceleration.

Construction

The digital accelerometer is basically a double-ended quartz tuning fork. Crystalline quartz is used as sensing element and is excited at its mechanical resonant frequency. (The precise configuration and dimensions are determined by the range of acceleration to be measured and other environmental considerations.)

More Instruments for Learning Engineers

Mile 10.10

A **Bolometer** invented by Samuel Pierpont Langley (1878), measures the power of incident electromagnetic radiation via the heating of material with a temp.-dependent electrical resistance. It was. It can be used to measure power at microwave frequencies. A dc bias current is applied to the resistor exposed to microwave power to raise its temp. via Joule heating, so that resistance matches to impedance. To reject the effect of ambient temperature changes, the active element is in a bridge circuit with an identical element not exposed to microwaves; variations in temperature is calibrated. The average response time of the bolometer permits convenient measurement of the power of a pulsed source.

Bolometer

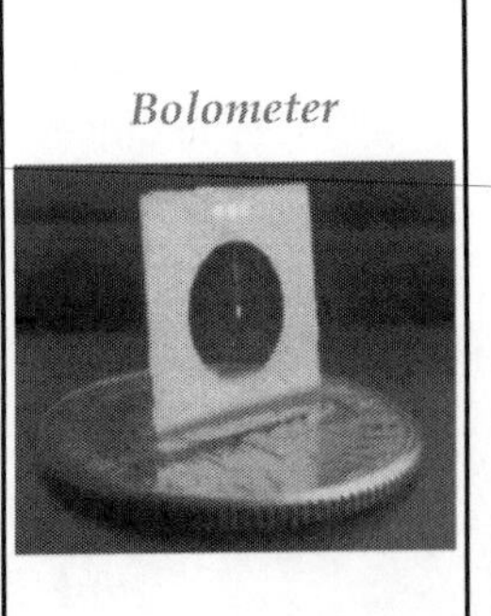

Two quartz resonators are used in each instrument. Both the resonators are fastened to the instrument case and to an individually suspended proof mass as shown in Fig. 10.16.

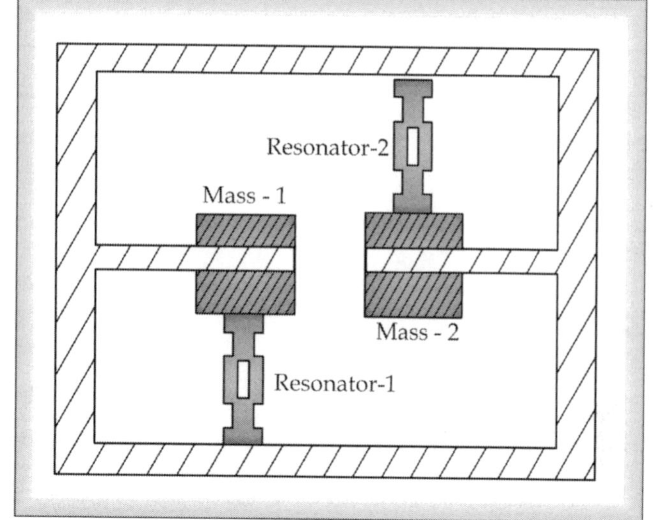

Fig. 10.16 Resonator Type Digital Accelerometer

Operation

Acceleration components along the axis of sensitivity will keep one resonator in tension and the other in compression. The resultant stresses cause precisely defined changes in the resonant frequency. The change in frequency as a result of applied acceleration is measured directly as a digital output.

Merits

- Linear digital accelerometers have high reliability
- These show accuracy
- They are smaller in size.
- The problem of normal analog to digital conversion is eliminated.

Demerits

- High cost
- More sensitive

Applications

Digital accelerometers are generally used in

- Inertial guidance systems for aircrafts, and
- Guidance and telemetry systems for missiles.

More Instruments for Learning Engineers

A **clapometer** or **applause meter** measures and displays the volume of clapping or applause made by an audience. It can be used to indicate the popularity of contestants and decide the result of competitions based on audience popularity. Clap-o-meters were popular in talent shows and television realty game shows in the 1950s and 60s, but more sophisticated methods are now available to measure audience response.

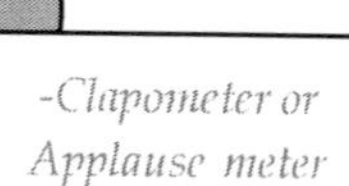

Know Instruments More? – Accelerometer, Gravity gradiometer

An **accelerometer** is a device that measures 'proper acceleration'. The proper acceleration measured by an accelerometer is not necessarily the coordinate acceleration (rate of change of velocity). Instead, the accelerometer sees the acceleration associated with the phenomenon of weight experienced by any test mass at rest in the frame of reference of the accelerometer device. For example, an accelerometer at rest on the surface of the earth will measure an acceleration g= 9.81 m/s^2 straight upwards, due to its weight. By contrast, accelerometers in free fall or at rest in outer space will measure zero. Another term for the type of acceleration that accelerometers can measure is g-force acceleration. Single- and multi-axis models of accelerometer are available to detect magnitude and direction of the proper acceleration (or g-force), as a vector quantity, and can be used to sense orientation (because direction of weight changes), coordinate acceleration (so long as it produces g-force or a change in g-force), vibration, shock, and falling in a resistive medium (a case where the proper acceleration changes, since it starts at zero, then increases). Micro machined accelerometers are increasingly present in portable electronic devices and video game controllers, to detect the position of the device or provide for game input.

Pairs of accelerometers extended over a region of space can be used to detect differences (gradients) in the proper accelerations of frames of references associated with those points. These devices are called gravity gradiometers, as they measure gradients in the gravitational field. Such pairs of accelerometers in theory may also be able to detect gravitational waves. Accelerometers have multiple applications in industry and science. Highly sensitive accelerometers are components of inertial navigation systems for aircraft and missiles. Accelerometers are used to detect and monitor vibration in rotating machinery. Accelerometers are used in tablet computers and digital cameras so that images on screens are always displayed upright. Accelerometers are used in drones for flight stabilization.

SELF ASSESSMENT QUESTIONS-10.4

1. Write the seven factors affecting the choice of transducers.
2. Explain the various types of calibration methods for vibration transducers.
3. Draw the circuit diagram of servo accelerometer. Explain its working principle.
4. Describe the principle of digital accelerometer with a neat diagram.
5. What is the sensing element in digital accelerometer?

SUMMARY

A vibration is a continuously repetitive variation in the displacement with time. The quantities measured in a vibrating system are usually amplitude or displacement, velocity, acceleration, frequency and modes of vibration at a particular frequency which are inter-related. Using a suitable transducer circuit, one of the quantities may be measured and converted to other or calibrated to measure the vibration. The three essential elements in such transducer circuit are Transducer (Fixed reference or Seismic type), Signal conditioning element and Display. In fixed reference type transducer the motion of the moving object is measured relative to the fixed datum (earth). If a fixed datum is not available, Seismic type is used in which a mass mounted on a spring and damper works as displacement transducer.

The principle, construction, operation, merits/demerits and applications of various instruments used to measure vibration/acceleration are discussed in this chapter. The piezoelectric accelerometer is based on the piezo-electric effect i.e. when a piezoelectric crystal is subjected to a mechanical forces or stresses due to acceleration or vibration along specific planes, a voltage is generated across the crystal. A Seismic (Linear/Rotational) Displacement Sensing Accelerometer works on the principle that the mass is displaced proportionally when a spring-mass-damper system is subjected to acceleration. In Strain - Gauge Accelerometer a cantilever beam, attached with a mass at its free end when subjected to vibration, displacement of the mass takes place that deflects the beam causing strain, which is proportional to the displacement of the mass and hence the vibration/ acceleration.

The output of a parallel plate capacitor due to the change in gap between its movable and fixed plates is the measure of vibration in Capacitance Vibration Accelerometer. In Variable Induction Type Accelerometer, measure of vibration is the output voltage (magnitude of inductance) from a field coil, proportional to the displacement due to vibration by which the magnet is dipped into the field coil. The differential output voltage of the secondary winding of the LVDT is proportional to the displacement experienced by the sensing mass (core) in LVDT - Accelerometer. The vibrations in the reed are matched with those of vibrating body with Reed Type Vibrometer. The reflection of the flash light by the vibrating pointer is measured by the stroboscope. Servo Accelerometer measures the force required to prevent the mass moving relative to the instrument frame which is proportional to the acceleration/ vibration. Digital Accelerometer uses the principle of variation of the natural resonant frequency of a structural membrane, with an applied force to measure the acceleration in a digital format.

KEY CONCEPTS

Vibration: If the displacement-time variation is of a generally continuous form with some degree of repetitive nature it is thought of as being a vibration.

Seismic Instrument (Seismic Type Transducer): Consists a mass mounted on a spring, damper and the relative motion of with respect to the frame of the instrument by displacement transducer is a measure of the unknown vibration.

Piezo Electric Accelerometer: Operates on piezo-electric effect i.e., when a piezo electric crystal loaded using a spring-plate, between two electrodes is subjected to a mechanical force or stress along specific planes, a voltage generated across the crystal becomes a measure of the acceleration when calibrated.

Seismic Displacement Sensing Accelerometer: When a spring-mass-damper system is subjected to acceleration, the mass is proportionally displaced becomes a measure of acceleration.

Strain - Gauge Accelerometer: In a cantilever beam a mass at its free end is subjected to vibration experiences displacement and beam deflects thence strain is proportional to the displacement being measured by calibration.

Capacitance Vibration Sensor/Accelerometer: The output of a parallel plate capacitor depends on the gap between its movable and fixed plates due to vibration and this change in capacitance becomes a measure of vibration/acceleration.

Variable Induction Type Accelerometer: The output voltage (magnitude of inductance) from a field coil is proportional to the amount by which a magnet is dipped due to vibration and causes displacement into the field coil, becomes the measure of acceleration/vibration when calibrated.

LVDT: Linear Voltage Differential Transformer.

LVDT – Accelerometer: The differential output voltage of the secondary winding of the LVDT proportional to the displacement experienced by the sensing mass (core) between two flexible reeds caused due to vibration/ acceleration, becomes the measure when calibrated.

Reed Type Vibrometer: The length of flexible reed adjusted until the frequency of the reed is equal to that of the vibrating structure by a knob which reads on a calibrated scale.

Vibration Measurement Using Stroboscope: Frequency of flashing light from a stroboscope falls on pointer attached to the vibrating structure is adjusted until a stationary image of the pointer is obtained and the flash frequency becomes a measure of frequency or amplitude of vibration.

Servo Accelerometer: A displacement transducer produces a signal proportional to the relative movement of the mass with respect to the instrument frame and the signal is amplified and fed back as dc to the force coil suspended in a magnetic field and the current needed to constrain the mass is a measure of the input acceleration.

REVIEW QUESTIONS

SHORT ANSWER QUESTIONS

1. A seismic type accelerometer is relatively rugged compared to a seismic type vibrometer comment.
2. What is vibrometer?
3. What are the advantages of piezoelectric type accelerometer?
4. Why vibrations has to be measured? How vibrations are measured?
5. What is the basic difference in design and application between vibrometer and accelerometer?
6. Write short notes on seismic instruments.
7. Discuss the principle of working of a vibrometer.
8. Explain the calibration procedure for an accelerometer.
9. Explain how a vibrometer is calibrated to measure acceleration.

LONG ANSWER QUESTIONS

1. Explain the general theory of seismic instruments need for vibration/acceleration measurement.
2. Explain the working of a bonded strain gauge accelerometer with neat sketch.

3. How are seismic instruments used for measuring acceleration? Explain in detail.
4. Compare the working of a servo & digital accelerometers.
5. Explain the different methods adopted for the calibration of accelerometers.
6. How seismic instruments are used for measuring acceleration. Explain in detail
7. Explain the working principle of a simple accelerometer and a vibrometer.
8. Sketch and explain the general theory of the seismic instrument.
9. Explain the principle of seismic instruments. Derive an expression to measure velocity & acceleration using this instrument.
10. What is a vibrometer? Explain anyone of them.
11. What are the different methods of converting vibration into a voltage? Explain anyone detail.
12. Name the different vibration sensing systems used in practive. Explain anyone such system for the measurement of vibration.
13. Explain the working of piezoelectric accelerometer with neat sketch.
14. How is measurement of vibrations on large structures done? Explain the method in detail.
15. Name any two types of velocity transducers used for the measurement of vibration and explain anyone type with neat sketch.
16. Explain the working of servo accelerometer with neat sketch.

MULTIPLE CHOICE QUESTIONS

1. The units of vibration is_____.
   (a) Ohms (b) Meters
   (c) Hertz (d) Ampere
2. The units of acceleration is____.
   (a) m (b) m/s
   (c) m^2/s (d) m/s^2
3. Rate of change of velocity is called _____.
   (a) Acceleration (b) Vibration
   (c) Linear velocity (d) Displacement
4. In fixed reference type transducer the amplitude is given as______.
   (a) $a = x \tan \theta/2$ (b) $a = x \tan 2\theta$
   (c) $a = x \tan 3\theta$ (d) $a = x \tan \theta^2$
5. For the measurement of vibration of a moving vehicle ________ type of transducer is used
   (a) Scale type fixed reference
   (b) Drum type fixed reference
   (c) Seismic
   (d) Vibrometer
6. In seismic transducer ________ is used for the measurement of vibration
   (a) Seismic mass
   (b) Seismic spring
   (c) Seismic thermocouple
   (d) Seismic Resistor
7. Which of the following is wrong about Piezo-Electric Accelerometer
   (a) Size of this accelerometer is small
   (b) It is a cheaper instrument
   (c) Crystal has low output impedance
   (d) Sensitivity of the instrument is high
8. In strain gauge accelerometer, strain is proportional to _____.
   (a) Clockwise rotation of the mass
   (b) Anti-clockwise rotation of the mass

(c) Displacement of the mass
(d) Deformation of the mass

9. A flash light source is related to which of the following instrument_____.
(a) Stroboscope
(b) Reed type vibro meter
(c) LVDT
(d) Variable induction type accelerometer

10. For finding vibration in large structures ______ instrument is used
(a) Reed type vibro meter
(b) Servo accelerometer
(c) LVDT
(d) Variable induction type accelerometer

11. The undamped natural frequency is given as
(a) $w_n = \sqrt{\frac{k+FG}{m}}$
(b) $w_n = \sqrt{\frac{k+F}{m}}$
(c) $w_n = \sqrt{\frac{k+FG}{2m}}$
(d) $w_n = \sqrt{\frac{k+2FG}{m}}$

12. Digital accelerometer uses_____ principle
(a) Induction
(b) Wheat stone bridge
(c) Natural resonant frequency
(d) Emf

13. The full form of LVDT is
(a) Linear value of differential transducer
(b) Linear voltage digital transducer
(c) Linear voltage differential transducer
(d) Lighter value of digital transducer

14. When a force or stress is applied on Piezo electric crystal ____________ is generated
(a) Voltage (b) Frequency
(c) Mass (d) Amplitude

15. In variable induction type accelerometer the amount by which a magnet is dipped into the field coil is proportional to the
(a) Voltage (b) Length
(c) Resistance (d) ampere

16. In the strain gauge accelerometer the strain gauges are attached to
(a) The body of the accelerometer
(b) Beam having a mass
(c) To a string
(d) to a motor

17. In which of the following instrument both primary and secondary winding are used
(a) Capacitance type
(b) Reed type vibrometer
(c) Variable induction
(d) LVDT accelerometer

18. The Vibration of gap between parallel plane capacitor subjected to ________.
(a) Capacitance vibration sensor
(b) Reed type vibrometer
(c) Variable induction
(d) LVDT accelerometer

19. In which of the following instrument a core is moved between primary and secondary winding
(a) Capacitance type
(b) Reed type vibrometer
(c) Variable induction
(d) LVDT accelerometer

20. Among which of the following instrument inertia efforts should be minimized
(a) Seismic type
(b) Reed type vibro meter
(c) LVDT
(d) Variable induction type accelerometer

ANSWERS

1.	(c)	2.	(d)	3.	(a)	4.	(a)	5.	(c)
6.	(a)	7.	(c)	8.	(c)	9.	(a)	10.	(b)
11.	(a)	12.	(c)	13.	(c)	14.	(a)	15.	(a)
16.	(b)	17.	(d)	18.	(a)	19.	(d)	20.	(a)

CHAPTER INFOGRAPHIC

11 Glimpses of Measurement of Force, Torque and Power

S.No	Instrument	Principle	Construction	Operation/Measurement	Application	Merits	Demerits/ Limitations
1. Direct method: Comparison is made with known gravitational force acting on a standard mass.							
1(a)	Analytical or equal arm balance	Moment comparison $M_2L_2 = M_1L_1$	A beam with a pointer resting on the fulcrum of a knife-edge	A known mass is at one end an unknown mass at other end can be determined by balancing		Very easy: useful for smaller forces	Cannot be used for measuring large force
1(b)	Unequal arm balance	Moment comparison $FL_2 = FL_1$	Graduated beam pointed to knife edge. Known mass attached to right of beam and unknown mass on left.	Position of known weight is adjusted so that pointer reads zero which indicates unknown force.	To measure force	Very easy, can measure higher loads	Needs adjusting known weight each time
1(c)	Pendulum scale (Multi lever type)	Moment comparison	Unknown force is converted to torque and then balanced by torque as pendulum	Unknown weight is applied to load rod.; movement of rack by equalizer beam indicates force.		Can measure higher loads, directly reads	Long use tapes expands leading to recalibration
2. Indirect method: Effect of force on a body is measured. (Force is that which changes or tends to change the state of rest or of uniform motion in a straight line. Force can be measured directly or indirectly)							
2(a) Applying force to known mass and measuring the resulting acceleration							
2(b) Transducing the force to a fluid pressure and then measuring the resulting pressure.							
2(b) (i)	Hydraulic load cell	Increase in pressure of fluid in confined space indicates force	A diaphragm above which piston is placed to apply force and from below liquid under pressure is applied	Force on piston deflects diaphragm to increase pressure of liquid underneath which varies with applied pressure	To measure force	Heavy loads can be measured	Leakages must be avoided
2(b) (ii)	Pneumatic load cell	Force on one side of diaphragm and air pressure on other indicates force.	A corrugated diaphragm with the top surface attached with arrangement to apply force.	Force on diaphragm closes bleed valve and increases back pressure and lifts diaphragm. Air pressure indicates force		Working medium available freely	Rrecalibrated frequently
2(c) Applying the force to same elastic member and measuring the resulting deformation or deflection							
2(c) (i)	Provide ring	Steel ring under force across its dia., deflects proportionally	A steel ring attached with external bosses to apply force	Force to be measured is applied to the external bosses; change in diameter indicates force	To measure force	Reading can be obtained quickly	Higher forces cannot be measured
2c. (ii)	Strain gauge	Strain of gauges on steel cylinder indicates force.	A steel cylinder mounted with four identical strain gauges	When load is applied the strain gauges get strain causing resistance change as a measure of applied force.	To measure force	Very effective, quick readability, can automate	Small loads cannot be measured.

Contd...

S.No	Instrument	Principle	Construction	Operation/Measurement	Application	Merits	Demerits/ Limitations
3	**Dynamometers (d/m):** Torque transmitted is estimated by angular twist or shear stress; Shear stress = $f_s = 16\tau/\pi d^3$; Shear strain = $y = f_s/G = 16\tau/G\pi d^3$; Angular twist = $\theta = 32\tau l/G\pi d^4$						
3(a)	Mech. Prony Brake dynamometer	Torque (τ) = Cross product of force and radius ($F \times R$)	Rotating fly wheel attached to test m/c with rope or band/brake shoes	Force reqd to bring down speed is measured and multiplied to get torque	For performance test	Very easy to use	Lot of heat is generated
3(b)	Hydraulic Dynamometer		Stator is freely pivoted on bearing and rotor hydraulically coupled	Varying braking effect produced between rotor & stator is measured indicates torque	If separate cooling arrangement cannot be used		Cradling must be done carefully
3(c)	Eddy current Dynamometer		A metallic disc is rotated in a magnetic field	Eddies produced as disc rotates in magnetic field and rotate d/m in trunnion bearings. attached arm measures torque	Electrical dynamometer	Load can be varied by varying the field current	Cooling is needed
3(d)	Electrical generator type		Cradle type d/m and is generator driven by test m/c	Mech. energy is converted elec. energy by generator is dissipated thru' variable resistance grid.	Electrical generator type		Rarely used
3(e)	Driving Dynamometer		Cradle type DC motor	Cradle type DC motor drives test m/c and reaction torque on stator is measured in eddy current dynamometer	For performance air compressor and pumps	Measures torque and gives energy to operate test m/c	Only for testing air compressors and pumps
4.	**Torsion Meters:** (Torque estimated angular twist are called torsion meters and Torque estimated by shear stress is called torque meters)						
4(a)	Mechanical torsion meter	Mechanical means	A shaft with 2 drums and 2 flanges one acting as driven and other driver	Angle of twist b/n 2-flanges varies with torque on shaft measured by stroboscope	Can be measured during power transmission	Simple and inexpensive	Sensitivity is reduces even by small speed variations
4(b)	Optical torsion meter	Optical means	2 castings with tension strip and mirrors, a light beam falls on mirror	When shaft transmits torque mirrors change position and angular deflection of light rays indicates torque	Steam turbines and internal combustion engines	Very accurate	Calibration must be done carefully
4(c)	Electric torsion meter	Time lapse b/n 2 pulses on slotted discs $\alpha\ \theta$	2 slotted discs fitted to driving and driven and a unit for pulse generation	As shaft rotates, relative displacement of slotted discs due to twist and time lapse b/n pulses of 2- discs varies twist $\alpha\ \theta$	Can be automated	No signal leakage and noise problems	Initial alignment of teeth on both discs must be perfect
4(d)	Strain gauge torsion meter	Resistance change of strain gauge $\alpha\ \tau$	4bounded wire strain gauges on a 45° helix to Wheatstone bridge	When shaft rotates 2- strain gauges get tension and other 2- compression. Imbalance of ckt indicates torque	Used if automatic torque measurement is required	Sensitivity is high	Difficult to connect power source and display unit to bridge ckt

Chapter 11

Measurement of Force, Torque and Power

STARTERS

To study this chapter, you should have awareness on the following concepts. For a better understanding, it is always a good idea to revise these prerequisites.

- Definitions of terms and basics of linear motion and acceleration due to gravity
- Equations of motion such as $v = u + at$, $s = ut + \frac{1}{2} at^2$, $v^2 - u^2 = 2as$ etc.
- Fundamentals of mechanics, Newton's laws, fundamentals of rotatory motion, harmonic motion, vibrations and wave motion.
- A study on relation between Force, mass and acceleration i.e., $F = ma$ or $a = F/m$; $W=mg$ or $g = W/m$ also the relation between force and stress.
- A fair idea on various types of forces, i.e., tensile-compressive, gravitational, elastic, magnetic, electric, mechanical, centripetal, centrifugal forces and so on.
- Definitions of terms and equations of angular motion, angular velocity, angular acceleration
- Relation between the terms of linear/angular motions such as $F = mr\grave{u}^2 = mv^2/r$
- Relation between force and electrical parameters such as resistance, capacitance etc.
- Conversion of mechanical to light, electrical, magnetic and piezo-electric systems, concept of load cell and strain gauge
- Awareness on torque, torsion, power, moment, and couple and their expressions and conversions from one to the other.
- Some mechanisms of motion conversion (translatory to rotatory and vice-versa)
- Simple gear mechanisms for calibration into suitable scale and
- First four chapters of this book.

LEARNING OBJECTIVE

After studying this chapter you should be able to

- Discuss the force,
- Classify the methods of measurement of force,
- List out and explain the direct methods of measurement of force
- List out and explain the indirect methods of measurement of force.
- Define the torque and understand how it is related to Power,
- Classify Dynamometers and explain them and
- Classify Torsion Meter and discuss them.

11.1 INTRODUCTION

The physical quantity, which changes or tends to change the state (motion/rest) or direction or shape of a body on which it is applied is called force, and its unit is NEWTON.

It is a vector quantity. Therefore, its measurement involves the determination of both magnitude and direction.

According to Newton's second law of motion, Force, 'F' is proportional to the product of mass 'm' (in kg), and acceleration, 'a' it produces in (m/s^2), i.e.

$$F = ma$$

Force in addition to an effect along its line of action (push or pull) may also cause of a turning effect relative to an axis. Such turning effect is called moment, torque or couple (depending on the manner in which it is produced), which can cause bending a beam or torsion in a shaft.

The weight of a body is the force exerted on the body by the gravitational acceleration. Weight will vary from point to point to the earth's surface.

$$W = mg$$

where 'g' is the acceleration due to gravity, Mass 'm' is invariable.

11.2 METHODS OF MEASUREMENT OF FORCE

Force measurement can be done by the methods shown in Figure 11.1.

11.2.1 Direct Method

In this method, a direct comparison is made with known gravitational force acting on a standard mass. This can be achieved directly or indirectly using a system of levers. Devices in this category are

1. Analytical balance or equal arm balance
2. Unequal arm balance
3. Pendulum Scale (Multi-lever type)

More Instruments for Learning Engineers

Mile 11.1

The **wattmeter** measures the electric power (or the supply rate of electrical energy) in watts of any given circuit. Electromagnetic wattmeters are used for measurement of utility frequency and audio frequency power; other types are required for radio frequency measurements.

Wattmeter

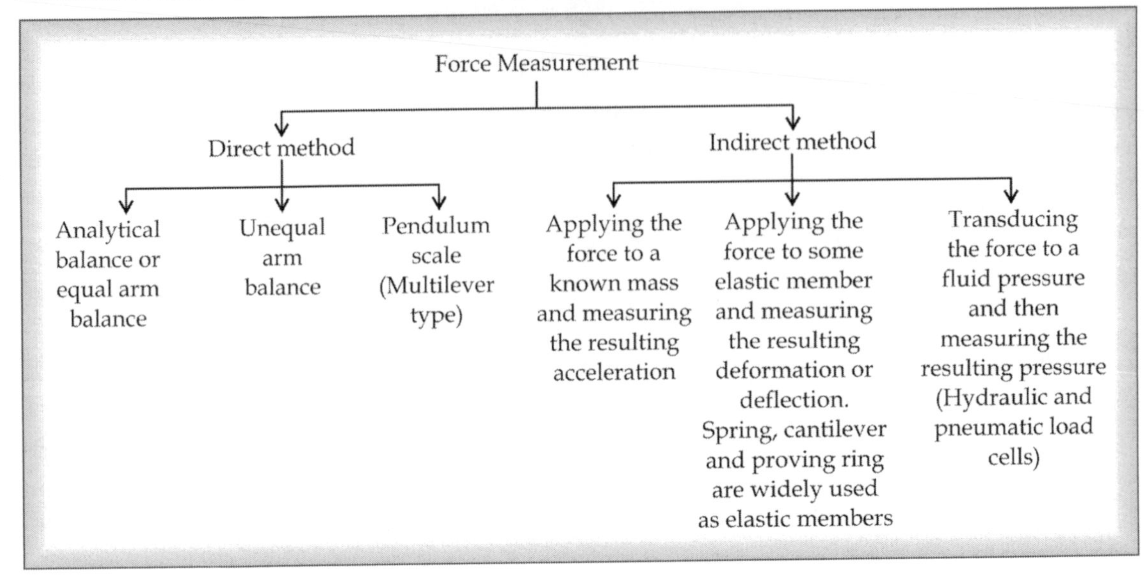

Fig. 11.1 Methods of Measurement of Force

11.2.2 Indirect Method

These involve measurement of the effect of force on a body.

1. Applying the force to a known mass and measuring the resulting acceleration.
2. Applying the force to some elastic member and measuring the resulting deformation or deflection. Spring, cantilever and proving ring are widely used as elastic members.
3. Transducing the force to a fluid pressure and then measuring the resulting pressure (Hydraulic and pneumatic load cells).

Self Assessment Questions-11.1

1. Define force. List the various methods for the measurement of force.
2. How the direct method is used for the measurement of force? Explain it.
3. Give the differences between direct and indirect method for the measurement of force.

11.3 EQUAL ARM BALANCE

Perhaps! This is the simplest and widely used instrument for measurement of force or weight. The main parts of this instrument are:

1. A beam
2. Knife edge fulcrum
3. A pointer
4. Calibrated scale
5. Hooks or a provision to hang masses on either side of beam

Principle

It works on principle of equilibrium of moments. The beam of equal arm balance is in equilibrium position when:

(Clockwise rotating moment) = (Anticlockwise rotating moment)

i.e., $m_2 L_2 = m_1 L_1$

where, m_1, m_2, L_1, L_2, are the masses and lengths about fulcrum in clock wise and counter clock wise directions respectively.

Thus the unknown force is balanced against the known force.

Construction

A beam with center is pointed rests on the fulcrum of a knife edge. Either side of the beam is of equal length with respect to the fulcrum (i.e., $L_1 = L_2 = L$). A pointer is attached to the center of the beam. This pointer would point vertically downwards when the beam is in equilibrium. On either side of the beam hooks or a provision is given to place masses one known and the other unknown at either end of the beam as shown in the Figure 11.2.

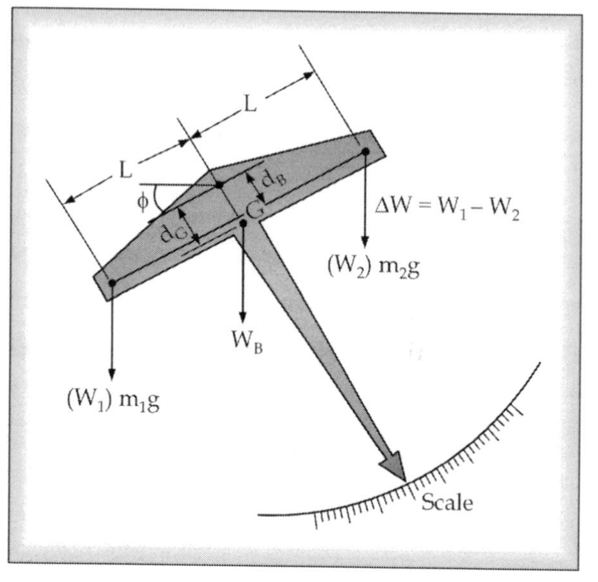

Fig. 11.2 Equal Arm Balance or Analytical Balance

Operation

A known standard mass (m_1) is placed at one end of the beam and an unknown mass (m_2) is placed at its other end. Now for Equilibrium condition, we know that

(Clock wise rotating moment) = (Anti Clock wise rotating moment)

i.e., $m_1 L_1 = m_2 L_2$

Since $(L_1 = L_2)$; $m_1 = m_2$

Acceleration due to gravity is constant at a place. Hence the equilibrium condition can also be written as

$$m_1 g = m_2 g$$

$$W_1 = W_2$$

Now when an unknown weight is to be measured, it is hanged at the provision and based on the position of the pointer, when calibrate its weight can be found.

11.4 UNEQUAL ARM BALANCE

The main parts of this instrument are listed below:

1. A graduated beam
2. Two knife edge supports
3. Known mass (movable on beam)
4. Unknown mass (to be measures)
5. Leveling pointer and Graduated scale for null balancing
6. Calibrated scale on beam

Principle

An unequal arm balance also works on the principle of equilibrium of moments. The beam of the unequal arm balance is in equilibrium position when:

(Clock wise rotating moment) = (Anti clockwise rotating moment)

i.e., $F.L_2 = F_x.L_1$

where F = Known force or weight

F_x = Unknown force or weight

L_1 = distance between knife edge 'z' & knife edge 'y'

L_2 = distance between knife edge 'y' and moving mass 'm'.

Construction

A graduated beam is simply supported by two knife edges 'y' & 'z' as shown in Figure 11.4. A known movable mass (m proportional to F) which can slide on the right side of the beam is attached and a provision is given for placing an unknown mass (F_x) in the left side of the beam. A leveling pointer is attached to the beam for null balancing.

Operation

F_x is applied on the left side of the beam through knife edge "z" as shown in figure 11.3. Now the position of mass "m" on the right side of the beam is adjusted until the leveling pointer reads null balance position. When leveling pointer is in this position, the beam is in equilibrium.

Hence, Clock-wise rotating moment = Anti clockwise rotating moment

$$F_x \cdot L_1 = F.L_2$$

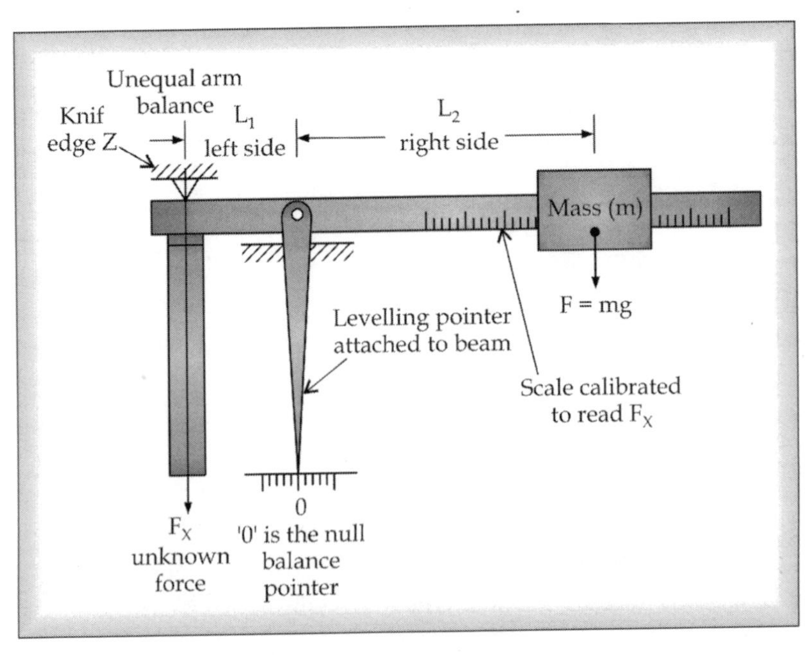

Fig. 11.3 Unequal Arm Balance

Therefore $$F_x = \frac{F}{L_1} L_2 = \frac{mg}{L_1} L_2$$

$$= \text{Constant} \times L_2$$

$$F_x = \mu \, L_2$$

Thus the unknown force "F_x" is proportional to the distance "L_2". The right hand side of the beam, which is graduated, is calibrated to get a direct measure of "F_x". This type of instrument is mostly used to measure the weight of steel rods.

11.5 PENDULUM SCALE (MULTI-LEVER TYPE)

The main parts of this instrument are as follows:

1. Scale frame
2. Supporting ribbons
3. Sectors (two)
4. Pivot
5. Equalizer beam
6. Counter weights
7. Rack and Pinion
8. Loading ribbon tapes
9. Load rod
10. Weighing platform

Principle

It also works on the principle of equilibrium of moments. The unknown force is converted to torque which is then balanced by the torque of a fixed standard mass arranged as a pendulum.

Construction

As shown in Figure 11.4 a scale frame carries two support ribbons, which are attached to the sectors on either side. Two loading ribbons are attached to these sectors and load rod. The load rod is in turn attached to the weighing platform. An equalizer beam is pivoted to the two sectors on either side. The sectors carry counter weights. A rack and pinion arrangement is at the center of the equalizer beam for which a pointer is attached. This pointer moves over a scale calibrated for weight or force.

Operation

When an unknown force is applied to the load rod, the loading tapes are pulled downwards, rotating the sectors. As the sectors rotate about the pivots, it moves the counter weights outwards. This movement increases the counter weight that causes moment. An equilibrium is established when

More Instruments for Learning Engineers

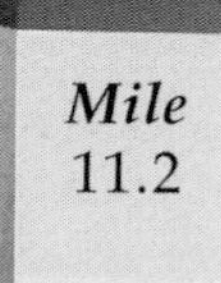

Zymoscope or **zymometer** or **zymosimeter** (Zymo- from Ancient Greek æõìüù : to leaven) measures fermentation efficiency of yeast by measuring the amount of carbon dioxide produced from a given quantity of sugar.

Zymnoscope

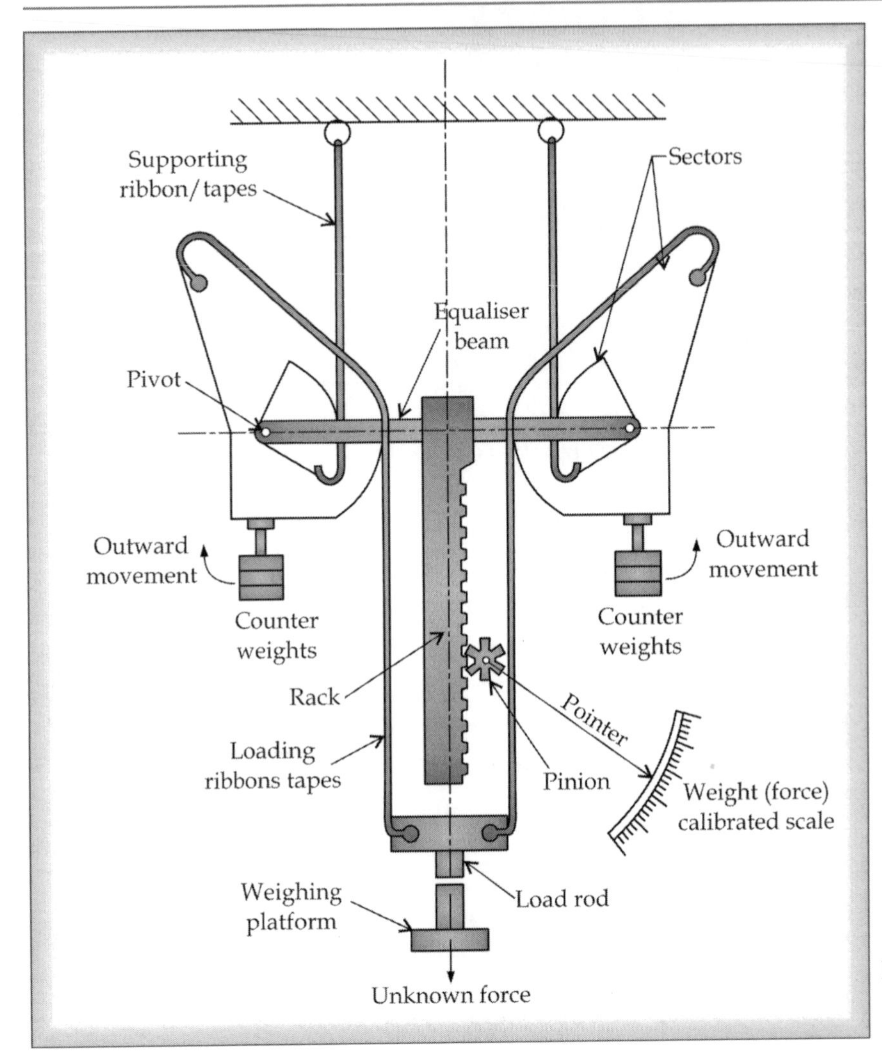

Fig. 11.4 Pendulum Scale

the torque produced by the force applied to the load rod and the moment produced by the counter weight balance each other. Due to this, the equalizer beam would be displaced downwards. This displacement of the equalizer beam rotates the rack which in turn rotates pinion and hence the pointer attached to it moves on a scale. The scale is calibrated to read weight or force directly. Thus the force applied on the load rod is measured.

Self Assessment Questions-11.2

1. Explain the principle of equal arm balance
2. Describe the main parts of equal arm balance with the help of a diagram
3. Differentiate between the unequal balance arm and equal balance arm.
4. How does the equal arm balance work? Explain its operation.
5. Describe the operation of unequal arm balance
6. Elucidate the operation and principle of pendulum scale (multi-lever type)

11.6 ELASTIC FORCE METER (PROVING RING)

The main parts of a proving ring are as follows:

1. Steel circular ring of rectangular cross section
2. External bosses (two)
3. Micrometer
4. Vibrating reed

Principle

When a steel ring is subjected to a force (tensile or compressive) across its diameter, it deflects. This deflection is proportional to the applied force when calibrated.

Construction

A steel circular ring of rectangular cross-section is attached with external bosses to enable to apply force. A precision micrometer with one of its ends mounted on a vibrating reed as shown in Figure 11.5.

Operation

The force to be measured is applied to the external bosses of the proving ring. Due to this force, the ring changes in diameter by deflection proportional to the applied force. At this stage, i.e., when the ring deflects, the reed is plucked to obtain a vibrating motion. When the reed is vibrating, the micrometer wheel is turned until the micrometer contact moves forward and makes a noticeable damping of the vibrating reed. Thus the micrometer measures the deflection of the ring, which is calibrated to get a measure of force.

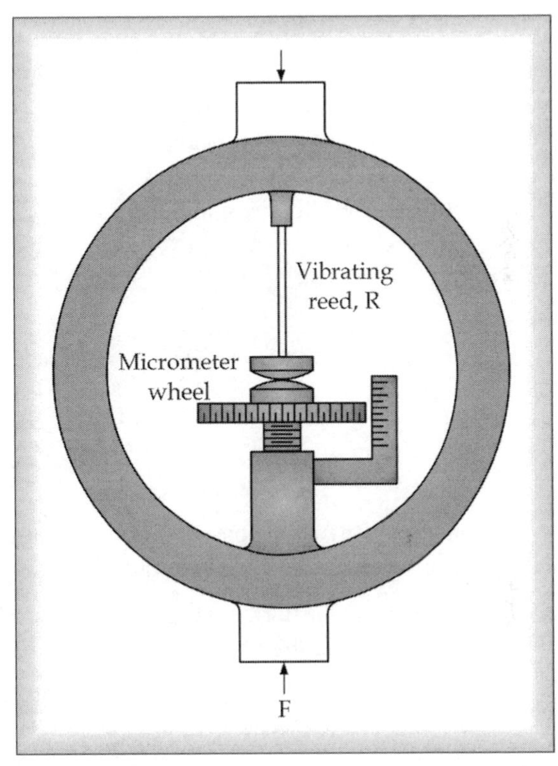

Fig. 11.5 Elastic Force Meter (Proving Ring)

11.7 STRAIN GAUGE LOAD CELL

The main parts of this instrument are as follows:

1. Load cell with 4-strain gauges
2. Wheatstone bridge circuit

Principle

The strain gauge load cell operates on the principle that when strain gauges are subjected of forces, their dimensions change, thence change the resistance which is calibrated to indicate applied load.

Construction

A simple load cell consists of a steel cylinder, which has four identical strain gauges mounted upon it as shown in Figure 11.6. The gauges R_1 and R_4 are fixed in the direction of applied load and the gauges R_2 and R_3 are attached circumferentially at right angles to gauges R_1 and R_4. These four gauges are connected eclectically to the four limbs of a Wheatstone bridge circuit.

Operation

At no load condition on the cell, all the four gauges have the same resistance. Therefore, the terminals B and D are at the same potential, the bridge is balanced and the output voltage is zero.

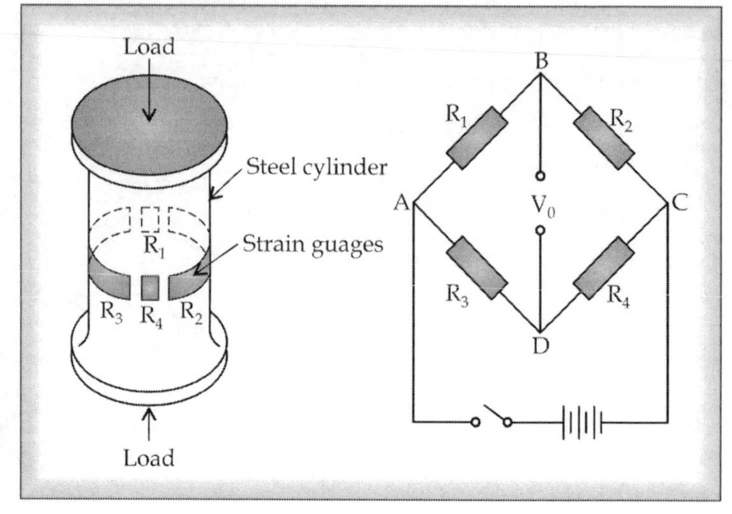

Fig. 11.6 Strain Gauge Load Cell

Thus at this condition the voltage drop from A to B (i.e. V_1) is same as the voltage drop from A to D, (i.e.., V_s) also equal to the drops from B to C (i.e.., V_2) and D to C, (i.e., V_4) each is equal to half of the battery voltage (V_E).

Hence $V_1 = V_3 = V_{E/2}$; when output voltage is zero then, $V_3 - V_1 = 0 = V_0$

$$V_1 = V_3 = V_{Es}/2$$

$$V_o = V_3 - V_1 = 0.$$

When a compressive load is applied to the unit, the vertical gauges ($R_1$and R_4) undergo compression and hence decrease in resistance.

Simultaneously the horizontal gauges R_2 and R_3 undergo tension and hence increase in resistance. Thus, when strained, the resistances of the various gauges are:

$$R_1 = R_4 = R - dR \text{ (Compression)}$$

$$R_2 = R_3 = R + dR \text{ (Tension)}$$

Potential at terminal B is

$$V_1 = \frac{R_1}{R_1 + R_2} V_E = \frac{R - dR}{(R - dR) + (R - \mu dR)} \times V_E \; \frac{R - dR}{2R - dR(1 - \mu)} \times V_S$$

Potential at terminal D is,

$$V_3 = \frac{R_3}{R_3 + R_4} V_E$$

$$= \frac{R + \mu dR}{(R - \mu dR) + (R - dR)} \times V_E$$

$$= \frac{R + \mu dR}{2R - dR + (1 - \mu)} V_E$$

The changed output voltage is,

$$V_o + dV_o = \frac{R - dR}{2R - dR(1-\mu)} \times V_E - \frac{R + \mu dR}{2R - dR1 - \mu} V_E$$

$$\frac{dR(1+\mu)}{2R} \times V_s = 2(1-\mu)\left(\frac{dR}{R}\frac{V_E}{4}\right)$$

The output voltage $V_o = 0$ under unloaded conditions, and therefore change in output voltage due to applied load becomes.

$$dV_0 = 2(1+\mu)\left(\frac{dR}{R}\frac{V_E}{4}\right)$$

Hence, it is clear that, this

is a measure of the applied load. The use of four identical strain gauges in each arm of the bridge provides full temperature compensation and also increases the bridge sensitivity $2(1 + \mu)$ times.

Merits

- The strain gauge load cells are excellent force measuring devices, particularly when the force is dynamic
- They are generally stable, accurate

Applications

- They are extensively employed in industrial applications such as draw bar and tool-force dynamometers, crane load monitoring, and load vehicle weighing devices etc.

11.8 HYDRAULIC LOAD CELL

The main parts of this -instrument cell are as follows:

1. Load to be measured
2. Piston
3. Diaphragm
4. Seals
5. Hydraulic oil
6. Base plate
7. Pressure gauge

More Instruments for Learning Engineers

Magnetometers measures the magnetization of a ferromagnetic material, or the strength or direction of the magnetic field at a point in space. Magnetometers are used to measure the earth's magnetic field and also to detect magnetic anomalies, submarines. They can also be used as metal detector: as they detect only magnetic (ferrous) metals up to a depth (at 10's of meters) than conventional metal detectors (rarely more than 2 m).
The first magnetometer was invented by Carl Friedrich Gauss in 1833 and developed in the 19th century, included the Hall Effect which is still used. Today they have been miniaturized to the extent that they can be incorporated in integrated circuits at very low cost and are finding increasing use as compasses in mobile phones and tablet etc.

Magnetometer

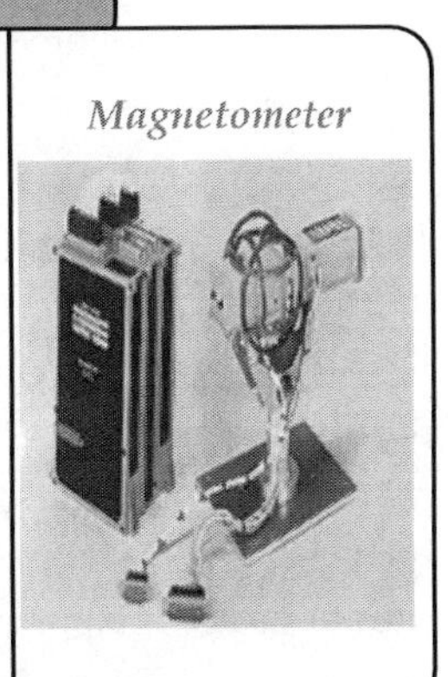

Principle

The pressure of fluid increases proportionately, when a force is applied on the fluid medium in a container of definite space. Hence, measure of increased pressure of the fluid becomes the measure of the applied force when calibrated.

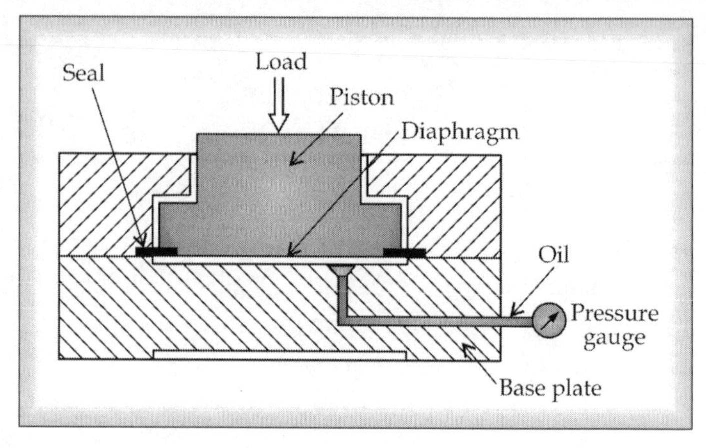

Fig. 11.7 Hydraulic Load Cell

Construction

On a diaphragm a piston is placed on which force (load) can be applied as shown in Figure 11.7. A preloaded liquid medium is present on the other side of the diaphragm which is connected to a pressure gauge.

Operation

The force (or load) to be measured is applied on the piston as shown in Figure 11.7. This load pushes the piston downwards, deflecting the diaphragm. This deflection of the diaphragm enhances the pressure in the liquid medium. The increase in the pressure is measured by the connected pressure gauge. If the pressure is calibrated in force units, then the indication in the pressure gauge becomes a measure of the force applied on the piston.

Precaution

The load cell should be adjusted to zero setting before using, as it is sensitive to pressure changes.

Merits

- These cells have accuracy of the order of 0.1% of its scale and
- They can measure heavy loads up to 2.5×10^6 N.

11.9 PNEUMATIC LOAD CELLS

The chief parts of this instrument are as listed below:

1. Force (Load) to be measured
2. Load plate
3. Corrugated diaphragm
4. Bleed valve
5. Air supply pipe with nozzle, flapper and regulator
6. Pressure measuring device

Principle

This works on the same principle as that of hydraulic load cell, except for the difference, here the fluid is the air instead of oil. The force is applied on one side of a diaphragm and air pressure is applied on the other sides, which are exactly balanced by a valve. This pressure is proportional to the applied force.

Construction

To a cell, a corrugated diaphragm is fitted with a load plate on top surface so that a load (force) can be applied on it. An air supply regulator, nozzle and a pressure gauge are arranged as shown in Figure 11.8. A flapper may be arranged above the nozzle. A bleed valve and a pressure measuring device (pressure gauge) are fitted to the cell.

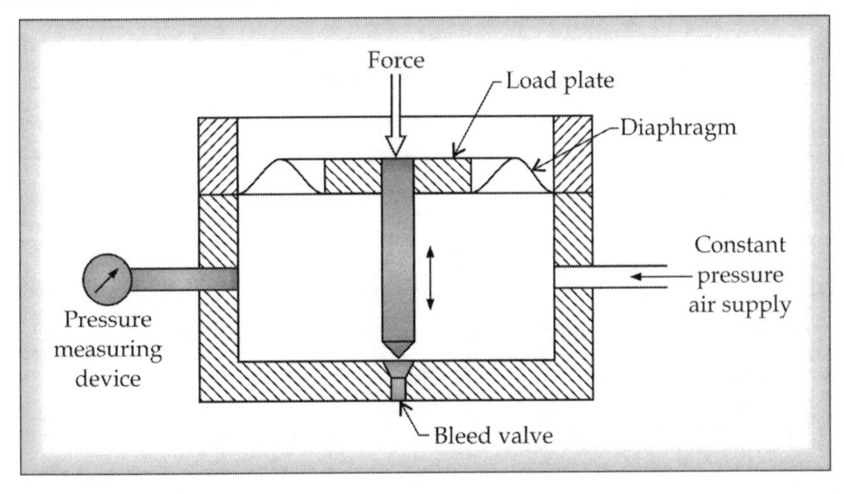

Fig. 11.8 Pneumatic Load Cell

Operation

The force to be measured is applied on load plate at the top of the diaphragm as shown in Figure 11.8. The diaphragm deflects due to the load and causes the flapper to shut off the nozzle opening. Now an air supply is provided as shown which results in back pressure underneath the diaphragm as the flapper has closed the bleed valve. This back pressure produces an upward force. Air pressure is regulated until the diaphragm returns to the preloaded position, which is indicated by air starting to come out of the nozzle. The corresponding pressure shown by the pressure measuring device becomes a measure of the applied force when calibrated.

SELF ASSESSMENT QUESTIONS-11.3

1. Describe the principle and operation of elastic force meter.
2. Draw the circuit diagram of strain gauge load cell. Explain its principle and operation.
3. Give the differences between the strain gauge load cell and hydraulic load cell.
4. With the aid of a diagram explain the operation of hydraulic load cell.
5. What are the advantages of hydraulic load cell? Give its precautions.
6. Explain the principle of pneumatic load cell.
7. Sketch and explain the main parts of pneumatic load cell.

11.10 MEASUREMENT OF TORQUE AND POWER

The main purpose of torque measurement is to determine the mechanical power required or developed by a machine. Torque measurement also helps in obtaining load information necessary for stress or strain analysis. In some cases other variables are determined by measuring torque. For example, in the case of rotating cylinder viscometer, measurement of torque developed at the fixed end of the stationary cylinder helps in determining the viscosity of the fluid between the movable and stationary cylinder.

11.10.1 Definitions of Torque and Power

Torque: Torque represents the amount of twisting moment, and is given by

$$\tau = F \times R$$

where

τ = Torque in Nm,

F = Force in N

R = perpendicular distance from the point of rotation to the point of application of force.

It can also be defined as the power that a shaft transmits per radian per second of rotation.

Power: Power is defined as the work done per unit time or rate of doing work.

Mathematically, Power (P) = W/T

Or $P = dW/dT$

where

P = Power in Watts or KW

W = Work done in Joules or KJ

T = Time in seconds.

11.10.2 Relation between Torque and Power

Torque and power are related as

$$P = \omega T = 2\pi NT$$

where

P is the power (Watt)

T is the torque (Nm)

ω is the angular speed (radian/s)

N is the angular speed (revolution/second)

Thus, if torque and angular speed are known, power can be determined.

Torque measuring devices are commonly known as dynamometers and are classified as shown in Figure 11.9.

The other instruments, which are indirectly used for measuring torque, are called as Torque or/and Torsion meters and are classified as shown in Figure 11.10.

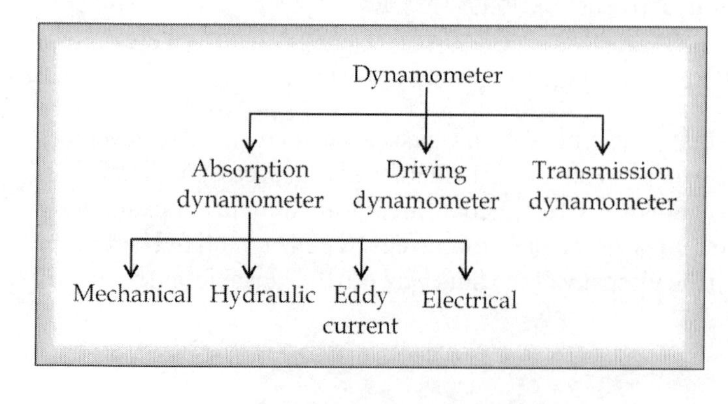

Fig. 11.9 Classification of Dynamometers

11.11 DYNAMOMETERS

Absorption Dynamometers

In these types of dynamometers the mechanical energy absorbed and dissipated as torque of the machine is measured. These are particularly useful for measuring power or torque developed by engines or motors. Absorption dynamometers can further be classified as follows (Refer Figure 11.9).

1. Mechanical type
2. Hydraulic type
3. Eddy current type
4. Electrical generator.
5. Driving
6. Transmission

11.11.1 Mechanical/Prony Brake Dynamometer

One of the most familiar and simplest mechanical dynamometers is prony brake dynamometer. The main parts of this instrument are:

1. Fly wheel (attached to test machine)
2. Tightening arrangements for brake shoes
3. Brake wooden blocks
4. Arm
5. Spring balance

Principle

Force required to stop rotating shaft is proportional to the power and is calibrated to read the power directly.

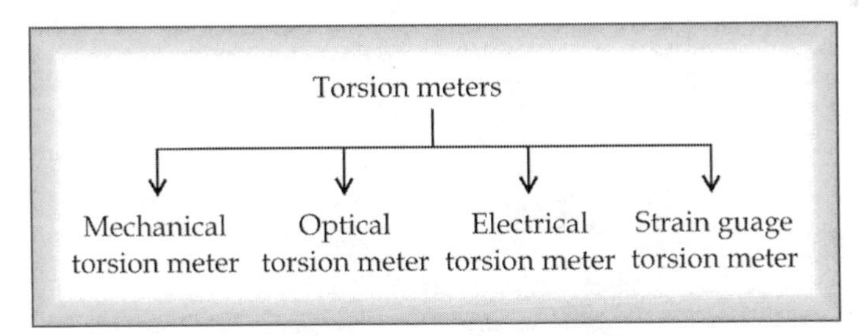

Fig. 11.10 Classification of Torsion Meters

More Instruments for Learning Engineers

Mile 11.4

A **load cell** is a transducer that converts a force into an electrical signal. This conversion happens in two stages. First, the force deforms a strain gauge which in turn changes the resistance and so an electrical signal. A load cell is configured in a Wheatstone bridge circuit with four strain gauges (full bridge) or two strain gauges (half bridge) or one strain gauge (quarter bridge). The electrical signal output is in the order of a few millivolts and requires amplification by an instrumentation amplifier before using. The output of the transducer can be calibrated to measure force, torque, energy, power etc. Various types of load cells are Hydraulic, Pneumatic and Strain gauge load cells.

Load Cell

The force ‘F’ can be measured by some force measuring device.

Torque $\tau = F \times l$

Power $P = 2\pi N\tau$

Where N is the speed in rps

$$P = 25\pi \text{ß} NFl$$

Construction

The wooden brake blocks are lined along the rope or band and surrounded on the fly wheel from top and bottom sides. The one end of band/rope is tied which includes some arrangement to vary the frictional resistance, e.g. tightening of rope or wooden brakes while the other to the arm as shown in Figure 11.11. The arm is connected to the spring balance.

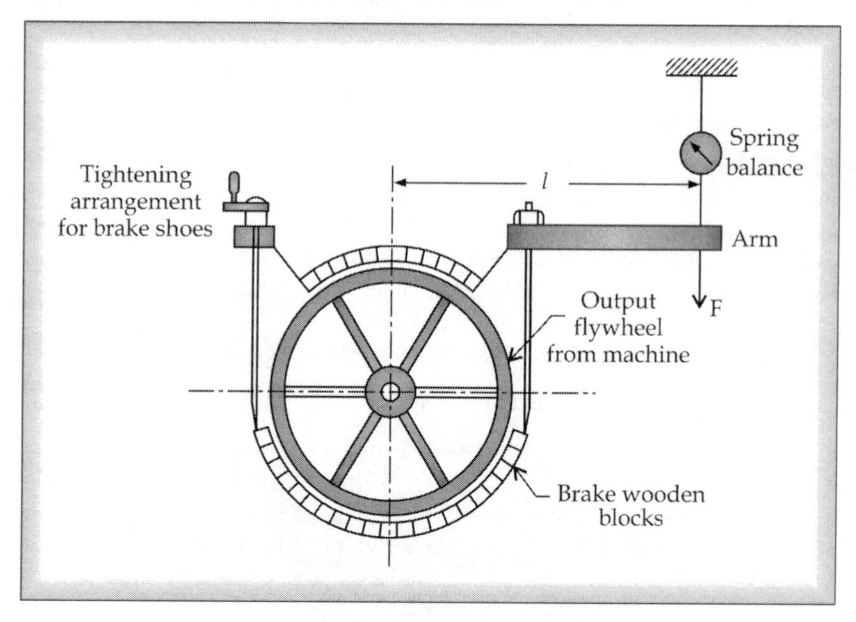

Fig. 11.11 Prony Brake Dynamometer

Operation

When brakes are applied on a rotating shaft, it offers resistance called friction. In these dynamometers, frictional resistance is created between a rotating fly wheel attached to the test-machine and a rope or band or brake shoes. This friction can be measured and is calibrated. The heat generated is dissipated by circulating cooling water.

11.11.2 Hydraulic Dynamometer

The main parts of this instrument are:

1. Shaft to be coupled to test machine
2. Rotor
3. Half sectors (two) of stator
4. Semi-elliptical grooves on rotor
5. Casing

Principle

The braking effect can be varied between rotor and stator by controlling the amount of water flowing.

Construction

It consists of a stator freely pivoted on bearing (cradling) and a rotor hydraulically coupled. The stator is in two halves and placed on either side of the rotor as shown in Figure 11.12. Semi-elliptical grooves in the rotor match with that in the casing through which a flow of water is maintained.

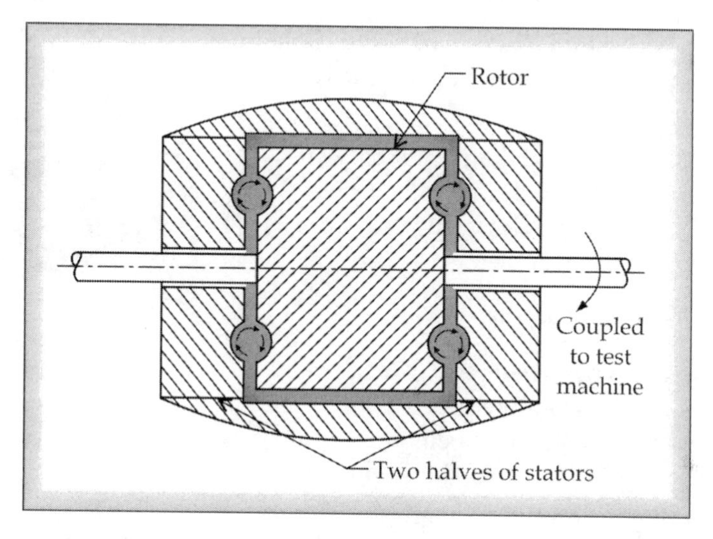

Fig. 11.12 Hydraulic Absorption Dynamometer

Operation

Water flows through helical paths creating vortices and eddy currents resulting in dissipation of energy. The weights applied to the moment arm on the stator will prevent stator from rotating. The braking effect can be varied by controlling the amount of water flowing by some mechanism between rotor and stator. There is no need to provide a separate cooling arrangement in this type of dynamometer because water itself acts as coolant.

11.11.3 Eddy Current Dynamometer

Construction

It is a cradle type electrical dynamometer. It consists of a metal disk, which is rotated in a magnetic field. The stator is mounted in trunnion bearings (cradling) and housed in the dynamometer which consists of field coils excited by an external source producing magnetic field.

Operation

As the disc rotates in the magnetic field, eddy currents are generated and the reaction with the magnetic field tends to rotate the dynamometer housing (stator) in the trunnion bearings. The torque is measured with the help of moment arm, attached to the stator. The load can be varied by varying the field current and the power absorbed is converted into heat which is removed by circulating cooling water.

11.11.4 Electrical Generator Type

It is also cradle type dynamometer and is essentially a generator driven by the test machine. The mechanical energy is converted into electrical energy by the electrical generator. The electrical energy is then dissipated through a variable resistance grid, which also serves to vary the load cradling in trunnion bearings permits the determination of reaction torque on the stator as in eddy current dynamometer.

11.11.5 Driving Dynamometer

It is essentially a cradle type dc motor, which drives the test machine and reaction torque on the stator is measured as in eddy current dynamometer.

These dynamometers measure the torque as well as provide the energy to operate the device to be tested. To study the performance characteristics of air compressors and pumps, these types of dynamometers are used.

11.11.6 Transmission Dynamometer

This can be considered as a passive device, which neither gives to nor absorbs energy from the test machine. The torque in the shaft is measured as it is being transmitted and the shaft itself acts as a load cell. The angular twist 'θ' of the shaft or the stresses becomes a measure of torque as explained below (Refer Figure 11.13).

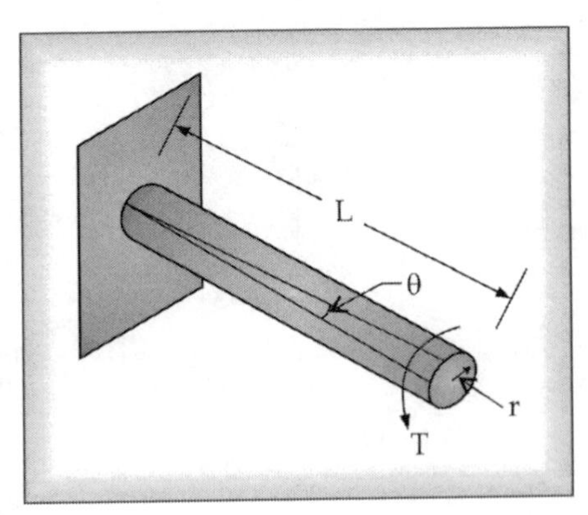

Fig. 11.13 Transmission Dynamometer

Shaft in Torsion-Theoretical Relations: We have torsion equation as

$$\frac{\tau}{J} = \frac{f_s}{R} = \frac{G\theta}{l}$$

where, J = polar moment of inertia of the section

R = distance from neutral axis to the fibre of interest

θ = angular twist

f_s = shear stress.

Maximum shear strain occurs at the surface,

$$R = \frac{d}{2} = r$$

$$J = \frac{1}{2}\pi R^4 = \frac{1}{32}\pi d^4$$

Hence $$\tau = \theta G\ \pi = \frac{d^4}{32}$$

$$\tau = \theta \times a\ \text{constant}$$

So, Torque is proportional to angular twist of the shaft i.e., ô µ q

Also, $$\tau = f_s\left(\frac{\pi}{16}d^3\right)$$

$$\tau = f_s \times \text{ a constant.}$$

Torque α shear stress at the surface for the shaft

Shear stress $$f_s = \frac{16\tau}{\pi d^3}$$

Shear strain $$y = \frac{f_s}{G}$$

Shear strain $$y = \frac{16\tau}{G\pi d^3}$$

Angular twist $$\theta = \frac{32\tau l}{G\pi d^3}$$

We can observe from the above relations that the torque transmitted by a shaft is proportional to its angular twist θ. The angle 'θ' can be measured *by some optical means or electromechanical* methods. Such devices are called *torsion meters.*

Also, it can be seen that torque transmitted is proportional to shear stress and strain developed in the shaft. Devices based on this concept are called *torque meters.*

ILLUSTRATION-11.1

A torque is applied to a torque tube made of steel ($G = 8 \times 10^{10}$ N/m^2) and of physical dimensions are $r_0 = 16 \pm 0.011$mm; $r_1 = 12 \pm 0.01$ mm and the length $l = 125 \pm 0.03$ mm. The angle of twist θ is 1.50° and its uncertainty is ± 0.05°.

(a) Calculate the nominal value of the impressed torque and its uncertainty.

(b) Calculate the 45° strains.

More Instruments for Learning Engineers

Mile 11.5

A **viscometer** (also called **viscosimeter**) is an instrument used to measure the viscosity of a fluid. For liquids with viscosities which vary with flow conditions, an instrument called a rheometer is used. Viscometers only measure under one flow condition.
In general, either the fluid remains stationary and an object moves through it, or the object is stationary and the fluid moves past it. The drag caused by relative motion of the fluid and a surface is a measure of the viscosity. The flow conditions must have a sufficiently small value of Reynolds number for there to be laminar flow.
At 20.00 degrees Celsius the viscosity of water is 1.002 mPa·s and its kinematic viscosity (ratio of viscosity to density) is 1.0038 mm^2/s. These values are used for calibrating certain types of viscometer.

Viscosimeter

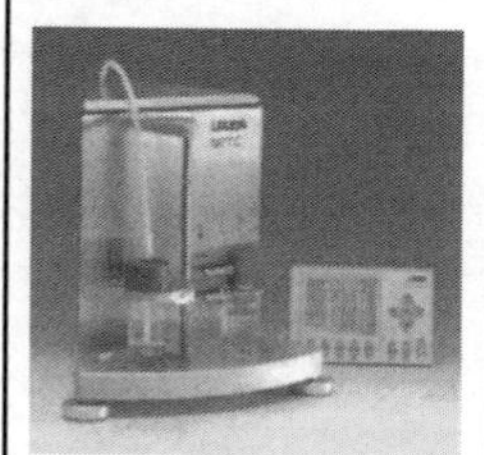

Solution:

Given $G = 8 \times 10^{10}\ N/m^2 = 8 \times \frac{10^{10}}{10^6}\ N/mm^2 = 8 \times 10^4\ N/mm^2$

$r_0 = 16 \pm 0.01$ mm

$r_1 = 12 \pm 0.01$ mm

$l = 125 \pm 0.03$ mm

$\phi = 1.5^0$ with $\pm 0.05^o$

To find

(a) $T_{nominal}$ uncertainty

We have torsion equation $\frac{\tau}{J} = \frac{f_s}{R} = \frac{G\theta}{l}$

$$\frac{\tau}{J} = \frac{G\theta}{l} \Rightarrow \tau = \frac{JG\theta}{l} = \frac{\pi\left(r_0^4 - r_1^4\right)G\theta}{2l}$$

$$\therefore \qquad \text{Torque} = \frac{\pi\left(r_0^4 - r_1^4\right)G\theta}{2l}$$

Nominal torque at $\tau_{10} = 16$ mm, $\tau_{11} = 12$ mm, $l = 125$ mm

$$\theta = 1.50° \times \frac{\pi}{180} = 0.26179\ G$$

is
$$\tau = \frac{\pi\left[16^4 - 12^4\right] \times 8 \times 10^4 \times 0.26179}{2 \times 125} = 11795210\ \text{N-mm}$$

Torque at $r_o = 16 + 0.01 = 16.01$ mm

$r_1 = 12 + 0.01 = 12.01$ mm

$l = 125 + 0.03 = 125.03$ mm

$$\theta = 1.50° \times \frac{\pi}{180} \neq 0.26179954$$

$$\tau = \frac{\pi\left[16.01^4 - 12.01^4\right] \times 8 \times 10^4 \times 0.26179954}{2 \times 125.03}$$

$$= 11817746.87\ \text{N - mm (or) } 11817747\ \text{N-mm}$$

Torque at $r_0 = 16 - 0.01 = 15.99$ mm

$r_1 = 12 - 0.01 = 11.99$ mm

$l = 125 - 0.03 = 124.97$ mm

$$\theta = 1.50° \times \frac{\pi}{180} - 0.05^6 = 0.026179923$$

$$\tau = \frac{\pi\left[15.99^4 - 11.99^4\right] \times 8 \times 10^4 \times 0.26179923}{2 \times 124.97}$$

$$= 1173530.12 \text{ N-mm (or) } 11773530$$

$$\text{Uncertainty} = 11817747 - 11773530$$

$$= 44217 \text{ N-mm or } 44216.75$$

(b) Strains at 45° $\in_{45^0}$ at $\tau = 1181300.613$N-mm

$$\in_{45^0} = \frac{f_s}{G} = \frac{\tau R}{GJ}\left[\because f_s = \frac{\tau R}{J}\right]$$

$$= \frac{1181300.613 \times 16.01}{8 \times 10^4 4 \times \pi\left[16.01^4 - 12.01^4\right]} = 0.00167616$$

$\in_{45^0}$ at $\tau = 1176879.319$ Nmm

$$\in_{45^0} = \frac{1176879.319 \times 15.99}{\pi \times 8 \times 10^4 \times \pi\left[15.99^4 - 11.99^4\right]} = 0.00167616$$

Self Assessment Questions-11.4

1. Define torque. Give the relation between the torque and power.
2. Give the classification of dynamometers.
3. Name the four types of absorption dynamometers. Explain each of them.
4. Classify various types of torsion meters.
5. State the working of Prony Brake Dynamometer
6. Describe the construction of Hydraulic Dynamometer. Explain the torque measurement using this dynamometer.
7. Explain the working principle of following dynamometers
   (a) Eddy Current Dynamometer (b) Electrical Generator Type
   (c) Driving Dynamometers (d) Transmission Dynamometers

11.12 TORSION METER

11.12.1 Mechanical Torsion Meter

The main parts of the mechanical torsion meter are:

1. Driving engine
2. Shaft
3. Drum with pointer (Driven side)
4. Drum with torque calibrated scale (Driving side)
5. Flanges (two)
6. Stroboscope

Principle

When a shaft is connected between a driving engine and driven load, a twist occurs on the shaft between its ends, which is measured and calibrated in terms of torque.

Construction

Two drums and two flanges one each on the driving side and another each on driven side are mounted on a shaft at its ends as shown in Figure11.14. One drum carries a pointer and the other drum has a torque calibrated scale. A stroboscope is arranged to take readings of pointer on a rotating shaft.

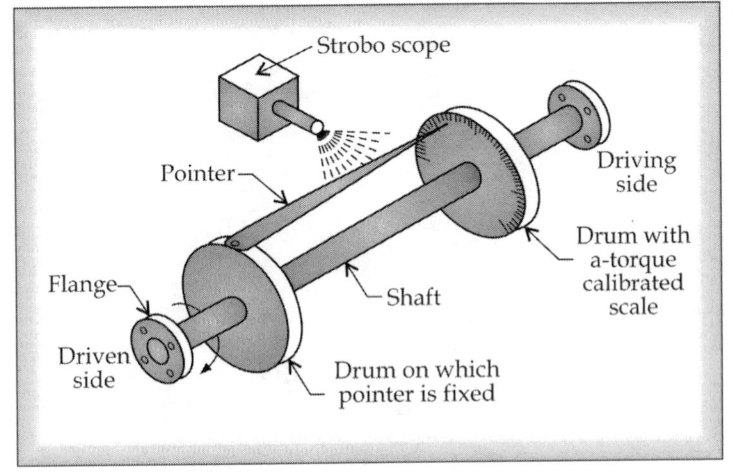

Fig. 11.14 Mechanical Torsion Meter

Operation

One end of the shaft of the torsion meter is connected to the driving engine and its other end to the driven load. An angle of twist is experienced by the shaft along its length between the two flanges, which is proportional to the torque applied to the shaft.

The resulted angular twist is noticed on the torque calibrated scale corresponding to the position of the pointer. As the scale on the drum is rotating, readings cannot be taken directly. Therefore, a stroboscope is used. The flash light of stroboscope is made to fall on the scale and the flashing frequency is adjusted till a stationary image is obtained. Then the scale reading is noted.

Merits

- Simple and inexpensive method.
- Power of shaft also can be calibrated by knowing the speed which given by flashing frequency.

Demerits

- Due to small displacement of the pointer, the accuracy of the instrument is small.
- Sensitivity is reduced even due to small variations in speed.
- This method can be used only when shafts are rotating at a constant speed.

11.12.2 Optical Torsion Meter

The main parts of an optical torsion meter are listed below:

1. Shaft
2. Castings (two)
3. Tension strip
4. Flanges (two)
5. Two mirrors
6. Optical system
7. Calibrated scale

Principle

Due to torque, an angular twist occurs on the shaft between its ends, which is measured using optical means where in the angular deflection of light rays is proportional to the twist and thence the torque.

Construction

Two castings M and N are mounted on a shaft at a known distance as shown in Figure 11.15 with a tension strip linking the two castings. Two mirrors are fitted and aligned with the castings in such a way that a light beam falling on the mirror, passes through an optical system that can measure the torque on a calibrated scale.

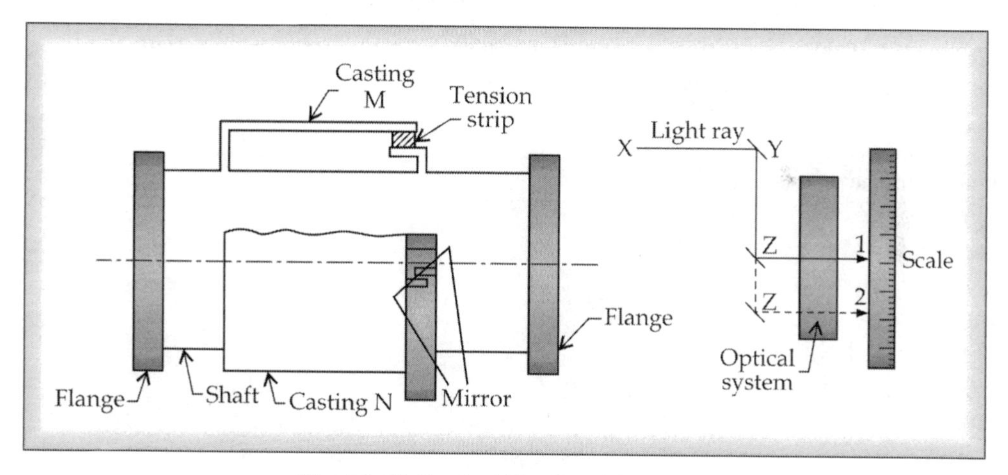

Fig. 11.15 Optical Torsion Meter

More Instruments for Learning Engineers

A **potentiometer**, or simply **pot**, is a 3-terminal resistor with a sliding contact that forms an adjustable voltage divider. If only two terminals are used, (one end and the wiper) it acts as a ***variable resistor*** or ***rheostat***. Potentiometers are used to control electrical devices such as volume controls on audio equipment. Pot mechanism can be used as position transducer like in a joystick. Pots are rarely used to directly control significant power (more than a watt), since the power dissipated in the pot would be comparable to the power in the controlled load.

Potentiometer - Rheostat

Operation

A relative movement occurs between castings M and N when the shaft transmits torque, due to which the positions of mirrors attached to the castings will change. Since, the mirrors are constantly made to reflect a light beam on the torque calibrated scale, and due to the changed position of the mirrors, there will be an angular deflection of the light rays measured from the calibrated scale. This angular deflection of the light rays is proportional to the twist on the shaft and hence the torque of the shaft.

Applications

This is widely employed to measure torque in

- Steam turbines and
- I.C. engines.

11.12.3 Electrical Torsion Meters (Torque Measurement using Slotted Discs)

An electrical torsion meter consists of the following main parts:

1. Driving engine
2. Driven load
3. Shaft
4. Slotted discs (two)
5. Pickup coils
6. Electronic units

Principle

A relative displacement between the two slotted discs due to torque applied causes a phase shift between the pulses generated by the transducers and an electronic unit connected shows a time lapse between the two pulses, which is proportional to the twist of the shaft and hence the torque of the shaft.

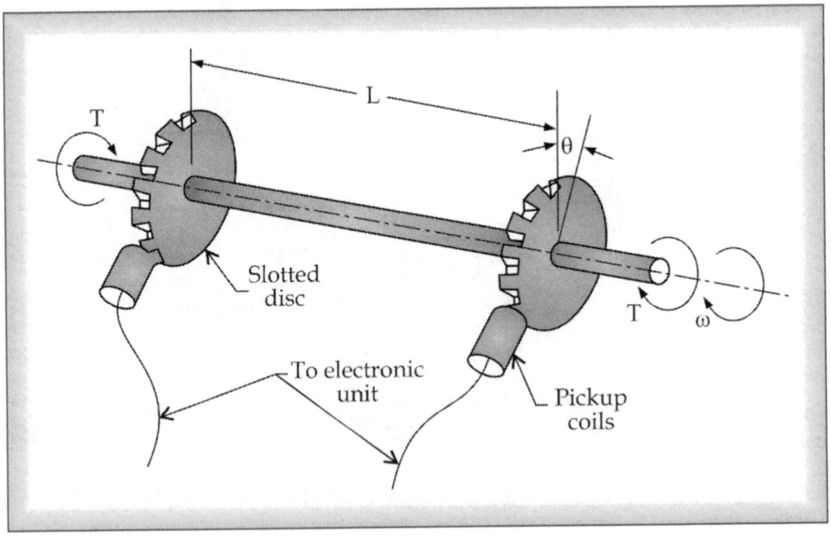

Fig. 11.16 Electrical Torsion Meter or Torque Measurement using Slotted Discs

Construction

A shaft is connected between a driving engine and a driven load. Two slotted discs are attached to the shaft on either side as shown in Figure 11.16. This set up is connected with magnetic or photo electric transducer to count pulses from the slotted discs.

Operation

The teeth of slotted discs produce voltage pulses in the transducers. When torque is not applied on the shaft, the teeth of both the discs perfectly align with each other and hence the voltage pulses produced in the transducer have zero time difference. But, as the torque is applied on the shaft, there will be a relative displacement of the slotted discs due to twist experienced by the shaft and hence the teeth of both the discs will not align with each other and so the voltage pulses produced in the transducers will have a time difference. This time lapse between the pulses of the two discs is proportional to the twist of the shaft and hence the torque of the shaft. A measure of this time lapse becomes a measure of torque when calibrated.

Merits

- There is no noise creation problem.
- There is no signal leakage.

11.12.4 Strain Gauge Torsion Meter (Strain Gauges on Rotating Shafts)

The arrangement consists of the following:

1. Shaft
2. Strain gauges (four)
3. Wheatstone bridge circuit
4. Calibrated scale

Principle

When the strain gauge is stretched or compressed, its resistance changes due to the change in its length and diameter.

The strain gauges bonded to a shaft, get strained when subjected to torque and hence twist. Thus the resistance of the strain gauges changes proportional to the torque in the shaft.

Construction

A strain gauge is a metallic conductor. Four bonded wire strain gauges mounted on a 45° helix with the axis of the shafts rotation. These gauges are placed in pairs diametrically opposite to each other as shown in Figure 11.17. All the four strain gauges are connected to a Wheatstone bridge circuit used to measure change in the resistance of the strain gauges. The system is temperature compensated and a change in resistance of the strain gauges will occur only due to the twist.

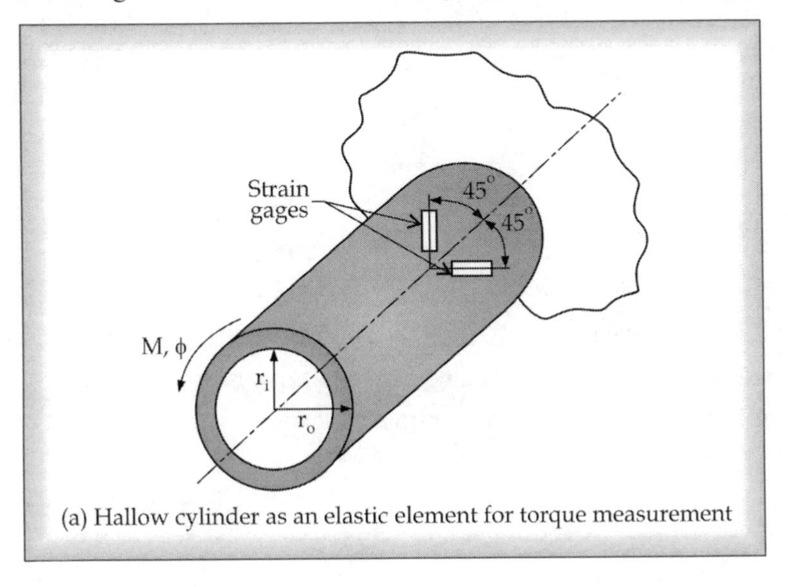

(a) Hallow cylinder as an elastic element for torque measurement

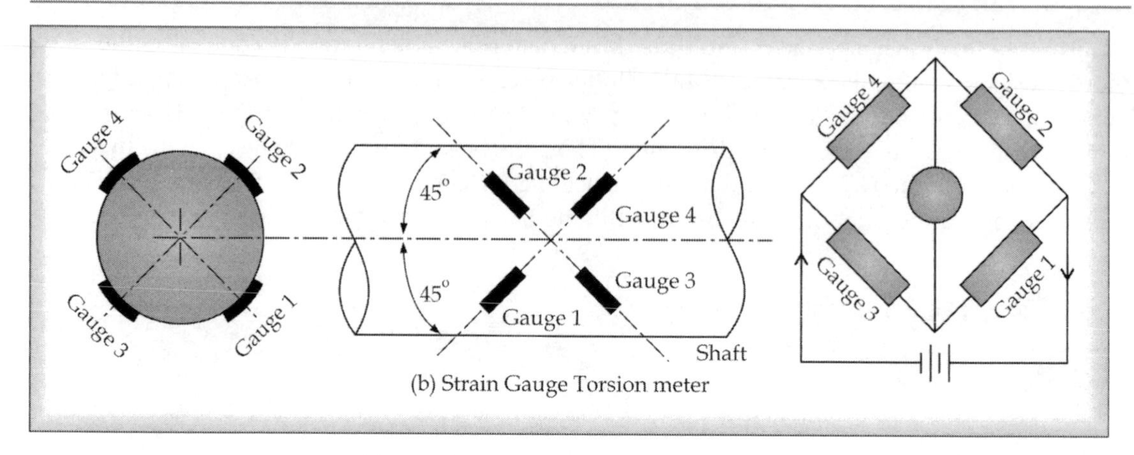

Fig. 11.17 Strain Gauge Torsion Meter

Operation

When the shaft is subjected to torque, strain gauges 1 and 4 will elongate due to tension and strain gauges 2 and 3 will contract due to compression. Due to the tensile and compressive effects on the strain gauges, their resistance will change. This change in resistance of the strain gauge, which is proportional to the torque subjected on the shaft, is measured using the Wheatstone bridge circuit. Thus a measure of this change in resistance of the strain gauges becomes a measure of torque when calibrated.

Merits

- Fully temperature compensated
- Sensitivity is high

Demerits

- Generally, it is difficult to connect the bridge circuit to the power source.
- It is difficult to connect the display (galvanometer) to the bridge circuit.

Self Assessment Questions-11.5

1. What are the different types of torsion meters? Explain each of them.
2. Describe the principle and operation of mechanical torsion meter with the help of a diagram.
3. What is the need of stroboscope used in mechanical torsion meter? Explain its function.
4. Give the advantage and limitation of mechanical torsion meter.
5. Discuss the construction and operation of optical torsion meter
6. How the torque is measured using strain gauge torsion meter? Explain the principle behind it.
7. Explain the torque measurement using electrical torsion meters
8. List the advantages and limitations of strain gauge torsion meter.

SUMMARY

The physical quantity, which changes or tends to change the state (motion/rest) or direction or shape of a body is called force. Force may also cause of a turning effect relative to an axis called moment, torque or couple which can cause bending a beam or torsion in a shaft. Torque, the amount of twisting moment ($\tau = F \times R$) and power, the rate of doing work ($P = W/T$) are related as $P = 2\pi N\tau$

Force can be measured by direct comparison making use of gravitation concept in equal/ unequal arm balance and Pendulum Scale (Multi-lever type). More precisely, they work on principle of equilibrium of moments i.e., clockwise and anticlockwise rotating moments are equal. In the indirect methods use the applications of elastic members (springs, cantilever and proving ring etc.) which deform or deflect. Further, hydraulic and pneumatic load cells use the concepts and characteristics of fluid.

Proving ring measures deflection proportional to the force, while load cells (with 4-strain gauges) operate on the principle of dimension strain converted into resistance and measured by a Wheatstone bridge. Hydraulic/pneumatic load cells use fluid properties to obtain strain.

Absorption Dynamometers (Mechanical/ Hydraulic/ eddy current/ Electrical generator type) measure absorbed and dissipated torque. Prony Brake Dynamometer measure force to stop rotating shaft is proportional to the power. Hydraulic Dynamometer measures the braking effect varied between rotor and stator by controlling the amount of water flowing. Eddy Current Dynamometer and electrical generator are cradle type electrical dynamometers that use field coils excited by an external source producing magnetic field. Transmission Dynamometer is a passive device, measures torque in the shaft as it is being transmitted and the shaft itself acts as a load cell.

The torque transmitted by shaft is proportional to its angular twist q measured by optical or electromechanical methods in torsion meters. They use the concept of torque transmitted proportional to shear stress and strain developed in the shaft. Torsion meters (Mechanical/ Optical) measure the twist occurring on the shaft between its ends. Torque is measured using slotted discs in electrical torsion meters. Torsion meter can also be constructed by mounting strain gauges on rotating shafts.

KEY CONCEPTS

Force: The physical quantity, which changes or tends to change the motion or shape of a body to which it is applied.

Units of Force: Dyne (Metric), Newton (SI), Pound (British), gf, kgf, lbf (Gravitational).

Newton's second law of motion: The rate of change in momentum is directly proportional to the force applied and the body moves in the direction of force. $F = ma$.

Weight: The force exerted on the body by the gravitational acceleration. $W = mg$.

Acceleration due to gravity 'g': 981 cm/sec^2; 9.81 m/sec^2; 32 ft/sec^2 .

Direct Method of measuring force: A direct comparison is made with known gravitational force acting on a standard mass.

Equal Arm Balance: Works on principle of moment equilibrium. The beam of equal arm balance is in equilibrium position when the unknown force is balanced against the known force.

Unequal Arm Balance: Works on the principle of moment equilibrium. The beam of the unequal arm balance is in equilibrium position when: (Clock wise rotating moment) = (Anti clockwise rotating moment)

Pendulum Scale (Multi-Lever Type): Works on the principle of moment equilibrium. The unknown force is converted to torque which is then balanced by the torque of a fixed standard mass arranged as a pendulum

Elastic Force Meter (Proving Ring): When a steel ring is subjected to a force (tensile or compressive) across its diameter deflects proportional to the applied force and measured on calibrated scale.

Strain Gauge Load Cell: Works on the principle of strain gauge that when strain gauge is subjected of force, its dimension changes which in turn change the resistance, used as an indication of applied load.

Hydraulic Load Cell: When a force is applied on a fluid medium contained in confined space, the pressure of the fluid increases proportional to the applied force and the increase in pressure of the fluid can indicate the applied force when calibrated.

Pneumatic Load Cell: Force is applied to one side of a diaphragm and the air pressure on the other side that exactly balances is the measure of applied force when calibrated.

Torque: The amount of twisting moment, and is given by $\tau = F \times R$ measured in Nm

Power: Rate of doing work ($P = W/T$)

Relation between Torque and Power: Torque and power are related as $P = \omega T = 2\pi NT$

Dynamometer: Device that measures the Torque or Power

Absorption Dynamometer: Measures torque of the machine in terms of mechanical energy absorbed and dissipated.

Mechanical/ Prony Brake Dynamometer: Frictional resistance is created between a rotating fly wheel attached to the test machine with a rope or band or brake shoes and the heat generated/ dissipated by circulating cooling water is the measure of torque/power.

Hydraulic Dynamometer: Consists stator freely pivoted on bearing (cradling) in two halves and placed on either side of rotor hydraulically coupled. Water flows through helical paths of semi elliptical grooves in the rotor creating vortices and eddy currents results in dissipation of energy that measures torque/power when calibrated.

Eddy Current Dynamometer: A cradle type electrical dynamometer, consisting of a metal disk rotated in a magnetic field and the stator mounted in bearings (cradling). As the disc rotates in the magnetic field, eddy currents generated are calibrated to measure torque.

Electrical Generator Type Dynamometer: A cradle type dynamometer generator driven by the test machine that converts mechanical energy into electrical energy and then dissipated through a variable resistance grid becomes the measure by calibrating.

Driving Dynamometers: A cradle type dc motor drives the test machine and reaction of torque on the stator is measured as in eddy current dynamometer.

Transmission Dynamometers: The torque in the shaft is measured as is being transmitted and the shaft itself acts as a load cell. The angular twist 'θ' of the shaft or the stress becomes a measure of torque.

Torsion-Equation: $\frac{\tau}{J} = \frac{f_s}{R} = \frac{G\theta}{l}$

Shear Stress: $f_s = \frac{16\,\tau}{\pi d^3}$

Shear strain: $y = \frac{f_s}{G}$

Shear strain: $y = \frac{16\,\tau}{G\pi d^3}$

Angular twist: $\theta = \frac{32\,\tau l}{G\pi d^3}$

Torque Meter: Devices based on the concept that torque transmitted is proportional to shear stress and strain developed in the shaft.

Torsion Meter: Devices based on the concept that the torque transmitted by a shaft is proportional to its angular twist 'θ' measured by some electromechanical or optical methods.

Mechanical Torsion Meter: When a shaft is connected between a driving engine and driven load, a twist occurs on the shaft between its ends. This angle of twist is measured and calibrated.

Optical Torsion Meter: Due to torque, an angular twist occurs on the shaft between its ends which is measured using optical means where in the angular deflection of light rays is proportional to twist and hence the torque.

Electrical Torsion Meters (Torque Measurement Using Slotted Discs): A relative displacement between the two slotted discs due to torque causes a phase shift between the pulse generated by the transducers connected to an electronic unit, that shows a time lapse between the two pulses, proportional to the twist of the shaft and hence the torque of the shaft.

Strain Gauge Torsion Meter (Strain Gauges on Rotating Shafts): When strain gauge bonded to a shaft subjected to torque is stretched or compressed, its resistance changes as both length and diameter of the strain gauge changes, which is proportional to the torque in the shaft.

REVIEW QUESTIONS

Short Answer Questions

1. Define force.
2. What is the difference between weight and mass?
3. Explain how a pendulum scale works.
4. How a proving ring is used to measure force?
5. What are load cells? Name the application of load cells.
6. Define torque.
7. Write about strain gauges on rotating shafts.
8. How a strain gauge is used for measuring torque?

LONG ANSWER QUESTIONS

1. Explain the method of measuring force using (a) strain gauge load cell (b) hydraulic load cell (c) pneumatic load cell.
2. How will you use an elastic transducer to measure force? Give at least three different configurations for such measurement and write down the relationship between the force and resulting deformation in each care.
3. What is a proving ring? How is it used to measure force?
4. Explain the working of a hydraulic load cell for the measurement of force.
5. A load cell is proposed in the form of a cantilever with one resistance strain gauge on the top and the underside; the force being applied at the free end of the cantilever. Explain the arrangement and set up a relation for the bridge output voltage in terms of the applied load, the gauge factor and the bridge supply voltage.
6. A strain gauge load cell consists of a solid steel cylinder, which has 4 identical strain gauges mounted upon it in the Poisson's arrangement. For each gauge the nominal resistance is 100 ohm, gauge factor is 2.0 and the gauges are connected electrically to the four arms of a Wheatstone bridge circuit. The applied compressive load is stated to produce a stress equivalent to the limit of proportionality stress in compression. What would be the output voltage if the bridge is energized with a supply voltage of 4 volt? For the steel cylinder: Poisson's ratio = 0.3, modulus of elasticity = 200 GN/m^2, and limit of proportionality stress = 200 MN/m^2.
7. Explain the different methods used for measuring force.
8. Explain how torque can be measured using a stroboscope.
9. Explain how torque can be measured using an optical torsion meter.
10. Explain how torque can be measured using slotted discs.
11. Write briefly on torque measurement. A torque is applied to a torque tube made of steel ($G = 8 \times 10^{11}$ N/m^2) and a physical dimensions $r_o = 16 \pm 0.01$ mm; $r_l = 12 \pm 0.01$ mm & the length $l = 125 \pm 0.03$ mm. The angle of twist θ is 1.50° and its uncertainty is ± 0.05°. Calculate the nominal value of the impressed torque and its uncertainty. Calculate the 45° strains.
12. Describe the working principle of strain gauge bridge with neat sketch. Indicate their arrangement for measurement of torque on a circular shaft.
13. If a shaft of diameter 4 cm & length 1.5 m is transmitting torque of 800 Nm, find the angular deflection of the shaft assuming G = 8 × 1010 N/m^2.
14. For measuring torque transmitted by a shaft, a single strain gauge of resistance 170 ohms and gauge factor 2.0 was fixed at 45° to the axis the shaft. If the strain measured is 833 mm/m, find the torque transmitted, assuming, diameter of the shaft d = 3 cm, shear modulus of the shaft material = 8×10^{10} N/m^2.
15. Explain the principle of working of the following:
    (a) Prony brake dynamometer. (b) Hydraulic dynamometer.

MULTIPLE CHOICE QUESTIONS

1. The unit of force is ______,
   (a) N (b) N/m
   (c) N/m^2 (d) Nm
2. Newton's second law tells us ______
   (a) Force is proportional to the momentum
   (b) Law of conservation of momentum
   (c) The concept of Inertia
   (d) The concept of friction.
3. Equal Arm Balance method is based on the principle of equilibrium of _______.
   (a) Forces (b) Couples
   (c) masses (d) moments
4. The deflection in the proving ring is proportional to_______.
   (a) Diameter of the ring
   (b) Mass of the ring
   (c) Applied Force
   (d) Pressure inside the ring
5. In strain gauge load cell, if there is no load on the cell, all the four gauges have the ________ resistance.
   (a) Same (b) Zero
   (c) Infinite (d) Different
6. The unit of torque is_____.
   (a) N/m (b) Nm
   (c) N^2/m (d) N/m^2
7. Nm is also called ________.
   (a) Joule (b) Erg
   (c) Watt (d) Kilowatt
8. With usual notations, Power is given by _____.
   (a) $2\pi Nt$ (b) $2\pi t$
   (c) $4\pi Nt$ (d) $2\pi N^2 t$
9. Torque is proportional to _______ at the surface of the shaft.
   (a) Diameter (b) Shear stress
   (c) Shear strain (d) Length
10. In Prony Brake Dynamometer, the force to stop rotating shaft is proportional to___
    (a) Diameter of the shaft
    (b) Length of the shaft
    (c) Power
    (d) Energy
11. Transmission Dynamometer is used to find _____.
    (a) Power (b) Energy
    (c) Torque (d) Momentum
12. One pound is equal to ____ kg.
    (a) 0.4535 (b) 1.4535
    (c) 2.4535 (d) 3.4535
13. The torsion equation (with usual notations) is given as_______.
    (a) $\frac{\tau}{J}=\frac{f_s}{R}=\frac{G\theta}{l}$ (b) $\frac{\tau}{J}=\frac{f_s}{R}=\frac{\theta}{l}$
    (c) $\frac{\tau}{J}=\frac{f_s}{R}=\frac{\theta}{l}$ (d) $\frac{\tau}{l}=\frac{f_s}{R}=\frac{G\theta}{J}$
14. In the torsion equation, 'y' is equal to ______.
    (a) $\frac{16\,\tau}{G\pi d^3}$ (b) $\frac{6\,\tau}{G\pi d^3}$
    (c) $\frac{16\,\tau}{G\pi d^2}$ (d) $\frac{10\,\tau}{G\pi d^3}$
15. In electrical torsion meters, the voltage pulses produced in the transducer have ______ time difference.
    (a) Maximum (b) Minimum
    (c) Zero (d) Unit
16. Acceleration due to gravity 'g'= _____ ft/sec^2
    (a) 981 (b) 32
    (c) 10 (d) 9.8
17. The advantage of strain gauge torsion meter over electrical torsion meter is
    (a) Low cost
    (b) Wide range
    (c) High Sensitivity
    (d) Occupy less space

18. The accuracy of the Mechanical torsion meter instrument is small due to _____.
    (a) Small displacement of pointer
    (b) Corrosion
    (c) Design incapability
    (d) Vibrations
19. The type of grooves used in hydraulic dynamometer are
    (a) Circular
    (b) Elliptical
    (c) Semi-elliptical
    (d) Oval
20. The mechanism used in pneumatic load cell is _____.
    (a) Nozzle-Flapper
    (b) Single lever
    (c) Double lever
    (d) Crank & Pinion

ANSWERS

1.	(a)	2.	(a)	3.	(d)	4.	(c)	5.	(a)
6.	(b)	7.	(c)	8.	(a)	9.	(b)	10.	(c)
11.	(c)	12.	(a)	13.	(a)	14.	(a)	15.	(c)
16.	(b)	17.	(c)	18.	(a)	19.	(c)	20.	(a)

CHAPTER INFOGRAPHIC

12

Glimpses of Strain and Stress Measurement

Strain is the change in length by original lenght $e = (l_2 - l_1)/l_1$	**Stress is load per unit area $s = P/A$**	**Young's modulus is ratio of stress to strain. $E = s/e$**	**Poisson's ratio(v) is ratio of lateral strain to liner strain**
$e_x = 1/E\,[s_y + _v(s_y + s_z)$; $e_y = 1/E[s_z + s_x)$; $e_x = 1/E[s_z + s_y)$	**Strain measurment must be made over a finite lenght of the work piece**		**The basic relation for resistance type strain guage is F(gauge factor)= 1 + 2v + (dp/p)/(dl/l)**
If L increases and its area of cross section decreases (A), it is said to be positively strained; If L decreases and its A increases, it is said to be negatively strained			

The reisistance of Cu and Fe wire change when subjected to mechanical strain and these are called electrical resistance strain gages

S.No	Instrument	Transducer	Construction	Operation	Application	Merits	Limitations
1.	Unbounded strain gage (Notpasted)	Stain gauge principle	Two frames carrying rigidly fixed insulated pins and fine wire resistance strain gage is stretched around the insulated pins, which is connected to Wheatstone brodge	Due to force two frames move relative to each other and strain gauge changes in L and so resistance which indicates applied force and change in dimensions of the structure	Detachable; used to find force pressure and accelerarion	Accuracy is very high	Occupies more space
2.	**Bonded strain gauge (Pasted onto the surface of the structure)**						
2(a)	Fine wire strain gage	Strain gauge principle	Resistance wire dia. 0.025 mm is bent to increase length of wire placed between papers/bakelite or Teflon and leads are provided for electrically connected strain gauge wheatstone bridge	Due to force structure will change and hence dimension of the strain gauge and so resistance changes which indicates applied force	Uniform stress distribution is possible as no. of turns are more	Very high accuracy	High cost
2(b)	Metal foil strain gage	Strain gauge principle	Leads are soldered to metal foil 62 mm thick produced by using printed technique and pasted on of plastic backing	Strain gauge is pasted to a structure on which a force is applied	Used where stability and sensitivity and at high temp.	Made in any shape, very high fatigue life	High cost
2(c)	Semi-conductor or Piezo Resistive Strain Gages	Strain gauge principle	Sensing element made of doped crystals, is pasted to the structure on which force is applied. Two types of doped crystals are Resis-tance decreases (n-type) or increases (p-type) with tensile strain	force is applied on gauge pasted to structure changes resistivity of semiconductor sensing element which indicates applied force	Used for fatigue life is more	Very high Gauge factor and measures very small strains	Can't measure Large strain, poor linearity, sensitivity to change in temp

The important gauge parameters: configuration, backing material, bonding material, gauge protection and electrical circuitry

The important properties of grid material are High electrical stability, yield strength, gauge factor, resistivity, endurance limit, Good workability, corrosion resistance, Low hysteresis, thermal emf, temperature sensitivity	After preparing the surface, the strain gauges are mounted on it with bonding material

Resistive type strain gauges are sensitive to temp. & can be compensated by (i) Use of adjacent arm balancing or compensating gauge or gauges; Method I - Using a dummy gauge b) Method II - Poisson's method c)Method III- Use of two active gauges on a cantilever (ii) Self temperature compensation (Selected Melt or Dual element Gauge) (iii) Use of external control circuitry

ROSETTE is a strain gauge measurement system consisting of a combination of strain gauges usually 2or 3 gauges placed at specific angles with respect to each other in order to measure the values of principal strains & stresses without actually knowing direction of principal stresses. The common types of rosettes are: Two element, Rectangular & Delta rosette

Wheatstone bridge is used to measure change in resistance, most common being Null balanced Bridge	Applications of strain gauges are Force, vibration & pressure

Chapter 12

STRAIN AND STRESS MEASUREMENT

STARTERS

To study this chapter, you should have awareness on the following concepts. For a better understanding, it is always a good idea to revise these prerequisites.

- A fair idea on various types of forces, i.e., tensile-compressive, gravitational, elastic, magnetic, electric, mechanical, centripetal, centrifugal forces and so on.
- Elasticity, elastic, plastic and rigid bodies, definitions of terms and equations of elastic bodies
- Relation between stress and strain, types stresses, strains, modulus of elasticity and Hook's law
- Positive and negative strains, concept of strain gauge
- Relation between the terms of linear/angular motions such as $F = mrw^2 = mv^2/r$
- Some mechanisms of motion conversion (translatory to rotatory and vice-versa)
- Simple gear mechanisms for calibration into suitable scale
- Relation between force and electrical parameters such as resistance, capacitance etc. particularly, Whetstone's bridge circuit and rosettes.

LEARNING OBJECTIVE

After studying this chapter you should be able to

- Understand the fundamentals of Stress and Strain,
- Classify and discuss the strain gauges,
- Discuss about bonded and unbonded strain gauges
- Describe the temperature compensation and methods
- Describe the rosettes and their applications
- Narrate the application of Wheatstone bridge circuits
- Explain the operation and applications of strain gauges

12.1 INTRODUCTION

It is a very commonly known fact that when a body is subjected to stress, it gets strained. As described by Hook's law, this stress is linearly proportional to strain up to certain limit, called Hook's limit or elastic limit. The stress-strain relationship can be intelligently used as a principle of transduction in instrumentation to measure several physical quantities such as force, resistance, acceleration etc. A thorough awareness on the material behaviour with reference to stress-strain relation can help the instrumentation engineers to design the measuring instruments with a greater ease. Therefore, let us have a quick revision of these concepts.

To have a better understanding let us consider Low carbon steel, (Refer figure 12.1), which exhibits a very linear stress–strain relationship up to a well defined yield point. This linear portion can be named as the elastic region and the slope is the modulus of elasticity or Young's Modulus.

As we know, stress α strain (within the elastic limit) and hence, we have

Stress/Strain = Constant (E), called the Young's Modulus (or Modulus of Elasticity)

After the yield point, the curve typically decreases slightly (as dislocations escape from Cottrell atmospheres). As deformation continues, the stress increases due to strain hardening till the ultimate tensile stress is reached. Up to this point, the area of cross-section decreases uniformly and randomly due to Poisson contractions. Further, the actual fracture point is in the same vertical line as the visual fracture point.

However, after this point the local cross-sectional area becomes significantly smaller than the original and a *neck* forms. The ratio of the tensile force to the true cross-sectional area at the narrowest region of the neck is called the *true stress*. The ratio of the tensile force to the original cross-sectional area is called the *engineering stress*. If the stress–strain relation is plotted in terms of *true stress* and *true strain,* we can observe the stress continuing to rise until failure. Eventually the neck becomes unstable and the specimen fractures.

If the specimen is subjected to gradually increasing tensile force it reaches the ultimate tensile stress and then necking and elongation occur rapidly until fracture. As the specimen is subjected to

More Instruments for Learning Engineers

Mile 12.1

An **extensometer** invented by Charles Huston, measures changes in the length of an object. It is used for stress-strain measurements and tensile tests. *Clip-on extensometers (Contact type)* used for high precision strain measurement (< 1 mm to over 100 mm) are low cost, easy to use and can influence small/delicate specimens. *Laser extensometer* (Non-contact type) can measure strain or elongation on certain materials subjected to loading in a tensile testing machine. It works by illuminating the specimen surface with a laser, the reflections of which are received by CCD camera & processed by complex algorithms. Resolutions up to 0.1 im and elongation up to 900 mm can be achieved, thence suitable for the most complex testing. A *video extensometer* can measure stress/strain by capturing continuous images of the specimen during test, using frame grabber or digital video camera attached to PC.

Extensometers

gradually increasing length, we can notice the progressive necking and elongation. Civil structures such as bridges are mostly loaded in this pattern.

Less ductile materials such as medium to high carbon steels do not have a well-defined yield point. Usually, we find two types of yield points - upper and lower yield points. For such materials the yield stress is determined by "offset yield method", in which a line is drawn parallel to the linear elastic portion of the curve and intersecting the abscissa at some arbitrary value (usually between 0.1-0.2percent). The point of intersection of this line and the stress–strain curve is referred to as the yield point. The elastic region is the portion of the curve where the material regains its original shape when the load is withdrawn. The plastic region is the portion where the permanent deformation occurs, even if the load is withdrawn and failure point is when the specimen fractures.

12.2 FUNDAMENTALS OF STRESS AND STRAIN

Any machine part or a structural member deforms to some extent when subjected to external loads or forces. The deformations result in relative displacements that may be normalized as percentage displacements or strain.

Let us consider a simple bar of diameter d_1 and length l_1 subjected to load P resulting in increase in length to l_2 i.e., a change in length $D_l = (l_2 - l_1)$ and decrease in diameter to d_2 i.e., change in diameter $D_d = (d_1 - d_2)$ as shown in Fig. 12.1.

We know that

Strain is defined as the change in length (or elongation) divided by the original length.

So, Strain in the direction of load

$$\varepsilon = \frac{\Delta l}{l_1} = \frac{l_2 - l_1}{l_1}$$

Similarly, the strain in the transversal direction is

$$\varepsilon' = \frac{\Delta d}{d_1} = \frac{d_1 - d_2}{d_1}$$

Again we know that

Stress is defines as the force or load per unit area

Stress $$\sigma = \frac{P}{A} = \frac{P}{\pi d_1^2 / 4}$$

According to Hook's law, stress is directly proportional to strain within elastic limits.

Thus, stress and strain under uni-axial condition are related as,

$$E = \frac{\text{Stress}}{\text{Strain}} = \frac{\sigma}{\varepsilon}$$

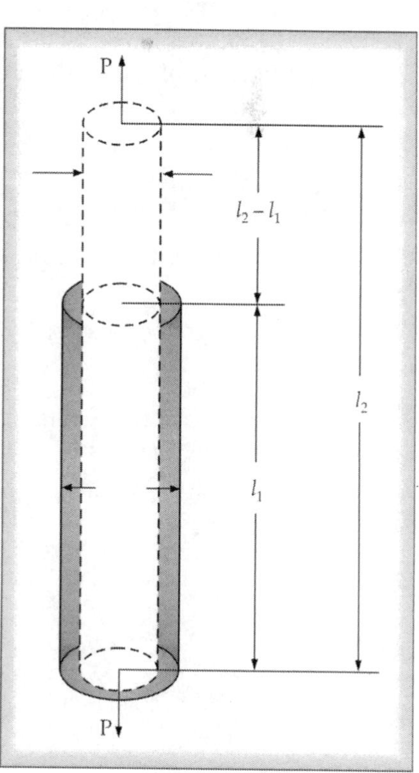

Fig. 12.1 Relations for Axial and Lateral Strain

Where, E is Young's modulus of elasticity of the material and is constant till stress is below the proportional limit.

Change in transverse direction, i.e., change in diameter D*d* with reference to the axial direction i.e. change in length D*l* in the above condition is related as

$$\nu = \frac{\text{Lateral Strain}}{\text{Axial Strain}}$$

$$\nu = \frac{-\dfrac{\Delta d}{d_1}}{\dfrac{\Delta l}{l_1}}$$

We call the above ratio as "Poisson's Ratio (*v)*" defined as the ratio of lateral stain to axial (linear) strain.

Observe and note the 'Negative Sign', in the above relation due to the fact that diameter decreases as length increases.

Now, let us consider an element subjected to orthogonal stresses σ_x and σ_y as shown in Fig. 12.2.

Suppose that the stresses σ_x and σ_y exist at the same time. If σ_x is applied first, there will be a strain in the x-direction equal to (σ_x/E). At the same time, because of Poisson's ratio there will be a strain in the y-direction equal to ($-\nu\,\sigma_y$/E).

Now suppose that the stress in the y-direction σ_y is applied. This stress will result in a y-strain of (σ_y/E) and an x-strain equal to ($-\nu\sigma_x$/E).

The net strains are then expressed by the relations.

$$\varepsilon_x = \{\sigma_x - \nu\sigma_y\}/E \text{ and}$$

$$\varepsilon_y = \{\sigma_y - \nu\sigma_x\}/E$$

Solving the above two relations for σ_x and σ_y, we get the equations

$$\sigma_x = \frac{E(\varepsilon_x + \nu\varepsilon_y)}{1-\nu^2} \text{ and}$$

$$\sigma_y = \frac{E(\varepsilon_y + \nu\varepsilon_x)}{1-\nu^2}$$

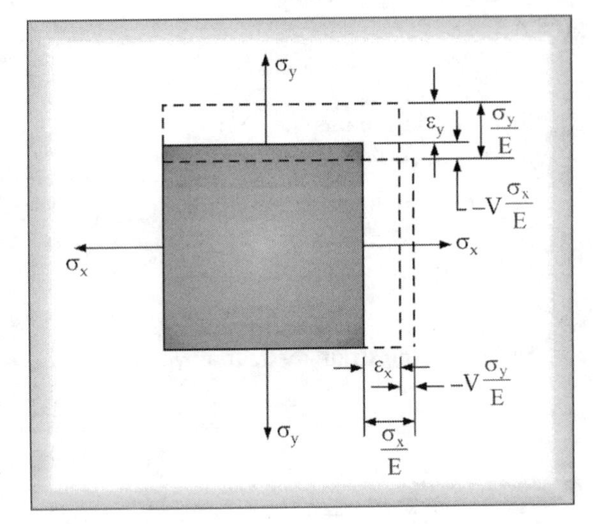

Fig. 12.2 Element Subjected to Biaxial Stress

When a stress σ_z is also considered to exist, acting in the third orthogonal direction, we can establish the more general three dimensional relations as follows:

$$\varepsilon_x = \{\sigma_x - \nu(\sigma_y + \sigma_z)\}/E$$

$$\varepsilon_y = \{\sigma_y - \nu(\sigma_z + \sigma_x)\}/E$$

$$\varepsilon_z = \{\sigma_z - \nu(\sigma_x + \sigma_y)\}/E$$

12.3 STRAIN MEASUREMENT

Strain measurement should be made over a definite length of the work piece. The smaller this length, the more nearly the measurement will approximate the unit strain at a point. The length over which the average strain measurement is taken is called the *base length*. The deformation sensitivity is defined as the minimum deformation that can be indicated by the approximate gauge.

Strain sensitivity is the minimum deformation that can be indicated by the gauge per unit base length.

Strain measurement can be done in a simple way by placing some type of grid marking on the surface of the work piece under zero load conditions and then the deformation of this grid when the specimen is subjected to a load. The grid may be scribed on the surface, drawn with a fine ink pen, or photo etched. Rubber threads have also been used to mark the grid. The sensitivity of the grid method depends on the accuracy which the displacement of the grid lines may be measured. A micrometer microscope is frequently employed for such measurements. An alternative method is to photograph the grid before and after the deformation and to make the measurements on the developed photograph. Photographic paper can have appreciable shrinkage so that glass photographic plates are preferred for such measurements. Grid methods are usually applicable to materials and processes having appreciable deformation under load. These methods are applicable in sheet-metal-forming processes. The grid could be installed on the flat sheet of metal before it is formed. The deformation of the grid after forming gives the designer an indication of the local stresses induced in the material during the forming process.

Brittle coatings offer a convenient means for measuring the local stress in a material. The specimen or work piece is coated substance having brittle properties. When the specimen is subjected to a load, small cracks appear in the coating. These cracks appear when the state of tensile stress in the work piece reaches a certain value and thus may be taken as a direct indication of a local stress. The brittle coatings are particularly useful for determination of stresses at stress concentration points that are too small or inconveniently located for installation of electrical resistance or other types of strain gauges.

12.3.1 Positive Strain

- When a strain gauge is subjected to tension, its length increases and its area of cross-section decreases then it is said to be positively strained.
- As the resistance of a conductor is proportional to its length and inversely proportional to its area of cross-section, the resistance of the gauge increases with positive strain.

12.3.2 Negative Strain

- When a strain gauge is subjected to compression, its length decreases and its area of cross section increases then it is said to be negatively strained.
- As the resistance of a conductor is proportional to its length and inversely proportional to its area of cross-section, the resistance of the gauge decreases with negative strain.

12.4 THE ELECTRICAL RESISTANCE STRAIN GAUGE

Lord Kelvin demonstrated that the resistances of copper wire and Iron wire change when the wires are subjected to mechanical strain. He used a Wheatstone bridge circuit with a galvanometer as the indicator.

Rauge of M.I.T. conceived the idea of making a preassembly by mounting wire between thin pieces of paper as shown in Figure 12.3.

In 1950's the foil type gauge, soon replaced the wire gauge. The common form consists of metal foil element on a thin Epoxy support and is manufactured using printed-circuit techniques. A few are shown in Fig. 12.4.

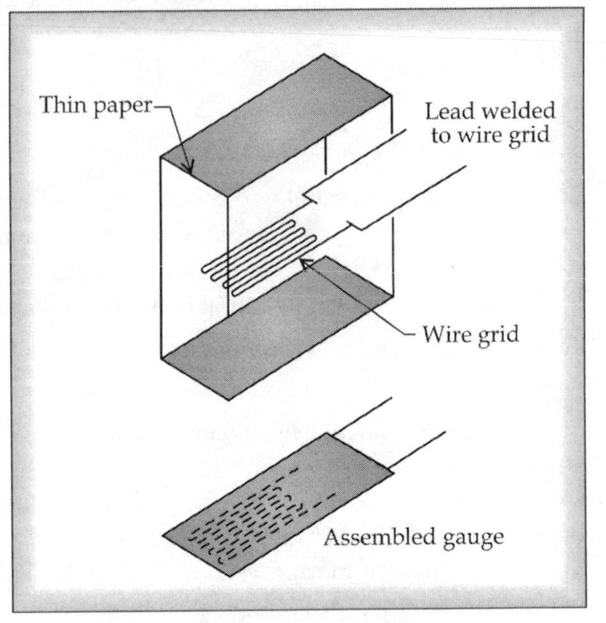

Fig. 12.3 Bonded - Wire - Type Strain Gauge

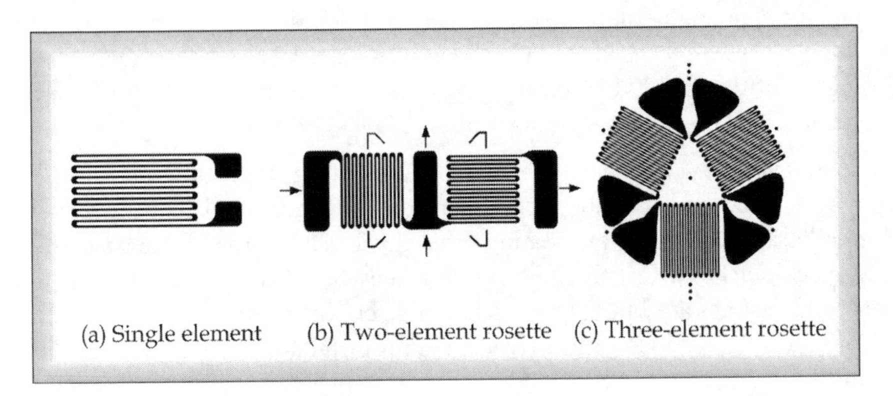

Fig. 12.4 Typical Foil Gauges

12.4.1 The Metallic Resistance Strain Gauge

Expression Relating Change in Resistance and Strain in Strain Gauge

The operation of the metallic resistance strain gauge is as follows. When a length of wire foil is mechanically stretched a longer length of smaller sectioned conductor results, hence the electrical resistance changes. If the length of resistance element is intimately attached to a strained member in such a way that the element will also be strained, then the measured change in resistance can be calibrated in terms of strain.

A general relation between the electrical and mechanical properties may be derived as follows:

Assume an initial conductor length of l having a cross sectional area of Cx. (In general the section need not be circular; hence x will be a sectional dimension and C will be a proportionality constant. If the section is square, then C = 1; If it is circular, the C = π/4 etc.). If the conductor is strained axially in tension, thereby causing an increase in length, the lateral dimension should reduce as a function of Poisson's ratio.

We know that
$$R = \frac{\rho l}{A} = \frac{\rho l}{Cx^2} \quad .....(12.1)$$

If the conductor is strained, we may assume that each of the quantities in Eqn.12.1 except for C may change.

Differentiating, we have
$$dR = \frac{Cx^2(ld\rho + \rho dl) - \rho l(C.2x.dx)}{(Cx^2)^2} \quad .....(12.2)$$

$$= \frac{1}{\left(Cx^2\right)}\left((ld\rho + \rho dl) - 2\rho l\frac{dx}{d}\right)^2$$

Dividing Eq (12.2) by Eq. (12.1) yields
$$\frac{dR}{R} = \frac{dl}{l} - 2\frac{dx}{d} + \frac{d\rho}{\rho}$$

Which may be rewritten as
$$\frac{dR/R}{dl/l} = 1 - 2\frac{d(x)/x}{dl/l} + \frac{d\rho/\rho}{dl/l}$$

Now $\frac{dl}{l} = \varepsilon_a$ axial strain,
$$\frac{d(x)}{d} = \varepsilon_L = \text{Lateral strain, and } \nu = \text{Poisson's ratio} = \frac{d\rho/\rho}{dl/l}$$

These substitutions give us the basic relation of gauge factor, for which we shall use the symbol F:

So,
$$F = \frac{dR/R}{dl/l} = \frac{dR/R}{\varepsilon_a} = 1 + 2\nu + \frac{d\rho/\rho}{dl/l}$$

This relation is the basic relation for resistance-type strain gauge.

ILLUSTRATION-12.1

An electrical resistance strain gauge has a gauge factor of 2.0 and resistance of 125Ω. Find the change in resistance of the gauge caused by strain of mm/m.

Solution:

Gauge Factor,
$$F = \frac{\Delta R/R}{\epsilon_a}$$

Substituting $F = 2.0,\ R = 125W,\ \epsilon_a = 2 \times 10^{-6}$ mm/m

$$2 = \frac{\Delta R}{125 \times 2 \times 10^{-6}}\ \Delta R = 0.0005\,\Omega.$$

$$\text{Percentage change in resistance} = \frac{0.0005}{125} \times 100 = 0.0004\ \%$$

ILLUSTRATION-12.2

An electric resistance strain gauge of 100 Ω resistance and gauge factor 2 is bonded to a specimen of steel. Calculate the change in resistance of the gauge when a tensile stress of 60 MN/m^2 is applied on the speciment.

Solution:

$$\text{Gauge factor} \quad (F) = \frac{dR/R}{dl/l}$$

$$2 = \frac{dR/100}{dl/l}$$

$$2 = \frac{dR/100}{\epsilon_a} \qquad .....(12.3)$$

$$\because \quad \text{Axial strain} = \left(\frac{\text{Change in Length}}{\text{Original Length}}\right)$$

$$\text{But,} \quad \text{Strain} = \frac{\text{Stress}}{\text{Young's modulus}} = \frac{60 \times 10^6\ N/m^2}{2 \times 10^{11}\ N/m^2} \quad \left[\because\ E = 2 \times 10^{11} N/m^2\right]$$

$$= 30 \times 10^{6-11}$$

$$= 30 \times 10^{-5}$$

$$E_a = 0.00030$$

Substituting E_a in (12.3)

$$2 = \frac{dR/100}{0.0003}$$

$$0.0006 = \frac{dR}{100}$$

$$dR = 0.0006 \times 100 = 0.06\ \Omega$$

SELF ASSESSMENT QUESTIONS-12.1

1. Define strain. Explain a simple method for measuring strain.
2. How do you differentiate between positive and negative strain. Brief some points.
3. Explain the importance of strain measurement. State some applications.
4. Derive an expression relating change in resistance and strain in strain gauge.
5. Describe the working of metallic resistance strain gauge
6. Explain the operation of bonded - wire - type strain gauge
7. Explain about the electric resistance strain gauge.

12.5 TYPES OF STRAIN GAUGES

The common types of strain gauges are as follows.

1. Unbonded strain gauge
2. Bonded strain gauge
   (a) Fine wire strain gauge
   (b) Metal foil strain gauge
   (c) Semi-conductor or piezo resistive strain gauge.

12.5.1 UNBONDED STRAIN GAUGE

These strain gauges are not directly pasted onto the surface of the structure. Therefore, they are termed as unbonded strain gauges.

The main parts of this instrument are:

1. Two frames
2. Insulated pins
3. Unbonded strain gauge (wire resistance)
4. Wheatstone bridge circuit

Principle

The applied load varies the dimensions of a strain gauge, hence proportionally changes the electrical resistance which can be calibrated.

Construction

The arrangement of this strain gauge consists of two frames say, *P* and *Q* which carry rigidly fixed insulated pins as shown in Fig. 12.5. These two frames can move with respect to each other and they are held together by a spring loaded mechanism. A fine wire resistance strain gauge is stretched around the insulated pins, which is connected to a Wheatstone bridge.

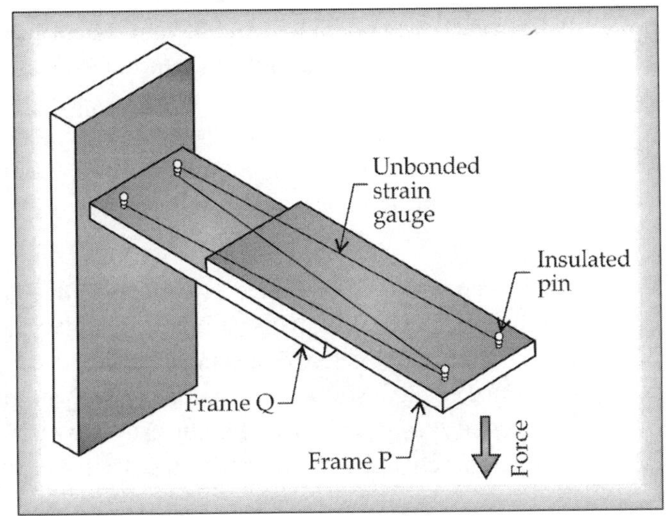

Fig. 12.5 Un-bonded Strain Gauge

Operation

When a force is applied on the structure, frame P moves relative to frame Q. This strains the strain gauge i.e., varies the length and cross section of the strain gauge. This strain changes the resistance of the strain gauge which is measured by using a Wheatstone bridge. This change in resistance when calibrated becomes a measure of the applied force and change in dimensions of the structure.

Merits

- The range of this gauge is ± 0.15% strain
- The accuracy of this gauge is very high.

Demerits

- Occupies more place.

Applications

- Suitable to places where the gauge can be detached and used again and again.

These strain gauges are used in force, pressure and acceleration measurement.

12.5.2 Bonded Strain Gauges

These gauges are directly pasted on the surface of the structure under study. Hence they are termed as bonded strain gauges. The three types of bonded strain gauges are:

(a) Fine wire strain gauge (b) Metal foils strain gauge (c) Semi-conductor gauge

Principle

Same principle is applicable here also. The applied load varies the dimensions of a strain gauge, hence proportionally changes the electrical resistance which can be calibrated.

(a) Fine Wire Strain Gauge

Construction

The arrangement of these strain gauges is as follows.

A resistance wire of diameter 0.025 mm is bent again and again so as to increase the length of the wire distributed permits a uniform distribution of stress as shown in Fig. 12.6. This resistance wire is placed between the two carrier bases made of Paper, Bakelite or Teflon connected to each other. The carrier base protects the gauge from damages. Leads are drawn for electrically connecting the strain gauge to a measuring instrument such as Wheatstone bridge.

Operation

As the name suggests (bonded), the strain gauge is stuck to the structure with the help of an adhesive material. Now, the structure is subjected to a force either tensile or compressive, due to which the structure dimension changes. As the strain gauge is bonded to the structure, the strain gauge also undergoes change both in length and cross section, which thence changes the resistance of the strain gauge, which can be measured using a Wheatstone bridge. This change in resistance of the strain gauge proportional to force applied becomes the measure when calibrated.

Fine-wire Strain Gauge Materials

Material	Composition	Gauge Factor (F)
Nichrome	80%Ni ; 20%Cr	2.0
Constantan or Advance	43%Ni; 57%Cu	2.0
Nickel	100% Ni	(-12)
Iso-elastic	36%Ni, 8%Cr, 0.5%Mo, 55.5%Fe	3.5
Karma	74%Ni, 20%Cr, 3%Al, 3%Fe	2.4

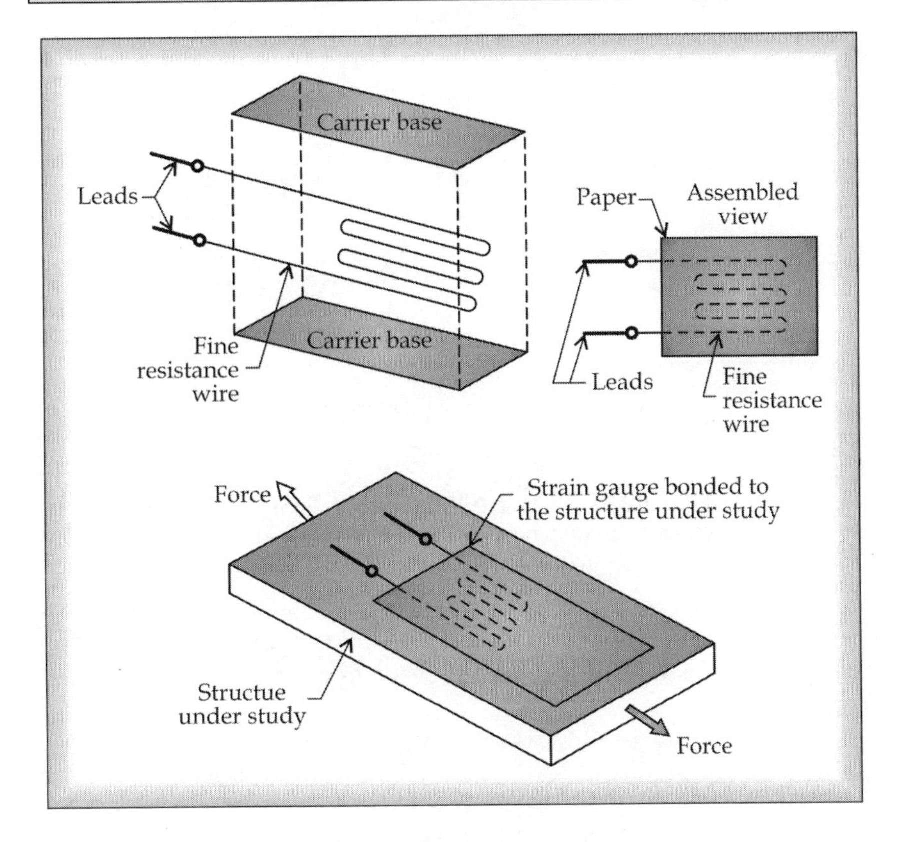

Fig. 12.6 Bonded Strain Gauge

Merits

- The range of this gauge is ± 0.3% of strain.
- The accuracy of this gauge is very high.
- It has a linearity of ± 1%

Demerits

- As these gauges are bonded to the structure, these cannot be detached and used again (i.e., not reusable)
- Cost of these gauges is usually very high.

(b) Metal Foils Strain Gauge

Construction

The metal foil of 0.02 mm thick is used to fabricate the strain gauge in the form of printed circuit board (PCB) which may be produced on one side of a plastic/ebonite/bakelite backing. Leads are soldered to the metal foil for electrically connecting the strain gauge to a Wheatstone bridge as shown in Figure12.7.

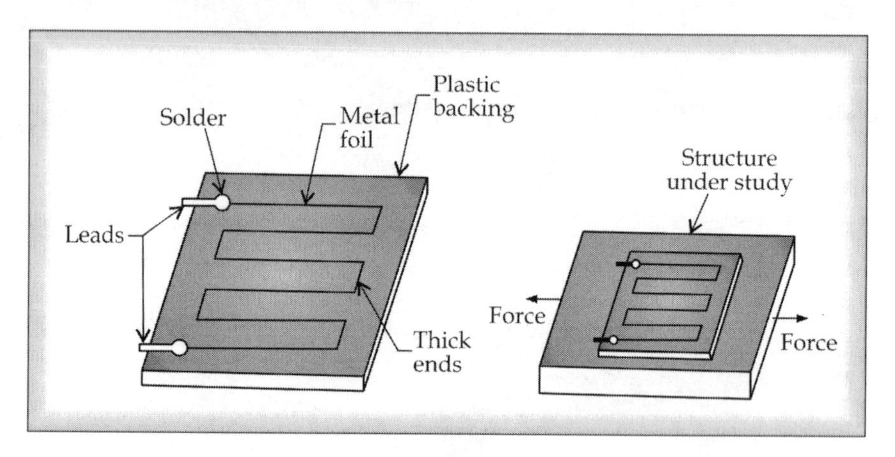

Fig. 12.7 Metal Foil Strain Gauge Structure

Operation

As suggested by the name (bonded-metal foil strain gauge) the metal foil (strain gauge) is pasted to the structure with the help of an adhesive material. Now the structure when subjected to a force either tensile or compressive will change in dimension. As the strain gauge is bonded to the structure, the strain gauge also undergoes change both in length and cross-section. This strain change, thence the variation of resistance of the strain gauge can be measured using a Wheatstone bridge, becomes a measure of the applied force when calibrated.

Merits

- These can be manufactured in any shape.
- Perfect bonding of the strain gauge is possible with the structure.
- The backing can be peeled off and the metal foil with leads can be used directly. In such cases, a ceramic adhesive is to be used.
- Fatigue life of these gauges is very high.
- Sensitivity and stability is very good at high temperatures.

(c) Semi-conductor or Piezo Resistive Strain Gauge

Construction

Here, the sensing element is a rectangular filament with a plastic or stainless steel backing made like a wafer from silicon or germanium crystals doped with boron to get some desired properties. Leads are drawn out from the sensing element for electrically connecting the strain gauge to a measuring instrument as shown in Figure 12.8.

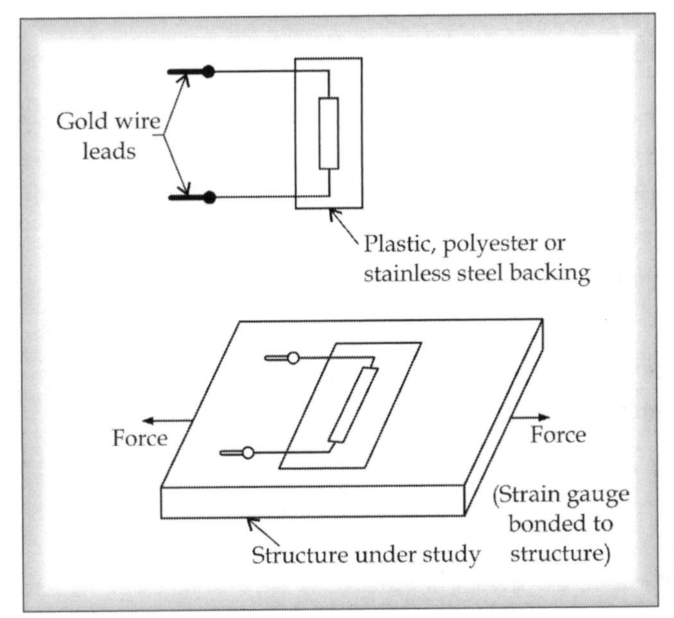

Fig. 12.8 Semi-conductor Strain Gauge

There are two types of sensing elements namely;

Negative or N -type (Resistance decreases with respect to tensile strain)

Positive or P -type (Resistance increases with respect to tensile strain).

Operation

As suggested by the name (bonded), the semiconductor (piezo-resistive) strain gauge is bonded to the structure with the help of an adhesive material. Now the structure when subjected to a force either tensile or compressive will change in dimension. As the strain gauge is bonded to the structure, the strain gauge also undergoes change both in length and cross-section. This strain change, thence the variation of resistance of the strain gauge can be measured using a Wheatstone bridge, becomes a measure of the applied force when calibrated.

More Instruments for Learning Engineers

A **Falling Weight Deflectometer** (FWD) evaluates the physical properties of pavement. FWD data is used to estimate pavement structural capacity for overlay design and to know if a pavement is overloaded at highways, local roads, airport pavements, harbor areas and railway tracks etc. It is also used to calculate stiffness-related parameters of a pavement structure, the degree of load transfer between adjacent concrete slabs, and to detect voids under slabs. FWD is usually contained in a trailer and towed by another vehicle. Deflection sensors (geophones; force-balance seismometers) mounted radially from centre of the load plate measure the deformation of the pavement in response to the load. A Light Weight Deflectometer (LWD) is portable, used to test in-situ base and sub-grade moduli during construction. A Heavy Weight Deflectometer (HWD) uses higher loads to test airport pavements. A Rolling Weight Deflectometer (RWD) can gather data at much higher speed (55 mph) than FWD. It is a specially designed tractor-trailer with laser measuring devices mounted on a beam under the trailer. Also RWD can gather continuous deflection data.

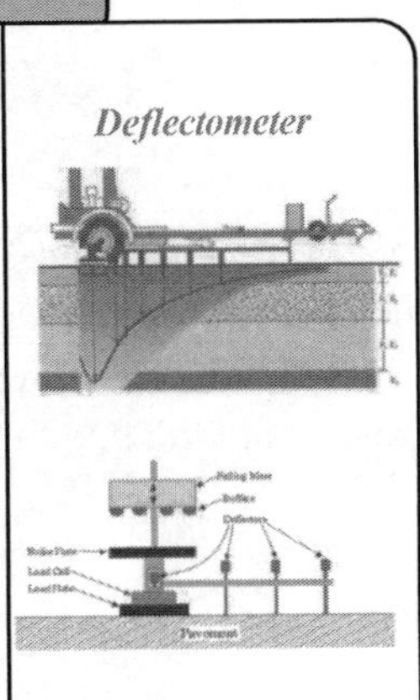

Merits

- They have an accuracy of 2.3 %.
- The gauge factor is very high; hence they can measure very small strains also.
- Fatigue life is very good.
- They have a good frequency of response.
- They can be manufactured in very small sizes.
- Hysteresis characteristics of these gauges are very good.

Demerits

- These gauges are brittle and hence they cannot be used for measuring large strain.
- The gauge factor does not remain constant.
- These gauges have poor linearity.
- These gauges are very costly
- The crystals are difficult to be bonded onto the structures.
- These gauges are sensitive to changes in temperature

Self Assessment Questions-12.2

1. Classify various types of strain gauges.
2. What are the different types of bonded strain gauges? explain each of them
3. What is an un-bonded Strain Gauge? Explain its construction and operation with a neat sketch.
4. List the applications, limitations and advantages of un-bonded strain gauge
5. Name the materials used for fine-wire strain gauge. Give the gauge factor for each material.
6. Give the advantages and limitations of fine-wire strain gauge.
7. Describe the construction and operation of metal foils strain gauge with a neat sketch.
8. Explain the working principle of piezo resistive strain gauges.
9. Give the advantages and limitation for following strain gauges.
   (i) Metal foils strain gauge (ii) Piezo resistive strain gauges

12.6 SELECTION AND PREPARATION OF BONDED STRAIN GAUGE

A strain gauge is too small to act directly, therefore it is fixed (bonded) on a material called backing material or base or carrier. For this fixing, the strain gauge surfaces as well as the backing material surface have to be prepared. Both surfaces have to be made sufficiently rough, and the adhesive is pasted and the bonding techniques have to be followed. This process is discussed here below.

12.6.1 Surface - Preparation and Mounting of Strain Gauges

The steps involved in preparing a surface to mount a strain gauge are listed below:

- Appropriate material is to be selected and the shaped into predetermined structure and fabricated/manufactured accordingly.

- Then structure under study is made even and free from dust and dirt by rubbing with an emery sheet or by sand blasting.
- The even surface is then cleaned by a volatile solution (acetone) using a cloth to remove oil/grease
- The bottom side of the backing (gauge carrier) is also cleaned by a solvent using a cloth.
- Then the strain gauge is mounted on the structure under study.

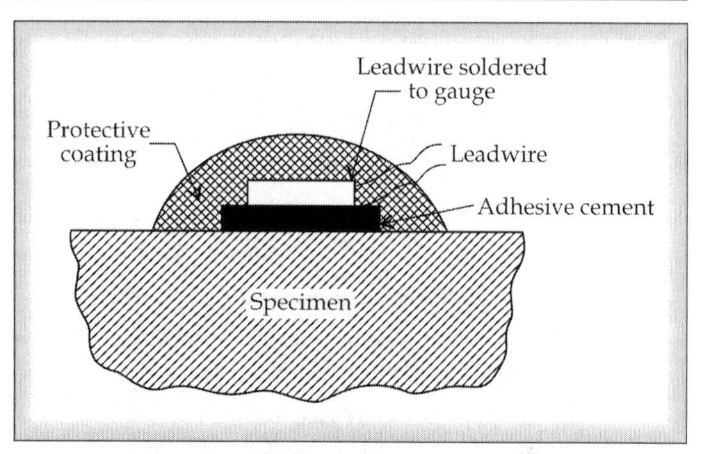

Fig. 12.9 Bonded Resistance Gauge Mounted on a Specimen

- Thus the performance of a strain gauge depends on appropriate selection, preparation and installation of three types of materials. These are
  1. Grid material
  2. Backing or base or carrier material.
  3. Bondage material or cement.

12.6.2 Grid Materials

The grid materials should possess the following properties:

- High electrical stability
- High yield strength
- High gauge factor F
- High resistivity, p
- High endurance limit
- Good workability
- Good corrosion resistance
- Good solder-ability or weld-ability
- Low hysteresis
- Low thermal emf when joined to other materials and
- Low temperature sensitivity

12.6.3 Backing, Base or Carrier Material

The purpose of providing the carrier/ backing material is to provide

- Support the resistance wire (grid) of the strain gauge arrangement.
- Protection to the sensing resistance wire of the strain gauge arrangement.
- Dimensional stability for the resistance wire of the strain gauge arrangement.

Characteristics required for a backing material are that it should

- be easy to apply
- spread easily and provide good bond strength (adhesion)
- be an insulator of electricity.
- have a high creep resistance.
- have good shear strength to transmit the force to the sensing resistance wire.
- go along with the backing material so that the backing material is fixed rigidly
- not absorb humidity, i.e., non-hygroscopic.
- not be affected by temperature changes.

12.6.4 Bonding Materials or Cements (Adhesive)

The strain gauge has to be bonded on the structure under study using an adhesive. These adhesives are called bonding materials or cements. Various adhesives, their composition and the temperatures for which they can be used are shown in following table.

Adhesive	Composition	For Temperatures
Bonding Material	Celluloid dissolved	-
Thermo-plastic cement	acetone	up to 75 °C
Thermo- Setting cement	-	From 150 °C to 210 °C
Special ceramic cement	phenol resin	Above 175 °C

Peper Backed gauge Backlite Guage Epoxy resin guage

Characteristics required for a bonding material is that it should be same as those listed under the backing materials.

Backing type	Bonding cement or Adhesive	Procedure to mount the strain gauge
Paper Backed gauge	Nitro cellulose cement	- Spread a thin layer of adhesive on the cleaned even surface of the structure - The strain gauge with backing is placed on the surface, pressed gently & the excess adhesive is wiped with a cloth. - The gauge is covered with a soft pad and pressure is applied by placing a weight about 2 kg. The required bondage is achieved after 8 hours. - The electrical continuity is checked and leads are soldered to the sensing resistance wire of the gauge.
Backlite Gauge	Phenolic resin	- The same procedure as in paper-back gauges is adopted. **Note:** To attain the required bondage, a clamping pressure of 2 kgf/cm^2 is applied and the arrangement is heated to 80°C and held for 2 hours. Then the pressure is removed and the arrangement is heated to 150°C for one hour.
Epoxy resin Gauge	Epoxy Cement	The same procedure as in paper-backed gauges is adopted

Summary of strain gauge selection, fabrication and mounting procedure

Note: To attain the required bondage, a clamping pressure 0.5 kgf/cm^2 is applied and the arrangement is heated to 75 °C and held for 3 hours.

12.6.5 Factors Affecting Selection of Bonded Metallic Strain Gauges

Thus summarily, the selection of material, fabrication, installation, mounting and performance of bonded metallic strain gauges largely depend on six important parameters as given below. (Remember the acronym 'BEG-BIG' formed by the first letters of the following words)

1. Bonding material
2. Electrical circuitry
3. Gauge protection
4. Bonding technique
5. Installation method & Backing material
6. Grid material and configuration

12.7 TEMPERATURE COMPENSATION

An important point to be noted in the strain gauges is that the strain measured is based on change in the resistance of the gauge. But this change in resistance need not only be due to the strained dimensional change but also may occur due to increase in temperature because the resistance is proportional to the dimensional change as well as temperature change. Thus resistive type strain gauges are sensitive to temperature. So, it is essential to account for the variations in the resistance of strain gauge occurring due to temperature changes. If the variations in resistance due to temperature changes are not accounted for or not compensated, the results could be an error. This is because the resistance of the strain gauge changes both with strain as well as with temperature.

When the ambient temperature changes following changes occur:

1. The gauge grid either elongates or contracts,

$$\Delta l = l\alpha\Delta T$$

where l = gauge lengths

α = Thermal coefficient of expansion of gauge material

ΔT = Ambient temperature change

2. The base material upon which gauge is fixed also elongates or contracts,

$$\Delta l = l\beta\Delta T$$

where β = Thermal coefficient of expansion of base material

3. The Resistance *(R)* of gauge material changes due to change of resistivity.

$$\Delta R = Ry\Delta T$$

where y = Temperature coefficient of resistivity of the gauge material

4. The strain due to the differential expansion between the grid and the base material,

$\varepsilon = \dfrac{\Delta l}{l} = (\beta - \alpha)\Delta T$ causes a change in resistance,

$$\frac{\Delta R}{R} = F(\beta - \alpha)\Delta T$$

where, F is gauge factor and the change in resistance due to change in resistivity.

$$\left(\frac{\Delta R}{R}\right) = y\Delta T$$

Temperature induced change in the resistance of the gauge having gauge factor *F*,

$$\left(\frac{\Delta R}{R}\right)\Delta T = F(\beta - \alpha) + y\Delta T$$

Thus a change in resistance of the gauge occurs, although the base material has not been subjected to any mechanical loading. The temperature effects may be handled by

1. Compensation or cancellation.
2. Evaluation as a part of the data reduction problem.

The first method is extensively used for both metallic as well as semiconductor strain gauges.

Compensation may be provided by

(i) Use of adjacent arm balancing or compensating gauge or gauges
   (a) Method I - Using a dummy gauge
   (b) Method II - Poisson's method
   (c) Method III- Use of two active gauges on a cantilever
(ii) Self temperature compensation
   (a) Selected Melt Gauge
   (b) Dual element Gauge
(iii) Use of external control circuitry

12.8 TEMPERATURE COMPENSATION USING AN ADJACENT ARM COMPENSATING GAUGE

Method - I Temperature Compensation using a Dummy Gauge

One of the ways in which temperature error can be eliminated by using adjacent arm compensating gauge is to use dummy gauge in adjacent arm. The arrangement is shown in Figure 12.10.

Gauge-1 is installed on the test specimen, while gauge-2 is installed on a like piece of material that remains unstrained throughout the tests but at the same temperature as the test piece. Any changes in the resistance of gauge-1 due to temperature are thus cancelled out by similar changes in the resistance of gauge-2 and the bridge circuit detects an unbalanced condition resulting only from the strain imposed on gauge-1 of course, care must be exerted to ensure that both gauges are installed in exactly the same manner on their respective work pieces.

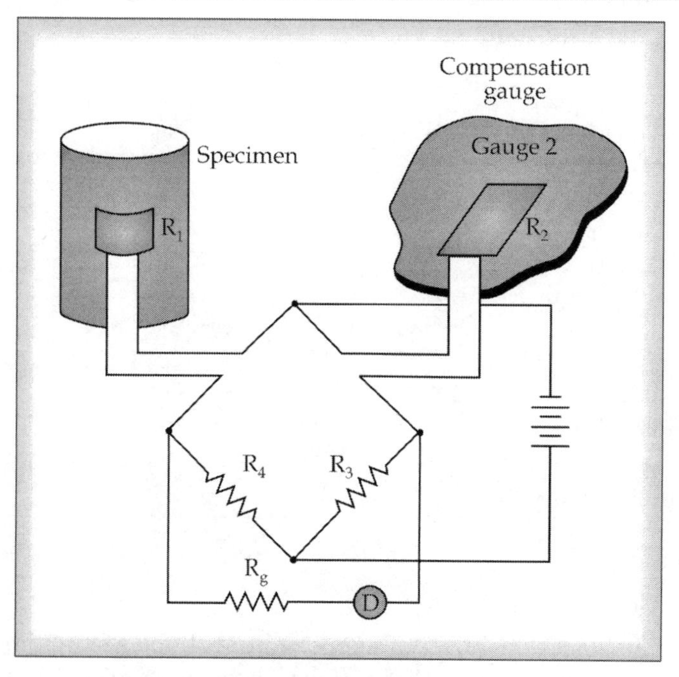

Fig. 12.10 Temperature Compensation Arrangement for Electrical- Resistance Strain Gauge

Initially when the bridge is balanced

$$\frac{R_1}{R_2} = \frac{R_3}{R_4} \Rightarrow \frac{R_1 R_4}{R_3} = R_2$$

If there is a change in temperature, the resistance R_1 and R_2 change by an amount ΔR_1 and ΔR_2 respectively.

$\therefore$ For balance
$$\frac{R_1 + \Delta R_1}{R_2 + \Delta R_2} = \frac{R_3}{R_4}$$

$$\Rightarrow \frac{R_4}{R_3}[R_1 + \Delta R_1] = R_2 + \Delta R_2$$

$$\Rightarrow \frac{R_4 R_1}{R_3} + \frac{R_4}{R_3}\Delta R_1 = R_2 + \Delta R_2$$

But
$$\frac{R_4 R_1}{R_3} = R_2$$

$\therefore$
$$R_2 + \frac{R_4}{R_3}\Delta R_1 = R_2 + \Delta R_2$$

$$\frac{R_4}{R_3}\Delta R_1 = \Delta R_2$$

Suppose $R_4 = R_3$

Then $\Delta R_1 = \Delta R_2$

This implies that for the bridge to remain insensitive to variations in temperature the resistance R_1 and R_2 should have their resistances change by equal amount when subjected to variation in temperature. Therefore, the active gauge R_1 and the dummy gauge R_3 should be identical. The use of dummy gauge for temperature compensation is simple and effective and should be employed whenever possible.

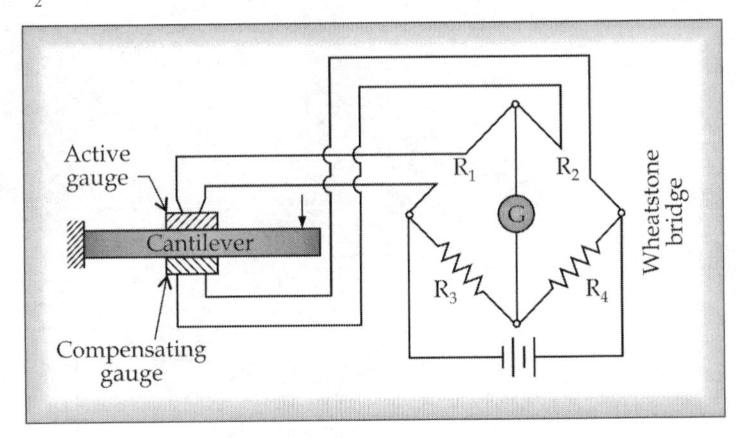

Fig. 12.11 Use of Two Active Gauges on a Cantilever

Method - II Poisson's Method

This arrangement eliminates the effect of temperature with the use of "Poisson's gauge" in the adjacent arm of the Wheatstone bridge. The arrangement consists of an active strain gauge installed on the structure of interest. A Poisson's gauge is placed at right angle with respect to the active gauge on the same piece (that is structure of interest). As both the gauges lie on the same piece, they are influenced equally by the ambient temperature changes.

Let R_1 = Resistance of the active strain gauge.

R_2 = Resistance of the Poisson's gauge

R_3, R_4 = Standard resistances.

Now, due to the applied force, the active gauge is in tension and the Poisson's gauge is in compression (depending on the Poisson's ratio of the material). In accordance with the change in temperature, the active gauge will undergo change in resistance in a certain direction and simultaneously the Poisson gauge will undergo change in resistance in the opposite direction and thus they nullify the effect of temperature changes.

Method III- Use of Two Active Gauges on a Cantilever

Here, a compensating gauge in the adjacent arm of the Wheatstone bridge eliminates the effect of temperature. This arrangement contains an active strain gauge bonded to the top surface of a cantilever. On the other hand, a compensating gauge is bonded to the bottom surface. As both the gauges are bonded to the same piece, they are influenced equally by the ambient temperature changes.

Let R_1 = Resistance of the active strain gauge

R_2 = Resistance of the compensating gauge

R_3, R_4 = Standard resistances.

Due to the applied force, if the active gauge is in tension, then the compensating gauge will be in compression. When there is a change in temperature, both the gauges are affected equally and hence no error is caused by resistance change due to temperature variation.

12.8.1 Self Temperature Compensation

The adjacent arm compensating methods cannot be used for temperature compensation in some cases due to the following reasons:

- Measuring gauge and compensating gauge cannot be kept at the same temperature
- When it is necessary to use the ballast circuit instead of the bridge circuit.

In the above situations, the "self temperature compensating gauges" is the better alternative, which is of two types, namely the "selected melt gauge" and the "dual element gauge". These are described in the paragraphs to follow.

12.8.1.1 Selected Melt Gauge

In this method, the alloy and processing (usually cold working) are manipulated and a control over the temperature sensitivity of the strain gauge grid materials is obtained. Thus the strain gauge grid is produced in such a way that it gives a very low strain variation with temperature, over a certain range of temperature.

12.8.1.2 Dual Element Gauge

In this method, two wire elements are joined in series to get one strain gauge assembly. The two wire elements have different temperature characteristics. The two wire elements are selected so that the net strain produced due to temperature changes is minimized when the gauges are fixed on the surface under study.

Illustration-12.3

An electrical resistance strain gauge of resistance 130 &! and gauge factor 2.5 is bonded to a specimen of steel. Calculate the resistance change of the gauge, (a) due to a tensile stress of 40 MN/m^2 in the specimen, (b) an increase of 40 °C in temperature.

For the gauge

Temperature coefficient of resistance = 20×10^{-6}/°C

Thermal coefficient of linear expansion = 16×10^{-6}/°C

For the steel

Modulus of Elasticity $\varepsilon = 200$ GN/m^2

Thermal coefficient of linear expansion = 12×10^{-6}/ °C

Solution:

(a) **Due to stress:** $\sigma = 40$ MN/m^2 = 40×10^6 N/m^2

Strainmace, $$\varepsilon = \frac{\sigma}{\in} = \frac{40 \times 10^6}{200 \times 10^9} = 2.0 \times 10^{-4}$$

$$F = \frac{\Delta R / R}{\varepsilon}$$

$$\Delta R = F\, \varepsilon\, R = 2.5 \times 2.0 \times 10^{-4} \times 130 = +\,0.065\ \Omega$$

(b) **Due to temperature:** It consists of two components,

Due to temperature coefficient of resistance

$$(\Delta R) = R\ y\ \Delta T = 130 \times 20 \times 10^{6} \times 40 = +\ 0.104\Omega$$

Due to differential expansion of gauge material and steel specimen

$$(\Delta R) = FR\ (\alpha_s - \alpha_g)\ \Delta T = 2.5 \times 130\ (12 - 16)\ 10^{-6} \times 40$$

$$= (-)\ 0.052\ \Omega \qquad (\alpha_g > \alpha_s)$$

Here the relative magnitude of resistance changes. Positive "R" is an increase and negative one is decrease in resistance. Further, the two components of resistance change due to temperature have opposite signs.

Self Assessment Questions-12.3

1. Give the factors that depend on the performance of bonded metallic strain gauges
2. What are the desirable properties of grid materials?
3. What is the purpose of backing material? List the required characteristics for backing material.
4. How the surface preparation is done for bonded metallic strain gauges? Explain in detail.
5. What are the various adhesives used in bonded metallic strain gauge also give the composition and range of temperature.
6. Give the eight characteristics required for a bonding material.
7. Explain the surface preparation for mounting of strain gauges.
8. What is the need of temperature compensation?
9. What are the different methods provided for compensation of temperature in strain gauges? Explain each of them.
10. With the help of circuit diagrams describe the temperature compensation in strain gauges by the following methods
    (a) Using a dummy gauge
    (b) Poisson's method
    (c) Use of two active gauges on a cantilever
11. What are the different methods used in self-temperature compensation of strain gauge. Explain each in detail.

12.9 GAUGE ROSETTES (MULTIPLE GRIDS)

When the direction of unit directional loading is know, such as tension or compression or bending then a single strain gauge is used. In such a case the results of the strain gauges is oriented exactly with the principal stress axis.

In practice, the test piece may be subjected to stresses in any direction and is impossible to find the exact direction of the principal stresses. Thence it is not possible to orient the strain gauges, along the direction of principal stresses.

To overcome this difficulty, a strain gauge measurement system called "ROSETTE" is used which measures the values of principle strains and stresses without actually knowing their direction.

Definition

ROSETTE is a strain gauge measurement system consisting of a combination of strain gauges usually two or three gauges placed at specific angles with respect to each other in order to measure the values of principal strains and stresses without actually knowing the direction of principal stresses.

In a rosette, the strain gauges are insulated from touching each other, and all the strain gauges all bonded to the same backing material.

The common types of rosettes are

1. Two element rosette
2. Rectangular rosette (three element)
3. Delta rosette (three element)

The arrangements of all these rosettes have been shown in Fig. 12.12.

In a two element rosette, the two strain gauges are placed at 90° to each other as shown in Fig. 12.12 (a).

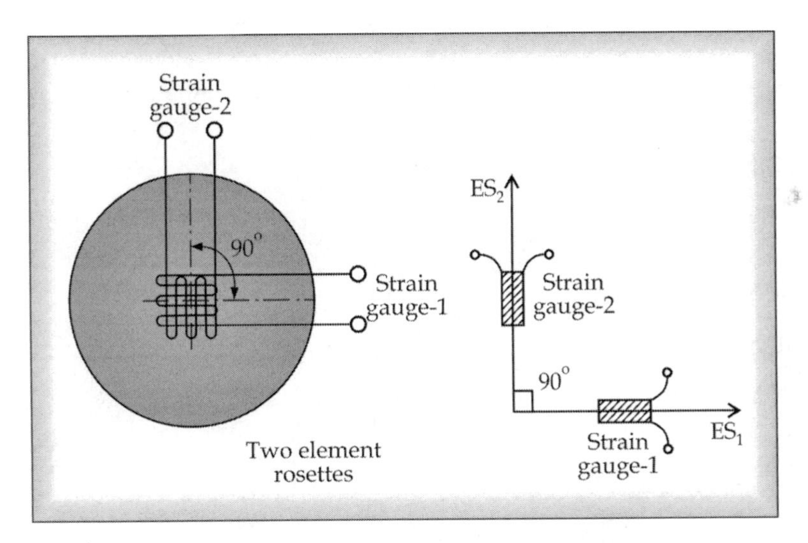

Fig. 12.12(a) Two Elements Rosette Principal Stresses

Principal strain = $\in_1$ or $\in_2$

Principal Stresses $\sigma_1 = \dfrac{E}{1-\nu^2}(\in_1 + \nu \in_2)$

$$\sigma_2 = \frac{E}{1-\nu^2}(\nu \in_1 + \in_2)$$

Maximum shear stress

$$\sigma_{max} = \frac{E}{2(1-\nu)}(\in_1 - \in_2)$$

where ν = Poisson's ratio

In a rectangular rosette, the three strain gauges are oriented as shown in Figure 12.12(b).

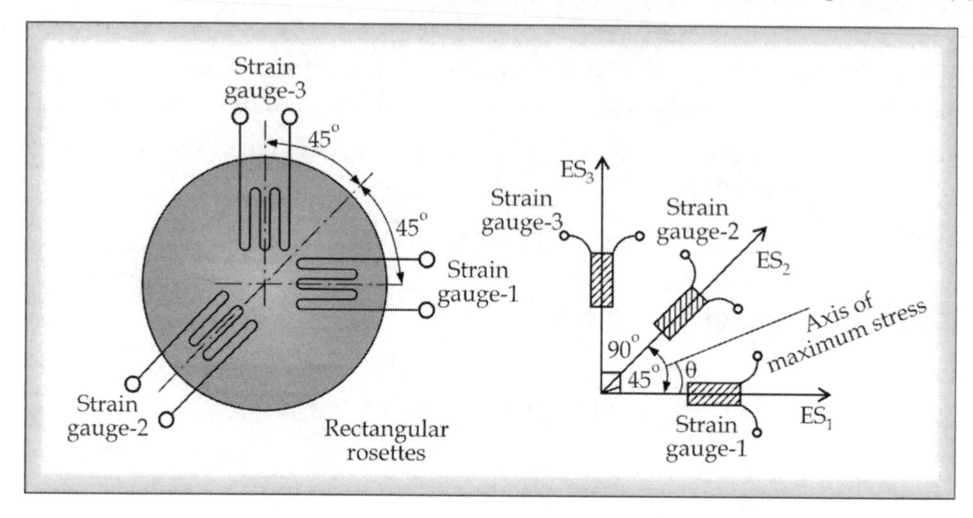

Fig. 12.12(b) Rectangular Rosette

The strains measured by the three strain gauges are $\in_1, \in_2$ and $\in_3$

Principal strains $\quad \in_{max}, \in_{min} = \dfrac{\in_1 + \in_3}{2} \pm \dfrac{1}{\sqrt{2}} Q$

where $\quad Q = \sqrt{(\in_1 - \in_2)^2 + (\in_2 - \in_3)^2}$

Principal Stress, $\quad \sigma_{max}, \sigma_{min} = \dfrac{E(\in_1 - \in_2)}{2(1-\nu)} \pm \dfrac{E}{\sqrt{2(1+\nu)}}.Q$

Max shear stress, $\quad \tau_{max} = \dfrac{E}{\sqrt{2(1+\nu)}}.Q$

More Instruments for Learning Engineers

Mile 12.3

Universal Testing Machine (UTM) is used to test/measure the tensile strength and compressive strength of materials. It is so named because it can perform many standard tensile and compression tests on materials, components, and structures. Further it can test, measure, record the deformation, strength of compression, tension, twist, torsion, shear etc.

These machines range from very small table top systems to ones with over 53 MN (12 million lbf) capacities.

Universal Testing Machine (UTM)

Orientation φ of principal Stresses,

$$\tan\ 2\varphi = \frac{2\in_2 - \in_1 - \in_3}{\in_1 - \in_3}$$

In a delta rosette, the three strain gauges are oriented as shown in the Fig. 12.12(c).

Principal Strains

$$\in_{max}, \in_{min} = \frac{1}{3}[(\in_1 + \in_2 + \in_3) \pm P]$$

where $$P = [2(\in_1 - \in_2)^2 + 2(\in_2 - \in_3)^2 + (\in_3 - \in_1)^2]^{1\ 2}$$

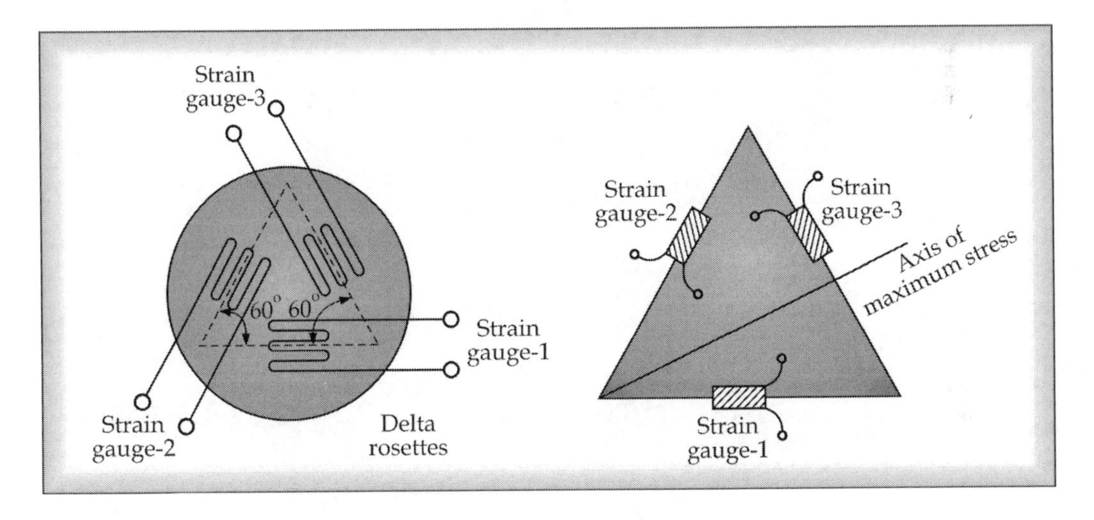

Fig. 12.12(c) Delta Rosette

Principal Stresses $$\sigma_{max}, \sigma_{min} = \frac{E}{3}\left[\frac{\in_1 + \in_2 + \in_3}{(1-\nu)} \pm \frac{1}{(1-\nu)}.P\right]$$

Maximum shear stress $$\tau_{max} = \frac{E}{\sqrt{2(1+\nu)}}.Q$$

Orientation τ of principal stress, $$\tan\varphi = \frac{[3(\in_3 - \in_2)]^{1/2}}{2\in_1 - \in_2 - \in_3}$$

ILLUSTRATION-12.4

Following observations were recorded for a rectangular rosette mounted on steel specimen; $\in_1 = (+)\ 800\ \mu m/m$, $\in_2 = (-)\ 100$ m/m, and $\in_3 = (-)\ 900\ \mu m/m$. Determine the principal strains, principal stresses and the location of principal planes. Assume $E = 200$ kN/mm^2 and $\nu = 0.3$.

Solution:

$$\in_{max}, \in_{min} = \frac{1}{2}[(\in_1 + \in_3) \pm \sqrt{2}\{(\in_1 - \in_2)^2 + (\in_2 - \in_3)^2\}^{1/2}]$$

$$= \frac{1}{2}[(800 - 900) \pm \sqrt{2}\{(800 + 100)^2 + (-100 + 900)^2\}^{1/2}]$$

$$= \frac{1}{2}[(-100 \pm \sqrt{2}\{(900)^2 + (800)^2\}^{1/2}]$$

$$= \frac{1}{2}[(-100 \pm 1702.9386]$$

$\in_{max}$ and $\in_{min}$ = 801.46932 and - 901.4693. $[(800 + 100) + (-100 + 900)^2]^{\frac{1}{2}} \times 10^{-6}$

$$\sigma_{max}, \sigma_{min} = \frac{200 \times 10^3 (800 - 900)}{2(1 - 0.3)} \pm \frac{200 \times 10^3}{\sqrt{2}(1 + 0.3)}$$

$$= -14285714 \pm 108785.66 \; [(900) + (800)]^{\frac{1}{2}} \times 10^{-6}$$

σ_{max} & σ_{min} = 116.7 N/mm^2 – and – 145.28 N/mm^2

$$\tan 2\theta = \frac{2\in_2 - \in_1 - \in_3}{(\in_1 - \in_3)} = \frac{2(-100) - (800) - (-900)}{800 - (-900)}$$

$$= \frac{-200 - 800 + 900}{800 + 900} = \frac{-1000 + 900}{1700}$$

$$\theta_{max}, \theta_{min} = -1.68^\circ \text{ or } 88.32^\circ$$

ILLUSTRATION-12.5

A delta rosette was used to determine the stress situation and the following observations were recorded = + 800, = 400, = 0 Determine the principle strains, principal stresses and the location of principal planes, assume elastic constants for aluminum specimen, E = 680 × 10^2 N/mm² and ν = 0.33.

Solution:

$$\in_{max}, \in_{min} \; \frac{(\in_1 + \in_2 + \in_3)}{3} \pm \frac{\sqrt{2}}{3}[(\in_1 - \in_2)^2 + (\in_2 - \in_3)^2 + (\in_3 - \in_1)^2]^{\frac{1}{2}}$$

$$= \frac{800 + 400 + 0}{3} \pm \frac{\sqrt{2}}{3}[(800 - 400)^2 + (400 - 0)^2 + (0 - 800)^2]^{\frac{1}{2}}$$

$$= \frac{1200}{3} \pm \frac{\sqrt{2}}{3}[(400)^2 + (400)^2 + (400 - 0)^2 + (0 - 800)^2]^{\frac{1}{2}}$$

$$= 400 \pm \frac{\sqrt{2}}{3}[160000 + 160000 + 640000]^{\frac{1}{2}}$$

$$= 400 + 461.81$$

$$\in_{max} \text{ and } \in_{min} = 861.81 \text{ and } -61.81$$

$$\tau_{max} \text{ and } \tau_{min} = \frac{E(\in_1 + \in_2 + \in_3)}{3(1-\nu)} \pm \frac{E}{(1+\nu)}\frac{\sqrt{2}}{3}[\in_1 - \in_2)^2 + \in_2 - \in_3)^2 + \in_3 - \in_1)^2]^{\frac{1}{2}}$$

$$= \frac{680 \times 10^2 (800 + 400 + 0)}{3(1-0.33)} \pm \frac{680 \times 10^2}{(1+0.33)} \times 461.81 \times 10^{-6}$$

$$= (40597015 \pm 23614917) \times 10^{-6}$$

$$= 64211932 \times 10^{-6} \text{ and } 16982098 \quad 10^{-6}$$

$$\therefore \sigma_{max} \ \& \ \sigma_{min} = 64.2 \text{ N/mm}^2 \text{ and } 16.88 \text{ N/mm}^2$$

$$\therefore \theta = -15^0 \text{ or } 75^0$$

ILLUSTRATION 12.6

A rectangular rosette was used to determine the stress situation in a certain experiment and following observation were recorded; $\in_1 = 900$ mm/m $\in_2 = 300$ mm/m, $\in_3 = 200$ mm/m, Determine the principal strains, principle stresses and the location of principal planes. Assume the elastic constants for steel, E = 200 Gpa, and $\nu = 0.3$.

Solution:

$$\in_{max},_{min} = \frac{1}{2}[\in_1 + \in_3 \pm \sqrt{2(\in_1 - \in_2)^2 + 2(\in_2 - \in_3)^2}]$$

$$\in_{max},_{min} = \frac{1}{2}\left[900 + (-200) \pm \sqrt{2(900-300)^2 + 2(300-200)^2}\right]$$

$$= \frac{1}{2}[700 \pm \sqrt{2(600)^2 + 2(500)^2}]$$

$$= \frac{1}{2}[700 \pm 100)\sqrt{72+50} = \frac{1}{2}[700 \pm 1104]$$

$$\in_{max} = \frac{1}{2}[700 + 1104] = -902 \text{ mm/m}$$

$$\in_{min} = \frac{1}{2}\ [700 - 1104) = -202 \text{ mm/m}$$

$$\tan 2\varphi = \frac{2(\in_2 - \in_1 - \in_3)}{\in_1 - \in_3} = \frac{2(300) - 900 + 200}{900 + 200}$$

$$\tan 2\varphi = \frac{-100}{+1100} = -0.0909$$

$$\tan 2\varphi = -5.19^{o}$$

$$\varphi = -206 \text{ or } 92.6^{o}$$

$$\sigma_{max}, \sigma_{min} = \frac{E}{2}\left[\frac{\epsilon_1 + \epsilon_3}{(1-v)} \pm \sqrt{2(\epsilon_1 - \epsilon_2)^2 + 2(\epsilon_2 - \epsilon_3)^2}\right]$$

$$= \frac{200 \times 10^9}{2}\left[\frac{700}{(1-0.3)} \pm \frac{1}{1+0.3}(1104)\right] \times 10^{-6}$$

$$= 100 \times 10^3 [1000 \pm 849]$$

$$\sigma_{max} = 1849 \times 10^5 \text{Pa}$$

$$\sigma_{min} = 151 \times 10^5 \text{Pa}$$

ILLUSTRATION-12.7

A delta rosette was used to determine the stress situation in a certain experiment and following observations were recorded. $\epsilon_1 = 600$ μm/m, $\epsilon_2 = 300$ μm/m, $\epsilon_3 = 400$ μm/m, determines the principal strains, principal stresses and the location of principal plane. Assume elastic constants for steel E = 200 Gpa and $v = 0.3$.

Solution:

$$\epsilon_{max}, \epsilon_{min} = \frac{}{3}\left[[(\epsilon_1 + \epsilon_2 + \epsilon_3)] \pm \sqrt{2(\epsilon_1 - \epsilon_2)^2 + 2(\epsilon_2 - \epsilon_3)^2 + 2(\epsilon_3 - \epsilon_1)^2}\right]$$

$$= \frac{1}{2}[(-600 + 300)] + 400 \pm \sqrt{2(-600 - 300)^2 + 2(300 - 400)^2 + 2(400 - 600)^2}$$

$$= \frac{1}{2}[100 \pm 100]\sqrt{162 + 2 + 200} = \frac{1}{3}[100 \pm 1900]$$

$$\epsilon_1 = \frac{1}{2}[(100 \pm 1900)] = 666.7 \ \mu\text{m/m}$$

$$\epsilon_2 = \frac{1}{2}[(100 - 900)] = -600 \ \mu\text{m/m}$$

$$\tan^2 \varphi = \frac{\sqrt{3}(\epsilon_3 - \epsilon_2)}{(2\epsilon_1 - \epsilon_2 - \epsilon_3)} = \frac{\sqrt{3}(400 - 300)}{(-2 \times 600 - 300 - 400)} = \frac{-\sqrt{3} \times 100}{1900}$$

$$\tan 2\varphi = -5.2^{o}$$

$$\varphi = +2.6^{o} \text{ as } \epsilon_c > \epsilon_b >>, 0 < \varphi < 90^{o}$$

$$\sigma_{max}, \sigma_{min} = \frac{E}{3}\left[\frac{(\in_1 + \in_2 + \in_3)}{(1-v)} \pm \frac{1}{(1+v)}\sqrt{2(\in_1 - \in_2)^2 + 2(\in_2 - \in_3)^2 + 2(\in_3 - \in_1)^2}\right]$$

$$= \frac{200\times10^9}{3}\left[\frac{100}{(1-0.3)} \pm \frac{1}{1.3}\times1900\right]\times10^{-6} \text{ micro strain.}$$

$$= \frac{200\times10^3}{3}[142.8 \pm 1461.5]$$

$\sigma_{max} = 1609 \times 10^5 Pa$

$\sigma_{min} = -878.7 \times 10^5 Pa$

12.10 WHEATSTONE BRIDGE CIRCUIT

Wheatstone bridge circuit is the best way of measuring change in resistance (i.e., indirectly strain in this case). The most common Wheatstone bridge is the "Null balanced bridge", which is explained as below:

The null balanced bridge arrangement is shown in Figure 12.13. It consists of a resistance R_1 (strain in this case) a variable resistance R_2 (Rheostat) and two standard resistance R_3 and R_4 whose resistance are known. (R_1 is the strain gauge in this case). "When Resistance R_1 is not strained", the resistance arrangement is such that the potential at B equals the potential at D, and the galvanometer gives zero deflection, i.e., no current is indicated by the galvanometer connected between the bridge points.

i.e., $$\frac{R_1}{R_2} = \frac{R_3}{R_4}$$

Therefore, $$R_1 = R_2\frac{R_3}{R_4}$$

"When Resistance R_1 is Strained", its resistance will change by an amount dR_1. This change in resistance (dR_1) will unbalance the bridge causing a deflection of the galvanometer. Now the resistance R_2 (Rheostat) is adjusted by an amount $dR_2$to bring back the bridge to its balanced position (Null position).

The rebalance condition gives: $$\frac{R_1 + dR_1}{R_2 + dR_2} = \frac{R_3}{R_4}$$

i.e., $$R_1 + dR_1 = (R_2 + dR_2)\frac{R_3}{R_{4'}},\ R_1 + dR_1 = R\frac{R_3}{R_4}2 + dR_2\frac{R_3}{R_4}$$

But, $$R_1 = R_2\frac{R_3}{R_4}$$

Therefore, $$R_1 + d R_1 = R_1 + dR_2 \frac{R_3}{R_4}$$

R_1 gets cancelled on both sides

Hence, $$dR_1 = dR_2 \frac{R_3}{R_4}$$

If all resistances are equal in the bridge (that is, $R_1 = R_2 = R_3 = R_4$) then, $dR_2 = dR_1$. Thus the change in the value of resistance R_2 is a direct measure of strain. The null balance bridge is used to measure static strain and is an accurate method to measure changes in resistance.

12.10.1 One Active Strain Gauge (Quarter Bridge) or Tension Measurement using Strain Gauge

Consider a single strain gauge mounted on a cantilever beam. When a force is applied to a beam the gauge is subjected to tensile stress. Therefore the resistance of the strain gauges increases. This increases in resistance is measured by using a Wheatstone bridge as shown in Fig. 12.13.

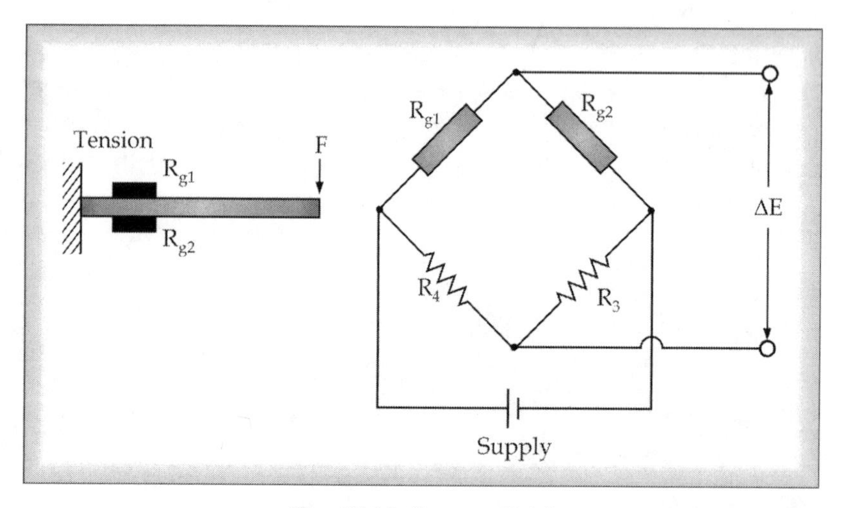

Fig. 12.13 Quarter Bridge

The output voltage ΔE due to tensile strain ε causing R_1 to change to $(R_1 + DR_2)$ is,

$$\Delta E = \frac{VR_1R_2}{(R_1 + R_2)}\left(\frac{\Delta R_1}{R_1} - \frac{\Delta R_2}{R_2} + \frac{\Delta R_3}{R_3} - \frac{\Delta R_4}{R_4}\right)$$

where, $R_1 = R_g = R_2 = R_3 = R_4$

and $DR_2 = DR_3 = DR_4 = 0$

$$\Delta E = \frac{VR_1^2}{4R_1^2}\frac{\Delta R_1}{R_1} = \frac{V}{4}\frac{\Delta R_1}{R_1}$$

Gauge factor, $F = \frac{\Delta R / R}{\varepsilon}$, $\Delta E = \frac{V}{4} F \in$

If R_2 used is dummy gauge temperature compensation is achieved.

12.10.2 Two Active Strain Gauges (Half Bridge) or Tension and Compression Measuring Strain Gauges

Two active strain gauges are mounted on a cantilever as shown in Fig. 12.14.

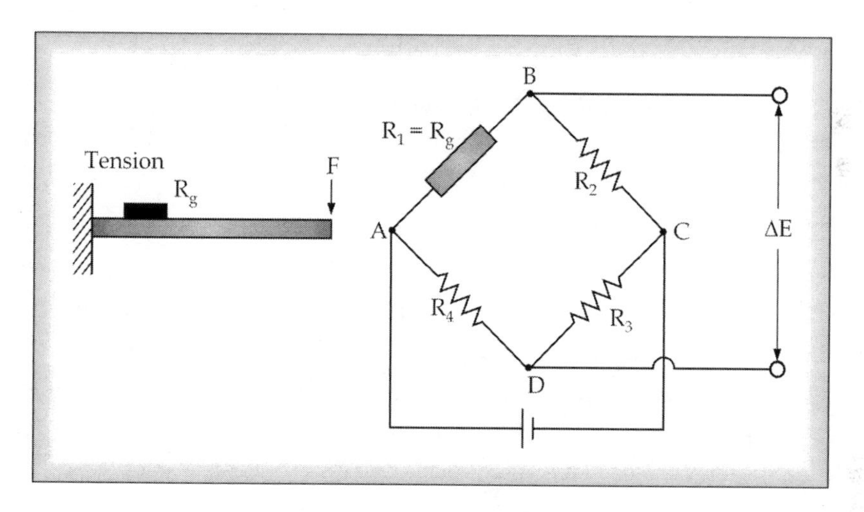

Fig. 12.14 Half Bridge

Each gauge is situated at an equal distance from the neutral axis is subjected to strain e of equal magnitude but of opposite sign. The gauge R_{g1} being in tension and R_{g2} in compression and are connected in adjacent arms. Two fixed resistances are connected in the arm 3 and 4 to complete the bridge.

$$R_1 = R_2 = R_3 = R_4 = R, \Delta R_3 = DR_4 = 0$$

$$\Delta E = \frac{V}{4}\left[\frac{\Delta R}{R} - (-)\frac{\Delta R}{R}\right]$$

As $\qquad DR_1 = DR_2, \qquad DR_1 = \Delta R, \; \Delta R_2 = DR$

$$\Delta E = \frac{V}{4}\frac{(2\Delta R)}{R}; \quad \Delta E = \frac{V(\Delta R)}{2R}$$

Gauges R_{g1} and R_{g2} are active gauges, their output due to mechanical strain add up being of opposite sign and connected in adjacent arms, whereas, temperature induced strains cancel, being of same sign and equal in magnitude.

12.11 CALIBRATION

A theoretical objection to the use of strain gauges is that, in most applications, it is impossible to check the accuracy of the readings obtained from them. Once the strain gauge, is bonded in its measuring position, it cannot be removed or transferred and subjected to a known strain for the calibration. Strain gauges are often used in applications where no other form of strain measurement is possible. Since no check on performance can be made, the value of the gauge factor as specified by the manufacturer has to be relied upon. However, this leads to errors because the value of gauge factor is specified with certain tolerance limits and hence each gauge should be individually calibrated.

The procedure for calibration of strain gauges: Null balancing bridges used in strain measurement employ a variable resistor having a scale calibrated in terms of strain. The amount of apparent strain to balance the bridge may then be read directly from the scale. The bridge is usually balanced by the methods of apex and shunt balancing.

Apex balancing: The arrangement has a potentiometer connected at one apex of the bridge circuit as shown in Figure 12.15. The resistance of this potentiometer is very small compared with the resistance value of the strain gauge. When the wiper of the pentameter is moved, the resistance of one limb of the bridge increases with a simultaneous decrease in the resistance of the adjoining limb. This aspect helps to balance the bridge and compensate for the difference in the resistance values of the gauges and the associated fixed resistors.

Shunt Balancing: The arrangement employs resistors placed in parallel (shunted across) with the strain gauges as shown in Fig. 12.15. The resistance values of these resistors are very high compared to the strain gauge resistance values. Evidently these parallel resistors produce only a small change in the total resistance of the bridge arm in which they are located.

More Instruments for Learning Engineers

Brinell scale, proposed by Swedish engineer Johan August Brinell (1900), measures hardness of materials through a scale of penetration of an indenter, loaded on a material test-piece. The test uses a 10 mm (0.39 in) steel ball as an indenter with 3,000 kgf (29.42 kN; 6,614 lbf) force. For softer materials, a smaller force is used; for harder materials, a tungsten carbide ball is used. The indentation is measured and hardness calculated as: $BHN = \dfrac{2P}{\pi D(D - \sqrt{D^2 - d^2})}$ where: P = applied force (kgf); D & d are diameters of indenter and indentation (mm) respectively. The two other equipment used for measuring hardness are Rockwell and Vicker's Hardness tester. The Hardness is measured in hardness numbers of respectives scales named after the founders, BHN or RHN and VHN.

Hardness Tester

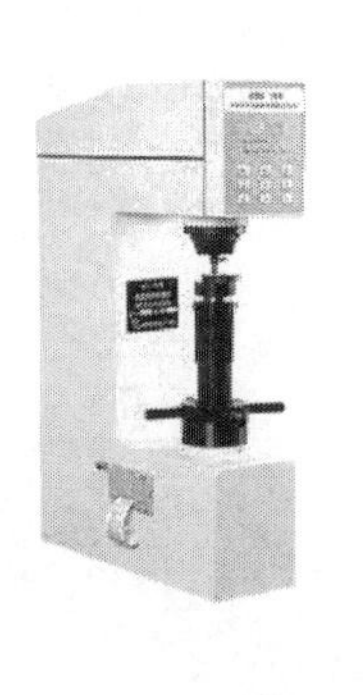

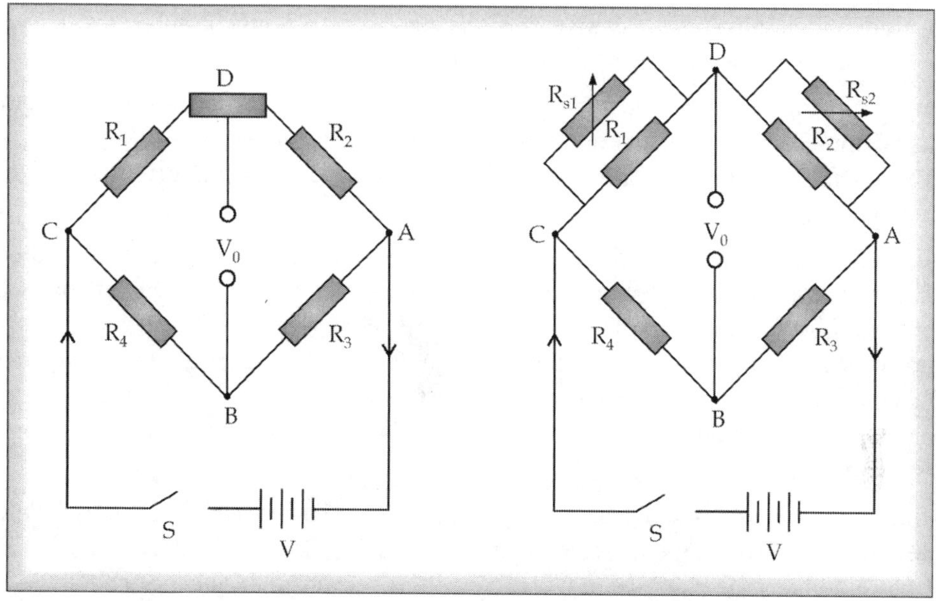

Fig. 12.15 APEX and Shunt Balancing

12.12 APPLICATIONS OF STRAIN GAUGES

The strain gauges are used in

1. Force measuring devices (such as strain gauge load cell)
2. Measurement of vibration / acceleration (strain gauge accelerometer)
3. Measurement of pressure.

SELF ASSESSMENT QUESTIONS-12.4

1. What is rosette? Describe different types of rosettes with neat sketches.
2. Explain the procedure for calibration of strain gauges using null balancing techniques.
3. Describe the calibration of strain gauges using following methods.
   (a) Apex balancing (b) Shunt Balancing
4. Give the applications of strain gauges.
5. Draw the circuit diagram of One Active Strain Gauge (Quarter Bridge) and explain its operation.
6. With the help of circuit diagram explain the principle of Two Active Strain Gauges (Half Bridge)

SUMMARY

Strain is the change in length per the original length while stress is defines as the force or load per unit area. According to Hook's law, stress is proportional to strain within the elastic limits and the ratio of stress to strain is modulus of elasticity. Depending on the type of stress and strain there are three (Young's, Rigidity and Bulk) modulus. Also, the ratio of lateral to linear strain is Poisson's ratio. Positive Strain occurs if strain gauge is subjected to tension while it is negative strain in case of compression.

Kelvin demonstrated the electrical resistance strain gauge using Wheatstone bridge circuit with a galvanometer as the indicator. In the resistance strain gauges, when a length of wire/ foil is mechanically stretched, a longer length of smaller cross section results, hence the electrical resistance changes.

In un-bonded strain gauges (not directly pasted on the structure), or in fine wire/ metal foil/ semi-conductor bonded strain gauges (directly pasted on the structure) the applied load varies the dimensions of a strain gauge, hence proportionally changes the electrical resistance. The selection of material, fabrication, installation, mounting and performance of bonded metallic strain gauges largely depend on 6-parameters (BEG-BIG) viz, Bonding material, Electrical circuitry, Gauge protection, Bonding technique, Installation method & Backing material, Grid material configuration.

The strain measured is based on change in the resistance of the gauge, which may be due to change in dimension or temperature. So, to account for the variations in the resistance due to temperature, compensation may be provided by using adjacent arm balancing or compensating gauge(s) such as using a dummy gauge, Poisson's method or two active gauges on a cantilever. Another method is self temperature compensation with selected melt gauge or dual element gauge or by using external control circuitry.

Rosettes is a combination of 2-element or 3-element (Rectangular/Delta) insulated strain gauges placed at specific angles to measure the values of principal strains and stresses without actually knowing the direction of principal stresses.

Wheatstone bridge circuit is the best way to measure change in resistance (thence strain), operation based on. The most common is the null balanced bridge. One/two active strain gauge (Quarter/half Bridge) is mounted on a cantilever beam to find the increase in resistance. It is impossible to check the accuracy (change due to gauge factor) of the bonded strain gauge as it cannot be removed or transferred and subjected to a known strain for the calibration. So each gauge should be individually calibrated. Null balancing bridges used in strain measurement employ a variable resistor having a scale calibrated in terms of strain. Apex balancing uses potentiometer connected at one apex of the bridge circuit while Shunt balancing employs resistors placed in parallel (shunted across) with the strain gauges.

KEY CONCEPTS

Elongation: The change in length due to stress.

Strain: The change in length (or elongation) divided by the original length $\varepsilon = \dfrac{\Delta l}{l_1}$.

Stress: The force or load per unit area Stress $\sigma = \dfrac{P}{A} = \dfrac{P}{\pi d_1^2/4}$.

Hook's Law: Stress is directly proportional to the strain within the elastic limits.

Young's Modulus: The constant of elasticity and is the ratio of stress to strain within elastic limits.

$$E = \frac{\text{Stress } \sigma}{\text{Strain } \varepsilon}$$

Poisson's Ratio: The ratio of lateral strain to axial strain. $\nu = \frac{\text{Laternal strain}}{\text{Axial strain}}$

Base length: The length over which the average strain measurement is taken.

Deformation sensitivity: The minimum deformation that can be indicated by the gauge.

Strain sensitivity: The minimum deformation that can be indicated by the gauge per unit base length.

Positive Strain: When a strain gauge is subjected to tension (length increases and its area of cross-section decreases).

Negative Strain: When a strain gauge is subjected to compression (length decreases and its area of cross section increases).

Electrical Resistance Strain Gauge: The resistances of copper wire and iron wire change when the wires are subjected to mechanical strain and is measured using a Wheatstone bridge circuit with a galvanometer as the indicator.

Metallic Resistance Strain Gauge: When a length of wire/foil is mechanically stretched, a longer length of smaller cross section results, hence the electrical resistance changes.

Unbonded Strain Gauge: Not directly pasted onto the surface of the structure. The applied load varies the dimensions of a strain gauge, and hence the electrical resistance.

Bonded Strain Gauges: Directly pasted on the surface of the structure under study; load varies the dimensions, hence the electrical resistance of a strain gauge.

Fine Wire Strain Gauge: A resistance wire of diameter 0.025 mm is bent again and again is subjected to load and the change is dimension, hence the resistance is measured.

Metal Foils Strain Gauge: Metal foil of 0.02 mm thick is used and is subjected to load and the change is dimension, hence the resistance is measured.

Semi-Conductor or Piezo Resistive Strain Gauges: The sensing element is a rectangular filament with a plastic or stainless steel backing made like a wafer from silicon or germanium crystals doped with boron

Grid/backing/bonding Materials properties: High electrical stability, yield strength, gauge factor, resistivity, endurance limit, workability, corrosion resistance, solderability or weld-ability and low hysteresis, thermal emf and temperature sensitivity.

Backing, or Carrier Material purpose: Providing support, protection and dimensional stability to the resistance wire (grid) of the strain gauge arrangement.

BEG-BIG factors of strain gauge materials: **B**onding material, **E**lectrical circuitry, **G**auge protection, **B**onding technique, **I**nstallation method & Backing material, **G**rid material configuration.

Temperature Compensation: Compensation for the change in resistance due to increase in temperature.

Temperature Compensating methods: Dummy gauge, Poisson's method, two active gauges on a cantilever, self temperature compensation with selected melt gauge or dual element gauge, and external control circuitry.

Rosette: Combination of two or three strain gauges placed at specific angles with respect to each other to measure the values of principal strains and stresses without actually knowing the direction of principal stresses e.g. Two element and 3-element (Rectangular/Delta) rosette.

Wheatstone bridge principle: $\frac{R_1}{R_2} = \frac{R_3}{R_4}$

Calibration of strain gauges: Null balancing bridges used in strain measurement employ a variable resistor having a strain calibrated scale.

Apex balancing: The potentiometer connected at one apex of the bridge circuit.

Shunt Balancing: Employs resistors placed in parallel (shunted across) with the strain gauges.

REVIEW QUESTIONS

Short Answer Questions

1. Name the different types of strain gauges used.
2. What is the need for rosette?
3. Mention the applications of the strain gauges.
4. Draw a Wheatstone bridge circuit for the strain measurement.
5. Define gauge factor of a strain gauge.
6. Define the terms gauge factor of strain gauge, and strain gauge, rosette. How are they useful in stress measurement?
7. What is the purpose of strain rosette?
8. Write short notes on rosettes.
9. How resistive strain gauges are calibrated?

Long Answer Questions

1. Deduce an expression relative change in resistance and strain in a strain gauge.
2. Discuss the salient features and applications of electrical strain gauges.
3. List the various factors that are considered for the selection of metallic strain gauges.
4. Describe the tension measurement using strain gauge with neat diagram.
5. Describe the working principle and theory of operation of a strain gauge.
6. Describe the working principle of strain gauge bridge with neat sketch. Indicate their arrangements for measurement of torque on a circular shaft.
7. How can a strain gauge be used for measuring torque? Explain.
8. Explain the method of calibration of a resistance strain gauge.
9. How temperature effects are eliminated in the measurement of strains using resistance wire gauges? Discuss.
10. Derive an expression for the gauge factor. Draw the sketches showing the arrange-ment of gauges to measure: (i) Unidirectional stress, and (ii) Bidirectional stress.
11. What are the requirements of materials for strain gauges?
12. Explain the method of measuring force using strain gauges.

13. Why bridge circuit is necessary for a strain gauge? Explain how the bridge circuit is used with a strain gauge?
14. Mention the important requirements of a strain gauge material. Name some of the materials used for making strain gauges and their desirable properties.

MULTIPLE CHOICE QUESTIONS

1. The unit of stress is __________.
   (a) N (b) N/mm^2
   (c) mm (d) no units
2. The units of strain is__________.
   (a) N (b) N/mm^2
   (c) mm (d) no units
3. Hooks law is applied___________.
   (a) within the elastic limit
   (b) beyond the elastic limit
   (c) up to the elastic limit
   (d) after the breaking point
4. Poisson's ratio is defined as _______.
   (a) the ratio of Lateral Strain to Linear Strain
   (b) the ratio of Linear Strain to Lateral Strain
   (c) sum of Lateral Strain And Linear Strain
   (d) difference of lateral and linear Strain
5. The length over which the average strain measurement is taken is called_________.
   (a) base length (b) average length
   (c) total length (d) zero length
6. Positive strained is said to be occurred on a specimen when __________.
   (a) length decreases and area of cross section increases
   (b) length increases and area of cross section decreases
   (c) it has no effect on change in length
   (d) it has no effect on change in cross section area
7. In un-bounded strain gauges the change in Dimensions of Strain gauge is calibrated with the change in_____.
   (a) EMF (b) temperature
   (c) resistance (d) pressure
8. In N-Type of sensing element. the resistance _____ with respect to tensile strain
   (a) decreases
   (b) increases
   (c) has no effect
   (d) first decreases then increases
9. Which of the following properties are desirable in the grid material
   (i) high electrical stability
   (ii) high yield strength
   (iii) low hysteresis
   (a) only i (b) i & ii
   (c) i, ii & iii (d) ii & iii
10. With the change in temperature the change in length is given as ________.
    (a) $l\alpha DT$ (b) $l\alpha$
    (c) $l\alpha - DT$ (d) $l\alpha + \Delta T$
11. The stress on the Principal Plane is called ______ Stress
    (a) maximum Tensile
    (b) maximum Compressive
    (c) principal
    (d) shear
12. While measuring pure torsion, the strain gauges are bonded to the shaft _______.
    (a) in the direction of the applied torque
    (b) longitudinally along the axis of the shaft.
    (c) transversally along the axis of the shaft.
    (d) at 45° to axis of rotation

13. Dummy strain gauge
    (a) is standard strain gauge connected in limbs that lie opposite to active gauges.
    (b) undergoes resistance change due to temperature as well as mechanical strain
    (c) is bonded to an unstrained component whose resistance changes due to temperature only.
    (d) does not have any change in resistance with temperature.
14. Mechanical strain gauges can measure_______
    (a) static strains only
    (b) dynamic strains only.
    (c) static and quasi-static strains.
    (d) static and dynamic strains.
15. Which of the following gauges can be detached from the test specimen and used again?
    (a) semi-conductor gauges
    (b) metal foil gauge
    (c) bonded metal wire gauges
    (d) unbounded metal wire gauge
16. Any change in the resistance of strain gauge due to temperature variation is compensated by____ which are connected in limbs adjacent to those containing ______
    (a) active gauges, dummy gauges
    (b) active gauges, active gauges only
    (c) dummy gauges, active gauge
    (d) dummy gauges, dummy gauges only
17. The ratio of fractional change in resistance to mechanical strain is called ______
    (a) bulk modulus
    (b) strain Rosette
    (c) gauge factor
    (d) temperature Gradient
18. The bridge constant (sensitivity) of a half-bridge arrangement when two strain gauges suffer equal and opposite strains, is
    (a) unity (b) doubled
    (c) quadrupled (d) halved
19. Strain gauge rosettes are used when the direction of _________
    (a) principal stresses are not known
    (b) circumferential stress is not known
    (c) principal stresses are known
    (d) circumferential stress are known
20. An electrical strain gauge is not sensitive to _________ strain
    (a) shear (b) compressive
    (c) tensile (d) any
21. The dimensional formula of stress is same as that of
    (a) torque (b) force
    (c) pressure (d) work
22. Which of the following pairs has same units and dimensions
    (a) torque & Force
    (b) shear Force & bending moment
    (c) pressure & stress
    (d) work and Force
23. Units of strain is
    (a) mm (b) cm
    (c) inches (d) no units
24. Hooke's limit is applicable up to
    (a) yield point (b) elastic limit
    (c) plastic limit (d) breaking point
25. Hooke's law is
    (a) stress a strain
    (b) pressure a 1/velocity
    (c) Pressure a 1/volume

ANSWERS

1.	(b)	2.	(d)	3.	(a)	4.	(a)	5.	(a)
6.	(b)	7.	(c)	8.	(a)	9.	(c)	10.	(a)
11.	(c)	12.	(d)	13.	(c)	14.	(d)	15.	(d)
16.	(c)	17.	(c)	18.	(b)	19.	(a)	20.	(a)
21.	(c)	22.	(c)	23.	(d)	24.	(b)	25.	(a)

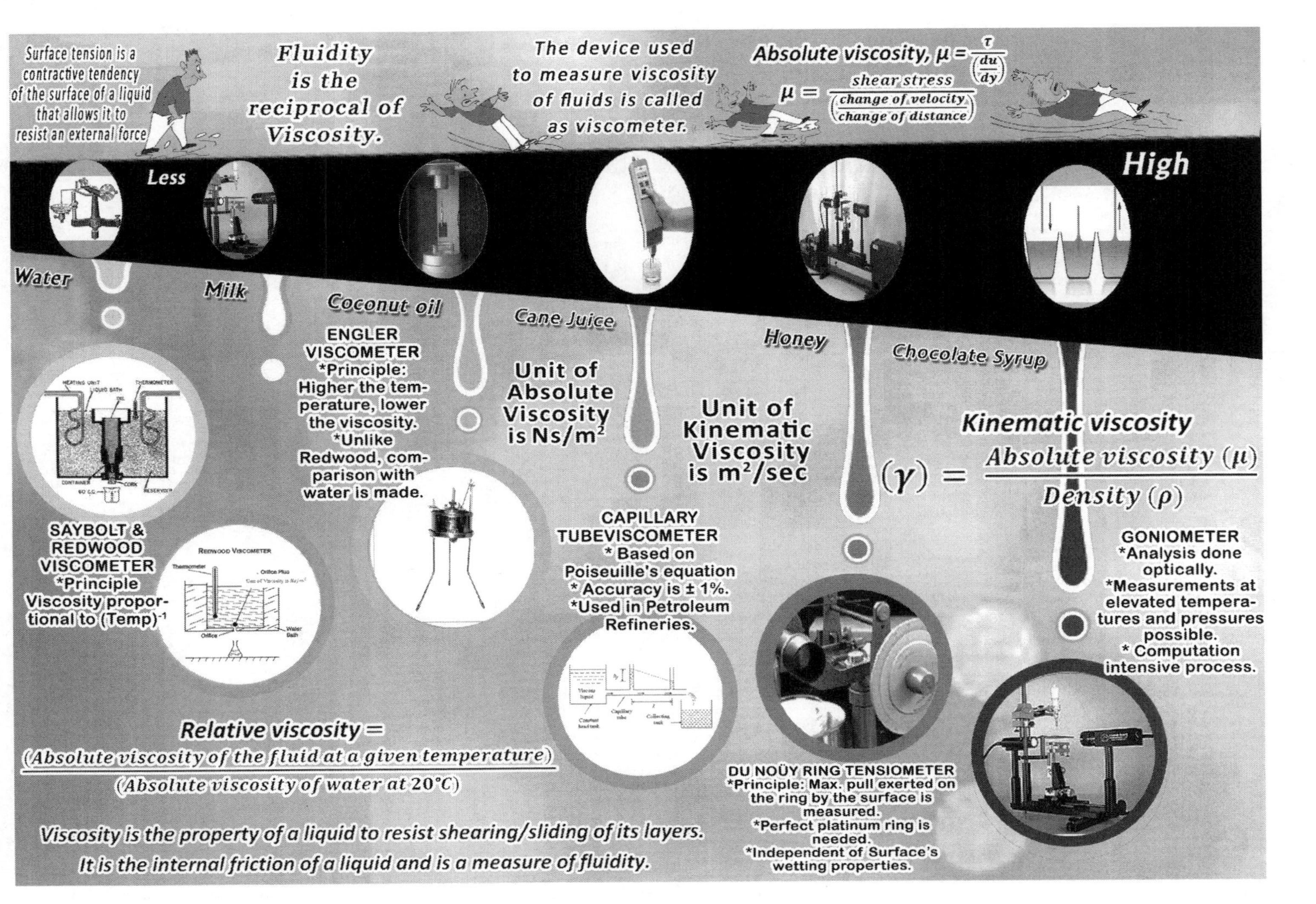

CHAPTER INFOGRAPHIC

Glimpses of Measurement of Viscosity and Surface Tension

MEASUREMENT OF VISCOSITY

Viscometers	Principle	Construction	Operation
Capillary tube viscometer	Poiseuille's equation		Liquid flowing through capillary tube drops at free end gives head loss (h_f) and time taken to fill flask is noted to compute flow rate
Efflux Viscometers: Redwood Viscometer	Flow Time at const. temp. varies with capillary	Level indicator flask, heating-coil, stirrer and thermometers are fitted to cylinder and bath to measure temperature constantly.	Level indicator filled with 50ml (test liquid) in Temp controlled bath is brought to t_r°C and heat transfers from bath to cylinder. When thermometers and bath show t_r°C, orifice is opened and time of flow thru' orifice indicates viscosity in Redwood seconds.
Efflux Viscometers: Saybolt Viscometer		60 ml collecting flask, heating-coil, stirrer amd thermometers are fitted on tank and bath to note temperature constantly	Temperature in controlled bath is brought to t_s°C; heat transfer takes place from the bath to tank. Thermometers immersed in liquid and bath shows t_s°C. Capillary is opened; time taken to collect liquid is noted indicates viscosity in Saybolt universal seconds.
Efflux Viscometers: Engler Viscometer		A shallow cup fitted with a standard orifice, level indicator, temperature bath	Temperature in bath and shallow cup are at t_e°C, orifice is opened and time taken for liquid to flow is noted. Temp. in bath is adjusted till a const. temp. is 20 °C in bath and shallow cup. Orifice is opened and time taken for 200 ml of water at 20 °C to flow is noted
Falling Sphere Viscometer	Steel ball fall Time varies with abs. viscosity	Graduated tube, funnel, temperature bath with electric heating-coil, thermometers and stirrer.	Thermometers immersed in liquid and bath show same temp. A steel ball is dropped to move through a distance of 150 or 200 mm to enable calculation of terminal velocity. Steel ball gets v_{max} when gravity force is balanced by buoyant and viscous drag
Rotating Cylinder Viscometers			
Rotating Cylinder Viscometer (Concentric Cylinders Type)	Torque varies with viscosity ($\tau \alpha \mu$)	Two concentric (inner and outer) inner cylinder is stationary and outer cylinder can rotate at a const. angular speed	Outer cylinder is rotated at a constant angular speed, a torque is developed at fixed end of the inner cylinder and is measured.
Rotating Cylinder Electrical Viscometer	Speed Lag of cylinder varies with viscosity ($\varpi \alpha \mu$)	Container with liquid flows continuously. Fitted with cylinder connected to calibrated spring & synchronous motor.	Cylinder connected to synchronous motor and rotates at const. angular speed. that lags as compared to that of synchronous motor which is directly proportional to the viscosity of liquid.

MEASUREMENT OF SURFACE TENSION

Instrument	Principle	Construction	Operation	Merits	Demerits
Du Noüy Ring Tensiometer	Maximum pull exerted on ring by surface is measured.	A platinum ring submersed in liquid.	As the ring is pulled out of the liquid, the tension required is precisely measured to determine the surface tension of the liquid.	Wetting properties are less	Inaccurate compared to the plate method
Du Noüy-Padday Tensiometer	Force required to pull the rod is proportional to surface tension	Small dia metal rod in test liquid connected to high sensitivity microbalance to record max. pull.	Rod is pulled out of liquid and the force required to pull the rod is precisely measured.	Very quick and minimal about 20 sec.	
Wilhelmy Plate Tensiometer	Force due to wetting is measured.	Vertical plate is attached to balance to make contact with liquid surface.	Force required to wet is proportional to surface tensions and so measures surface tension when calibrated.	Simple and more accurate	
Bubble Pressure Tensiometer	Max. pressure of each bubble is measured with Jaeger's bubble pressure method.	A capillary, bubble pressure is higher than in surroundings (water).	Air bubbles within liquids are compressed to rise and decrease bubble radius. Gas stream pumped in capillary makes bubble at tip continually bigger in surface gets smallest radius and forms hemisphere. Characteristic pressure pattern is evaluated	Easy handling and low cleaning effort	
Goniometer/ Tensiometer	Geometry of drop or angle of contact is analyzed optically		Profile of drop produced is captured by goniometer/tensiometer. Software analyses profile and makes a series of calculations using Young-Laplace equation	Measured in elevated temp. and pressures	

Chapter 13

Measurement of Viscosity and Surface Tension

STARTERS

To study this chapter, you should have awareness on the following concepts. For a better understanding, it is always a good idea to revise these prerequisites.

- Viscosity, its types (of expressing) and its related fundamental definitions
- Surface tension and its effects such as formation of spherical shape (in liquids), wine tears, fogging, mist, movement of water striders etc.
- Fluid properties such as fluidity, flow characteristics etc.
- Mass, Volume, Density, Specific gravity, acceleration due to gravity, force and related terms
- Meaning and concepts of buoyancy, capillarity, surfactants, adhesion, cohesion, meniscus etc.
- Awareness on floating, bubble formation, purging etc.
- Young-Laplace equation (surface tension) and its application
- Stoke's law, its applications and effects
- Contributions of Redwood, Saybolt, Engler and others
- Contributions of Pierre Lecomte du Nouy, Wilhelmy, Jaeger and others.
- Fundamentals of fluid mechanics and
- First four chapters of this book.

LEARNING OBJECTIVE

After studying this chapter you should be able to

- Define viscosity and various relevant terms
- Classify and explain various types of Viscometers,
- Define surface tension and its effects,
- Explain various types of Tensiometers and
- Discuss various measuring methods of surface tension.

13.1 INTRODUCTION

Why do we skid when oil is spilled on a floor surface? Why do some liquids (say water) have free flow while some other (like tar) do not? How can we measure this fluidity or flow rate? How does the density of fluids affect its fluidity? Why some objects though have higher density than the liquid, still float on the liquid?

Well! The above questions can be answered easily by understanding the terms viscosity and surface tension. The fluid flow rate and fluidity of a fluid through any part depends on several factors of which viscosity of the fluid is most significant. The flow rate of the fluid or fluidity can be analyzed and explained easily by using the concept viscosity. Particularly, the effects of temperature on fluidity can be well explained by understanding the concept that the viscosity of a fluid varies inversely with the temperature (i.e. lower the temperature, higher will be the viscosity and vice-versa). Similarly, the surface tension is yet another property of liquids that has several effects on the liquid behavior. The common terms with respect to viscosity and surface tension and their methods of measurement and the relevant instrumentation are discussed in this chapter.

13.2 THE JARGON OF VISCOSITY

1. *Viscosity*: Viscosity is the property of a liquid to resist shearing/sliding of its layers. Viscosity is the internal friction of a liquid and is a measure of fluidity.
2. *Fluidity*: Fluidity is the reciprocal of viscosity.
3. *Absolute Viscosity (Dynamic Viscosity or Viscosity Coefficient)*

According to Newton's law: $\tau \propto \frac{du}{dy}$

Therefore $\tau = \mu . \frac{du}{dy}$,

where du = change of velocity between two layers of liquid.

dy = change of distance between two layers.

More Instruments for Learning Engineers

Profilometer is used to measure a surface's profile, in order to quantify the roughness. The historical notion of a profilometer reveals a device similar to a phonograph that measures a surface as the surface is moved relative to its contact stylus. Development of non-contact profilometery changed this notion. The profilometer applications can also be extended to many engineering aspects such as to measure gear tooth profile, cam profile etc. Road pavement profilometers use a distance measuring laser along with odometer & accelerometer. The data is used to International Roughness Index (IRI) in inches/mile or mm/m [range: 0 (smoothest) to several hundred (roughest)]. IRI value is used for road management to monitor road safety &quality. Many road profilers can measure the pavement cross slope, curvature, longitudinal gradient and rutting.

Profilometer

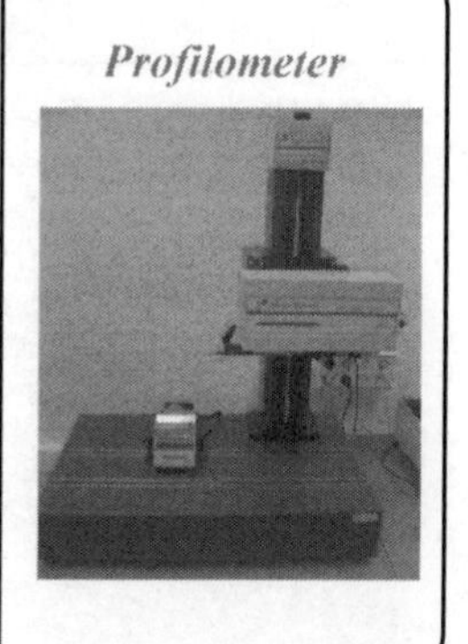

μ = Absolute viscosity (which is a constant of proportionality).

τ = Shear stress between the two layers of liquid.

Unit of absolute viscosity

We have, $\mu = \dfrac{\tau}{\left(\dfrac{du}{dy}\right)} = \dfrac{\text{Shear stress}}{\left(\dfrac{\text{Change of velocity}}{\text{Change of distance}}\right)}$

$$\frac{\left(\dfrac{\text{Force}}{\text{Area}}\right)}{\left(\dfrac{\text{Length}}{\text{Time}} \times \dfrac{1}{\text{Length}}\right)} = \frac{\left[\dfrac{\text{Force}}{(\text{Lenght}^2)}\right]}{\dfrac{\text{Length}}{\text{Time}} \times \dfrac{1}{\text{Length}}}$$

$$\frac{\text{Length} \times \text{Time}}{(\text{Length}^2)}$$

$$\frac{\text{Newton.second}}{\text{meter}^2}; \text{ or N s/m}^2$$

4. ***Kinematic Viscosity***

Kinematic viscosity $(\gamma) = \dfrac{\text{Absolute viscosity } (\mu)}{\text{Density } (\rho)}$

Unit of kinematic viscosity:

Kinematic viscosity

$$(\gamma) = \frac{\text{Absolute viscosity}}{\text{Density}} = \frac{\text{Force} \times \dfrac{\text{Time}}{\text{Length}^2}}{\dfrac{\text{Mass}}{\text{Length}^3}} = \frac{\text{Force} \times \text{Time}}{\text{Mass / Length}}$$

$$\frac{(\text{Mass} \times \text{Acceleration}) \times \text{Time}}{\text{Mass / Length}} = \frac{\text{Acceleration} \times \text{Time}}{1/\text{Length}} = \frac{\left(\dfrac{\text{Length}}{\text{Time}^2}\right) \times \text{Time}}{\dfrac{1}{\text{Length}}}$$

$$= \frac{\text{Length}^2}{\text{Time}}; \text{ or } \text{m}^2/\text{sec}$$

5. *Relative Viscosity*

$$\text{Relative viscosity} = \frac{(\text{Absolute viscosity of the fluid at a given temperature})}{\text{Absolute viscosity of water at 20 °C}}$$

6. *Specific Viscosity*

$$\text{Specific Viscosity} = \frac{(\text{Absolute viscosity of the fluid at a given temperature})}{(\text{Absolute viscosity of a standard liquid such as water at the same temperature at that of the fluid})}$$

7. *Stoke's Law*: If a body is passing through a liquid, a thin layer of the liquid at rest is in contact with the body and due to the viscosity of the liquid; a viscous drag force is exerted on the moving body. In order to keep the body moving inside the liquid with a uniform velocity, a steady force should be applied to the body to overcome the effect of the viscosity of the liquid. This is stoke's law.

   According to Stoke's Law, if a ball of radius "r" passes through the liquid with a velocity "V", the viscosity of the liquid being "p.", the force (F) applied to the ball which just balances the effect of drag due to viscosity is given by the equation:

   $$F = 6\pi r\mu V$$

8. *Viscometer*: The device used to measure viscosity of fluids is called as viscometer.

SELF ASSESSMENT QUESTIONS-13.1

1. What is viscosity?
2. What is fluidity?
3. Briefly discuss the below given terms
   (a) Absolute viscosity (b) Kinematic viscosity (c) Relative viscosity
4. Briefly discuss the following terms
   (a) Stroke's law (b) Viscometer (c) Specific viscosity

13.3 CLASSIFICATION OF VISCOMETERS

The following are the important viscometers:

1. Capillary tube viscometer.
2. Efflux viscometers.

   Redwood viscometer.

   Saybolt viscometer.

   Engler viscometer.
3. Falling sphere viscometer.
4. Rotating cylinder viscometer.
   (i) Concentric cylinders viscometer. (ii) Electrical viscometer.

13.4 CAPILLARY TUBE VISCOMETER

Principle: The viscosity (dynamic viscosity) for laminar flow condition is measured and it is based on Poiseuille's equation which is given as follows:

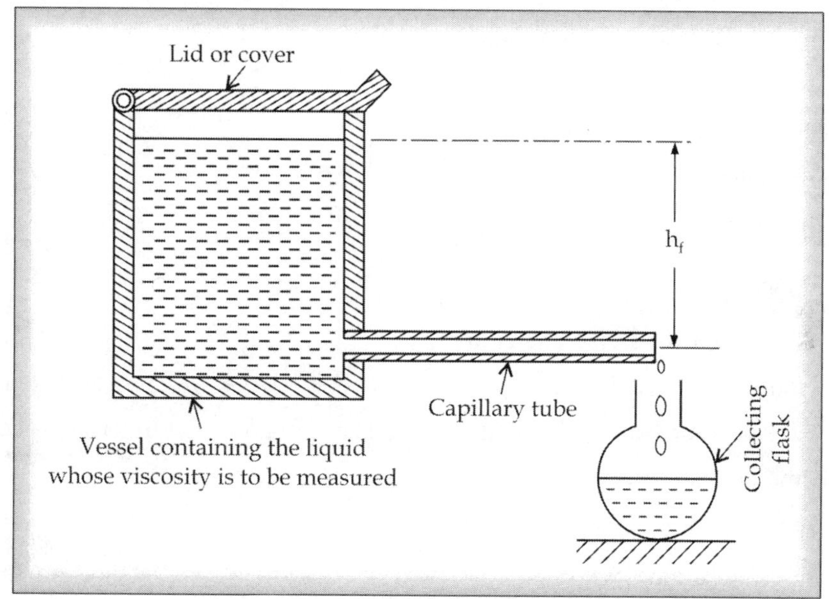

Fig. 13.1 Capillary Tube Viscometer

$$\text{Dynamic viscosity } (\mu) = \frac{\pi W d^4}{128\ QL} . h_f$$

where

W = Specific weight of liquid

Q = Flow rate of the liquid

h_f = Head loss over length L

d = Diameter of capillary tube

L = Length of capillary tube

Construction and Operation

To measure the dynamic viscosity of the liquid, two variables (Q and h_f) have to be determined.

So, take a flask (with lid) of known volume. The liquid flowing through the capillary tube drops at the free end which gives head loss (h_f) and the time taken to fill the known volume flask is noted. Using this time, the flow rate (Q) of the liquid is calculated.

Merit

- The accuracy is ± 1%

Demerit

- Error in measurement may arise due to variation in h_f because of sudden changes in inlet and outlet sections of the capillary tube and also due to kinetic energy of flow at outlet.

Application

- Used in refineries for measuring the viscosity of petroleum products.

13.5 EFFLUX VISCOMETERS

Principle: The time taken for a known quantity of fluid at constant temperature to flow through a standard capillary or orifice or nozzle is noted. This time becomes a measure of viscosity.

There are three Efflux viscometers namely:

1. Redwood viscometer.
2. Saybolt viscometer.
3. Engler viscometer.

All these viscometers use the same principle.

13.5.1 Redwood Viscometer

The main parts of a Redwood viscometer are:

1. A cylinder
2. Test liquid
3. Level indicator
4. A standard orifice and Ball valve
5. Collecting flask
6. Heating-coil and stirrer
7. Thermometers

Construction

A cylinder holds the test liquid whose viscosity is to be determined. A level indicator is fitted on the inner walls of the cylinder in which 50 ml test liquid can be filled easily. The temperature controlled bath is maintained around the cylinder. Collecting flask, heating-coil, stirrer and thermometers are also permanently fitted on the cylinder and bath to measure temperature constantly. A standard orifice and ball valve are fitted at the bottom of the cylinder. This set-up is shown in Figure 13.2.

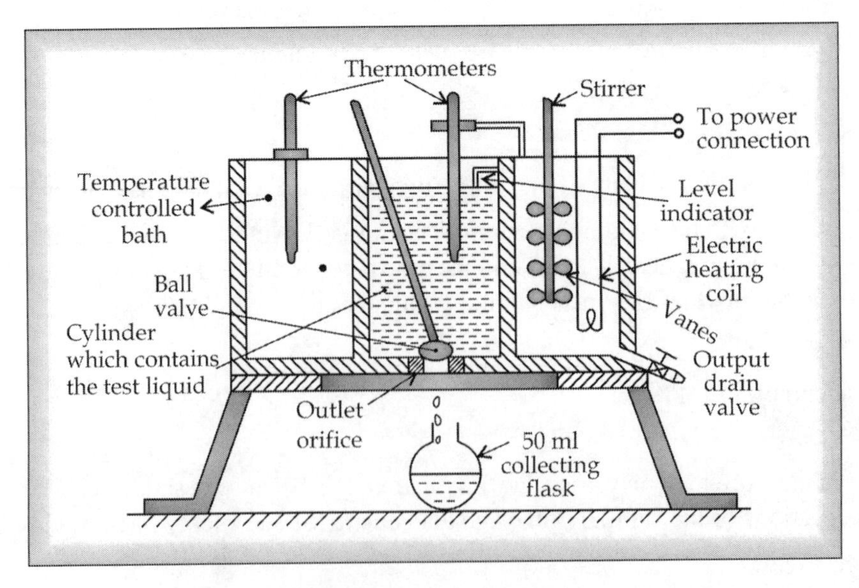

Fig. 13.2 Efflux (Redwood) Viscometer

Operation

We know that viscosity depends on temperature, i.e. higher the temperature, lower the viscosity. Hence, the temperature at which the viscosity of the liquid is to be measured has to be decided first, say, t_r°C. Fill the level indicator with 50 ml of the test liquid. Now, in the temperature controlled bath, the temperature is brought to t_r°C using the electric heating coil. Wait till the heat transfer takes place from the bath to the cylinder which contains the test liquid whose viscosity is being measured. The stirrer may be rotated frequently to enable the heat distribution proportionately in the bath.

When both the thermometers immersed in the test liquid as well as the bath show t_r°C, the orifice is opened by lifting the ball valve. The time taken for 50 ml of test liquid to flow through the orifice is noted. This time taken by 50 ml of the test liquid at t_r°C to flow through the orifice becomes a measure of viscosity of the liquid at t_r°C in Redwood seconds.

13.5.2 Saybolt Viscometer

The main parts of the Saybolt viscometer are:

1. A tank
2. Test liquid
3. A capillary and valve
4. A temperature controlled bath
5. 60 ml collecting flask
6. Heating-coil and stirrer
7. Thermometers

Construction

A tank contains the test liquid with a capillary attached to the bottom of the tank and is surrounded by a temperature controlled bath. A 60 ml collecting flask, heating-coil, stirrer and thermometers are permanently fitted on the tank and bath to measure temperature constantly as shown in the Figure 13.3.

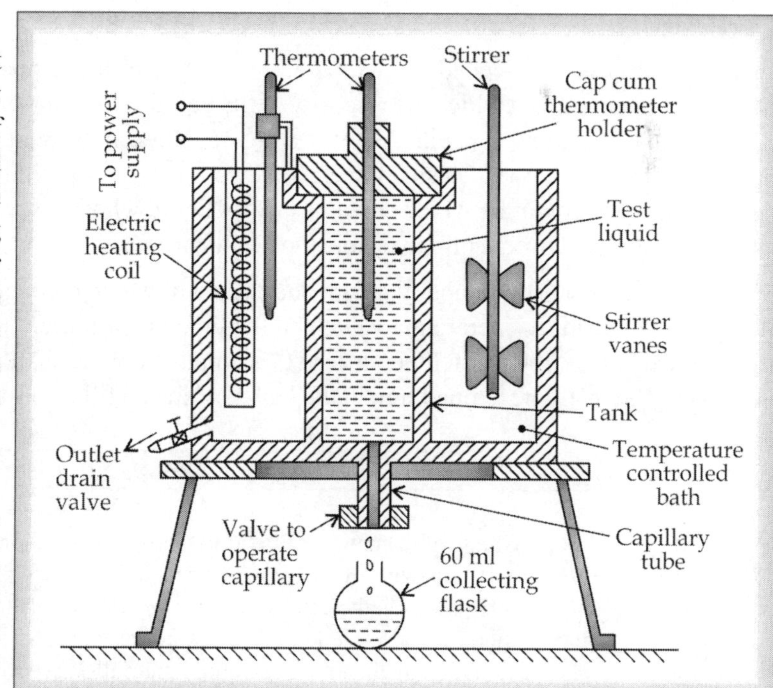

Fig. 13.3 Saybolt Viscometer

Operation

We know that, viscosity depends on temperature, that is, higher the temperature, lower the viscosity. Hence, the temperature at which the viscosity of the liquid is to be measured has to be decided first, say, t_s°C. Now the temperature in the temperature controlled bath is brought to t_s°C using the electric heating coil. Wait till heat transfer takes place from the bath to the tank which contains the test liquid whose viscosity is being measured. The stirrer may be rotated frequently so as to enable the heat distribution proportionately in the bath.

When both the thermometers immersed in the test liquid as well as the bath show t_s°C, the capillary is opened by operating the valve and the time taken to collect 60 ml of test liquid in the collecting flask is noted. This time taken by 60 ml of the test-liquid at t_s°C to flow through the capillary becomes a measure of viscosity of the liquid at t_s°C in Saybolt universal seconds.

Note: *Observe the basic difference between the above two viscometers – i.e., Orifice (in Redwood Viscometer) and Capillary (in Saybolt Viscometer).*

13.5.3 Engler Viscometer

The main parts of an Engler viscometer are:

1. Shallow cup
2. Test liquid
3. Level indicator
4. A standard orifice and Ball valve
5. Collecting flask
6. Heating-coil and stirrer
7. Thermometers

Construction

As shown in Figure 13.4, a shallow cup fitted with a standard orifice at its bottom. A level indicator fitted to inner wall of the shallow cup in which 200 ml of liquid can be filled easily. A constant temperature bath is maintained surrounding the shallow cup. An electric heating-coil, stirrer and 200 ml collecting flask, and thermometers are permanently fitted to the shallow cup and the bath.

Operation

We know that viscosity depends on temperature, i.e. higher the temperature, lower the viscosity. Hence, the temperature at which the viscosity of the liquid is to be measured has to be decided first, say, t_e°C. Fill the shallow cup with 200 ml test liquid. Now, in the temperature controlled bath, the temperature is brought to t_e°C using the electric heating coil. Wait till the heat transfer takes place from the bath to the cylinder which contains the test liquid whose viscosity is being measured. The stirrer may be rotated frequently to enable the heat distribution proportionately in the bath.

When the temperature in both the bath and the shallow cup containing the test liquid are t_e°C, the orifice is opened by operating a valve and the time taken for 200 ml of the test liquid to flow through the orifice at the test-temperature t_e°C is noted. Now the shallow cup is cleaned and filled with water (200 ml). The temperature in the bath is adjusted till a constant temperature of 20 °C is

More Instruments for Learning Engineers

A **Rain Gauge** (also known as **Pluviometer**, or **Ombrometer,** or **Udometer**) is a type of instrument used by meteorologists and hydrologists to gather and measure the amount of liquid precipitation over a set period of time. George James Symons was the founder of the first systematic rainfall survey on a national basis. Vaious types of rain gauges include, Standard rain gauge, Pluviometer of intensities, Weighing precipitation gauge, Tipping bucket rain gauge, Optical rain gauge, Acoustic rain gauge.

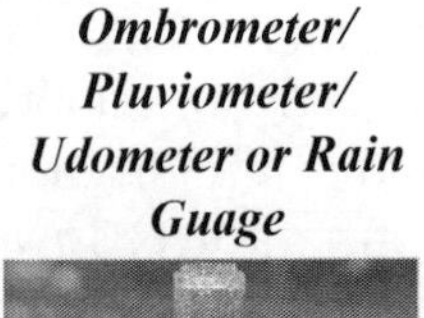

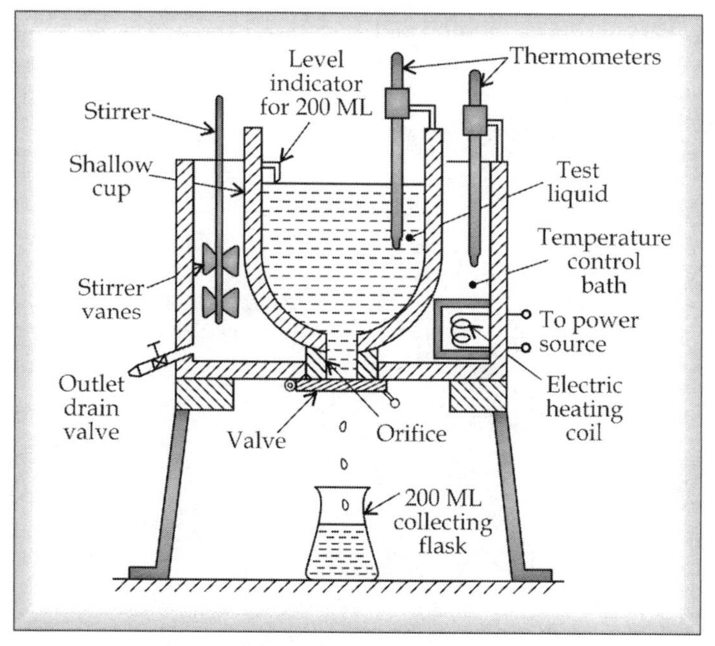

Fig. 13.4 Engler Viscometer

obtained in both the bath and the water in the shallow cup. Now the orifice is opened by operating the valve and the time taken for 200 ml of water at 20 °C to flow through the orifice is noted.

Viscosity is calculated as follows:

$$\left(\text{Viscosity in Engler degrees}\right) = \frac{\text{Time taken for 200 ml of test fluid to flow through the orifice at test temperature}}{\text{Time taken for 200 ml of water at 20 °C to flow through the orifice}}$$

Note: *Observe the difference between the Redwood and Engler viscometers. In Redwood we take an absolute value while in Engler, we compare with water.*

13.6 FALLING SPHERE VISCOMETER

The main parts of the falling sphere viscometer are:

1. Graduated Tube with guiding funnel
2. Test liquid
3. Steel ball sphere
4. A constant temperature bath
5. Heating-coil and stirrer
6. Thermometers

Principle

The time taken for a steel ball to fall through a distance in a test liquid in a tube at test temperature is proportional to absolute viscosity.

Construction

A graduated tube contains the test liquid whose viscosity is to be measured. A guiding funnel fitted at the top of the tube through which a steel ball/sphere falls. The tube is surrounded by a constant temperature bath with electric heating-coil and stirrer. Thermometers are permanently fitted to the tube containing the test liquid and the constant temperature bath to measure temperature constantly. The set-up is shown in Figure 13.5.

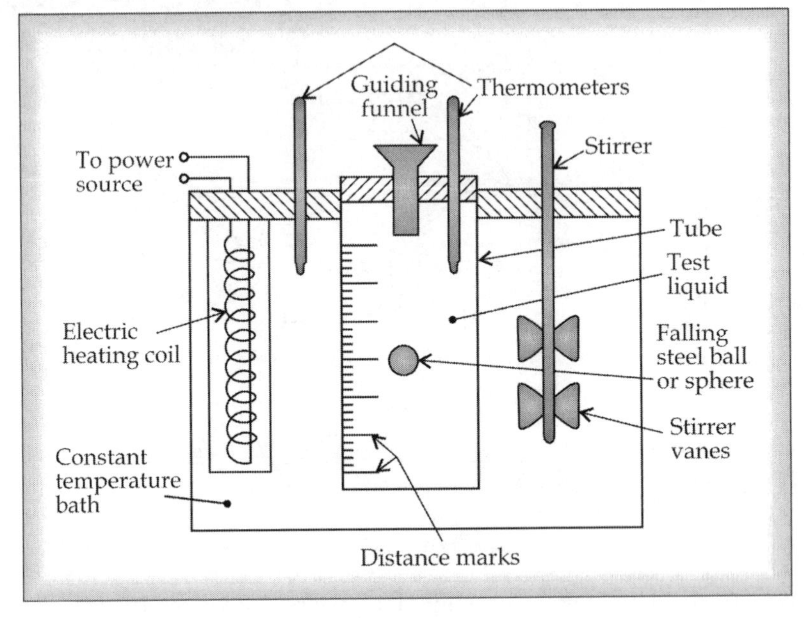

Fig. 13.5 Falling Sphere Viscometer

Operation

We know that viscosity depends on temperature, i.e. higher the temperature, lower the viscosity. Hence, the temperature at which the viscosity of the liquid is to be measured has to be decided first, say, t_f°C. Now, in the temperature controlled bath, the temperature is brought to t_f°C using the electric heating coil. Wait till the heat transfer takes place from the bath to the cylinder which contains the test liquid whose viscosity is being measured. The stirrer may be rotated frequently to enable the heat distribution proportionately in the bath.

When both the thermometers immersed in the test liquid as well as the bath show the same temperatue i.e. t_f°C, a steel ball is dropped through the guiding funnel to move in the test liquid. This steel ball is made to move through a distance of 150 or 200 mm to enable calculation of terminal velocity. The steel ball obtains a maximum velocity (v) when the gravity force is balanced by the buoyant and viscous drag forces.

The dynamic viscosity is calculated using the following expression:

$$\text{Dynamic viscosity } (\mu) = \frac{2}{9}\frac{r^2}{V}(\rho_s - \rho_t).g$$

where; r = Radius of the steel ball.

V = Velocity;

ρ_s = Density of the steel ball

ρ_t = Density of the liquid.

Note: *In this method, timing can be obtained with great accuracy by using field coils at the start and finishing points.*

Application

- Used to measure viscosity of high viscous oils in petroleum industries.

13.7 ROTATING CYLINDER VISCOMETERS

These viscometers are of two types namely:

(a) The rotating cylinder viscometer of concentric cylinders type

(b) The rotating cylinder viscometer of electrical type.

13.7.1 Rotating Cylinder Viscometer (Concentric Cylinders Type)

The following are the main parts of the arrangement

1. Two concentric cylinders (inner & outer)
2. Test liquid
3. Torque measuring device

Principle

The torque is proportional to the viscosity $(\tau \propto \mu)$

Construction

Two concentric (referred to as inner and outer) cylinders are mounted such that their axis is same. The inner cylinder is stationary and is equipped with a torque measuring device while the outer cylinder can rotate at a constant angular speed (w). There is a gap $(R_2 - R_1)$ between the inner and outer cylinders. This gap is filled with the test liquid whose viscosity is to be measured.

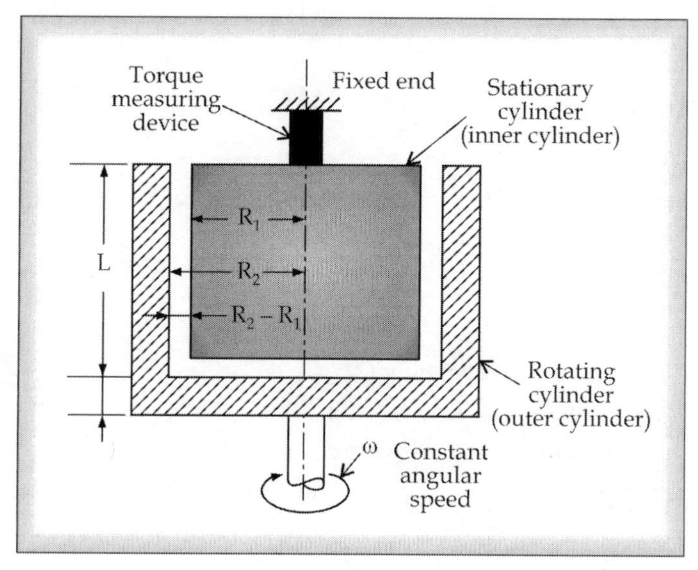

Fig. 13.6 Rotating Cylinder Viscometer

Operation

When the outer cylinder is rotated at a constant angular speed, a torque is developed at the fixed end of the inner cylinder as the inner cylinder is stationary. This torque developed at the fixed end of the inner cylinder is measured using a torque measuring device which is permanently attached to the fixed end of the inner cylinder. The viscosity can be determined when calibrated using the relationship-

$$\text{Viscosity } (\mu) = \frac{\tau}{\pi\omega R_1^2\left(\frac{R_1}{2a} + \frac{2LR_2}{R_2 - R_1}\right)}$$

where τ = Torque

ω = Constant angular speed.

R_1 = outer radius of inner cylinder

R_2 = inner radius of outer cylinder

a = base thickness of the outer cylinder and

L = inner height of outer cylinder

13.7.2 Rotating Cylinder Electrical Viscometer

The main parts of this arrangement are:

1. Flow container with inlet and outlet
2. Cylinder
3. Test liquid
4. Calibrated spring
5. Synchronous motor

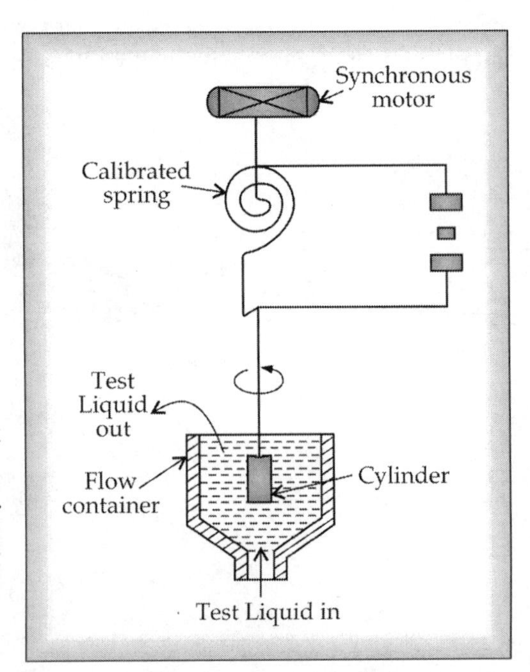

Fig. 13.7 Rotating Cylinder Electrical Viscometer

Principle

Lag in speed of cylinder is proportional to the viscosity ($v \propto \mu$)

Construction

A flow container with an inlet (at bottom) and outlet (at top) will have the test liquid that flows continuously. A cylinder is immersed in the test liquid. The cylinder is connected to a calibrated spring and a synchronous motor.

Operation

The test liquid whose viscosity is to be measured at a particular temperature is made to flow through the flow container continuously. The cylinder is then rotated using the synchronous motor. Since the cylinder is connected to the synchronous motor, both (the cylinder and the synchronous motor) are supposed to rotate at a constant angular speed. But the angular speed of the cylinder lags when compared to the speed of the synchronous motor since the cylinder is immersed in the test liquid. This lag in speed of the cylinder is directly proportional to the viscosity of the test liquid. The measure of this lag in the speed becomes the measure of the viscosity when calibrated.

As the drive is through a calibrated spring, the lag in speed of the cylinder with respect to the synchronous motor can be directly obtained as an electric signal by using a variable resistance or variable capacitance transducer.

Merits/Applications

- Can be used over wide ranges of viscosity and this is made possible by employing suitable cylinder type and speed.
- Can be used in open or closed vessels under pressure or vacuum.

Self Assessment Questions-13.2

1. List out the various types of viscometers.
2. Discuss capillary tube viscometer with a neat sketch.
3. What is the principle of Efflux viscometer? Name the three Efflux viscometers.
4. With a neat sketch explain Redwood viscometer.
5. Describe Saybolt viscometer with a neat diagram.
6. Describe Engler viscometer with a neat diagram.
7. What is the principle of Falling sphere viscometer? Explain its operation and construction with a neat sketch.
8. What are the rotating cylinder viscometers? Write their principles.
9. Discuss Rotating Cylinder Viscometer (Concentric Cylinders Type) with a neat sketch.
10. Explain Rotating Cylinder Electrical Viscometer with a neat sketch.

13.8 SURFACE TENSION

- Certain insects such as water-striders can float or walk or run on water surface. How?

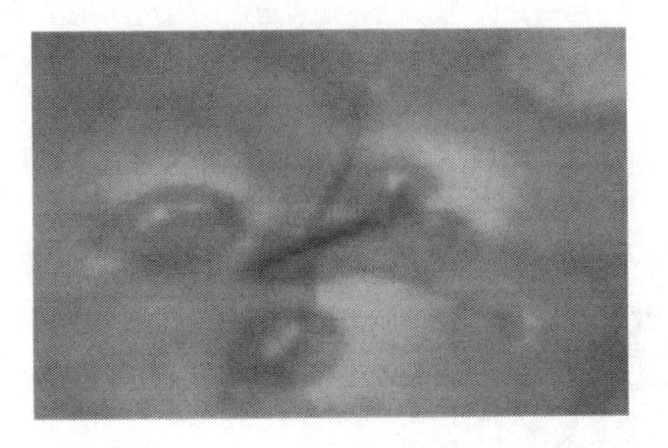

- How a gem clip or a pin can float on water (though steel is denser than water)?

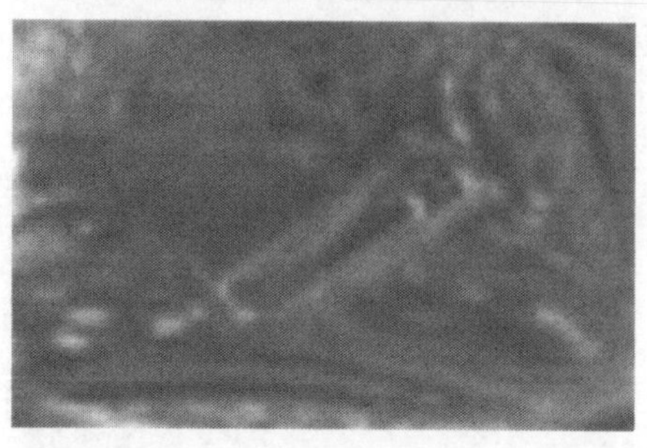

- A water drop on a leaf looks like a pearl (round sphere). Why?

- Why water raises in the tube when a tube is placed (capillarity)?

Well! The answer for all the above question is the special property of liquids, the surface tension. Surface tension helps an object or insect (e.g., water striders) though denser than water is able to float or run along the water surface. Surface tension is an important property that markedly influences the ecosystem. At liquid-air interfaces, surface tension results from the greater attraction of water molecules to each other (cohesion) than to the molecules in the air (adhesion).

In other words, the net effect of inward forces at its surface makes water to behave as if its surface were covered with a stretched elastic membrane. Due to the relatively high attraction of water molecules for each other, water has a high surface tension (72.8 milli Newtons per meter at 20 °C) as compared to that of most other liquids. Further, surface tension is an important factor in the phenomenon of capillarity.

Definition

Surface tension is a contractive tendency of the surface of a liquid that allows it to resist an external force.

Physical Units

Surface tension, usually represented by the symbol γ is measured in forces per unit length. Its SI unit is Newton per meter but the CGS unit of dyne per cm is also used.

$$\gamma = 1\frac{\text{dyn}}{\text{cm}} = 1\frac{\text{erg}}{\text{cm}^2} = 0.001\frac{\text{N}}{\text{m}} = 0.001\frac{\text{J}}{\text{m}^2}$$

Surface tension is also expressed as energy per unit area.

Though both (force per unit length or energy per unit area) are same, when referring to energy per unit of area, the term surface energy is preferred since it is a more general term that it can be applied also to solids and not just liquids.

13.9 MEASUREMENT OF SURFACE TENSION

Since surface tension exhibits various effects, it offers a number of ways and means to its measurement. The suitability of method for a given liquid and given situation depends on the nature of the liquid being measured, the conditions under which its tension is to be measured, and the stability of its surface when it is deformed. To choose the optimal method, one needs to be aware of the effects of surface tension. A few of these are as follows.

13.9.1 Effects of Surface Tension

Several effects of surface tension can be seen with ordinary water, some of which are given below. These effects can be made use to measure the surface tension.

1. **Beading of rain water on a waxy surface (such as a leaf):** Water clusters into drops as it adheres weakly to wax and strongly to itself. Surface tension gives them almost spherical shape, as sphere has the smallest possible surface area to volume ratio.
2. **Formation of drops when a mass of liquid is stretched:** The water adhered to a tap or a faucet does not fall unless it gains mass and stretches to a point where the surface tension can no longer bind it to the tap/faucet. Then, it separates and forms the drop into a sphere. If a stream of water runs from the faucet, the stream breaks up into drops during its fall. The gravitation force pulls to stretch the stream, and then surface tension pinches it into spheres.
3. **Floating of objects denser than water:** This occurs when the object is non-wetting type and its weight is sufficiently small to withstand the forces arising from surface tension. For instance, water striders make use of surface tension to walk on the surface. Here, the surface of the water acts like an elastic film so that the surface area stretches due to the indentation of insect's feet on the surface.
4. **Separation of oil and water (such as water and liquid wax):** This is caused in the surface between dissimilar liquids due to the so called "interface tension". However, its principle is

More Instruments for Learning Engineers

A **Tensiometer** measures surface tension (γ) of liquids or surfaces. They are used in research and development laboratories to determine the surface tension of liquids like coatings, lacquers or adhesives. A further application field of tensiometers is the monitoring of industrial production processes like part's cleaning or electroplating.

Tensiometer

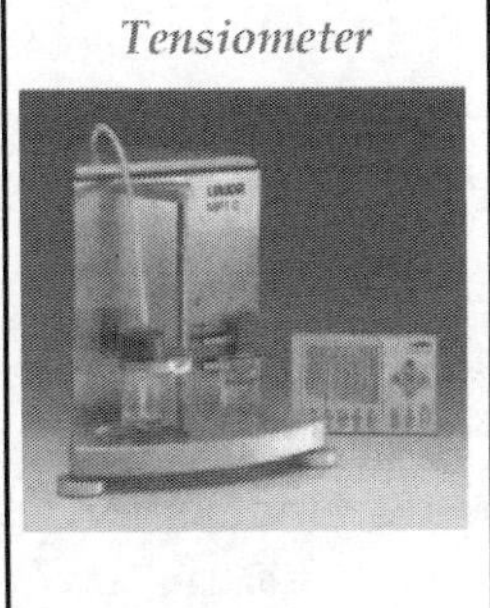

the same. Example: Lava lamp with interaction between dissimilar liquids; water and liquid wax.

5. **Tears of wine:** A complex interaction between the differing surface tensions of water and ethanol causes the formation of drops and rivulets on the side of a glass containing an alcoholic beverage. This is induced by a combination of surface tension modification of water by ethanol together with ethanol evaporating faster than water.
6. **Surfactants:** Surface tension is visible in other common phenomena, especially when surfactants are used to decrease it:

- Bubbles in pure water are unstable, but the addition of surfactants, such as soap, produces a stabilizing effect on the bubbles (Marangoni effect). Soap bubbles have very little mass but very large surface areas. The surfactants in fact reduce the surface tension of water threefold or more.
- Another good example we can notice the effect of surface tension is with emulsions. The small fragments of oil suspended in pure water will spontaneously assemble themselves into much larger masses. But the presence of a surfactant reduces surface tension, and thence permits stability of minute droplets of oil in the bulk of water (or vice versa).

13.9.2 Methods of Measuring Surface Tension

Making use of the above effects, we have several methods to measure the surface tension as listed below.

1. Du Noüy Ring method
2. Du Noüy-Padday method
3. Wilhelmy plate method
4. Spinning drop method
5. Pendant drop method
6. Bubble pressure method (Jaeger's method)
7. Stalagmometric method
8. Vibrational frequency of levitated drops method
9. Resonant oscillations of spherical/hemispherical liquid drop method

We shall now discuss most popular instruments, the methods used in the instruments, the basic principle and concept behind these methods to measure the surface tension in the sections to follow.

13.9.3 Instruments used to Measure Surface Tension

The instruments used to measure surface tension are called tensiometers. The popular tensiometer in use are

1. Du Noüy Ring Tensiometer
2. Du Noüy-Padday Method Tensiometer
3. Wilhelmy Plate Tensiometer
4. Bubble Pressure Tensiometer
5. Goniometer/ Tensiometer

These are briefly described as follows.

Self Assessment Questions-13.3

1. Define Surface Tension?
2. What are the physical units of surface tension?
3. Discuss various effects of surface tension.
4. List out the various methods of measuring surface tension.
5. What are the various instruments used to measure surface tension?

13.10 DU NOÜY RING TENSIOMETER

This instrument uses the traditional method to measure surface or interfacial tension.

Principle

Maximum pull exerted on the ring by the surface is measured.

Construction and Operation

This type of tensiometer (Fig.13.8) uses a platinum ring submersed in a liquid. As the ring is pulled out of the liquid, the tension required is precisely measured to determine the surface tension of the liquid.

Fig. 13.8 Du Noüy Ring Tensiometer

Merits

- Wetting properties of the surface or interface have little influence on this measuring technique.

Demerits

- This method requires that the platinum ring be nearly perfect; even a small blemish or scratch can greatly alter the accuracy of the results.
- A correction for buoyancy must be made.
- This method is considered inaccurate compared to the plate method (but is still widely used for interfacial tension measurement between two liquids).

13.11 DU NOÜY-PADDAY TENSIOMETER

It is a modified version of Du Noüy method and a small diameter metal needle instead of a ring as shown in Figure 13.9.

Fig. 13.9 Du Noüy-Padday Tensiometer

Principle

A small diameter metal needle is used in combination with a high sensitivity microbalance to record maximum pull.

Construction and Operation

This method uses a rod which is placed into a test liquid. The rod is then pulled out of the liquid and the force required to pull the rod is precisely measured. This is a rather novel method which is accurate and repeatable.

Merits

- The very small sample volumes (down to few tens of microliters) can be measured with very high precision, without the need to correct for buoyancy (for a needle or rather rod with proper geometry).

- The measurement can be performed very quickly, minimally in about 20 seconds.
- This can work well with liquids with a wide range of viscosities.
- First commercial multichannel tensiometers were recently built based on this principle.

13.12 WILHELMY PLATE TENSIOMETER

This is a universal method especially suited to check surface tension over long time intervals.

Principle

A vertical plate of known perimeter is attached to a balance, and the force due to wetting is measured.

Construction and Operation

The Wilhelmy Plate tensiometer requires a plate to make contact with the liquid surface. The force required to wet is proportional to surface tensions and hence becomes a measure of the surface tension when calibrated.

Merits

- It is the simplest and most accurate method.

13.13 BUBBLE PRESSURE TENSIOMETER

This is a measurement technique for determining surface tension at short surface ages.

Principle

Maximum pressure of each bubble is measured using Jaeger's bubble pressure method.

Construction and Operation

As shown in Figure 13.10, the bubble pressure method makes use of this bubble pressure which is higher than in the surrounding environment (water). Due to internal attractive forces of a liquid, air bubbles within the liquids are compressed. The resulting pressure (bubble pressure) rises at a decreasing bubble radius.

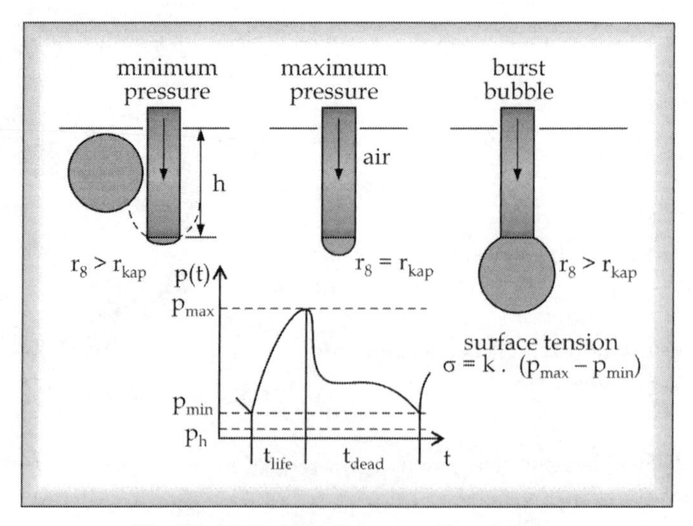

Fig. 13.10 Bubble Pressure Tensiometer

A gas stream is pumped into a capillary that is immersed in a fluid. The resulting bubble at the end of the capillary tip continually becomes bigger in surface; thereby, the bubble radius is decreasing.

The pressure rises to a maximum level. At this point the bubble has achieved its smallest radius (the capillary radius) and begins to form a hemisphere. Beyond this point the bubble quickly increases in size and soon bursts, tearing away from the capillary, thereby allowing a new bubble to develop at the capillary tip. During this process a characteristic pressure pattern develops, which is evaluated for determining the surface tension.

Merits/Demerits/Applications

- Because of the easy handling and the low cleaning effort of the capillary, bubble pressure tensiometers are a common alternative for monitoring the detergent concentration in cleaning or electroplating processes.
- Bubble pressure method can measure the dynamic surface tension of liquids.

13.14 GONIOMETER/ TENSIOMETER

Principle

Geometry of a drop (***Pendant drop method***) or the angle of contact (***Sessile drop method***) is analyzed optically.

Construction and Operation

Goniometer/ Tensiometer (as shown in Figure 13.11) can be used to measure contact angle, surface tension and interfacial tension of a liquid using the pendant or sessile drop methods. A drop is produced and the profile is then captured using a goniometer/ tensiometer. The software then analyses the profile of the drop and makes a series of calculations based on Young-Laplace equation of surface tension.

Fig. 13.11 Goniometer/Tensiometer

Merits/Demerits/Applications

- Surface and interfacial tension can be measured by this technique, even at elevated temperatures and pressures.
- It is a more difficult experimental measurement to accomplish and is not nearly as accurate as the Wilhelmy plate method.
- It is computation intensive process.

13.15 SOME MORE METHODS

In addition to the above methods, surface tension can also be measured by using the following principles.

13.15.1 Spinning Drop Method

This technique is ideal for measuring low interfacial tensions. The diameter of a drop within a heavy phase is measured while both are rotated.

13.15.2 Drop Volume Method

This is a method for determining interfacial tension as a function of interface age. Liquid of one density is pumped into a second liquid of a different density and time between drops produced is measured.

13.15.3 Capillary Rise Method

The end of a capillary is immersed into the solution. The height at which the solution reaches inside the capillary is proportional to the surface tension.

13.15.4 Stalagmometric Method

This is a method of weighing and reading a drop of liquid.

13.15.5 Sessile Drop Method

This is a method for determining surface tension and density by placing a drop on a substrate and measuring the contact angle.

13.15.6 Vibrational Frequency of Levitated Drops Method

The natural frequency of vibrational oscillations of magnetically levitated drops has been used to measure the surface tension of super fluid 2^{He^4}. This value is estimated to be 0.375 dyne/cm at T = 0 K.

13.15.7 Resonant Oscillations of Spherical/ Hemispherical Liquid Drop Method

This technique is based on measuring the resonant frequency of spherical and hemispherical pendant droplets driven in oscillations by a modulated electric field. The surface tension and viscosity can be evaluated from the obtained resonant curves.

Self Assessment Questions-13.3

1. Discuss Du Noiiy Ring Tensiometer.
2. Explain Du Noiiy-Padday Tensiometer.
3. What is the principle of Wilhelmy plate Tensiometer? Explain its operation and construction.
4. Discuss Bubble pressure Tensiometer with a neat sketch.
5. Explain Goniometer/ Tensiometer with its principle.
6. Briefly explain the following methods
   (a) Spinning drop method (b) Drop volume method
   (b) Capillary rise method (d) Stalagmometric method
7. Briefly explain the following methods
   (a) Sessile drop method (b) Vibrational frequency of levitated drops method
   (c) Resonant oscillations of spherical/hemispherical liquid drop method

SUMMARY

Viscosity is the property of a liquid to resist shearing/sliding of its layers. Due to internal friction of liquid and is a measure of fluidity (the reciprocal of viscosity) and measuring device known as viscometer. Absolute or Dynamic Viscosity or Viscosity Coefficient is the ratio of shear stress to change of velocity per length. Kinematic Viscosity is the ratio of absolute viscosity to density. Relative viscosity is the ratio of absolute viscosity of the fluid at a given temperature to that of water at 20 °C while the specific viscosity is measured with standard liquid.

A capillary tube viscometer works on Poiseuille's equation to measure the dynamic viscosity of the liquid by determining flow rate and head loss. There are three Efflux viscometers, namely, Redwood, Saybolt and Engler viscometer, which operate by measuring the time taken for a known quantity of fluid at constant temperature to flow through a standard capillary (e.g., Saybolt) or orifice (e.g., Redwood & Engler) or nozzle. The first two measure absolute while the third one measures relative viscosity. Falling sphere viscometer measures the time taken for a steel ball to fall through a distance in a test liquid in a tube at test temperature using the principle, dynamic viscosity $(\mu) = \frac{2}{9}\frac{r^2}{V}(\rho_s - \rho_t)$. Rotating cylinder viscometers use concentric cylinders to measure the torque is proportional to the viscosity whilst electrical viscometer measures the lag in speed of cylinder is proportional to the viscosity.

Surface tension is a contractive tendency of the surface of a liquid that allows it to resist an external force measured in N/m or J/m^2. Some of the effects of surface tension can be seen with beading of rain water on a waxy surface (such as a leaf), Formation of drops when a mass of liquid is stretched, Floating of objects denser than water, separation of oil and water (such as water and liquid wax), tears of wine, surfactant effects.

Surface tension can be measured using Du Noüy Ring, Du Noüy-Padday method, Wilhelmy plate, spinning drop, pendant drop, Jaeger's bubble pressure, stalagmometric, vibrational frequency of levitated drops and resonant oscillations of spherical/hemispherical liquid drop methods.

The instrument used to measure the surface tension is known as tensiometer. Du Noüy Ring Tensiometer uses a platinum ring submersed in a liquid is pulled out to measure force (tension). Du Noüy-Padday Method Tensiometer a modified version of Du Noüy method and uses a small diameter metal needle instead of a ring. Wilhelmy Plate Tensiometer is a universal method in which a vertical plate of known perimeter is attached to a balance, and the force due to wetting is measured. Bubble Pressure Tensiometer is a measurement technique for determining surface tension at short surface ages and the maximum pressure of each bubble is measured using Jaeger's bubble pressure method. Goniometer/ Tensiometer uses geometry of a drop (Pendant drop method) or the angle of contact (Sessile drop method) is analyzed optically.

Spinning drop method is ideal for measuring low interfacial tensions where the diameter of a drop within a heavy phase is measured while both are rotated. Drop volume method measures as a function of interface age. Capillary rise method makes use of the principle of capillary rise proportional to the surface tension. Stalagmometry goes by weighing a drop of liquid. Surface tension and density can be found by placing a drop on a substrate and measuring the contact angle in sessile drop method. Vibrational frequency of levitated drops or resonant oscillations of spherical/ hemispherical liquid drop can be measured to determine the surface tension.

KEY CONCEPTS

Viscosity: The property of a liquid to resist shearing/sliding of its layers due to the internal friction of a liquid and is a measure of fluidity.

Fluidity: The reciprocal of viscosity.

Absolute Viscosity or Dynamic Viscosity or Viscosity Coefficient:

$$\mu = \frac{\tau}{\left(\frac{du}{dy}\right)} = \frac{\text{Shear stress}}{\left(\frac{\text{Change of velocity}}{\text{Change of distance}}\right)} \quad \text{Ns/m}^2$$

Kinematic Viscosity: $(\gamma) = \dfrac{\text{Absolute viscosity } (\mu)}{\text{Density } (\rho)} \, m^2/s$

Relative Viscosity: $\dfrac{\text{(Absolute viscosity of the fluid at a given temperature)}}{\text{(Absolute viscosity of water at 20 °C)}}$

Specific Viscosity: $\dfrac{\text{(Absolute viscosity of the fluid at a given temperature)}}{\text{(Absolute viscosity of a standard liquid such as water at the same temperature at that of the fluid)}}$

Stoke's Law: To keep the body moving inside the liquid with a uniform velocity, a steady force should be applied to the body to overcome the effect of the viscosity of the liquid.

Viscometer: The device used to measure viscosity of fluids.

Capillary tube viscometer: Measures dynamic viscosity for laminar flow condition based on Poiseuille's equation (μ)

Poiseuille's Equation: Dynamic viscosity $(\mu) = \dfrac{\tau W d^4}{128\, QL} . h_f$

Efflux Viscometers: Operates on the principle that time taken for a known quantity of fluid at constant temperature to flow through a standard capillary or orifice or nozzle is a measure of viscosity.

Redwood Viscometer: The Efflux viscometer that measures absolute viscosity using an orifice.

Saybolt Viscometer: The Efflux viscometer that measures absolute viscosity using a capillary tube.

Engler Viscometer: The Efflux viscometer that measures relative viscosity using an orifice.

Falling Sphere Viscometer: Operates on the principle that the time taken for a steel ball to fall through a distance in a test liquid in a tube at test temperature is proportional to absolute viscosity.

Electrical Rotating Concentric Cylinders Viscometer: Operates on the principle that the torque is proportional to the viscosity ($v \propto \mu$).

The Rotating Cylinder Electrical Ciscometer: Operates on the principle that the lag in speed of cylinder is proportional to the viscosity ($v \propto \mu$).

Surface Tension: A contractive tendency of the surface of a liquid that allows it to resist an external force.

Du Noüy Ring Tensiometer: Maximum pull exerted on the ring by the surface is measured.

Du Noüy-Padday Tensiometer: A small diameter metal needle is used in combination with a high sensitivity microbalance to record maximum pull.

Wilhelmy Plate Tensiometer: A vertical plate of known perimeter is attached to a balance, and the force due to wetting is measured.

Bubble Pressure Tensiometer: Maximum pressure of each bubble is measured using Jaeger's bubble pressure method.

Goniometer/Tensiometer: Geometry of a drop (***Pendant Drop Method***) or the angle of contact (***Sessile Drop Method***) is analyzed optically.

Spinning Drop Method: The diameter of a drop within a heavy phase is measured while both are rotated.

Drop Volume Method: Liquid of one density is pumped into a second liquid of a different density and time between drops produced is measured.

Capillary Rise Method: Operates on the principle that the height at which the solution reaches inside the capillary is proportional to the surface tension.

Stalagmometric Method: A method of weighing and reading a drop of liquid.

Sessile Drop Method: A method of measuring surface tension and density by placing a drop on a substrate and measuring the contact angle.

Vibrational Frequency of Levitated Drops Method: The natural frequency of vibrational oscillations of magnetically levitated drops used to measure the surface tension of super fluid $^4He_2$. This value is estimated to be 0.375 dyne/cm at T = 0 K.

Resonant Oscillations of Spherical/Hemispherical Liquid Drop Method: Measuring the resonant frequency of spherical and hemispherical pendant droplets driven in oscillations by a modulated electric field. The surface tension and viscosity can be evaluated from the obtained resonant curves.

REVIEW QUESTIONS

SHORT ANSWER QUESTIONS

1. Define and explain briefly
   (a) Surface Tension (b) Viscosity
2. Distinguish between viscosity and fluidity.
3. Distinguish between absolute viscosity and kinematic viscosity.
4. Discuss the concept of capillarity.
5. Explain the principles behind the measurement of surface tension.
6. What is surface tension? Give some examples.
7. Explain the principle behind floating of a steamer though the steel/iron is denser than water.
8. How can a water-strider can walk on water? Discuss the principle behind it.
9. Why do we skid on oil spilled floor?

LONG ANSWER QUESTIONS

1. With a neat sketch explain the construction and working of saybolt viscometer.
2. Explain the construction and working of engler viscometer. How does it differ from Red wood viscometer?
3. Describe the construction, operation and measurement of saybolt viscometer. How does it differ from Red wood viscometer.
4. Describe the principle behind falling sphere viscometer. With a neat sketch explain its construction and method of measurement.
5. Explain the construction and working of any tensiometer.
6. Discuss the principle of operation and construction of Goniometer/tensiometer.
7. Wite short notes on:
   (a) Bubble pressure tensiometer (b) Rotating cylinder viscometer
8. Discuss the effects of viscosity and surface tensions in nature with at least two examples to each.
9. Wite notes on: (a) Dunouy tensiometer (b) Efflux viscometer.

MULTIPLE CHOICE QUESTIONS

1. Fluid is defined as_______.
   (a) Substance which has ability to flow
   (b) A substance which is capable of flowing /moving under the action of shear force
   (c) Any substance which is liquid is called Fluid
   (d) Any substance which is not solid is called fluid
2. Viscosity is defined as________.
   (a) The force required to flow liquid
   (b) The reciprocal of bulk modulus
   (c) The ratio of mass of fluid to its volume
   (d) The internal resistance offered by one layer opf fluid to another adjacent layer
3. Units of viscosity______.
   (a) Pa-sec (b) Pa-sec^2
   (c) Pa-sec^3 (d) Pa

4. Units of kinematic viscosity is______.
   (a) m^2/sec (b) m^2/sec^2
   (c) $m^2 - sec$ (d) $m^2 - sec^2$
5. The rise or fall of a liquid when a small diameter tube is immersed in it is known as_____.
   (a) Viscosity
   (b) Capillarity
   (c) Surface tension
   (d) Fluidity
6. The Capillary rise is due to ______ force
   (a) Viscous (b) Adhesion
   (c) Cohesion (d) Newtonian
7. The capillary fall is due to_____.
   (a) Viscous (b) Adhesion
   (c) Cohesion (d) Newtonian
8. In case of liquids viscosity _____ with increase in temperature
   (a) Increases
   (b) Does not change
   (c) Initially decreases then increases
   (d) Decreases
9. Capillary tube is used in____ Viscometer
   (a) Redwood (b) Saybolt
   (c) Engler (d) Falling sphere
10. ______ is defined a contractive tendency of the surface of a liquid that allows it to resist an external force.
    (a) Viscosity (b) Capillarity
    (c) Surface tension(d) Fluidity
11. Spherical shape of the droplet is due to_______.
    (a) Viscosity (b) Capillarity
    (c) Surface tension(d) Fluidity
12. The instrument used to measure the surface tension is known as____.
    (a) Tensiometer (b) Viscometer
    (c) Surface meter (d) Barometer
13. To check surface tension over long time intervals______ Tensiometer is used.
    (a) Du Noüy Ring
    (b) Du Noüy-Padday Method
    (c) Wilhelmy Plate
    (d) Bubble Pressure
14. For calculating surface tension weighing a drop of liquid is used in ____ method
    (a) Stalagmometric(b) Capillary rise
    (c) Drop volume (d) Spinning drop
15. A Capillary tube viscometer works on ___ equation to measure the dynamic viscosity of the liquid
    (a) Buoyancy (b) Poiseuille's
    (c) Pascal's (d) Hydrostatic
16. The device used to measure viscosity of fluids is called
    (a) Viscosity meter(b) Viscosensor
    (c) Viscometer (d) Viscomeasure
17. The viscometer that measures absolute viscosity using a capillary tube is __________.
    (a) Redwood viscometer
    (b) Saybolt viscometer
    (c) Engler viscometer
    (d) Falling sphere viscometer
18. The viscometer that measures absolute viscosity using an orifice is called
    (a) Redwood viscometer
    (b) Saybolt viscometer
    (c) Engler viscometer
    (d) Falling sphere viscometer
19. _________ is a contractive tendency of the surface of a liquid that allows it to resist an external force.
    (a) Viscosity
    (b) Fluidity
    (c) Surface tension
    (d) Liquidity
20. The viscometer that measures relative viscosity using an orifice is
    (a) Redwood viscometer
    (b) Saybolt viscometer
    (c) Engler viscometer
    (d) Falling sphere viscometer

ANSWERS

1.	(b)	2.	(d)	3.	(a)	4.	(a)	5.	(b)
6.	(b)	7.	(c)	8.	(d)	9.	(b)	10.	(c)
11.	(c)	12.	(a)	13.	(c)	14.	(a)	15.	(b)
16.	(c)	17.	(b)	18.	(a)	19.	(c)	20.	(c)

CHAPTER INFOGRAPHIC

Glimpses of Measurement of Humidity/Dampness and Density/Specific Gravity

Amount of water vapor contained in air/gas is Humidity	**RH or Ψ = (mass of water vapor actually present)/(mass of water vapor needed for saturation)**
Temp. of bulb directly exposed to air-water vapor mixture, is Dry bulb temperature	**Temperature of bulb covered by wet wick exposed to air-water vapor mixture is Wet-bulb temperature**
Dew Point Temperature is temp at which water vapor starts to condense	**Difference of dry and wet bulb temperatures is Wet bulb depression**

S.No	Instrument	Principle/Transducer	Construction	Operation	Applications	Merits	Limitations
1.	Sling Psychrometer	By measuring both dry and wet bulb temperature	2-thermometers in frame to measure temp of dry bulb and wet bulb covered by cotton wick	Frame is rotated at 3-10 m^{-1} then temp indicated by 2 thermometers are noted	To measure Humidity	Range 0-180 °C, hair hygro-metric is set	Continuous use not possible, medium gets disturbed in use
2. Absorption Hygrometer (AH)							
2(a)	Mechanical humidity sensing AH	Hygroscopic materials varies linearly with absorbed humidity	No. of hairs in parallel beam in air increase in length, magnified by leverage and connected to pointer/scale	Absorption of humidity from air causes the hair to expand and indicates humidity	To measure Humidity	Range 0-75 °C	Response time is slow; needs calibration if used constantly
2(b)	Electric humidity sensing AH	Resistance changes due to change in humidity	2 metal electrodes placed in hygroscopic salt (**LiCl**) connected to a null balance	Change in resistance by a Wheatstone bridge indicates humidity	To measure Humidity	Fast Response time and accuracy +25%	Should expose to 100% humidity and to use in constant temperature condn
3.	Dew point meter	At const pressure if air temperature is reduced, water vapor condenses	A shiny mirror surface, nozzle, light, source, photo voltaic cell	Light falls on mirror of therm-ocouple reflects, measured by photocell and air on mirror condenses to droplets; light deflected indicates humidity	To measure dew pt temp in cargos/ ships/ industries etc	Good response; protects cargoes from condensation damage	Limited to cooling fluids and light measurement

MEASUREMENT OF DENSITY/SPECIFIC GRAVITY

Instrument	Principle	Construction	Operation
Pycnometer	Specific gravity is ratio of wt of given liquid to that of boiled distilled water at given temp.	Straight walled glass bottle, a ground glass stopper	S.G = $\frac{W_2 - W_1}{W_2 - W_1}$ where W_1 = Weight of empty bottle; W_2 = Weight of pycnometer with boiled, distilled water; W_3 = Weight of pycnometer with liquid
Hydrometers			
Simple Hydrometer	Archimedes principle.	Container closed with a lid/cap, which contains a hole to introduce float. A float fitted with weight on its bottom end & a stem of constant diameter on its other end.	Weight is introduced into test liquid. Float has constant wt. and so its buoyant force is constant. Float sinks to depth inversely proportional to density of liquid. On calibrated stem, reading indicates density or specific gravity
Photo-Electric Hydrometer	Output of photo electric transducer is inversely proportional to density.	A chamber with inlet &outlet, a hollow glass float, a weight fitted at its bottom. light source & a photo-electric transducer are housed at top portion of chamber	Test-liquid when flows through chamber, hollow glass float immersed floats at bottom. Light ray from source at top slit received by the photo-electric transducer opposite to source indicates density
Differential Type bubbler Systems			
2-Vessels by Manometer and Reference Column	π = rgH i.e., for a constant depth of liquid, force of liquid is proportional to density.	2-vessels for test liquid and reference liquid, 2-bubbler tubes of same height connected by U-tube fitted with air pressure regulators.	Pressure of air to each bubbler tube is regulated; $\rho = \left(\frac{\rho_m.H_m}{H} + \rho_r\right)$ Where ρ_m & ρ_r are densities of manometric & reference liquids; H_m &H = heights
Single Vessel and Pressure Gauge	Pressure in vessel is proportional to density	A container, 2-bubbler tubes immersed in the test liquid. Both bubbler tubes are connected with separate air regulators.	Pressure of air to each bubbler tube is regulated, indicated differential pressure by pressure gauge indicates density

Chapter 14

Measurement of Humidity/ Dampness and Density/Specific Gravity

STARTERS

To study this chapter, you should have awareness on the following concepts.For a better understanding, it is always a good idea to revise these prerequisites.

- Humidity, moisture, dew, fog, wet and dry air etc. and their definitions
- Concepts of Wet-bulb and dry bulb temperatures, saturated/unsaturated vapour, vapour pressure, etc.
- Mass, Volume, Density, Specific gravity and related terms
- Basics of Psychrometry
- Different types of pressures such as atmospheric, absolute, vacuum, gauge pressure etc., and about manometers and pressure gauges
- Relation between vapour pressure and heat and mechanical energy
- The relation between pressure and temperature of gases i.e. Gas laws, particularly Charles law ($P \alpha T$) Boyle's law and Ideal gas equation and Kinetic gas equation
- The instrument components such as U-tube, Bourdon, bellows, diaphragms etc. and their shapes/utility
- Some mechanisms to convert one type of motion to the other such as rotational to translatory or reciprocatory motion etc. and Simple gear mechanisms for calibration into suitable scale
- Relation between force and electrical parameters such as resistance, capacitance etc.
- Awareness on float, bubbler, purge, buoyancy etc., and
- First four chapters of this book

LEARNING OBJECTIVE

After studying this chapter you should be able to

- Understand the terminology of humidity
- List out and explain the instruments used to measure humidity
- Know the fundamentals of density and specific gravity and
- Classify and discuss the instruments to measure specific gravity and density
- Differentiate between hydrometers and hygrometers

14.1 INTRODUCTION

The amount of water vapour present in air or gas is called humidity. It plays a significant role in many industrial processes such as chemical, textile, paper, food processing, leather, pharmaceutical industries in addition to manufacture of precision equipment. Before going into the details of measuring of humidity it is always a good idea to get awareness on the terms related to humidity measurement.

14.2 TERMINOLOGY

1. **Humidity:** The amount of water vapour contained in air or gas is called humidity.
2. **Dry Air:** If there is no water vapour present in the atmosphere, it is called dry air.
3. **Moist Air:** If there is water vapour present in the atmosphere, then the air is called moist air.
4. **Saturated Air:** Saturated air is the moist air where the partial pressure of water vapour equals the saturation pressure of steam corresponding to the temperature of the air.
5. **Humidity Ratio or Specific Humidity or Absolute Humidity or Moisture Content:** It is defined as the ratio of the mass of water vapour to the mass of dry air in a given volume of the mixture and is denoted by w.

   Humidity Ratio= Mass of water vapour/Mass of dry air
6. **Relative Humidity:** It is defined as the ratio of the mass of water vapour in a certain volume of moist air at a given temperature to the mass of water vapour in the same volume of saturated air at the same temperature and is denoted by **RH** or Φ.

   Relative Humidity (At a given temperature)

   = (Water vapour actually present)/(Water vapour required for saturation)

   Here a comparison is made between the humidity of air and the humidity of saturated air at the same temperature and pressure. It is to be noted that if relative humidity is 100%, it is saturated air, i.e., the air contains all the moisture it can hold.

   It should also be noted that the degree of saturation (percentage of relative humidity) of air keeps on changing with temperature.
7. **Dew Point Temperature:** By continuous cooling at constant pressure if the temperature of air is reduced, the water vapour in the air starts condensing at a particular temperature. This temperature at which the water vapour starts condensing is known as dew point temperature.

More Instruments for Learning Engineers

An **alcoholmeter**, also known as a proof & Tralles hydrometer (named after Johann Georg Tralles) determines the alcoholic strength of liquids. It only measures the density of the fluid with certain assumptions. It has a scale marked with volume percent of "potential alcohol", based on a pre-calculated specific gravity. Approximate alcohol content is determined by subtracting the post fermentation reading from the pre-fermentation reading.

Alcoholmeter

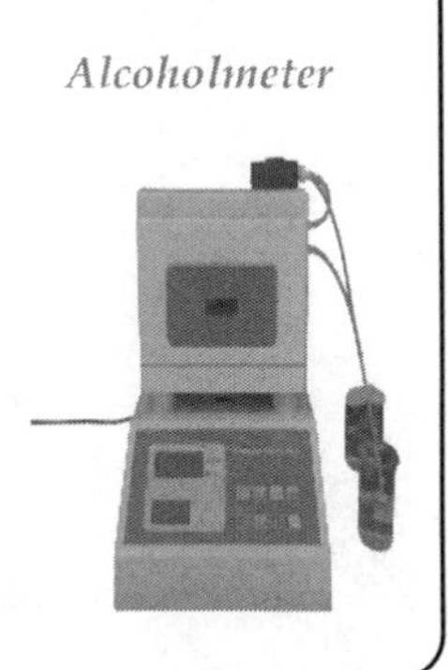

8. **Dry Bulb Temperature:** When a thermometer bulb is directly exposed to an air -water vapour mixture, the temperature indicated by the thermometer is the dry-bulb temperature. This dry-bulb temperate is not affected by the moisture present in the air i.e., the temperature of air is measured in a normal way by the thermometer.
9. **Wet Bulb Temperature:** When a thermometer bulb is covered by a wet wick, and if the bulb covered by the wet wick is exposed to air-water vapour mixture, the temperature indicated by the thermometer is the wet bulb temperature.
   When air passed on the wet wick present on the bulb of the thermometer, the moisture present in the wick starts evaporating and this creates a cooling effect at the bulb. The bulb now measures the thermodynamic equilibrium temperature reached between the cooling affected by the evaporation of water and heating by convection.
10. **Wet Bulb Depression:** It is the difference between dry bulb temperature and wet bulb temperature
    Wet bulb depression = (Dry bulb temperature) – (Wet bulb temperature)
    Always dry bulb temperature is higher than the wet bulb temperature.
    i.e., (Dry bulb temperature > Wet bulb temperature)
11. **Percentage Humidity:** It is defined as ratio of weight of water vapour in a unit weight of air to weight of water vapour in the same weight of air if the air were completely saturated at the same temperature.
    Percentage Humidity = (Weight of water vapour in a unit weight of air)/(Weight of water vapor in the same weight of air were completely saturated at the same temperature)

14.3 LIST OF INSTRUMENTS USED TO MEASURE HUMIDITY

The three most popular instruments used for measuring humidity are

1. Sling Psychrometer
2. Absorption hygrometer
   (a) Mechanical humidity sensing absorption hygrometer.
   (b) Electrical humidity sensing absorption hygrometer.
3. Dew point meter.

SELF ASSESSMENT QUESTIONS-14.1

1. What is humidity?
2. List out the terms which are related to humidity measurement.
3. Briefly explain the following terms.
   (a) Dry Air (b) Moist Air (c) Saturated Air
4. Discuss the following terms.
   (a) Specific Humidity (b) Relative Humidity (c) Percentage Humidity
5. Briefly explain the following terms
   (a) Wet Bulb Temperature (b) DewPoint Temperature
   (c) D ry Bulb Temperature (d) Wet Bulb Depression
6. List out the instruments used to measure humidity.

14.4 PSYCHROMETERS (WET AND DRY BULB THERMOMETERS)

1861 diagram of a psychrometer with wet bulb (a) and dry bulb (b). The wet bulb is connected to a reservoir of water.

The interior of a Stevenson screen showing a motorized psychrometer A psychrometer consists of two thermometers, one which is dry and one which is kept moist with distilled water on a sock or wick. The two thermometers are thus called the dry-bulb and the wet-bulb. At temperatures above the freezing point of water, evaporation of water from the wick lowers the temperature, so that the wet-bulb thermometer usually shows a lower temperature than that of the dry-bulb thermometer. When the air temperature is below freezing, however, the wet-bulb is covered with a thin coating of ice and may be warmer than the dry bulb. Relative humidity is computed from the ambient temperature as shown by the dry-bulb thermometer and the difference in temperatures as shown by the wet-bulb and dry-bulb thermometers. 3 can also be determined by locating the intersection of the wet and dry-bulb temperatures on a psychrometric chart. The two thermometers coincide when the air is fully saturated, and the greater the difference the drier the air. Psychrometers are commonly used in meteorology, and in the HVAC industry for proper refrigerant charging of residential and commercial air conditioning systems.

Fig. 14.1 Psychrometer

14.5 SLING PSYCHROMETER

The main parts of the instrument are:

1. Wet bulb thermometer (with wick)
2. Dry bulb thermometer
3. Frame
4. Glass casing
5. Swivel joint
6. Swivel handle

More Instruments for Learning Engineers

Mile 14.2

A **thermohydrometer** is a hydrometer having a thermometer enclosed in the float section. For measuring the density of petroleum products, like fuel oils, the specimen is usually heated in a temperature jacket with a thermometer placed behind it since density is dependent on temperature. Light oils are placed in cooling jackets, typically at 15°C. Very light oils with many volatile components are measured in a variable volume container using a floating piston sampling device to minimize light end losses.

Thermohydrometer

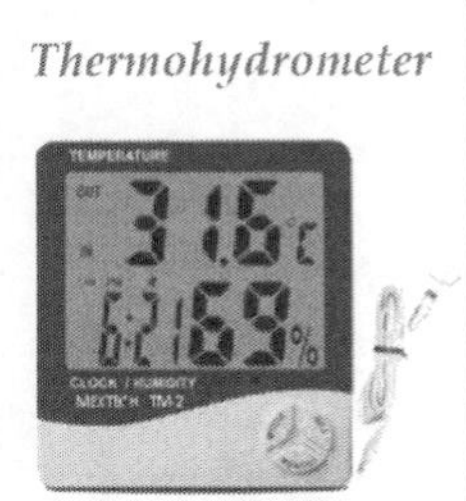

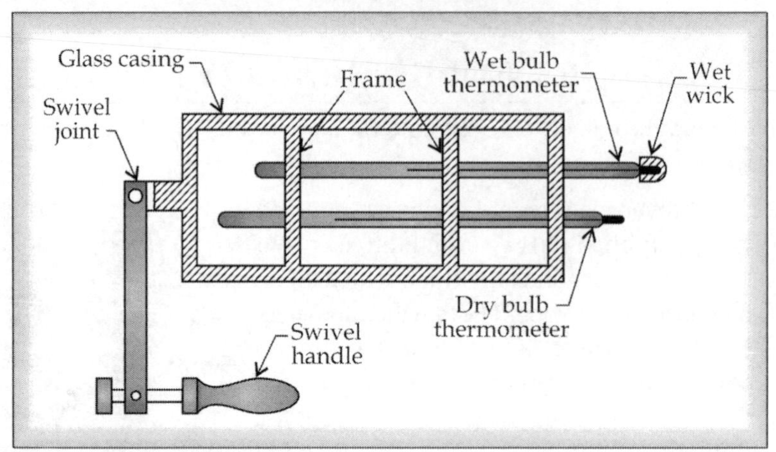

Fig. 14.2 Sling Psychrometer

Principle

This instrument is used to measure both the drybulb and wet bulb temperatures at a time and with these temperatures the humidity content in air can be measured.

Construction

A frame holds two mercury-in-glass thermometers, one to measure dry bulb temperature and the other to measure wet bulb temperature as shown in Figure 14.1. The frame carrying these thermometers is covered by a glass casing. A swivel handle is attached to this arrangement of frame-glass casing-thermometers to ensure that the air at the wet bulb is always in immediate contact with the wet wick.

Operation

To measure wet bulb temperature, the psychrometer arrangement of parts (frame-glass covering-thermometer) as shown in Figure 14.1 is rotated at 5 m/s to 10 m/s to get the necessary air motion. This is because the accurate measurement of wet bulb temperature is obtained only if air moves with a velocity around the wet wick.

The thermometer whose bulb is bare (without wick) contacts the air and indicates the dry bulb temperature. At the same time, the thermometer whose bulb is covered with the wet wick comes in contact with the air and when this air passes on the wet wick, the moisture present in the wick starts evaporating and a cooling effect is produced at the bulb. This temperature is the wet bulb temperature, and is obviously less than the dry bulb temperature.

Demerits

- Continuous recording of humidity is not possible.
- The evaporation process at the wet bulb adds moisture to the air, which disturbs the measured medium.
- Automation is not possible with these instruments.
- If the wick is covered with dirt, the wick becomes stiff and its water absorbing capacity reduces.

Applications

It is used to:

- Check humidity level in air conditioned rooms and installations.
- Set and check hair hygrometers.
- Measure range of 0 to 100% RH and can measure wet bulb temperatures between 0 °C to 180 °C.
- Measure wet bulb temperatures between 0 °C to 180 °C.

14.6 ABSORPTION HYGROMETERS

A **hygrometer** measures moisture content in the atmosphere. Humidity measurement instruments usually rely on measurements of some other quantity such as temperature, pressure, mass or a mechanical or electrical change in a substance as moisture is absorbed. By calibration and calculation, these measured quantities can lead to a measurement of humidity. Modern electronic devices use temperature of condensation (the dew point), or changes in electrical capacitance or resistance.

Fig. 14.3 Hygrometer

Principle

Humidity can vary the physical, chemical and electrical properties of several materials. We make use of this property by applying suitable transducers that are designed and calibrated to read relative humidity directly.

There are two types of absorption hygrometers namely.

1. Mechanical humidity sensing absorption hygrometer.
2. Electrical humidity sensing absorption hygrometer.

14.6.1 Mechanical Humidity Sensing Absorption Hygrometer

The main parts of the mechanical hair hygrometer are:

1. Humidity sensor (hair/animal membrane/wood etc.)
2. Arm
3. Link
4. Tension spring
5. Pointer
6. Calibrated scale

Principle

Hygroscopic materials such as human hair, animal membranes, wood, paper etc, undergo changes in linear dimensions when they absorb moisture from their surrounding air. This change in linear dimension is used for the measurement of humidity present in air.

Construction

This instrument uses human hair as the humidity sensor. The hair is arranged in parallel beam and they are separated from one another to expose them to the surrounding air. Numbers of hairs are placed in parallel to increase mechanical strength as shown in Figure 14.4. This hair arrangement is subjected to light tension by the use of a tension spring to ensure proper functioning.

The hair arrangement is connected to an arm and a link arrangement which in turn is fixed with a pointer, pivoted sweeps on a humidity calibrated scale.

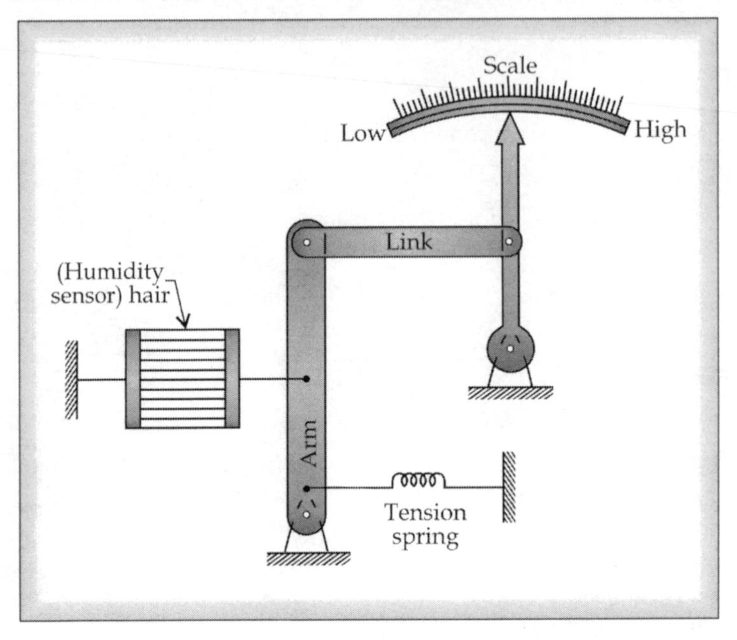

Fig. 14.4 Hair Hygrometer

Operation

As and when the humidity of air is to be measured, the hair arrangement is exposed to the air medium. Obviously, this absorbs the humidity from the surrounding air and expands or contracts in the linear direction.

The expansion/contraction of the arrangement moves the arm and hence the link and thus the pointer on the calibrated scale, indicating the humidity present in the atmosphere.

These hygrometers are called *'membrane hygrometers'* when the sensing element is a membrane.

Demerits

- Calibration tends to change if is it used continuously
- Response time is slow

More Instruments for Learning Engineers

Mile 14.3

Antifreeze Tester is used in automotives to test the quality of the antifreeze solution used for engine cooling. The degree of freeze protection can be related to the density (and so concentration) of the antifreeze; different types of antifreeze have different relations between measured density and freezing point.

Antifreeze Tester

Applications

- It is used in temperature ranges of 0 to 75 °C.
- Relative humidity (RH) range is 30 to 95%.

14.6.2 Electrical Humidity Sensing Absorption Hygrometer

The main parts of this instrument are:

1. Hygroscopic salt (Lithium chloride)
2. Electrodes
3. Base
4. Electrical leads (positive/negative)
5. Wheatstone bridge circuit
6. Calibrated scale

Principle

Humidity varies the resistance of hygroscopic material which can become the measure of humidity when calibrated.

Construction

The two metal electrodes, which are coated and separated by a humidity sensing hygroscopic salt (lithium chloride), are arranged in a base as shown in Fig. 14.5. The leads of the electrodes are connected to a null balance Wheatstone bridge circuit to measure the resistance.

Operation

The electrodes coated with lithium chloride are kept open to atmosphere, where humidity is to be measured. Then, due to the variation in humidity, the resistance of lithium chloride changes (increases/decreases) as the chemical absorbs or loses moisture. Higher the humidity (*RH*) in the atmosphere more will be the moisture absorbed by lithium chloride and lower will be the resistance (and higher will be the resistance in case of lesser humidity). The change in resistance can be measured using a Wheatstone bridge which becomes a measure of humidity (RH) present in the atmosphere.

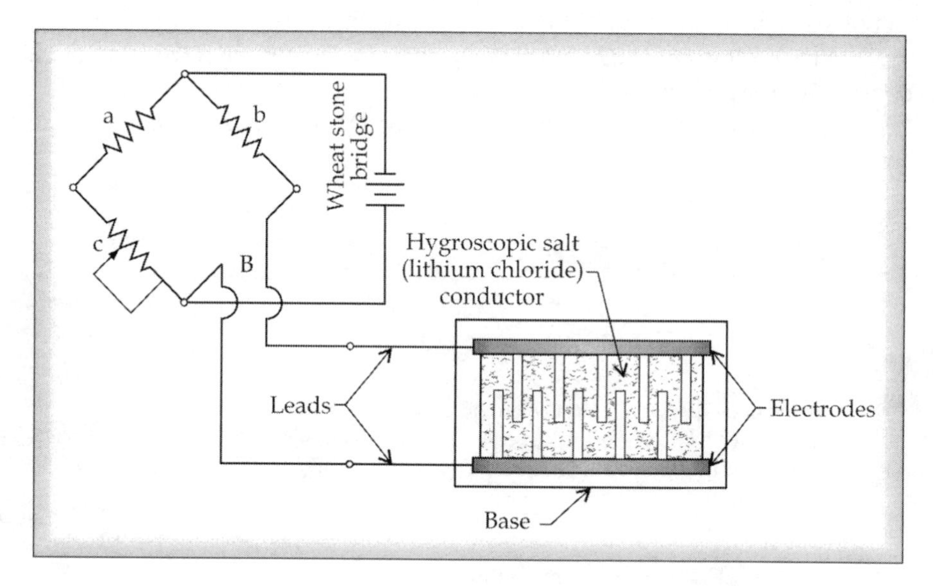

Fig. 14.5 Electrical Humidity Sensing Absorption Hygrometer

Merits

- The accuracy of this instrument is ± 25%
- The response is very fast, of the order of few seconds.

Demerits

- Temperature corrections must be made if they are not used at constant temperature conditions.
- This instrument should not be exposed to 100% humidity because this makes chemical absorb all the humidity and damage the instrument.

Applications

- These are useful in case if these are used under constant temperature conditions.
- These are useful when high response is needed

14.7 DEW POINT METER

The chief parts of this instrument are:

1. A shiny mirror surface
2. A thermocouple
3. A nozzle
4. A mirror.
5. A light source with a lens
6. A photo cell
7. Cup cylinder
8. Cooling medium
9. Refrigerant liquid
10. Stirrer
11. Heater

Principle

At constant pressure if the temperature of air is reduced, the water vapour in the air will start to condense at a particular temperature. This temperature is called dew point temperature.

Construction

A shiny mirror surface is fixed with a thermocouple as shown in Fig. 14.6. A nozzle is arranged to provide a jet of air on the mirror. A light is focused through a lens constantly on the mirror from a light source. A photo cell is attached to detect the amount of light reflected from the mirror. The mirror is constantly cooled by a cooling medium, which is maintained at a constant temperature. A line is taken to heater to heat the shiny surface i.e. mirror again and again to measure the dew point temperature repeatedly.

More Instruments for Learning Engineers

A **urinometer** is a medical hydrometer designed for urinalysis. A urinometer measures urine's specific gravity, found by its ratio of solutes (wastes) to water, and so, can assess a patient's overall level of hydration.

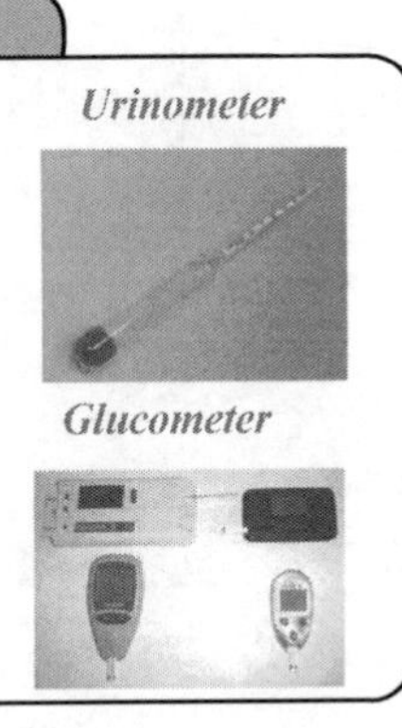

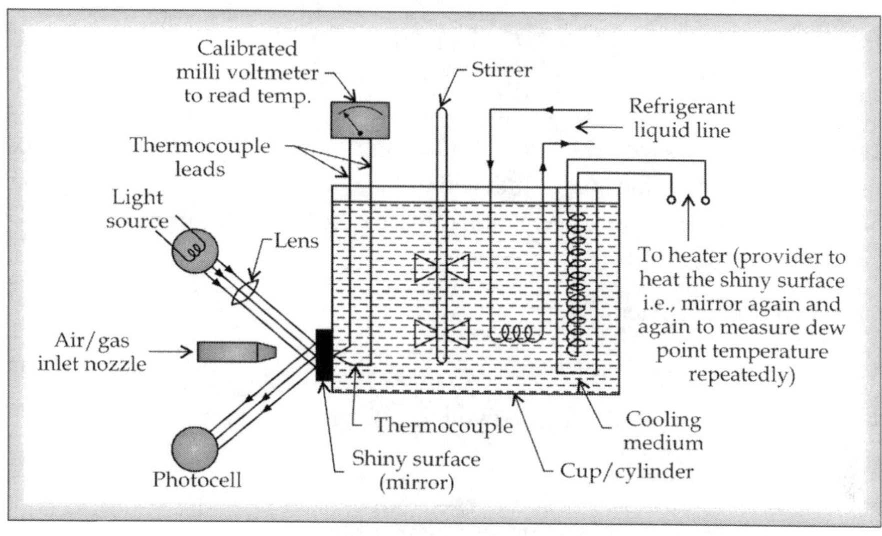

Fig. 14.6 Dew Point Meter

Operation

A thermocouple is attached to the mirror, whose leads are connected, to a milli voltmeter. Through a lens, a light from a source is made to fall constantly at an angle on the mirror and the amount of reflected light is sensed by a photo cell as shown in Fig. 14.6. Now an air jet is projected on the mirror and the water- vapour (moisture) content in the air starts condensing on the mirror and they appear as small drops (dews) on the mirror. This moisture formed on the mirror lowers the amount of light reflected, which is detected by the photocell. When for the first time, there is a change in the amount of transmitted light; it becomes an indication of dew formation. At this instance i.e., the temperature indicated by the thermocouple attached to the mirror becomes the dew-point temperature. Thus, this instrument is used to know the time at which the dew appears for the first time and to know the dew point temperature.

Demerits

- Effective light measurement is very difficult.
- Cooling fluids exists.

Applications

- Cargoes can be protected from condensation damage by this instrument by maintaining the dew point of air in holds lower than the cargo temperature.
- Used in industries for determining dew point.

Self Assessment Questions-14.2

1. With neat sketch explain the construction and operation of Sling Psychrometer.
2. What is absorption hygrometer? What are its types?
3. Discuss the mechanical humidity sensing absorption hygrometer with neat sketch.
4. Explain the Electrical Humidity sensing Absorption Hygrometer with its merits, demerits and applications with neat diagram.
5. With neat sketch discuss the Dew Point Meter. What are its applications?

14.8 FUNDAMENTALS OF DENSITY AND SPECIFIC GRAVITY

To determine and control concentrations of fluids in process control situations measurement of density and specific gravity is the most suitable method. Further, this fluid property can be used as a means for measuring products directly. Some important terms and definitions most useful in this context are given below:

(a) Specific Gravity (In General)

$$\text{Specific Gravity} = \frac{\text{Mass of a given volume of the substance}}{\text{Mass of an equal volume of some substance}}$$

(b) Specific Gravity of Liquids

$$= \frac{\text{Mass of the volume of liquid}}{\text{(Mass of an equal volume of distilled water at some standard reference temperature, usually 4 °C}}$$

(c) Specific Gravity of Gases

$$= \frac{\text{Mass of the volume of gas}}{\text{(Mass of equal volume of air 20 °C and at a pressure of 74.8 mm of Hg}}$$

(d) Specific Weight
Specific weight is the weight per unit volume. (Sp.Wt. = Weight/Volume)

(e) Density
Density is defined as the mass per unit volume (Density = Mass/Volume)

More Instruments for Learning Engineers

Battery Hydrometer: The status of charge of lead-acid battery is estimated from the density of the electrolyte (sulfuric acid solution). A hydrometer calibrated to read specific gravity relative to water at 60°F is a standard tool for servicing automobile batteries. Tables are used to correct the reading to the standard temperature.

Battery Hydrometer

14.9 INSTRUMENTS TO MEASURE SPECIFIC GRAVITY AND DENSITY

The following are the most popular instruments used to measure the specific gravity (or density) and/or relevant physical quantities.

1. Pycnometer
2. Hydrometers
   (a) Hydrometers – simple type
   (b) Hydrometers – photo electric type.
3. Bubbler systems of differential type.
   (a) Bubbler systems of differential type - two vessels attached by a manometer reference column method.
   (b) Bubbler system of differential type - single vessel attached with pressure gauge.

We shall now discuss the construction and operation of these instruments.

14.10 PYCNOMETER

Relative density, or **specific gravity**, is the ratio of the density (mass of a unit volume) of a substance to the density of a given reference material. Specific gravity usually means relative density with respect to water. The term "relative density" is often preferred in modern scientific usage.

If a substance's relative density is less than one then it is less dense than the reference; if greater than 1 then it is denser than the reference. If the relative density is exactly 1 then the densities are equal; that is, equal volumes of the two substances have the same mass. If the reference material is water then a substance with a relative density (or specific gravity) less than 1 will float in water. For example, an ice cube, with a relative density of about 0.91, will float. A substance with a relative density greater than 1 will sink.

Temperature and pressure must be specified for both the sample and the reference. Pressure is nearly always 1 atm equal to 101.325 kPa. Where it is not, it is more usual to specify the density directly. Temperatures for both sample and reference vary from industry to industry. In British brewing practice the specific gravity as specified above is multiplied by 1000. Specific gravity is commonly used in industry as a simple means of obtaining information about the concentration of solutions of various materials such as brines, sugar solutions (syrups, juices, honeys, brewers wort, must, etc.,) and acids.

This instrument consists of two units as

1. The sensitive analytical balance and
2. The pycnometer which is composed of the main parts as
   (a) Glass bottle of definite volume
   (b) Ground glass stopper
   (c) Hole for overflow

Principle

The ratio densities is equal to the ratio of masses when volume is same (constant) which is equal to the ratio of the weights when measured at the same place (acceleration due to gravity is constant). Thus the ratio of weight of the given liquid to that of boiled distilled water at a given temperature becomes the measure of the specific gravity of the liquid.

Construction

This instrument contains a straight walled glass bottle with definite volume, a ground glass stopper with an overflow hole at its centre. The stopper can exactly fit into the glass bottle. The sketch of a simple pycnometer is shown in the Figure 14.7.

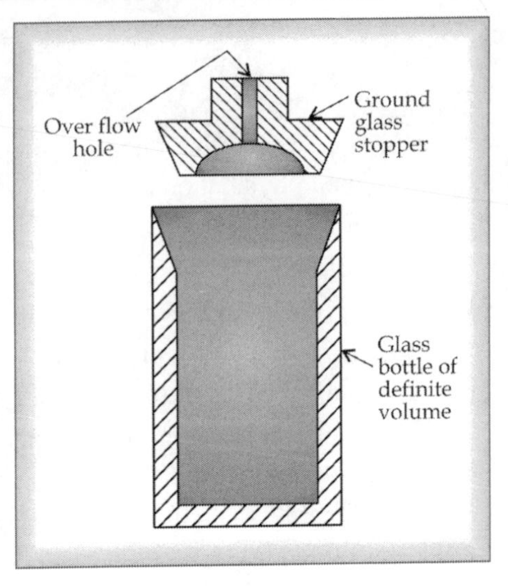

Fig. 14.7 Pycnometer

Operation

For measuring specific gravity of a liquid, the following procedure is adopted:

- First of all, the weight of the empty pycnometer with the stopper is determined using a sensitive analytical balance (say, W_1)
- Now the glass bottle of pycnometer is filled with boiled distilled water at the desired temperature. Then the ground stopper is fitted on to the glass bottle and the excess water drains out from the overflow hole. Wipe off the excess over flown water using a clean cloth. Now the weight of the pycnometer with water & stopper is determined using the same sensitive analytical balance (say, W_2)
- Now empty the glass bottle of pycnometer, wipe it cleanly using a cloth and fill it with the liquid whose specific gravity is to be determined. Fit the stopper and wipe the excess liquid which comes out from the overflow hole. Now the weight of the pycnometer with the liquid (whose specific gravity is required) and stopper is determined using the sensitive analytical balance (Say, W_3)

More Instruments for Learning Engineers

Lactometer is used to check purity of milk. The specific gravity of milk does not give a conclusive indication of its composition as milk contains a variety of substances either heavier or lighter than water. So, additional tests for fat content are necessary to determine overall composition. It is graduated into a hundred parts. Milk is poured in and allowed to stand until the cream has formed, then the depth of the cream deposit in degrees determines the quality of the milk. The device works on the principle of Archimedes's principle that a solid suspended in a fluid will be buoyed up by a force equal to the weight of the fluid displaced. If the milk is pure, the lactometer floats on it and if it is adulterated or impure, then the lactometer sinks.

AcidoMeter

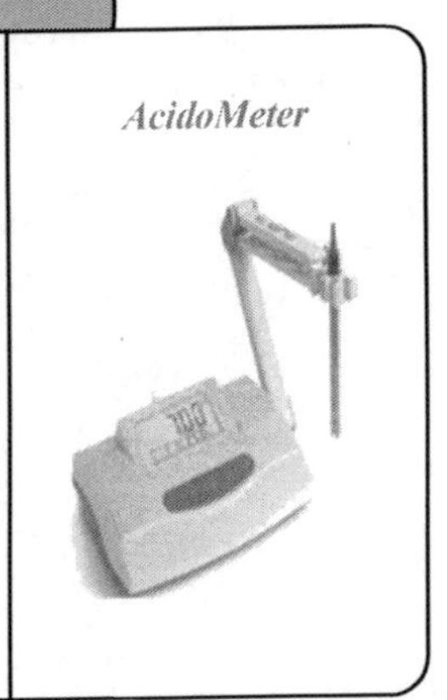

- Specific gravity is calculated as follows:

  Weight of empty bottle of pycnometer with ground glass stopper = W_1

  Weight of pycnometer with ground glass stopper and boiled, distilled water at a desired temperature = W_2

  Weight of boiled, distilled water at a desired temperature = $W_2 - W_1$

  Weight of pycnometer with ground glass stopper and liquid whose Sp.Gr. is to be found at the temperature = W_3

  Weight of liquid whose Sp.Gr. is to be found at the temperature = $W_3 - W_1$

  $$\text{Specific gravity of the liquid} = \frac{W_3 - W_1}{W_2 - W_1}$$

Applications

- To measure specific gravity of liquids and semi liquids of low melting points.
- To measure specific gravity of powder and granular solids.

14.11 HYDROMETERS

A **hydrometer** measures the specific gravity (or relative density) of liquids; that is, the ratio of the density of the liquid to the density of water. A hydrometer is usually made of glass and consists of a cylindrical stem and a bulb weighted with mercury or lead shot to make it float upright. The liquid to be tested is poured into a tall container, often a graduated cylinder, and the hydrometer is gently lowered into the liquid until it floats freely. The point at which the surface of the liquid touches the stem of the hydrometer is noted. Hydrometers usually contain a scale inside the stem, so that the specific gravity can be read directly. A variety of scales exist, and are used depending on the context.

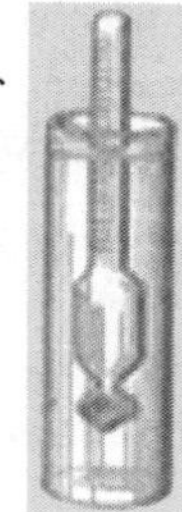

Hydrometers may be calibrated for different uses, such as a lactometer for measuring the density (creaminess) of milk, a saccharometer for measuring the density of sugar in a liquid, or an alcoholometer for measuring higher levels of alcohol in spirits.

Principle

Operation of the hydrometer is based on Archimedes' principle that a solid suspended in a fluid will be buoyed up by a force equal to the weight of the fluid displaced by the submerged part of the suspended solid. Thus, the lower the density of the substance, the farther the hydrometer will sink.

There are two types of hydrometers namely:

1. Simple Hydrometer
2. Photo-electric Hydrometer

14.11.1 Simple Hydrometer

The main parts of the simple hydrometer are:

1. Test liquid
2. Transparent glass container with drain tap
3. Cap/lid of container chamber with a hole for float-stem
4. Calibrated stem
5. Float - Weight

Principle

The depth to which the float sinks in a liquid depends on the density on the liquid. This is in accordance with Archimedes principle.

Construction

This instrument is composed of a container with a drain tap and holds the test liquid whose density is to be measured. It is closed with a lid/ cap, which contains a hole to introduce the float. A float fitted with a weight on its bottom end and a graduated and calibrated stem of constant diameter on its other end. When the float is put in the test liquid, the stem is always in an upright (vertical) position due to the presence of the weight at the bottom of the float as shown in the Figure 14.8.

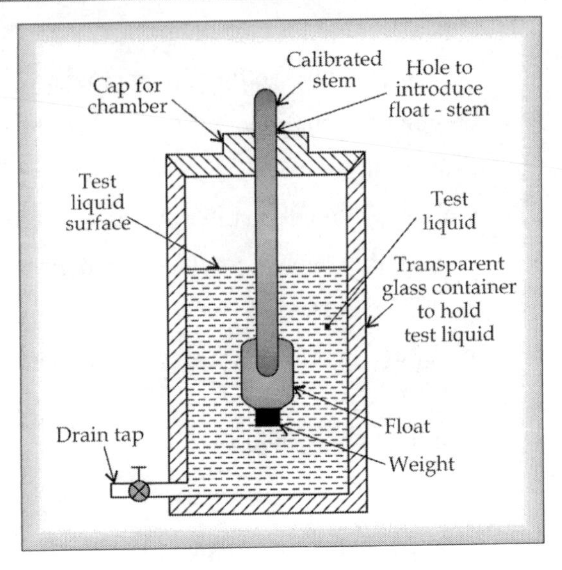

Fig. 14.8 Simple Hydrometer

Operation

The test liquid is partially filled in the container and cap is placed on it. The float carrying the stem and weight is introduced into the test liquid from the hole provided on the cap and see that the arrangement will float upright (vertically) in the test liquid. The float possesses a constant weight and so its buoyant force is constant. Thus, the float sinks to the depth inversely proportional to the density of the liquid. The higher the density of the test liquid, lesser is the depth to which the float sinks, i.e., more it will buoy up the float, and vice versa. On the calibrated stem, the reading corresponding to the surface of the test liquid becomes the measure of density or specific gravity of the test liquid.

14.11.2 Photo-Electric Hydrometer

The main parts of the photo electric hydrometer are:

1. Test liquid
2. Container/chamber with inlet and outlet

More Instruments for Learning Engineers

Mile 14.7

A **saccharometer**, invented by Thomas Thomson determines the amount of sugar in a solution. It is used primarily by winemakers and brewers and also finds use in making sorbets and ice-creams. The first brewers' saccharometer was constructed by Benjamin Martin (with distillation in mind) and initially used by James Baverstock Sr (1770). Henry Thrale adopted it but was later popularized by John Richardson (1784). It consists of a large weighted glass bulb with a thin stem rising from the top with calibrated marks. The sugar level is determined by reading the value where the surface of the liquid crosses the scale. It works by the principle of buoyancy. A solution with a higher sugar content is denser, causing the bulb to float higher. Less sugar results in a lower density and a lower floating bulb.

Saccharometer

3. Cap/lid
4. Opaque stem
5. Hollow glass float with weight
6. Light source (lamp)
7. Output (photo tube)

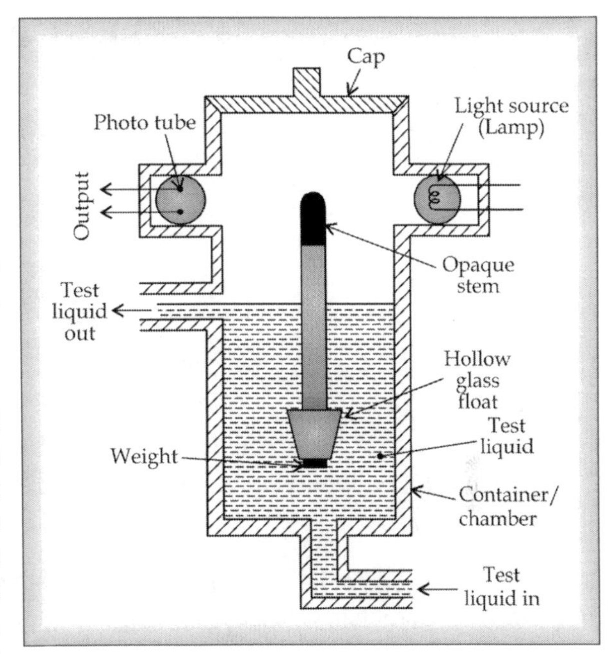

Fig. 14.9 Photo-electric Hydrometer

Principle

Output (amount of light received) of photo electric transducer is inversely proportional to density.

Construction

A chamber with inlet and outlet is designed to hold the test liquid whose density is to be measured. It should be noted that the test liquid flows continuously through the chamber. A hollow glass float with an opaque stem fitted on its top side and a weight fitted at its bottom side is placed inside the chamber. When the float is put in the test liquid, the opaque stem is always in an upright (vertical) position due to the presence of the weight at the bottom of the float. A light source (lamp) and a photo-electric transducer are housed at the top portion of the chamber exactly as shown in the Figure 14.9.

Operation

The test-liquid whose density is to be found is sent to flow through the chamber. Now, the hollow glass float with the opaque stem is immersed in the test liquid. Obviously it starts floating in an upright (vertical) position due to weight at the bottom. Now, a light source housed in the top portion of the chamber in a slit produces the light ray. This is received by the photo-electric transducer, provided at a position diametrically oppositely at the same height to that of the light source. The portion of the opaque stem will be between the light source and the photo-electric transducer as shown in the Figure 14.7.

The more the density of the test liquid is, the lesser will be the depth to which the float sinks, i.e., more it will buoy up the float, and vice versa. This means, for dense liquids, the stem rises to a greater height while for less dense liquids, the float sinks and the stem is lowered.

So, if density of denser liquids is measured, the opaque stem cuts off more light since it rises higher and hence the output of the photo-electric transducer will be lesser. On the other hand, if the density of the liquid is lesser, the stem comes down allowing more light to be transmitted to the photo-electric transducer which now gives a higher output. Thus, the output of the photo-electric transducer is inversely proportional to the density. The photo-electric transducer is calibrated so that its output becomes a measure of the density of the test liquid.

Application

- The photo-electric hydrometer is better applicable for continuous monitoring of the density of liquids or continuously flowing liquids.

14.12 DIFFERENTIAL TYPE BUBBLER SYSTEMS

There are two differential bubbler systems namely:

1. Differential type bubbler system with two vessels attached by a manometer and a reference column method
2. Differential type bubbler system with single vessel attached with pressure gauge.

14.12.1 Differential Type Bubbler System with Two Vessels Attached by a Manometer and Reference Column Method

The main parts of the arrangement are:

1. Test liquid
2. Standard/Reference liquid
3. Two vessels with inlet and outlets (one for test-liquid and other for reference liquid)
4. Manometer with pressure regulators
5. Bubbler tubes

Principle

This method uses the principle $p = \rho gH$, i.e., for a constant depth of the liquid, the force exerted by the liquid is proportional to density.

Construction

This instrument contains two vessels, one to hold the test liquid whose density is to be measured and a second to hold a standard reference liquid. Both the vessels are filled to equal height. Two bubbler tubes of same height are introduced into both the vessels for a height H. Both the bubbler tubes are connected with separate air pressure regulators. The bubbler tubes are interconnected by a U-tube manometer as shown in Figure 14.10.

Operation

The operation starts with filling first vessel with the test liquid whose density is to be measured and the standard reference liquid in the other vessel. Both the vessels are filled to equal height. Now the pressure of air to each bubbler tube is regulated to make air bubbles to be released slowly. At this stage, the height difference in the U-tube manometer is noted as H_m (indicated differential pressure). The density of the test liquid (ñ) can be calculated using the following expression:

More Instruments for Learning Engineers

Mile 14.8

A **barkometer** is calibrated to test the strength of tanning liquors used in tanning leather.

Sometimes spelled **acidimeter**, is a hydrometer used to measure the specific gravity of an acid.

Barkometer

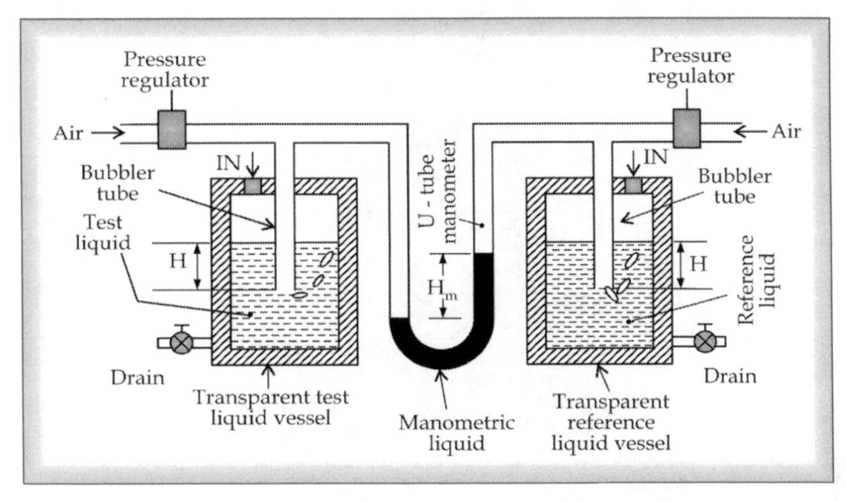

Fig. 14.10 Differential Type Bubbler System with 2-Vessels Attached by Manometer & Reference Column Method

$$\rho = \left(\frac{\rho_m . H_m}{H} + \rho_r\right)$$

Where ρ_m = Density of manometric liquid

ρ_r = Density of reference liquid.

H_m and H = heights of liquids and manometer reading.

14.12.2 Differential Type Bubbler System with Single Vessel Attached and Pressure Gauge

The salient parts of the arrangement are:

1. Test liquid
2. Container with inlet and drain
3. Pressure regulators (two)
4. Pressure gauge
5. Bubbler tubes (two)

Construction

This instrument contains a container to hold the test liquid whose density is to be measured. Two bubbler tubes with a fixed distance between their open ends are immersed in the test liquid. Both the bubbler tubes are connected with separate air regulators. Further, both the bubbler tubes are connected to a differential pressure gauge as shown in Figure14.11.

Operation

The test liquid is filled into the container so that both the bubbler tubes are immersed in the test liquid. The pressure of air to each bubbler tube is regulated to make air bubbles to be released slowly. At this point, the indicated differential pressure by the pressure gauge is a measure of the density of the test liquid on account of the following factors:

- Differential height, which is known
- The liquid specific gravity that determines the liquid density.

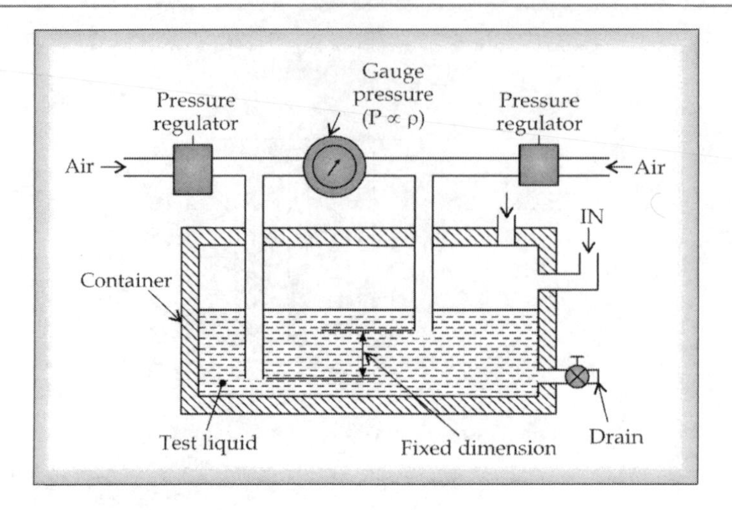

Fig. 14.11 Differential Type Bubbler System with Single Vessel Attached and Pressure Gauge

Self Assessment Questions-14.3

1. What are the instruments used to measure the specific gravity?
2. What is the principle of Pycnometer?
3. Discuss the Pycnometer with neat sketch.
4. With neat sketch explain simple hydrometer.
5. Explain the photo electric hydrometer with neat sketch.
6. What are differential bubbler systems?
7. Explain the differential type bubbler system with two vessels attached by a manometer and a reference column method with neat sketch.
8. With neat sketch discuss the differential type bubbler system with single vessel attached and Pressure Gauge.

14.13 SPECIAL APPLICATIONS OF HYDROMETERS

The principle of hydrometer is extensively used in various area with specialized applications. Some of these specific applications are listed below.

Alcohol meter: To measure the alcoholi strength in the liquid (beverages).

Acidometer (or Acidimeter): To measure the acid strength or pH.

Urinometer: For urinalysis (measures urine's Sp. gravity).

Battery hydrometer: To estimate status of change of lead-acid battery (by density of electrolyte).

Lactometer: To measure purity of milk.

Saccharometer: To measure sugar content in a solution.

Barkometer: To test the strength of of tanning liquors used tanning leather.

SUMMARY

Humidity refers to the water content in air. Saturated air is the moist air where the partial pressure of vapour equals the saturation pressure of steam. Humidity ratio or specific humidity is the ratio of the mass of vapour to that of dry air. Relative humidity is the ratio of the mass of vapour in a certain volume of moist air at a given temperature to that of vapour in the same volume of saturated air at the same temperature. Dew-point is the temperature at which the vapour starts condensing. It is temperature shown by the thermometer directly exposed to an air-water and if the bulb covered by the wet wick exposed to air, it is the wet bulb temperature. Wet Bulb Depression is the difference between these two.

Sling psychrometer can measure both dry and wet bulb temperatures at a time. Absorption hygrometers make use of the property that humidity can vary the physical, chemical and electrical properties. Under humidity sensing absorption hygrometers, the mechanical hygrometers use human hair/ animal membranes which undergo changes in linear dimensions when they absorb moisture from the air while electrical hygrometers operate on the principle that humidity varies the resistance of hygroscopic material. Dew-point meter works at constant pressure if the temperature of air is reduced, the water vapour in the air starts condensing at a particular temperature, called dew point.

Density is defined as the mass per unit volume. Specific gravity in general is the ratio of density of given to substances to that of the standard one (e.g. water for liquids, air for gases). Specific weight is the weight per unit volume.

The specific gravity is measured by Pycnometer, Hydrometers (simple/ photo electric types), Bubbler systems of differential type (two/ single vessels). Pycnometer works on the principle that the ratio of densities is equal to the ratio of masses (hence, weights) when volume is same (constant) measured at the same place. Thus the ratio of weight of the given liquid to that of boiled distilled water at a given temperature is the measure of the specific gravity of the liquid. Hydrometer measures the depth to which the float sinks in a liquid which depends on the density on the liquid. This is in accordance with principles of buoyancy and Archimedes. Photo-electric hydrometer measures output (amount of light received) of photo electric transducer as inversely proportional to density. Differential type bubbler system (with single/ two vessels) uses the principle $p = \rho g H$, i.e., for a constant depth of the liquid, the force exerted by the liquid is proportional to density.

KEY CONCEPTS

Humidity: The amount of water vapour contained in air or gas.

Dry Air: No water vapour contained in the atmosphere.

Moist Air: There is water vapour contained in the atmosphere.

Saturated Air: The moist air where the partial pressure of water-vapour equals the saturation pressure of steam corresponding to the air temperature.

Humidity Ratio or Specific Humidity or Absolute Humidity or Moisture Content: The ratio of the mass of water vapour to the mass of dry air in a given volume of the mixture and is denoted by *w*; Humidity Ratio = $M_{water\ vapour}/M_{dry\ air}$

Relative Humidity: The ratio of the mass of water vapour in a certain volume of moist air at a given temperature to the mass of water vapour in the same volume of saturated air at the same temperature and is denoted by **RH** or Φ

$$\text{Relative Humidity (At a given temp.)} = (WV_{actually\ present}) / (WV_{reqd.\ for\ saturation})$$

Dew Point Temperature: By continuous cooling at constant pressure if the temperature of air is reduced, the temperature at which the water - vapour in the air will start to condense.

Dry Bulb Temperature: When a thermometer bulb is directly exposed to an air -water vapour mixture, the temperature indicated by the thermometer

Wet Bulb Temperature: When a thermometer bulb is covered by a wet wick, and if the bulb covered by the wet wick is exposed to air-water vapour mixture, the temperature indicated by the thermometer

Wet - Bulb Depression: The difference between dry-bulb temperature and wet-bulb temperature. Wet bulb depression = $(T_{Dry\ bulb}) - (T_{Wet\ bulb})$

Percentage Humidity: The ratio of weight of water vapour in a unit weight of air to weight of water vapour in the same weight of air if the air were completely saturated at the same temperature.

$$\%\text{Humidity} = (W_{wvin\ unit\ wt.\ of\ air})/(W_{wvin\ the\ same\ wt.\ of\ completely\ saturated\ air\ at\ the\ same\ temp.})$$

Sling Psychrometer: A swivel handle is attached to the frame - glass casing - thermometer arrangement to ensure that the air at the wet bulb is always in immediate contact with the wet wick

Absorption Hygrometer: Measures humidity by absorption and works on fact that humidity changes the physical, chemical and electrical properties

Mechanical Humidity Sensing Absorption Hygrometer: Hygroscopic materials like human hair, animal membranes, wood, paper etc, undergo changes in linear dimensions when absorb moisture from their surrounding air which is sensed and measured.

Electrical Humidity sensing Absorption Hygrometer: Humidity changes the resistance of some material which is sensed and calibrated. The two metal coated electrodes and separated by a humidity sensing hygroscopic salt (lithium chloride) are connected to a null balance Wheatstone bridge.

Dew Point Meter: The reflection of light source focused constantly on the shiny mirror surface fixed with a thermocouple is detected by a photo cell.

Density: The ratio of mass to volume denoted by ρ ($\rho = m/V$)

Specific Gravity (In General): The ratio of mass of given volume of substance to that of an equal volume of some substance considered as standard

$$\text{Specific Gravity} = \frac{\text{Mass of a given volume of the substance}}{\text{Mass of an equal volume of some substance}}$$

Specific Gravity of Liquids: The ratio of mass of given volume of liquid to that of an equal volume of distilled water at some standard reference temperature usually at 4°C

$$\text{Specific gravity of liquids} = \frac{\text{Mass of the volume of liquid}}{\text{Mass of an equal volume of distilled water at some standard reference temperature, usually } 4^{\circ}\text{C}}$$

20. **Specific Gravity of Gases:** The ratio of mass of given volume of gas to that of an equal volume of air at 20°C and at a pressure 74.8 mm of Hg

    Specific gravity of gases =

$$\frac{\text{Mass of the volume of gas}}{\text{Mass of equal volume of air } 20^{\circ}\text{C and at a pressure of } 74.8 \text{ mm of Hg}}$$

21. **Specific Weight:** Specific weight is the weight per unit volume.
22. **Density:** Density is defined as the mass per unit volume
23. **Pycnometer:** A ground glass stopper exactly fit into the glass bottle with a over flow hole at its centre and a sensitive analytical balance measure the weights and the ratio is calculated
24. **Hydrometer:** Measures the density or specific gravity of liquids
25. **Simple Hydrometer:** The depth to which the float sinks in a liquid depends on the density on the liquid which is calibrated on a suitable scale.
26. **Photo-Electric Hydrometer:** Output of photo-electric transducer is inversely proportional to density. The test liquid flows continuously through a chamber with a hollow glass float (with a weight) at bottom and opaque stem on top along with a light source & photo-electric transducer. For dense liquids, the stem rises to greater height while for less dense, the float sinks and the stem is lower which may be calibrated.
27. **Differential Type Bubbler System (with 2-Vessels Attached by Manometer & Reference Column):** Works on $p = \rho gH$, i.e. for a constant depth, the force exerted by the liquid is proportional to density. 2-vessels, test liquid in 1st & standard reference liquid in 2nd filled to equal height and bubbler tubes of same height put in both for a height connected with separate air pressure regulators and interconnected by a U-tube manomete.
28. **Differential Type Bubbler System (with Single Vessel Attached & Pressure Gauge):** Two bubbler tubes connected with separate air regulators and differential pressure gauge with a fixed distance between their open ends are immersed in the test liquid.

REVIEW QUESTIONS

Short Answer Questions

1. Why humidity has to be measured?
2. Define humidity.
3. What are dry air, moist air, and saturated air?
4. What is absolute humidity and relative humidity?
5. What is humidity ratio?
6. Explain dew point temperature?
7. What is dry bulb temperature and wet bulb temperature?
8. What is wet bulb depression?
9. What is percentage humidity?
10. Explain the principle of a hygrometer.
11. Define specific gravity.

12. Define specific weight and density.
13. Give the procedure to measure specific gravity
14. Explain the basic principle of hydrometers.

LONG ANSWER QUESTIONS

1. Explain how a sling psychrometer is used to determine the dry and wet bulb temperatures?
2. Explain the working of any one of the absorption hygrometers.
3. What is the importance of humidity control in process industries?
4. How is relative humidity measured using hygrometer?
5. What are hygroscopic materials? Give examples.
6. Explain how a dew point meter is used to measure the dew point temperature?
7. Explain one application of finding due point temperature.
8. Name the different types of hygrometers used for measuring humidity?
9. How is dew point temperature measured?
10. Describe a method of measuring humidity of the atmospheric air.
11. What is a psychrometer? Explain the working of a psychrometer with neat sketch?
12. Name various methods for measuring humidity. Briefly compare their characteristic?
13. How absolute humidity is measured?
14. Explain how the density of a liquid is determined using a photo hydrometer.
15. Explain any one bubbler arrangement to measure density of liquids.

MULTIPLE CHOICE QUESTIONS

1. The amount of water vapour present in air is defined as ______.
   (a) density (b) viscosity
   (c) humidity (d) pressure
2. Saturated air is the moist air where the partial pressure of water vapour equals______.
   (a) atmospheric pressure
   (b) zero
   (c) the saturation pressure of steam
   (d) infinite
3. The ratio of the mass of water vapour to the mass of dry air in a given volume of the mixture is defined as_____.
   (a) relative Humidity
   (b) saturated Air
   (c) humidity Ratio
   (d) moist air
4. The temperature at which the water vapour starts condensing is known as_______.
   (a) dew point temperature
   (b) standard temperature
   (c) wet bulb temperature
   (d) dry bulb temperature
5. Wet bulb temperature is always_______ with reference to Dry bulb temperature
   (a) greater (b) lower
   (c) zero (d) more than Unity
6. Units of Specific weight is______.
   (a) newton/kg (b) Nm
   (c) N/m^3 (d) kg

7. Which instrument is used to measure Specific gravity________.
   (a) barometer (b) pycnometer
   (c) hydrometer (d) pitot tube
8. Dew point temperature is associated with__________.
   (a) phase change of a liquid at constant pressure
   (b) increase the temperature of a liquid at constant pressure
   (c) maintaining the same temperature at variable pressure
   (d) decreasing the temperature and decreasing the pressure
9. Which principle is used in hydrometer_____.
   (a) pascal law
   (b) buoyancy force
   (c) archimedes Principle
   (d) hydrostatic law
10. Hygrometers using membranes as sensing element for measuring humidity is called______.
    (a) hair hygrometer
    (b) membrane hygrometer
    (c) sling hygrometer
    (d) simple hygrometer
11. The difference of dry bulb temperature and wet bulb temperature is called________.
    (a) wet bulb depression
    (b) dry bulb depression
    (c) differential bulb temperature
    (d) critical temperature
12. Following is a hygroscopic material
    (a) plastic (b) hair
    (c) steel (d) moist air
13. The temperature indicated by a thermometer when it is exposed to an air-water vapour mixture is called __________ temperature
    (a) wet bulb
    (b) dry bulb
    (c) dew point
    (d) differential bulb temperature
14. If relative humidity is 100 % then air is called________air.
    (a) saturated (b) moist
    (c) dry (d) unsaturated
15. When there is water vapour contained in the atmosphere it is called ____________air.
    (a) dry (b) saturated
    (c) moist (d) unsaturated
16. Change in resistance refers to ________.
    (a) dew point meter
    (b) absorption hygrometer
    (c) sling psychrometer
    (d) simple hygrometer
17. Hygroscopic materials undergo change in ____________ when they absorb moisture
    (a) colour (b) dimension
    (c) temperature (d) odour
18. In __________ instrument humidity is measured by measuring both dry bulb and wet bulb temperature
    (a) hair hygrometer
    (b) membrane hygrometer
    (c) sling hygrometer
    (d) simple hygrometer
19. Which instrument among the following uses lithium chloride as sensing element?
    (a) absorption hygrometer
    (b) dew point meter
    (c) sling hygrometer
    (d) simple hygrometer
20. Light source is used among the following instrument
    (a) absorption hygrometer
    (b) dew point meter
    (c) sling hygrometer
    (d) simple hygrometer

ANSWERS

1.	(c)	2.	(c)	3.	(c)	4.	(a)	5.	(b)
6.	(c)	7.	(b)	8.	(a)	9.	(c)	10.	(b)
11.	(a)	12.	(b)	13.	(b)	14.	(a)	15.	(c)
16.	(b)	17.	(b)	18.	(c)	19.	(a)	20.	(b)

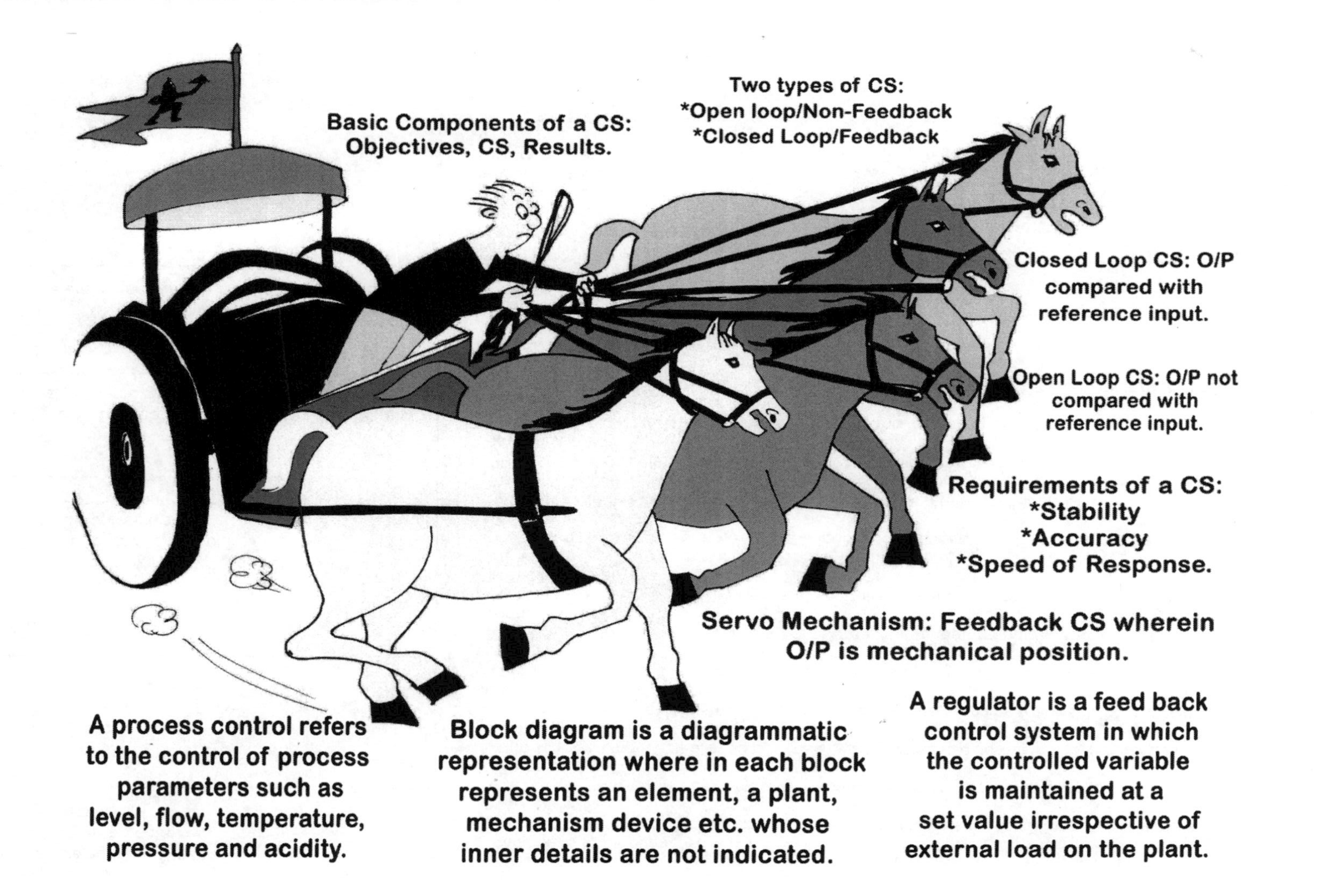

CHAPTER INFOGRAPHIC

15

Glimpses of Elements of Control Systems

Basic components of control systems

1. Objectives of control
2. System controls
3. Result

Requirements of control system

1. Stability
2. Accuracy
3. Speed of response

Examples of control systems

(a) Speed control and steering of automobile system
(b) Automobile tank level control
(c) Temperature control in a car
(d) Biological systems
(e) Position control system etc.

Negative feedback: Occurs when the feedback signal subtracts form the reference signal ($e = r - b$)

Positive feedback: If the feedback signal adds to the reference signal ($e = r + b$)

Speed of Response: The time taken by the system to respond to the give input and given that as output.

Ideal system: Perfectly stable, 100% accurate and has instantaneous speed of response.

Control Systems Terminology

1. **Processes, Plant or System:** A body, process or machine of which a particular quantity or condition is to be controlled
2. **Controlled Variable:** Quantity/condition that characterizes process through constant by controller or by certain law.
3. **Controlled Medium:** Process material in controlled system or flowing through it in which variable is to be controlled.
4. **Command:** An input established or varied by some means external to and independent of feedback control system.
5. **Set Point:** Signal established as standard of comparison for feedback control system by its relation to command.
6. **Manipulated Variable:** Quality/condition varied as function of actuating signal to change value of controlled variable.
7. **Actuating Signal:** An algebraic sum of reference input and primary feedback (also called error or control action)
8. **Primary Feedback Signal:** Function of controlled output compared with reference input to obtain actuating signal.
9. **Error-Detector:** An element that detects feedback
10. **Disturbance:** An undesirable variable applied to system, which tends to affect adversely on variable being controlled.
11. **Feedback Element:** An element that establishes functional relationship between controlled variable & feedback signal
12. **Control Element:** An element required to generate appropriate control signal (manipulated variable) applied to plane.
13. **Forward/Backward Paths:** Transmission path from actuating signal to controlled output is forward. Reverse is backward path.
14. **Process Control:** Refers to the control of process parameters such as level, flow, temperature, pressure and acidity.
15. **Regulator:** Maintains controlled variable at set value irrespective of ext. load & output at const. level even under disturbance

Types of Control Systems

Open loop system	Closed loop system
Def: Output has no influence on input and operates on time basis in a no-disturbance environment irrespective of feedback · have no feedback · response is sensitive · Accurate & expensive components have to be used · Immediate response is possible	Def: Maintains a prescribed relationship between the output and reference input by comparing them and using the difference as a means of control · have feedback · response is insensitive · Inaccurate & inexpensive components can also be used · Immediate response is not possible **SERVO MECHANISM:** The objective of system is to control the positions of an object.

Chapter 15

ELEMENTS OF CONTROL SYSTEMS

STARTERS

To study this chapter, you should have awareness on the following concepts. For a better understanding, it is always a good idea to revise these prerequisites.

- Different types of systems
- Ability to express a function or job in terms of a control system
- Feedback and its significance
- Identify various control systems in and around us and observe the components and mechanism in them
- Zeroth, first and second order systems
- System errors and corrective methods and
- First four chapters of this book.

LEARNING OBJECTIVE

After studying this chapter you should be able to

- Understand the control system and its types,
- Explain the requirements of a control system,
- Distinguish between open loop & closed loop systems,
- Describe the significance and advantages of feedback control systems,
- Narrate the Servomechanism,
- Understand the requirements of an ideal control system and,
- Observe, apply and explain various control systems found in daily life and work areas.

15.1 INTRODUCTION

There are several "objectives" that need to be addressed and performed in our daily life such as to regulate the temperature, pressure and humidity of homes and buildings for comfortable living. Similarly, an automobiles, trains, air-planes etc., have to be controlled to reach their destinations accurately and safely. Industrially, manufacturing process also has several objectives for products to satisfy the cost, safety and quality requirements. The means of achieving these "objectives" usually involve the use of control systems that implement certain control strategies.

In recent past, control systems have gained an increasingly significant role in the development and advancement of modern civilization and technology. Particularly, every aspect of life is affected by some type of control system. Control Systems have become an integral part of our day-to-day activities. We shall discuss some of such control systems.

15.2 BASIC COMPONENTS OF A CONTROL SYSTEM

A control system can be described by

1. Objectives of control (input)
2. System Controls
3. Results (outputs)

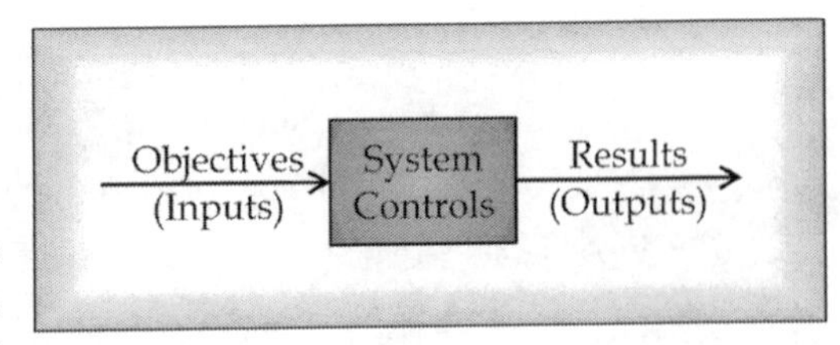

Fig. 15.1 Components of Control System

The basic relationship between the above three components is illustrated in Figure 15.1. More technically speaking, the objectives can be identified through *inputs* or *actuating signals* while the results are *the controlled variables* or *outputs*. In general, the objective of the control system is to control the outputs in some prescribed manner by the inputs through *the elements of the control system* or often called *system controls*.

15.3 CONTROL SYSTEMS TERMINOLOGY

A control system consists essentially of a process, error detector and control elements. Some of the terms related to these control systems are defined below and are shown in Figure 15.2.

More Instruments for Learning Engineers

A **breathalyzer** (a portmanteau of *breath* and *analyzer*) estimates blood alcohol content (BAC) from a breath sample. **Breathalyzer** is the brand name for the instrument developed by inventor Robert Frank Borkenstein. It was registered as a trademark on May 13, 1958, but many people use the term to refer to any generic device for estimating blood alcohol content. Public Breathalyzers are popular for testing the accused for the alcohol consumption. It is the most reliable tool used by police to control/ prevent drunken drive. Public Fuel Cell Breathalyzers are used in pubs, bars, restaurants, charities, weddings and all types of licensed events. They are now found in all licensed business. Particularly, it is inceasingly demanded tool in educational instituions, hostels and government/public offices.

Breathalyzer

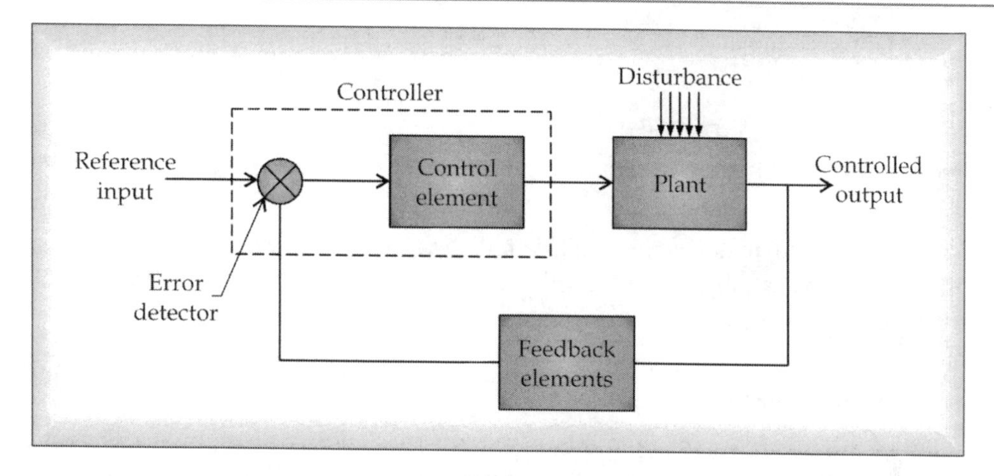

Fig. 15.2 Elements of a Control System

1. **Processes, Plant or System:** This is a body, process or machine of which a particular quantity or condition is to be controlled, e.g., a furnace, reactor or a space craft, etc.
2. **Controlled Variable:** This is the quantity or condition (temperature, level, flow rate, etc) characterizing a process or a system whose value is held constant or consistent by controller or is changed according to a certain law or principle.
3. **Controlled Medium:** The process material in the controlled system or flowing through it in which the variable is to be controlled is known as controlled medium.
4. **Command:** An input or instruction established or varied by some external means to and independent of the feedback control system is often termed as command.
5. **Set Point (or) Reference Input:** A signal established as a standard of comparison for a feedback control system by virtue of its relation to command. The set point either remains constant or changes with time according to a resent programme.
6. **Manipulated Variable:** The quality or condition that is varied as a function of the actuating signal so as to change the value of the controlled variable. The manipulated variable is applied to the plant g_2 by the control element $g_{1.}$
7. **Actuating Signal:** An algebraic sum of the reference input 'r' and the primary feedback 'b' The actuating signal is also called the error or control action.
8. **Primary Feedback Signal:** A primary feedback signal is the function of the controlled output c, which is compared with the reference input to obtaining the actuating signal.
9. **Error-Detector:** An element that detects the feedback, essentially it is a summing point which gives the algebraic summation of two or more signals. The direction of flow of information is indicated by arrows and the algebraic nature of summation by plus or minus signs.

   Negative feedback occurs when the feedback signal (b) subtracts from the reference signal (r).

$$e = r - b$$

If the feedback signal (b) adds to the reference signal (r), the feedback is said to be positive feedback.

$$e = r + b$$

Negative feedback tries to reduce the error, whereas positive feedback makes the error large.

10. **Disturbance:** An undesirable variable applied to the system, which tends to affect adversely the value of the variable being controlled. The process disturbance may be due to changes in set point supply, demand, environmental and other associated variables.
11. **Feedback Element:** An element of the feedback control system that establishes a functional relationship between the controlled variable and feed back signal.
12. **Control Element:** An element that is required to generate the appropriate control signal (manipulated variable) applied to the plant or system.
13. **Forward and Backward Paths:** The transmission path from the actuating signal to the controlled output constitutes the forward path. The backward path is the transmission path from the controlled output to the primary feedback signal.
14. **Process Control:** A process control means the monitoring and regulating the process parameters such as level, flow, temperature, pressure and acidity. The system is said to be under control when a chosen parameter is maintained at an optimum value (required point). Here, the external disturbance tries to vary this parameter but control system sees that the optimum set value is maintained. Thus by controlling these parameters, the process is controlled. However, usually these systems will have large time lags.
15. **Regulator:** A regulator is a feedback control system in which the controlled variable is maintained at a set value irrespective of external load on the plant. The output of the system is maintained at constant level even if the system is disturbed by either the change in load or change in environment or change in the system itself, e.g., thermostat control in an electric iron.

Self Assessment Questions-15.1

1. What are basic components of a control system?
2. List out some of the terms related to the control systems.
3. Briefly explain some elements of a control system with a neat diagram.
4. Briefly discuss the following terms
   (a) Plant or System (b) Controlled variable
   (c) Controlled medium (d) Command
5. Briefly explain the following terms
   (a) Reference input (b) Manipulated variable
   (c) Actuating signal (d) Primary feedback signal
6. Briefly explain the following terms
   (a) Error detector (b) Disturbance
   (c) Feedback element (d) Control element
7. Briefly explain the following terms
   (a) Forward and backward paths (b) Process control (c) Regulator

15.4 TYPES OF CONTROL SYSTEMS

There are two types of control system. They are

1. Non-Feedback Control Systems (open loop systems)
2. Feedback Control System (closed loop system).

We shall describe these in detail through the sections to follow.

15.4.1 Open Loop Control Systems (Non- Feedback Systems)

Observe the following examples

- A military officer gives a command to his subordinates without expecting any feedback and the subordinates blindly have to execute the command to perform the task.
- In a washing machine soaking, washing, and rinsing in the washer operate on a time basis. The machine does not measure the output signal, i.e., the cleanliness of the clothes.
- The traffic control by means of signals is operated on a time basis. Irrespective of the vehicle moving or waiting. The traffic signal changes after a particular preset time period. The vehicles have to just follow the signal.

In the above examples, we can notice some significant features as discussed below.

Features of Open-loop control systems

1. The output has no impact on the control action.
2. The output is neither measured nor fed-back for comparison with the input.
3. They are operated on time basis irrespective of feedback.
4. The control system will not perform desired task if there is a disturbance i.e., the system will function correctly if and only if there is neither internal nor external disturbance.
5. The accuracy of the system depends on calibration.

These systems are known as open-loop or non-feedback control systems. In any open-loop control system the output is not compared with the reference input that corresponds to a fixed operating condition and as a result, the accuracy of the system depends on calibration. In the presence of disturbance, an open-loop control system will not perform the desired task. Open-loop control can be used, in practice, only for a system of known input-output relationship without internal and external disturbances. Clearly such systems are not feedback control system. Note that any control system that operates on a time basis is open loop.

Definition

The control system in which the output has no influence on the input and operates on time basis in a no-disturbance environment irrespective of feedback is called open-loop control system (or non-feedback control system).

Working (Operation) of Open-loop control system

For the ease of study, an open-loop control system can usually be divided into two parts, viz. the controller and the controlled process, as shown in Figure 15.3.

- An input signal or command is given to the controller, whose output acts as the actuating signal.

- The actuating signal then controls the controller process so that the controlled variable will perform according to prescribed standards.

Reference input(x) → Controller → Actuating signal → Tank → Controlled variable (y)

Fig. 15.3 Elements of an Open-loop Control System

In simple cases, the controller can be an amplifier, mechanical linkage, filter or other control element(s) depending on the nature of the system. In more sophisticated cases, the controller can be a computer such as a micro processor. Because of the simplicity and economy of open loop control system, we find this type of system in many noncritical applications.

Illustration-15.1 (Traffic Control System)

Consider a traffic junction which is provided with time based traffic signals (Red, Yellow and Green to represent Stop, Look and Proceed). Describe the set up as a control system with the help of a block diagram.

Solution:

Input

Input in this system is the density of traffic (entering the system)

Control element/Actuating signal

Traffic signals by colors (Red to stop, Yellow to look and Green to proceed)

Output

The volume of traffic that moved away i.e., exiting the system

Feedback

No feedback

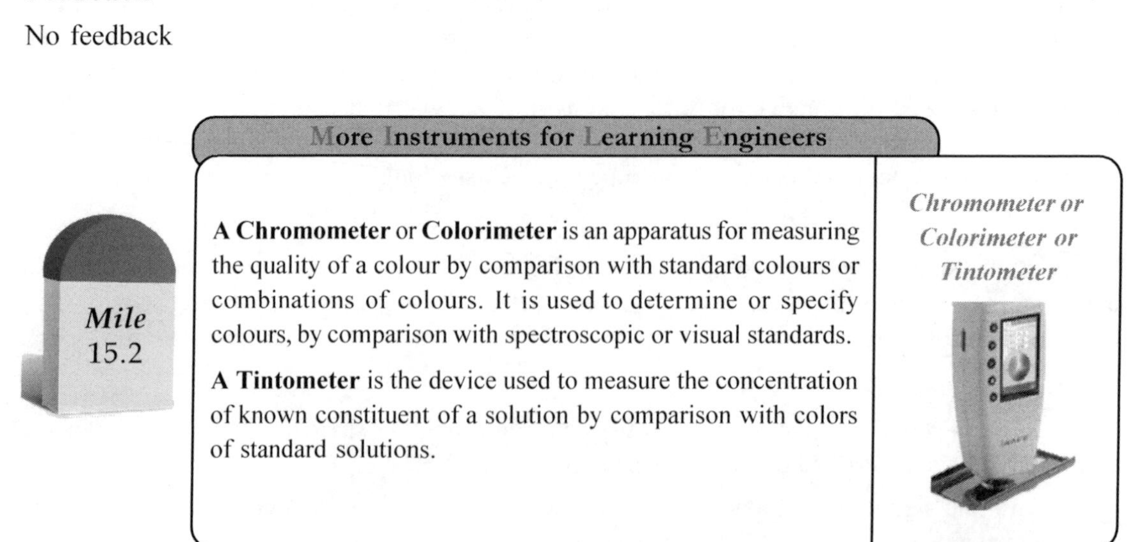

Type of control system

Open-loop system

Control Operation

The operation of ordinary traffic signal is an open-loop control. The variable (input) in this system is the density of traffic entering into the system. The volume or density of this traffic (entering) is not taken into consideration while switching Stop/Proceed of respective green/red lights. They are set by predetermined timing mechanisms and are no way influenced by the volume of traffic which is the output. The traffic stops when the red color signal is given and moves out of the system when green signal is given which is the output.

(The above system can be converted into closed loop-system by fixing sensors in the respective directions and knowing the actual density of traffic and allowing the traffic to flow from that direction which has highest density).

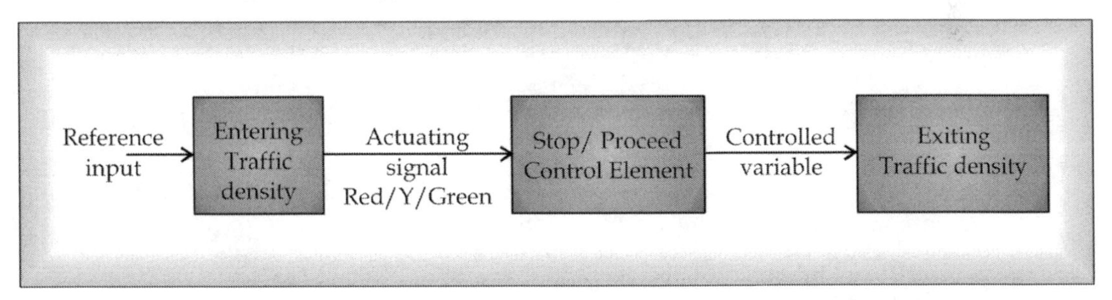

Fig. 15.4 Traffic Control System (Open-loop system)

15.4.2 Closed Loop or Feedback Control Systems

Let us now observe the following systems.

- In a room temperature control system, by measuring the actual room temperature and comparing it with, the reference temperature (desired temperature), the thermostat turns the heating or cooling equipment on or off in such a way as to a comfortable level regardless of outside conditions.
- In the human body, both body temperature and blood pressure are kept constant by means of physiological feedback.

In the above cases, the system input is essentially influenced by the output feedback. These are known as closed-loop or feedback systems. In fact, feedback performs a vital function in these systems. It makes relatively insensitive to external disturbance thus enabling to function properly even in a changing environment. The following features can be noticed in these systems.

Features of closed-loop control systems

1. Feedback is an essential feature which may be either positive or negative.
2. The output has an impact on input. In fact the input is regulated by the feedback given from the output.
3. The output signal is compared with the reference input (desired input) and controlled the variables accordingly.

4. The system can work in disturbance and changing environment also. The feedback makes the system insensitive to the disturbance.
5. The system accuracy depends on the quality of feedback but the system stability could be a major problem.

Thus we can define the closed-loop control system as follows.

Definition

A system that maintains a prescribed relationship between the output and the reference input by comparing them and using the difference as a means of control is called a feedback control system or a closed-loop control system.

Working (Operation) of Closed-loop Control Systems

A closed-loop control system often referred to as feedback control system can be studied by dividing into three parts, viz. controller, controlled processor and the error detector/ regulator.

- The reference input given to the controller device comes out as an actuating signal which is sent to the processor device.
- The actuating signal is processed in a processor device to produce the output signal which is fed back to the detector.
- A device detects and regulates the error by comparing the feedback with the reference input so as to minimize the error and sends as input.

In a closed-loop control system the actuating error signal, which is the difference between the input signal and the feedback signal (which may be the output signal itself or a function of the output signal and its derivatives and /or integral), is fed to the controller so as to reduce the error and bring the output of the system to a desired value. The term closed-loop control always implies the use of feedback control action in order to reduce system error.

A closed-loop idle-speed control system is shown in Figure 15.5. The reference input W_r sets the desired idling speed. The engine speed at idle should agree with the reference value, and any difference such as the load torque T_2 is sensed by the speed transducer and the error detector. The controller will operate on the difference and provide a signal.

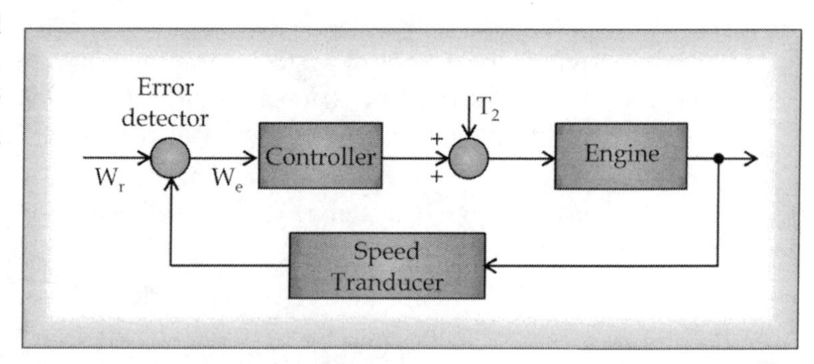

Fig. 15.5 Closed-loop Control System

15.4.3 Advantages of Feedback Control Systems

The following advantages can be sought through feedback control systems.

- The feedback systems are not only suitable but also desirable in the complex and fast acting or rapid systems beyond the physical abilities of a human.

- The human beings can be relieved from hard physical work, fatigue, boredom and drudgery that normally result from a continuous work.
- It is economic in the operating cost due to elimination of the continuous employment of a human operator.
- Feedback control system can increase the output or productivity due to reduced wastage.
- The quality and quantity of the products can be improved.
- Economy in the plant equipment, power requirement and in the processing material.
- Feedback control systems permit to initiate precise control by using relatively inexperienced components.
- These systems reduce the effect of non-linearity and distortion.
- There will be satisfactory response over a wide range of input frequencies.

Illustration-15.2 (Bathroom - Toilet Tank Control System)

Consider a bathroom toilet tank with the float valve/ball mechanism. Describe the control system with a block diagram and a neat sketch.

Solution:

The parts: The parts of this arrangement are

1. Pipe for water input
2. Valve (closing/opening)
3. Pivot
4. Lever
5. Hollow ball
6. Drain pipe

The Construction

The input pipe is fitted to a valve whose pivot is connected to a hollow ball (which floats in water) through a lever. This lever length may be chosen according the reference level and height at which the valve is fixed from the base of the tank. A drain pipe is fitted at the bottom of the tank. The arrangement of these parts of bathroom-toilet tank control system is shown in the Figure 15.6(a) and the block diagram is shown in 15.6(b).

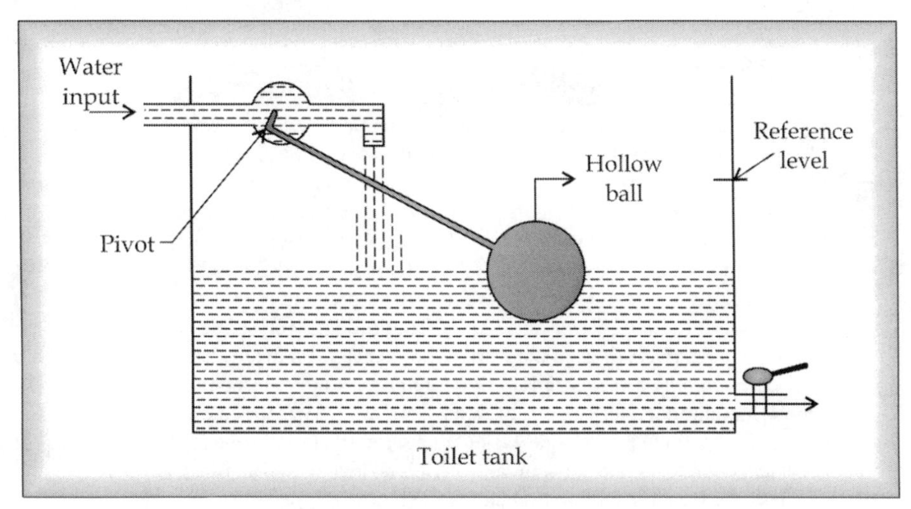

Fig. 15.6(a) Toilet Tank Control System

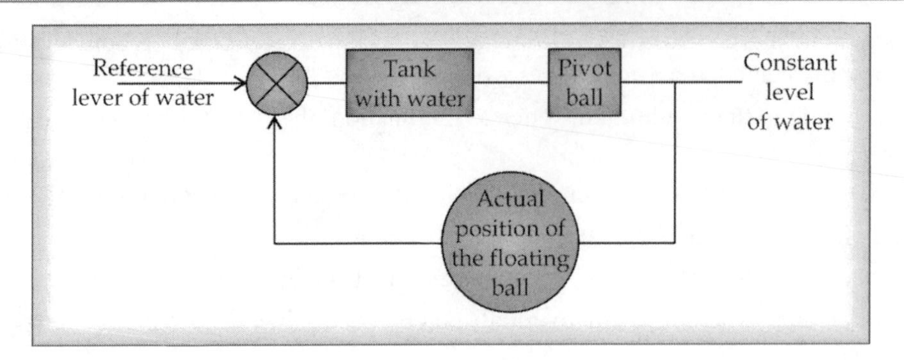

Fig. 15.6(b) Block Diagram of a Bathroom Toilet Control System

Components of control system

1. *Input:* Water and the reference level
2. *System controls:* The position of ball, the lever and hence the pivot
3. *Output:* Maintaining the water level
4. *Feedback:* Water level makes floating hallow ball to raise/fall when water filled/drained.

Type of Control System

Closed-loop or feedback control system

The Controlling Operation

The input to the system is the water and the control point is the reference level and whenever the level of water falls below this level the hallow ball which will be always in contact with the level also falls which in turn opens the valve and allowing water to flow into the tank which raises the level of the water and also the hallow ball. When water level touches the reference input the valve attached to the hallow ball closes completely. When the outflow knob is opened, the water level falls. This in turn opens the valve, thereby allows inflow of water. Hence here the output is continuously monitored.

More Instruments for Learning Engineers

A **Chronometer** is spring-driven escapement timekeeper like a watch, but parts are massive. Changes in the elasticity of Balance Spring caused by variations in temp are compensated. They often included other innovations (diamond, ruby, and sapphire were used as jewel bearings to reduce friction/ wear) to increase efficiency & precision. They also enjoyed usage of Au, Pt & Pd. Early, the term was used for specific timepiece tested & certified to meet certain precision standards. [In Swiss, only time pieces certified by Contrôle Officiel Suisse des Chronomètres (COSC) used the word 'Chronometer' on them]. It was coined by Jeremy Thacker, UK, (1714), referring to his invent clock. A marine chronometer (invent of John Harrison, 1730) for celestial navigation & finding longitude, was 1st of a series of chronometers that enabled accurate marine navigation. Now, it is no longer used as primary means of navigation, though needed as a backup.

Chronometer

Illustration-15.3 (Shower Water Temperature Control System)

Consider the case of two types of water (hot and cold) have to be mixed in a bathroom shower. With the help of block diagram design a control system.

Solution:

The parts are

1. Input water pipes (hot & cold)
2. Mixing tap
3. Output pipe
4. Nozzle (to spray)

The Construction

Figure 15.7(a) shows the arrangement for the control of shower bath water temperature.

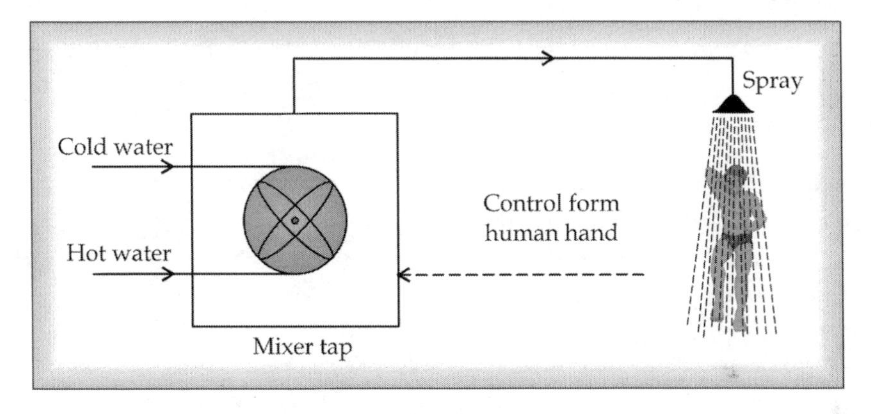

Fig. 15.7(a) Control of Shower Bath Water Temperature

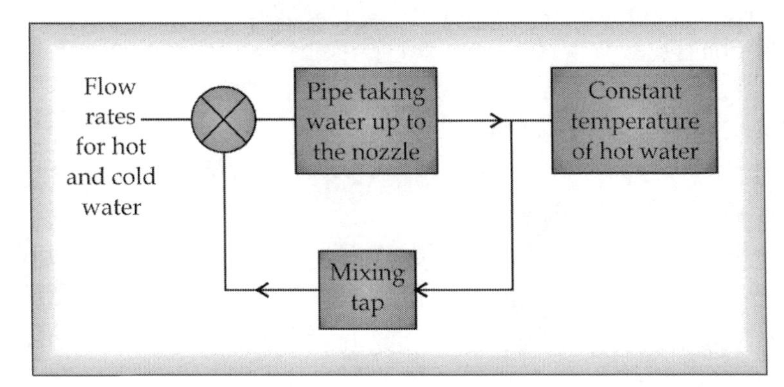

Fig. 15.7(b) Block Diagram for Shower Bath Water Temperature Control

Components of control system

Input to the system is the temperature and flow rates of hot & cold water.

Plant implies piping.

Output is the temperature of hot water flowing from the nozzle

Control element is the mixing tap

Feedback is by sensing and regulating by human hand

Type of Control System

Closed loop system (it includes sensing and regulating by hand)

The Controlling Operation

The block diagram given in Fig. 15.7(b) shows the control system for temperature control of a shower bath. Water coming from hot and cold pipes is mixed by a tap. This water is made to flow through pipes up to the nozzle. The hotness or coldness of the water is controlled by human hand. The adjustment is so made that the difference of the temperature between what the desired and the output temperature of the shower bath water is minimized.

15.4.4 Positive and Negative Feedback

In a system, wherever transduction takes place, there would be inputs and outputs. The inputs are the result of the environments influence on the system, while the outputs are the influence of the system on the environment.

We know that input and output are separated by some durations. The path from input to output is called forward path while that from output to input is reverse or backward path. In an open system, there is only forward path and no reverse (backward) path whereas in closed system both paths exist which is often referred to as feedback loop. In a feedback loop, information about the result of transduction is sent back to the input of the system in the form of input data as shown in Figure 15.8. This feedback may be of two kinds as described below:

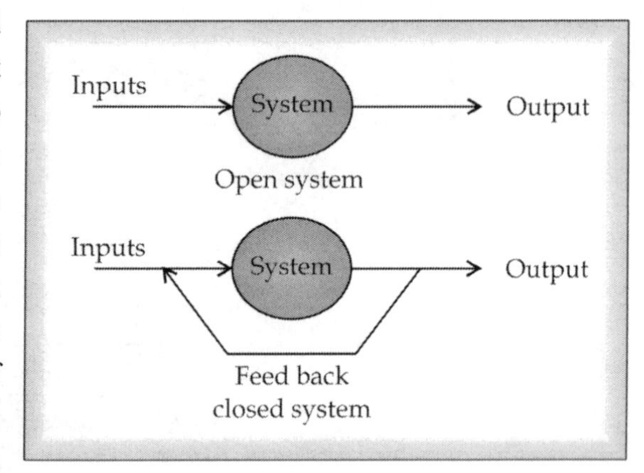

Fig. 15.8 Open and Closed Systems

1. **Positive feedback:** If the data fed to the input is in the same direction and facilitate or accelerate the input, the feedback is said to be positive-feedback
2. **Negative feedback:** If the data fed to the input is opposite to the input, the feedback is negative-feedback.

In positive feedback, may it be growth or decline, the system becomes unstable by exponentially as feedback accelerates the input whereas in negative feedback the system attains stability or equilibrium since the feedback opposes or restricts the input. Therefore, the positive feedback loop left unattended, can lead to the destruction of the system through explosion or through the blocking of all its functions.

For example, consider a positive feedback loop of chain reaction. Here the feedback neutrons accelerate the reaction and become disastrous over a period of time if it is not controlled. Similarly, population, explosion, industrial expansion, inflation etc. are some more examples for positive feedback systems.

In contrast, the negative feedback leads to adaptive or goal seeking behavior. For instance, the negative feedback loop is found in sustaining the same liquid level in a tank with a control valve mechanism. The control systems of temperature (in an air conditioner, refrigerator etc), speed or direction (such as in automobiles), concentration (in certain chemical reactions) etc. can be included in this category.

ILLUSTRATION-15.4 (BIOLOGICAL CONTROL SYSTEM TO GRIP AN OBJECT – WITH POSITIVE FEEDBACK)

When we grip an object, the brain, hand, eyes and senses work synchronously and signal system is established. Describe this natural biological activity as a control system with a block diagram. Is the feedback positive or negative? Discuss.

Solution:

System components

Input is signal from brain and

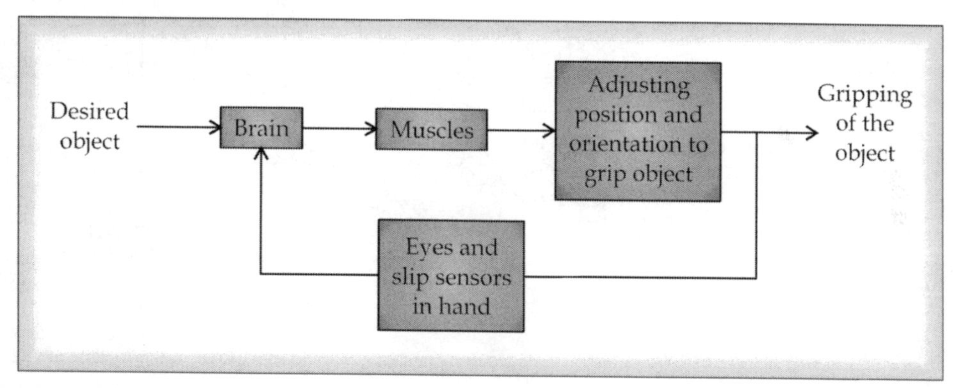

Fig. 15.9 Block Diagram of Biological Control System to Grip an Object by Human Hand

Plant or system is constituted by muscles.

Control elements are adjusting position and orienting to grip the desired object.

Output is gripping the object.

Feedback is provided by the eyes and slip sensors in hand.

System controlling operation

Figure 15.9 shows the biological control system to grip an object by human hand. Here the desired input is gripping of the object. The brain will send signal to muscles which in turn orient in such a way that the object is gripped perfectly. The feedback is received by eyes and slip sensors in hand. The output of the system is that object desired is gripped perfectly.

Discussion of feedback system

Here the feedback is encouraging the input to grip more, thus the feedback is positive type. This feedback may be used to enhance the grip to certain extent only after which the over gripping by excessive force will cause slip and even damage the object or the hand muscles. Suppose, a machine power (robot) is used instead of hand, the over gripping would damage the system also if the signal given to the machine is not monitored and controlled.

ILLUSTRATION-15.5 (CONTROL SYSTEM FOR REDUCING POLLUTION & IMPROVE MILEAGE OF AUTOMOBILE – WITH NEGATIVE FEEDBACK)

Construct a control system for reducing pollution by which the mileage can also be improved in an automobile. Check if the feedback is positive or negative and discuss its effect.

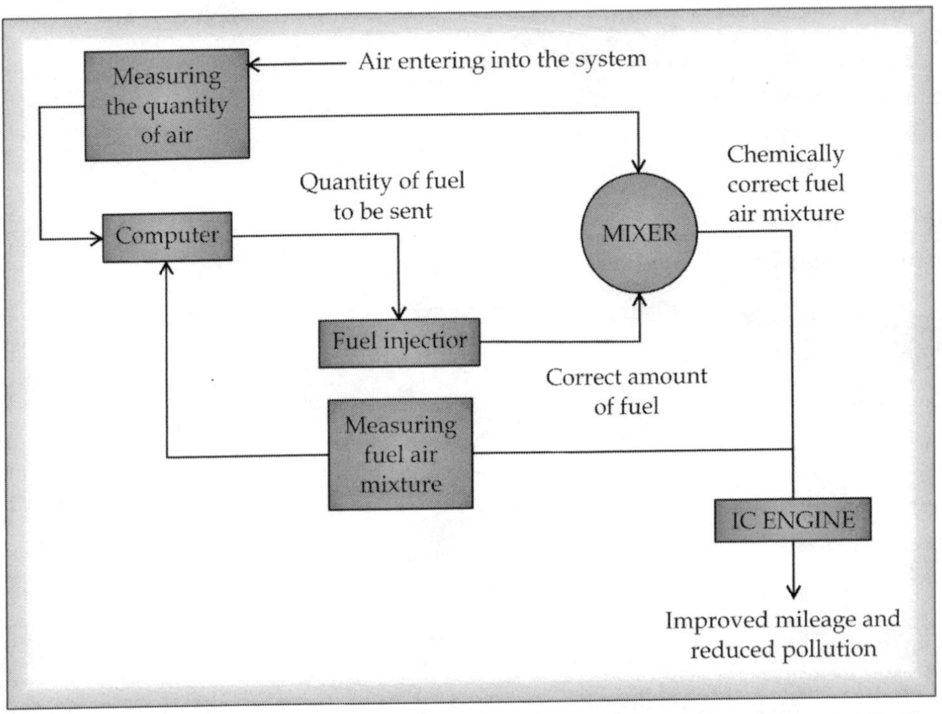

Fig. 15.10 Control System for Reducing Pollution & Improve Mileage of Automobile Engine

More Instruments for Learning Engineers

Scrubber systems are air pollution control devices used to remove some particulates and/or gases from industrial exhaust streams. The 1st air scrubber was designed to remove CO_2 from air of submarine, the Ictineo-I. Traditionally, the term "scrubber" has referred to pollution control devices that use liquid to wash unwanted pollutants from a gas stream. Today, the term is also used to describe systems that inject a dry reagent or slurry into a dirty exhaust stream to "wash out" acid gases. Scrubbers are the primary devices that control gaseous emissions, particularly acid gases. They can also be used for heat recovery from hot gases by flue-gas condensation. There are several methods (wet scrubber, dry scrubber, and absorber) to remove toxic or corrosive compounds from exhaust gas and neutralize it.

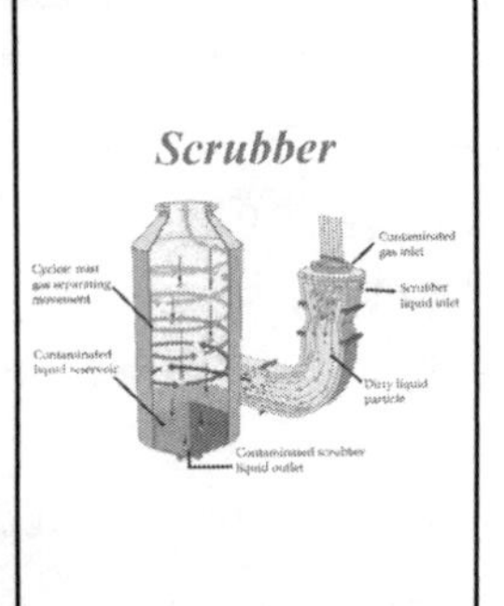

Solution: As shown in Figure 15.10 the air entering into the system is measured and this information is given to computer which in turn calculates the quantity of fuel required and actuates the fuel injector system so that fuel and air are mixed in the mixer and sent to the automobile engine thereby reducing pollution and thence improve mileage.

Here, the feedback discourages the pollution, thereby regulates the air-fuel ratio by continuously comparing the desired and actual results. If this difference is reduced it goes on improving thus the reduction of pollution is controlled by negative feedback. This is carried out till the difference between desired (set value) and actual result becomes zero.

(**Note:** *In the above illustration, the identification of system components and the detailed description about the control system, system elements, controlling operation etc., are left to the readers)*

SOME CASE STUDIES FOR PRACTICE

1. In an Automatic Washing Machine, the clothes are washed, dried automatically by certain controls. Prepare a block diagram of control system for this automatic washing machine. Is it an open loop or a closed loop system? Identify the inputs, outputs and the elements of control system.
2. Identify the type of control system and draw the block diagram for melting the material by an ordinary floor furnace with manual control and also explain the elements of control system.
3. As shown in the Fig. 15.11, describe the process of student learning with minimized error, as a control system. Discuss the system, assuming feedback is from student in the form of examination. Is the feedback positive or negative? What will be the effect if feedback is not monitored?

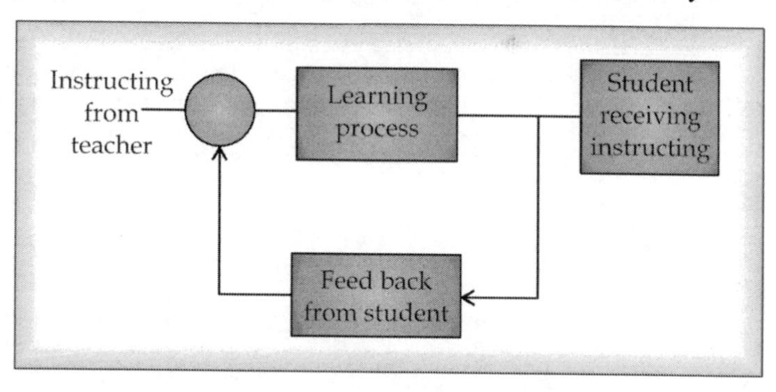

Fig. 15.11 Block Diagram of Learning Process

4. With the help of a block diagram, discuss the control system in automatic electric rice cooker.
5. With a block diagram describe TV operation (channel change/volume control) by a remote control. Is it an open-loop or closed loop?
6. Identify the control system and draw a block diagram for operations in flour mills. Discuss.

15.5 DIFFERENCES BETWEEN OPEN LOOP AND CLOSED LOOP SYSTEMS

The open-loop (non-feedback) and closed loop (feedback) control systems are clearly distinguished through a tabular form below.

Table 15.1 Differences between Open-loop and Closed-loop Control Systems

	Open - Loop Control Systems	Closed Loop Control Systems
1.	These systems have no feedback	These systems have feedback (either positive or negative)
2.	The output has no influence on the input and control elements	The output influences the input and the controlling elements through feedback
3.	The system response is sensitive to external disturbances and internal variations in system parameters	The system response is relatively insensitive to external disturbance and interval variations in system parameters.
4.	It is useful in systems where the inputs are known in advance and in which there are no disturbances	These are useful only when unpredictable disturbances and/or unpredictable variations in system components are present.
5.	There is no comparison between the actual (controlled) & the desired values of variable.	Always there is a comparison between the actual (controlled) and desired values of the variable.
6.	Accurate & expensive components have to be used to obtain the accurate control.	Inaccurate & inexpensive components can also be used to obtain the accurate control.
7.	The number of components used is less.	The number of components used is more.
8.	These are cheaper in cost and power	These are relatively higher in cost and power
9.	System stability is not a major problem.	System stability is a major problem as the feedback can over correct errors that can cause oscillations of constant or changing amplitude.
10.	Immediate response is possible	Immediate response is not possible
11.	Less accurate & somewhat unreliable	More accurate & reliable
12.	*Example:* In traffic control signals, the green/ yellow/ red lights are switched on a time basis irrespective of the feedback.	*Example:* In the human body both body temperature and blood pressure are kept constant by means of physiological feedback.

SELF ASSESSMENT QUESTIONS-15.2

1. What are the types of control systems? Give an example for each.
2. Give some suitable examples for Open Loop Control Systems (Non- Feedback Systems).
3. Define and explain the Open Loop Control Systems with a neat diagram.
4. List out some important features of Open-loop control systems.
5. Define and discuss Closed Loop or Feedback Control Systems with suitable example and a neat sketch.
6. What are the features and advantages of Closed Loop Systems?
7. What is positive feedback? Explain.
8. What is negative feedback? Explain.
9. Differentiate between open-loop and closed-loop control systems with some suitable examples.

15.6 SERVO MECHANISM

When the objective of the system is to control the positions of an object then the system is called servomechanism. It is a feedback control system in which the output is mechanical position or time derivatives of position e.g., velocity and acceleration.

The motors used in automatic control systems are called servomotors. The servomotors are employed to convert an electrical signal (control voltage) applied to them into an angular displacement of the shaft. They can either operate in a continuous or step duty depending on construction.

There is a wide variety of servomotors available for control system applications. A typical closed loop position control system is shown in Figure 15.12, which consists of a servomotor powered by a generator. The load whose position has to be controlled is connected to motor shaft through gear wheels. Potentiometers are used to convert the mechanical motion to electrical signals.

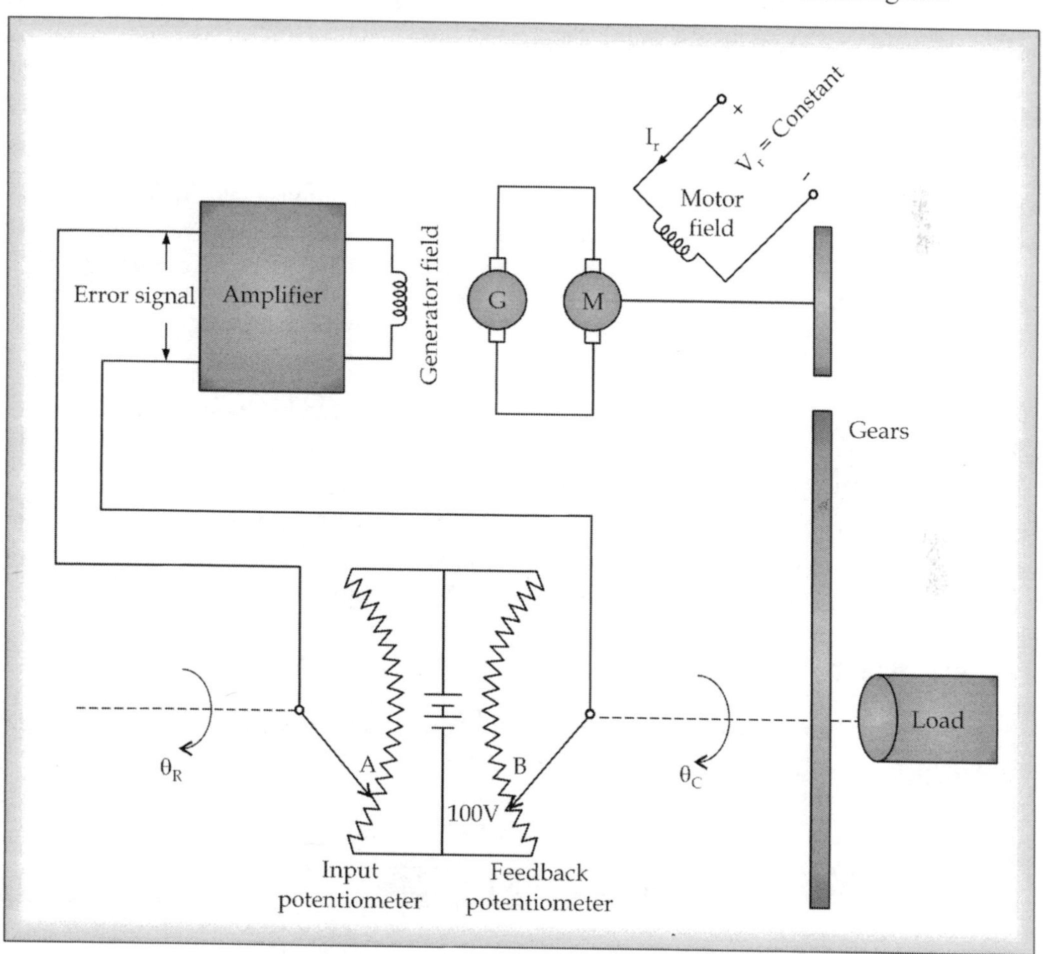

Fig. 15.12 Servo Mechanism

The desired load position (θ_R) is set on the input potentiometer and the actual load position (θ_e) is fed to feedback potentiometer. The difference between the two angular positions generates an error signal, which is amplified and fed to generator field circuit. The induced emf of the generator drives the motor. The motor stops rotating when the error signal is zero, i.e., when the desired load position is reached. This type of control system is called servomechanism. Thus the servomechanism is feedback control system in which the output is mechanical position (or time derivatives of position i.e., velocity and acceleration).

15.7 REQUIREMENTS OF A CONTROL SYSTEM

A control system should be consistent, correct and fast. In the technical terms, the three basic requirements that a control system essentially should possess are stability, accuracy and speed of response.

15.7.1 Stability

A system is said to be stable if the output of the system after fluctuations, variations, or oscillations, if any, settles at a reasonable value for any change in disturbance.

15.7.2 Accuracy

A system is said to be 100% accurate if the error (difference between accurate and output) is zero.

Usually, an accurate system is costly. So, it carries no meaning to go for a 100% accurate system if high degree of accuracy is not really needed. For instance, when variations say, 0.1°C cannot be sensed or differentiated by a human being there is no need to have a room air conditioner so highly accurate i.e., with temperature variation equal to zero.

15.7.3 Speed of Response

This refers to the time taken by the system to respond to the given input and gives as the output. Theoretically the speed of response should be infinity i.e., the system should have an instantaneous response. This requirement is of prime concern with follow up systems.

Obviously, an ideal system is perfectly stable, 100% accurate and has instantaneous speed of response. Unfortunately, the requirements are incompatible. Moreover, it may be unnecessary or impractical at times. Hence, there should be a compromise between these requirements.

15.8 CONTROL SYSTEMS - CASE STUDIES

As discussed at the beginning of this chapter, the control systems are widely employed everywhere at every moment of our lives. A few are illustrated below.

More Instruments for Learning Engineers

Mile 15.5

Dad-o-Meter is a social media created meter in the recent past to measure how better your dad is. This has become so popular that the greeting cards (including e-greeting) were are published and sent on Fathers' day. The software/ programme created quantify the answers given to the series of questionnaire about your dad and represents on a calibrated scale. The similar meters created in recent days include, love-o-meter, trust-o-meter, stress-o-meter, fraud-o-meter, mood-o-meter, legit-o-meter, bullshit-o-meter and so on.

Dad-o-Meter

15.8.1 Speed Control System in Automobiles

The difference between open & closed loop control is well explained by considering an example of driving system of an automobile.

The speed of the automobile is a function of the position of its accelerator.

Speed = *f*(position of accelerator).

Open-Loop Control: Controlling pressure on the accelerator pedal maintains the desired speed or a desired change in speed. This automobile driving system (accelerator carburetor and engine-vehicle) constitutes a control system. Figure 15.13 shows the general diagrammatic representation of a typical open-loop control system. For the automobile driving system the input signal is the force on the accelerator pedal, which through linkages causes the carburetor valve to open/ close so as to increase/ decrease fuel flow to the engine bringing the engine vehicle speed (controlled variable) to the desired value.

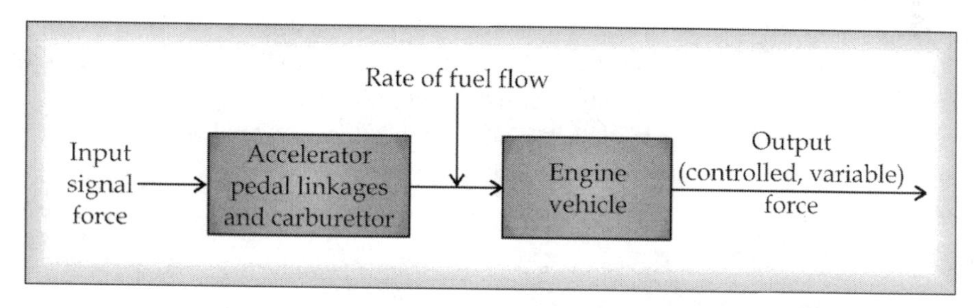

Fig. 15.13 Basic Open-loop Control System (Automobile System)

The diagrammatic representation in the above form is known as block diagram representation where in each block represents an element, a plant, mechanism device etc., whose inner details are not indicated. Each block has an input and output signal, linked by a relationship characterizing the block. It may be noted that the signal flow through the block is unidirectional.

Closed-Loop Control: Let us once again consider the same system. By observing traffic & road conditions, the route, speed and acceleration of the automobile are determined and controlled by properly manipulating the accelerator, clutch, gear-lever, brakes and steering wheel, etc., by the driver. Suppose the driver wants to maintain a speed of 40 km per hour, which becomes the desired output and can be achieved with the help of the accelerator and then maintained it by holding the accelerator steady. No error in the speed of the automobile occurs so long as there are no disturbances along the road. The actual speed of the automobile as measured by the speedometer and indicated on its dial is read by driver and compared with the desired one mentally. If they deviate from the desired speed, accordingly he takes the decision to increase or decrease the speed. The decision is executed by changing the pressure of his foot (muscular power) on the accelerator pedal.

These operations can be represented as shown in Figure 15.14. In contrast to the sequence of Figure 15.13, the events in the control sequence of Figure 15.14 follow a closed loop, i.e., the information about the instantaneous state of the output is feedback to the input and is used to modify it in such a manner as to achieve the desired output. It is on account of this basic difference that the system shown in Figure 15.13 is called an open-loop system, while the system of figure 15.14 is called a closed-loop system.

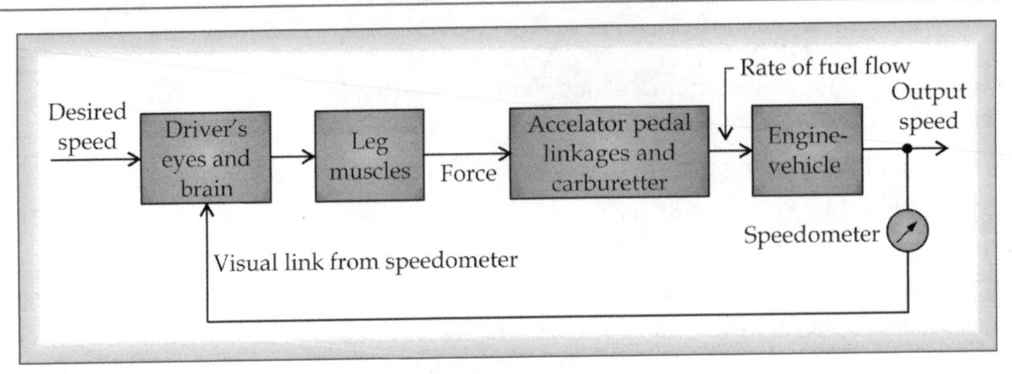

Fig. 15.14 Block Diagram of a Manually Controlled Closed-loop System

Let us consider the steering mechanism of an automobile. A simple block diagram of an automobile steering mechanism is shown in Figure 15.15. The driver senses visually and by tactile means (body movement) the error between the actual and desired directions of the automobile as in Figure 15.16. Additional information is available to the driver from the feel (sensing) of the steering wheel through his hand(s), this information constitute the feedback signals, which are interpreted by driver's brain, who then signals his hand to adjust the steering wheel accordingly. This again is an example of a closed-loop system where visual signal and tactile measurement constitute the feedback loop.

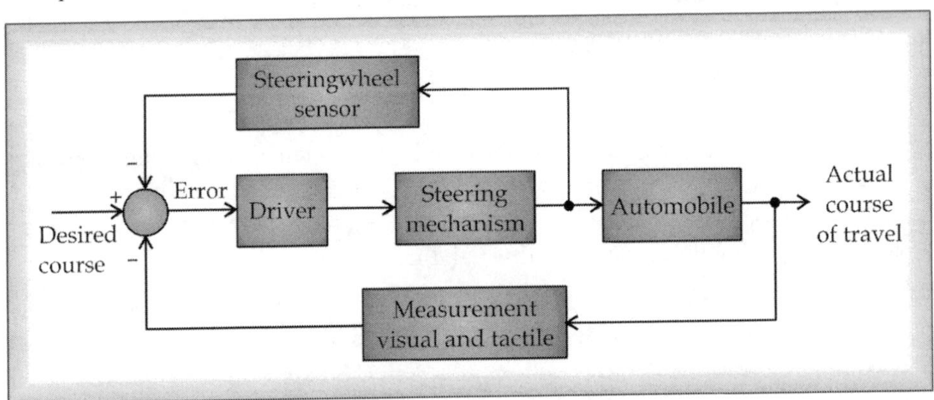

Fig 15.15 Automobile Steering Control System

In fact, unless human beings are not left out of in a control system study, practically all control system is a sort of closed-loop systems with intelligent measurement and sensing loops.

Systems of the type represented in Figure 15.14 and Figure 15.15 involve continuous manual control by a human operator. These are classified as manually controlled systems. The presence of human element in the control loop is undesirable in many complex and fast-acting systems. Because the system response may be too rapid for an operator to follow or the demand on operator's skill may be reasonably high. Furthermore, some of the systems that are self destructive such as missiles, human elements must be excluded. Even in situations where manual control could be possible, an economic case can often be made out for reduction of human supervision. Thus in most situations the use of some equipment, which performs the same intended function as a continuously employed human operator, is preferred.

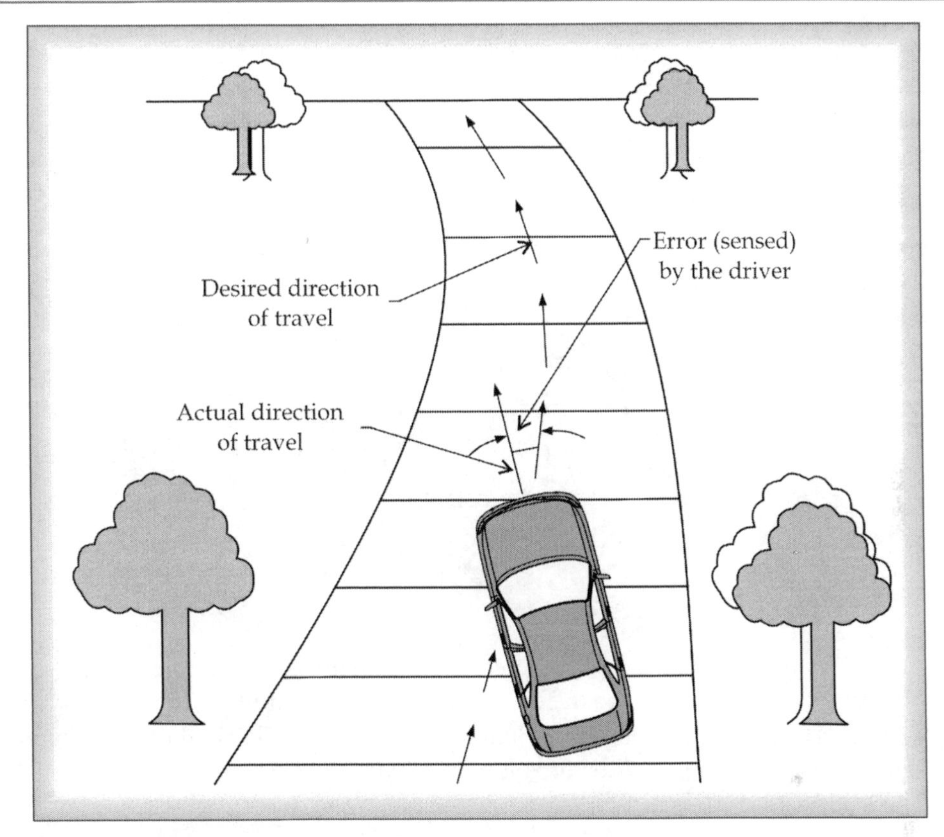

Fig. 15.16 The Driver uses the Difference between the Actual and Desired Direction of Travel to Adjust the Steering Accordingly

A system incorporating such equipment is known as automatic control system. In fact in most situations an automatic control system could be made to perform intended functions better than a human operator, and could further be developed to perform such functions, which are impossible for human operator.

15.8.2 Speed Control System in IC Engines

The basic principle of a Watt's speed governor for an engine is illustrated in the schematic diagram of Figure 15.17. The amount of fuel admitted to the engine is adjusted according to the difference between the desired and the actual engine speeds.

The sequence of actions may be stated as follows.

The speed governor is adjusted such that, at the desired speed, no pressured oil will flow into either side of the power cylinder. If the actual speed drops below the desired value due to disturbance, then the decrease in the centrifugal force of the speed governor causes the control valve to move downwards, supplying more fuel, and the speed of the engine increases until the desired value is reached. On the other hand, if the speed of the engine increases above the desired value, then the increase in the centrifugal force of the governor causes the control valve to move upward. This decreases the supply of fuel, and the speed of the engine decreases until the desired value is reached.

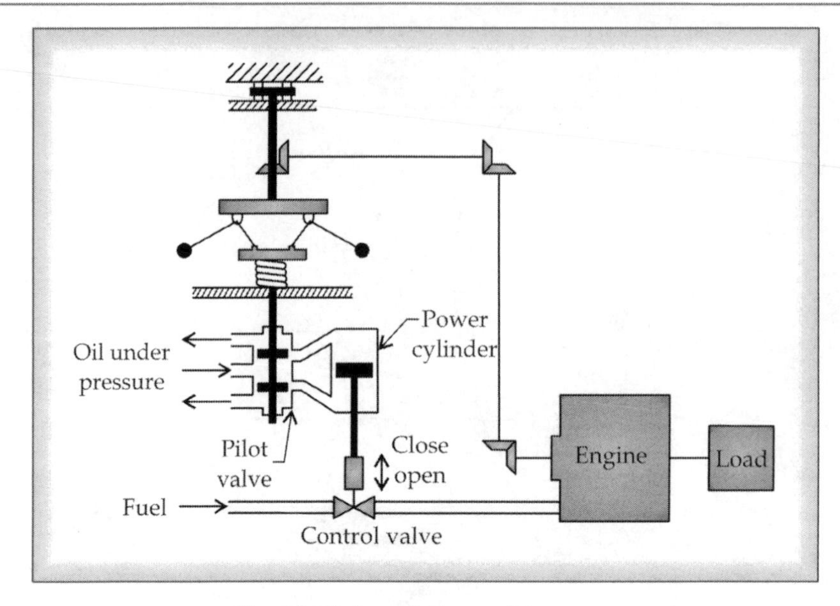

Fig. 15.17 Speed Control System

In this speed control system, the plant (controlled system) is the engine and the controlled variable is the speed of the engine. The difference between the desired speed and the actual speed is the error signal. The control signal (the amount of fuel) to be supplied to the plant (engine) is the actuating signal. The external input to disturb the controlled variable is the disturbance. An expected change in the load is a disturbance.

15.8.3 Temperature Control System of the Passenger Compartment of a Car

Figure 15.18 shows a functional diagram of temperature control of the passenger compartment of a car. Desired temperature, converted to voltage, is the input to the controller. The actual temperature of the passenger compartment is converted to voltage through a sensor and is feedback to the controller for comparison with input. The ambient temperature and radiation heat transfer from the sun which are not constant while the car is driven, act as disturbances. This system employs both

More Instruments for Learning Engineers

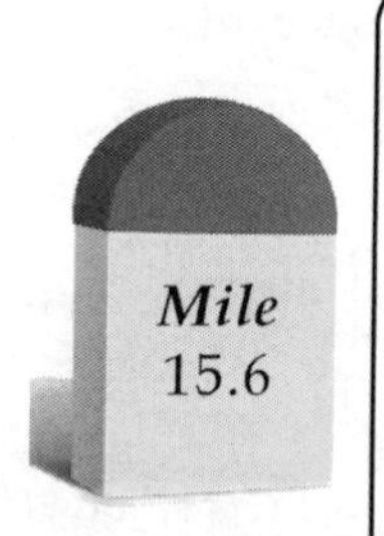

Anti-Plagiarism checking software is gaining popularity in the recent past, in honor to the intellectual property rights. This measures how much a text is having a copied matter.

A **Bullshit-o-meter** or **Bla-Bla-meter** is similar social media developed tool used to measure how much bullshit (useless matter) is coming out of mouth or in a write-up. PR-Experts, politicians, ad-writers or pseudo scientists need to be strong here! BlaBlaMeter unmasks without mercy how much bullshit hides in any text. It is a useful tool for everyone involved in speaking and writing! An app is available in the internes which works with English text up to 15000 characters. For a meaningful result a minimum length of 5 sentences is recommended.

Plagiarism and Bullshit-o-Meter or Bla-Bla meter

feedback control and feed forward control. (Feed forward control gives corrective action before the disturbances affect the output).

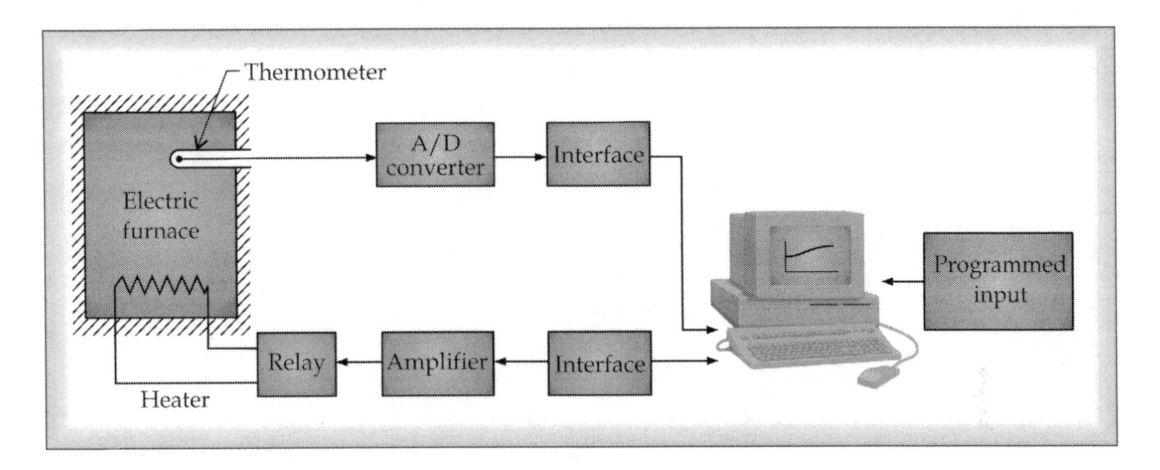

Fig. 15.18 Temperature Control Systems

The temperature of the passenger car compartment differs considerably depending on the place where it is measured. Instead of using multiple sensors for temperate measurement and over aging the measured values, it is economical to install a small suction blower at the place where passengers normally sense the temperature. The temperature of the air from the suction blower is an indication of the passenger compartment temperature and is considered as the output of the system (Refer Fig. 15.19).

The controller receives the input signal, and signals from sensors form disturbance sources. The controller seeds out an optimal control signal to the air conditioner or heater to control the amount of cooling air or warm air so that the passenger compartment temperature is about the desired temperature (refer Figure 15.19).

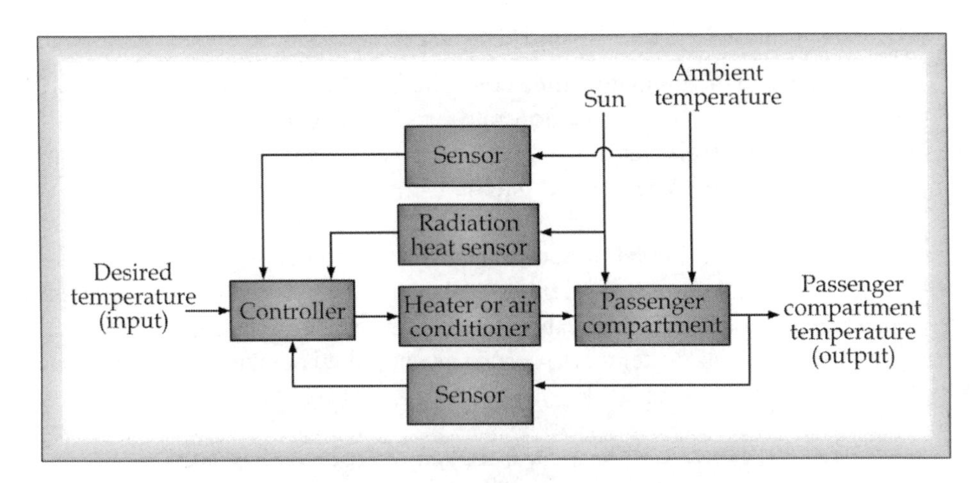

Fig. 15.19 Temperature Control of Passenger Compartment of a Car

15.8.4 Automatic Tank-Level Control System

In order to give a better understanding of the interactions of the constituents of control systems, let us discuss a simple tank level control system shown in Figure 15.20.

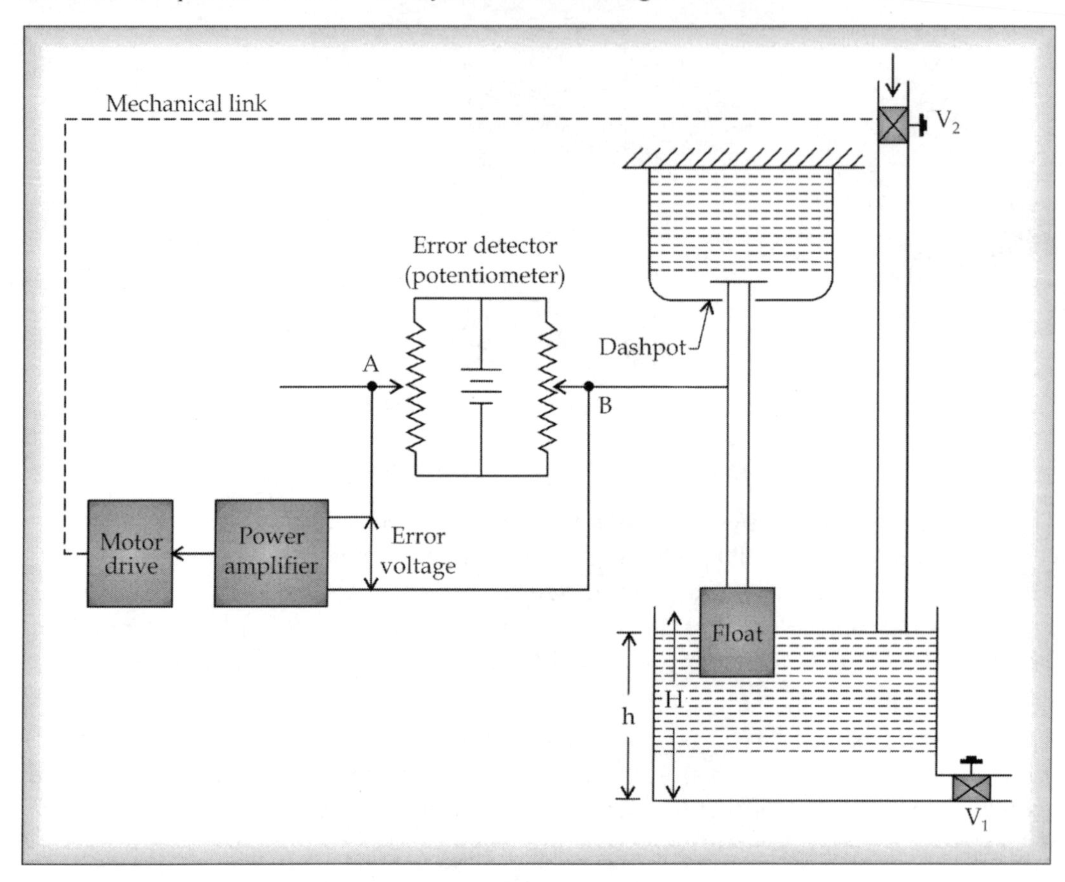

Fig. 15.20 Automatic Tank-Level Control System

This control system can maintain the liquid controlled output of the tank within accuracy of the desired liquid level even though the output flow rate through the valve V_1 is varied. The liquid level is sensed by a float (feedback element), which positions the slider arm B on a potentiometer. The slider arm A of another potentiometer is positioned corresponding to the desired liquid level H (The reference input). When the liquid level rises or falls, the potentiometer (error detector) gives as error voltage (error or actuating signal) proportional to the change in liquid level. The error voltage actuates the motor through a power amplifier (control elements), which in turn conditions the plant decreases or increases the opening of the valve V_2) in order to restore the desired liquid level. Thus the control system automatically attempts to correct any supplied deviation between the actual and desired liquid levels controlled in the tank.

15.8.5 Liquid Level Control System (Automatic and Manual)

Figure 15.21(a) is a schematic diagram of a liquid level control system. Here the automatic controller maintains the liquid level by comparing the actual level with desired level and correcting any error by adjusting the opening of the pneumatic valve. Figure 15.21(b) is a block diagram of the control system. The corresponding block diagram for a human operated liquid-level control system can be drawn in a similar way.

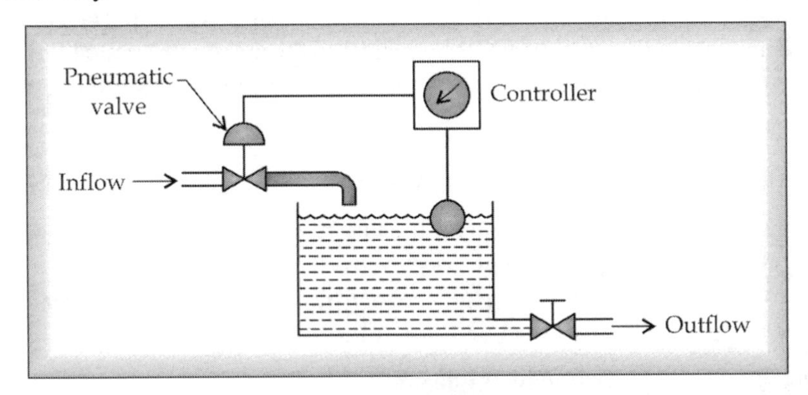

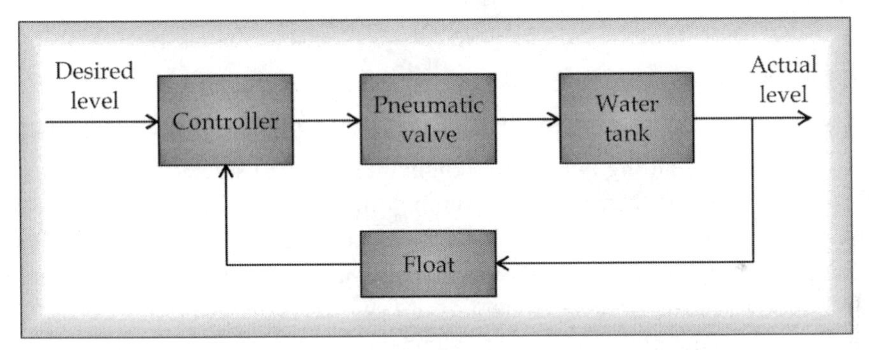

Fig. 15.21 Liquid Level Control System

In the human-operated system, the eyes, brain, and muscles correspond to the sensor, controller, and pneumatic valve, respectively. A block diagram is shown in Figure 15.22.

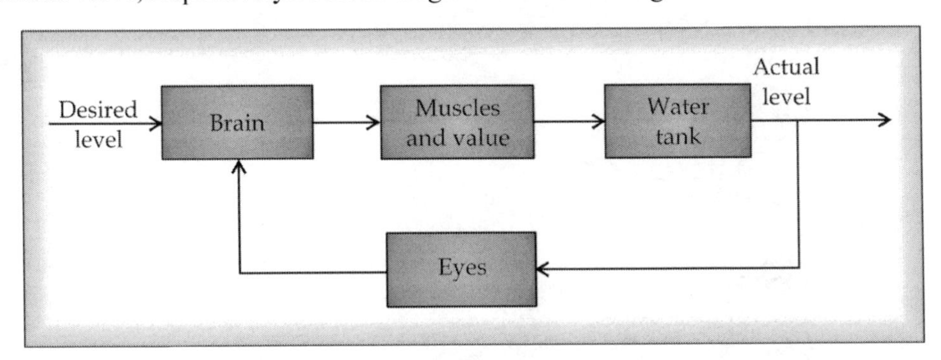

Fig. 15.22 Block Diagram of Human-operated Liquid-level Control System

15.8.6 Feedback System in Nature – Biological System

The temperature control in the human body is a very good example of feedback control in biological system. As we all know that normal or comfortable temperature of human body is 98.4° (or 36.8 °C). During summer, the outside temperature goes up by about 5°C, due to this increase in temperature;

the skin on the human body expands thus increasing the sweat pore diameter. From this the amount of moisture (sweat) exposed to atmosphere is more. The sweat absorbs the latent heat and evaporates there by maintaining the body temperature. The reverse happens in winter, is the sweat pores contract, decreasing the pore diameter. Hence the amount of sweat exposed to atmosphere also decreases, which in turn prevents the loss of body heat to the surrounding medium. In normal days the sweat pores are optimally opened so as the keep the body temperature optimum & constant within acceptable limits.

15.8.7 Position Control System – Missile Launching and Guidance System

In defense application of using missiles to hit the target enemy fighter planes, feed back systems are extensively used. The design of the control system ensures the probability of hitting the targets to be the highest. Figure 15.23 shows a typical missile control system. A beam riding missile continuously corrects its path after launching from the ground.

r(t) = reference input to the computed trajectory

θ(t) = computed trajectory

c(t) = error in the trajectory of the missile launching

The control system shown in Figure 15.23 shows radar positioned on the ground, with its antenna scanning the air-space by sending out electro magnetic signals of selected frequencies. As soon as an enemy fighter plane enters the region of scanning, the reflected signals travel backward and are received by the radar antenna. These signals (X) consist of several parameters such as *x, y,* z co-ordinates of the air craft, and the velocity components in those direction and all such necessary parameters required to locate precisely one air craft, which is also called the target. The ground computers process this data and evaluate a trajectory r(t), θ(t) and provide the reference input to the missile system. The missile has to travel along θ(t) so as to reach and collide with the target.

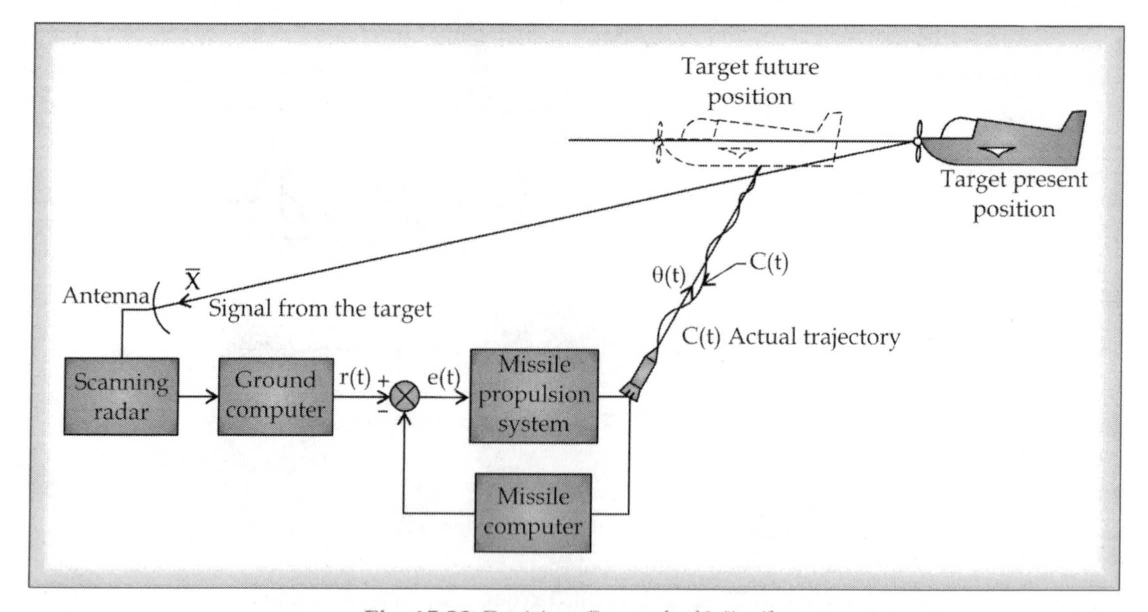

Fig. 15.23 Position Control of Missile

$\theta(t)$ also takes care of the time factor and is such that the aircraft moving along with its path and the missile along $\theta(t)$, meet at a common point, which will be the computed future position of the target. The propulsion system of the missile gets activated and the missile starts its journey along the trajectory $\theta(t)$. However due to the disturbances in the surrounding medium, or interval disturbances in tie propulsion system, the actual trajectory c(t) tend to deviate from $\theta(t)$. The antenna fixed to the missile head gets locked to the target and continuous to monitor the target location with respect to $\theta(t)$. The dedicated computer system in the missile, working in the control loop, helps in updating the feedback signal of c(t). The negative feedback ensures that the error e(t) in the trajectory of the missile is minimised along the course. Finally the missile gets close to the target and collides with it or explodes in the vicinity of the target.

SELF ASSESSMENT QUESTIONS-15.3

1. What is Servomechanism? Explain it with a neat diagram.
2. What are the requirements of a control system? Explain why they are required to a control system.

SUMMARY

A control system is described by its objectives of control (input), control system components and results (outputs). Stability, accuracy and speed of response are the three requirements of a control system. The two types of control systems are: Non-Feedback (open loop systems) and Feedback (closed loop system) Control System. In a transduction process, the inputs are the result of the environment influence on the system while the outputs are the influence of the system on the environment. In a closed loop, result of transduction is fed-back to the input of the system and if the data fed to the input is in the same direction and facilitate/accelerate the input, it is positive-feedback and if the data is opposite to the input, it is negative feedback. The servomechanism is a feedback control system in which the output is mechanical position or time derivatives of position such as velocity and acceleration.

The temperature control in the human body is the best example of feedback control in biological system. In defense, feedback systems are extensively used for missiles to hit the targets. The operation of traffic signal is an open-loop control system since the variable/input in this system (the density of traffic) is not considered while switching on/off of green/red lights as they are preset mechanisms irrespective traffic volume (output). This system can be converted into closed loop-system by fixing sensors to work as per the traffic density. The difference between open & closed loop control is explained by driving system of an automobile in which the speed of the automobile is a function of the position of its accelerator (open loop). The same system, by observing traffic & road conditions, the route, speed and acceleration of the automobile are determined and controlled by properly manipulating the accelerator, clutch, gear - lever, brakes and steering wheel, etc., by the driver (closed loop). The amount of fuel admitted to an IC engine (controlled system) is adjusted according to the difference between the desired and the actual engine speeds (controlled variable).

In an automatic Tank-Level Control System, the liquid level is sensed by a float (feedback element), which positions the arm on a potentiometer. In a Temperature Control System of a car, the actual temperature of the passenger compartment is converted to voltage through a sensor and is fed-back to the controller for comparison with input. In a bathroom - toilet tank control system, the reference level (input) to the system when falls, the hallow ball in contact opens the valve and allows water to flow into the tank to raise the level. In the control system for temperature control of a shower bath, the input to the system is the temperature and flow rates of hot/cold water while the output is the temperature of hot water flowing from the nozzle with control element as the mixing tap. In a Biological Control System, input (Gripping of the object) is signaled by brain to muscles to orient gap and positioning perfectly, and feedback is received by eyes and slip sensors in hand. In control system for automatic washing machine, the input is tuning for washing/drying based on the nature of clothes while the output of the system is washed/dried clothes. A floor furnace with manual control is an open loop control system as the temperature inside the furnace is controlled manually by seeing if the material is melted. For Reducing Pollution of Automobile Engineering, the air entering in to the system is measured and fed to computer to calculate the quantity of fuel required and actuates the fuel injector system so as to mix fuel and air correctly. In the learning process the input to the system is instructions by teacher whereas the output is student learning with minimized error and feedback is an exam from student.

KEY CONCEPTS

Basic Components of Control System: Objectives, Components and Results or outputs.

Processes, Plant or Control System: A body, process or machine of which a particular quantity or condition is to be controlled,

Controlled Variable: The quantity or condition (temperature, level, flow rate, etc) characterizing a process whose value is held constant by controller or is changed according to a certain law.

Controlled Medium: The process material in the controlled system or flowing through it in which the variable is to be controlled.

Command: An input that is established or varied by some means external to and independent of the feedback control system.

Set Point (or) Reference Input: A signal established as a standard of comparison for a feedback control system by virtue of its relation to command.

Manipulated Variable: The quality or condition that is varied as a function of the actuating signal so as to change the value of the controlled variable.

Actuating Signal: An algebraic sum of the reference input and the primary feedback (also called the error or control action)

Primary Feedback Signal: A function of the controlled output compared with the reference input to obtaining the actuating signal.

Error-Detector: An element that detects the feedback.

Negative feedback: Occurs when the feedback signal subtracts from the reference signal ($e = r - b$)

Positive feedback: If the feedback signal adds to the reference signal ($e = r + b$).

Disturbance: An undesirable variable applied to the system, which tends to affect adversely the value of the variable being controlled.

Feedback Element: An element of the feedback control system that establishes a functional relationship between the controlled variable and feed back signal.

Control Element: An element that is required to generate the appropriate control signal (manipulated variable) applied to the plant or system.

Forward Path: The transmission path from the actuating signal to the controlled output.

Backward Path: The transmission path from the controlled output to the primary feedback signal.

Open Loop Control Systems (Non- Feedback Systems): The systems in which the output has no effect on the control action.

Closed Loop/Feedback Control Systems: A system that maintains a prescribed relationship between the output and the reference input by comparing them and using the difference as a means of control.

Servomechanism: A feedback control system in which the output is mechanical position or time derivatives of position like velocity and acceleration.

Process Control: A process control refers to the control of process parameters such as level, flow, temperature, pressure and acidity.

Regulator: A regulator is a feed back control system in which the controlled variable is maintained at a set value irrespective of external load on the plant and the output of the system is maintained at constant level even the system is disturbed by either the change in load or change in environment or change in the system itself

Stability: The characteristic of system in which the output of the system after fluctuations, variations, or oscillations, if any, settles at a reasonable for any change in disturbance.

Accuracy: The degree of closeness to the true value.

Speed of Response: The time taken by the system to respond to the give input and given that as the output.

Ideal System: Perfectly stable, 100% accurate and has instantaneous speed of response.

REVIEW QUESTIONS

Short Answer Questions

1. What is a servo mechanism? Explain.
2. Draw a block diagram of the closed loop system and explain why negative feedback is invariably preferred in such systems.
3. State the advantages of a closed loop system.
4. Describe the working of one automatic control system used in practice, out line functional elements of that system.
5. Define the terms:

   (a) Closed loop and open loop system. (b) Servo mechanisms

6. Explain a position control system suitable for the table of N. C machine tool (Otherwise any other type familiar to you).
7. Why use of block diagrams are preferred for closed loop systems?
8. Sketch & explain a temperature control system.
9. What are the basic *elements* of a control system? Explain.
10. Briefly explain different types of control systems.
11. Explain briefly the difference between positive and negative feedback.
12. Write short notes on position control system
13. Write short notes on servomechanism.
14. How is water level in a boiler controlled?
15. State and explain a temperature control system. (or) Write short notes on temperature control system.

Long Answer Questions

1. Describe the operation of a driver driving an automobile on the road and identify the components input and output of the human system.
2. A constant water level is to be maintained in a boiler suggest a suitable automatic level control system with a block diagram and explain its working.
3. Explain the working of a system used for automatic control of temperature of a room
4. (a) Distinguish between open and closed loop systems with the help of suitable diagram.
   (b) Illustrate you answer using block diagram schematics.
   (c) Identify the system parameters and components in each case.
5. With suitable examples, bring out the advantages of systems over open loop systems. Draw a block diagram of the closed loop considered above. Explain why negative feed back is invariably preferred in such systems.
6. What are feed forward - feed back control systems? give some examples.
7. Explain with the help of block diagram anyone feed forward – feed back control system.
8. Explain briefly the difference between servomechanism, process control and regulations.
9. Explain with the help of block diagram the working of the variable speed d. c. drive control systems. State its characteristics and applications.
10. Describe the operation of a driver driving an automobile on the road and identify the components, input and output of a human system.
11. Distinguish between open-loop and closed loop control systems with the help of suitable diagram.
    (a) Illustrate you answer using block diagram schematics.
    (b) Identify the system parameters and components in each case.
12. Describe the working of one automatic control system used in practice. Outline functional elements of that system.
13. With the help of block diagrams explain the function of all the intents of a bathroom toilet tank control systems.
14. A constant water level is to be maintained in a boiler. Suggest a suitable automatic level control system with a block diagram and explain its working.

15. Describe the control system used for steering antenna. Explains control variables and the servomechanism.
16. Explain the operation of ordinary traffic signal. Why the system is called open loop? How can traffic be controlled more efficiently?
17. With the help of block diagrams explain the functions of all the ingredients of a bathroom toilet tank control system. How may this be made an automatic closed loop system? Explain with the help of a block diagram.
18. A computer controlled fuel injection system that automatically adjusts the fuel-air mixture ratio could improve gas mileage and reduce unwanted polluting emissions significally. Sketch a block diagram for such a system for an automobile and explain its working.
19. Describe a control system to fill a tank with water after it is emptied through an output at the bottom. This system automatically stops the inflow of water when the tank is filled. Draw the block diagram of the system.

MULTIPLE CHOICE QUESTIONS

1. The controller acts on the ___________ signal and produces an output, which actuates the ___________ element.
   (a) error, final control
   (b) final, control
   (c) input, final control
   (d) output, final control
2. Negative feedback tries to ___________ the error, where positive feedback makes the error ________________.
   (a) larger, reduce
   (b) reduce-large
   (c) constant, reduce
   (d) reduce, constant
3. A control system that has no connection between the output and the input is known as ___________ control system, whereas an error actuated control system is a ___________ control system.
   (a) semi open loop, closed loop
   (b) open loop, closed loop
   (c) closed loop, open loop
   (d) open loop, semi closed
4. A control system that regulates a variable in response to a fixed command signal is known as ___________ system, and a control system that follows changes in command signal is known as a ___________ system.
   (a) open loop, closed loop
   (b) follow up, regulator
   (c) regulator, follow up
   (d) closed loop, open loop
5. The ________________ determines the differences between the reference of set point and the measured variable. This difference is the or ___________ signal.
   (a) controller, error, actuating
   (b) error, controller, actuating
   (c) actuating, controller, error
   (d) none of the above.
6. For open control system which of the following statements is incorrect?
   (a) less expensive
   (b) recalibration is not required for maintaining the required quality of the output.
   (c) construction is simple and maintenance is easy
   (d) errors are caused by disturbances.
7. A control system in which the control action is some how dependent on the output ___________.
   (a) open - system
   (b) closed -loop system'
   (c) semi - closed system
   (d) none of the above

8. In an open ~ loop control system ...
   (a) output is dependent on control input
   (b) output is independent of control input
   (c) only system parameters have effect on the control output
   (d) none of the above
9. A control system working under unknown random actions is called ___________.
   (a) digital data system
   (b) stochastic control system
   (c) computer control system
   (d) adaptive control system
10. ___________ is a closed loop system.
   (a) direct current generator
   (b) electric switch
   (c) car starter
   (d) auto - pilot for an aircraft
11. ___________ is a part of the human temperature control system
   (a) leg movement
   (b) ear
   (c) digestive system
   (d) perspiration system
12. A car running at a constant speed of 60 kmph, which of the following is the feedback element for the driver?
   (a) steering wheel
   (b) eyes
   (c) clutch
   (d) needle of the speedometer
13. In open-loop system ...
   (a) the control action is independent of the output
   (b) the control action depends on system variables
   (c) the control action depends on the size of the system
   (d) the control action depends on the input signal
14. A good control system has all the following features except ___________.
   (a) good stability
   (b) good accuracy
   (c) slow response
   (d) sufficient power handling capacity
15. ___________ system has tendency to oscillate.
   (a) open-loop
   (b) closed-loop
   (c) booth (a) & (b)
   (d) neither (a) nor (b)
16. A closed-loop system in distinguished from open-loop system by which of the following
   (a) output pattern
   (b) input pattern
   (c) feedback
   (d) servo-mechanism
17. The initial response when the output is not equal to input is called ___________.
   (a) error response
   (b) transient response
   (c) dynamic response
   (d) any of the above
18. Any externally introduced signal affecting the controlled output is called as
   (a) signal (b) gain control
   (c) stimulus (d) feedback
19. The output of feedback control system must be a function of
   (a) input and feedback signal
   (b) output and feedback signal
   (c) reference and output
   (d) reference and input.

ANSWERS

1.	(a)	2.	(b)	3.	(b)	4.	(c)	5.	(a)
6.	(b)	7.	(b)	8.	(b)	9.	(b)	10.	(d)
11.	(d)	12.	(d)	13.	(a)	14.	(c)	15.	(b)
16.	(c)	17.	(b)	18.	(c)	19.	(c)		

Appendix

INSTRUMENTS AND THEIR MEASUREMENTS

Instrument	Measurement/Operation
Absorptiometer	Solubility of gases in liquids
Accelerometer	Acceleration or vibrations
Acetimeter	Strength of vinegar
Acidimeter	Concentration of acids
Actinograph	Calculates time of photographic exposure
Actinometer	Incident radiation
Aerometer	Weight or density of gas
Aethrioscope	Temperature variations due to sky conditions
Alcoholometer	Proportion of alcohol in solutions
Alcovinometer	Strength of wine
Algometer	Sensitivity to pain
Alkalimeter	Strength of alkaline
Altimeter	Altitude
Ammeter	Electrical current
Anemograph	Pressure and velocity of wind
Anemometer	Wind velocity
Areometer	Specific gravity
Arthroscope	Examines interior of a joint
Atmometer	Evaporating capacity of air
Audiometer	Acuity of hearing
Auriscope	Examines the ear

Contd...

Instrument	Measurement/Operation
Auxanometer	Growth of plants
Auxometer	Magnifying power
Ballistocardiograph	Detects body movements caused by heartbeat
Barograph	Records air pressure
Barometer	Air pressure
Baroscope	Weather-glass
Bathymeter	Records contours of deep oceans
Bathythermograph	Records water temperature as compared to depth
Bolometer	Radiant energy or infrared light
Bronchoscope	Examines the windpipe
Calorimeter	Absorbed or evolved heat
Cardiograph	Records movements of the heart
Cathetometer	Short vertical distances
Ceilometer	Height of cloud ceiling above earth
Ceraunograph	Records thunder and lightning
Chlorometer	Amount of chlorine in a solution
Chromatograph	Performs chromatographic separations
Chromatoptometer	Eyes' sensitivity to colour
Chronograph	Records the moment of an event
Chronometer	Time
Chronoscope	Very short time intervals
Clinometer	Slopes and elevations
Coercimeter	Coercive force
Colonoscope	Viewing the colon
Colorimeter	Determines color
Colposcope	Views the neck of the uterus
Coronagraph	Views the corona of the sun
Coulombmeter	Electric charge
Coulometer	Amount of substance released in electrolysis
Craniometer	The skull
Cratometer	Power of magnification
Crescograph	The growth of plants
Cryometer	Low temperatures
Cryoscope	Determines freezing points of substances
Cyanometer	Blueness of the sky or ocean

Contd...

Instrument	Measurement/Operation
Cyclograph	Describes arcs of circles without compasses
Cyclometer	Revolutions of a wheel
Cymograph	Traces the outline of mouldings
Cymometer	Frequency of electrical waves
Cystoscope	Examines the bladder
Cytometer	Counts cells
Decelerometer	Deceleration
Declinometer	Magnetic declination
Dendrometer	Trees
Densimeter	Closeness of grain of a substance
Densitometer	Optical or photographic density
Diagometer	Electrical conductivity
Diagraph	Enlarges or projects drawings
Diaphanometer	The transparency of air
Dichroscope	Examines crystals for dichroism
Diffractometer	Determines structure of crystal through light diffraction
Dilatometer	Expansion
Dioptometer	Focus or refraction of the eyes
Dipleidoscope	Moment when an object passes a meridian
Diplograph	Writes two lines of text at once
Dosimeter	Dose of radiation
Dromometer	Speed
Drosometer	Dew
Durometer	Hardness of substances
Dynamograph	Records mechanical forces
Dynamometer	Mechanical force
Ebullioscope	Boils point of liquids
Effusiometer	Compares molecular weights of gases
Eidograph	Copies drawings
Elatrometer	Gaseous pressure
Electrocardiograph	Records unusual electrical fluctuations of the heart
Electrodynamometer	Electrical current
Electroencephalograph	The brain's electrical impulses
Electrograph	Records electrical potential
Electrometer	Electrical potential

Contd...

Instrument	Measurement/Operation
Electromyograph	Diagnoses neuromuscular disorders
Electroretinograph	Electrical activity in the retina
Electroscope	Detects electrical charges in the body
Ellipsograph	Describes ellipses
Encephalograph	Records brain images
Endoscope	Visualizes interior of a hollow organ
Endosmometer	Osmosis into a solution
Epidiascope	Projects images of objects
Episcope	Projects images of opaque objects
Ergograph	Measures/Records muscular work
Ergometer	Work performed
Eriometer	Very small diameters
Eudiometer	Air purity
Evaporimeter	Rate of evaporation
Extensometer	Deformation in object due to forces applied
Fathometer	Underwater depth using sound
Fiberscope	Uses fiber-optics to examine inaccessible areas
Floriscope	Inspects flowers
Flowmeter	Properties of flowing liquids
Fluorimeter	Fluorescence
Fluoroscope	Uses X-rays to examine internal structure of opaque object
Focimeter	Focal length of a lens
Galactometer	Specific gravity of milk
Galvanometer	Electrical current
Galvanoscope	Detects presence and direction of electric current
Gasometer	Holds and measuring gases
Gastroscope	Examines interior of the stomach
Geothermometer	Subterranean temperatures
Goniometer	Angles between faces
Gradiometer	Gradient of a physical quantity
Gravimeter	Variations in gravitational fields
Gyrograph	Counts a wheel's revolutions
Haptometer	Sensitivity to touch
Harmonograph	Draws curves representing vibrations
Harmonometer	Harmonic relations of sounds

Contd...

Instrument	Measurement/Operation
Helicograph	Draws spirals on a plane
Heliograph	Intensity of sunlight
Heliometer	Apparent diameter of the sun
Helioscope	Observes sun without injury to the eyes
Hemacytometer	Counts blood cells
Hippometer	Height of horses
Hodoscope	Traces paths of ionizing particles
Hydrometer	Specific gravity of liquids
Hydroscope	Views under water
Hydrotimeter	Water hardness
Hyetograph	Records rainfall
Hyetometer	Rainfall
Hyetometrograph	Records rainfall
Hygrograph	Record variations in atmospheric humidity
Hygrometer	Air moisture
Hygroscope	Displays changes in air humidity
Hypsometer	Height of trees through triangulation
Iconometer	Finds size of object by measuring its image
Idiometer	Motion of observer in relation to transit of the heavens
Inclinometer	Inclination to the horizontal of an axis
Interferometer	Analyzes spectra of light
Iriscope	Exhibiting the prismatic colours
Katathermometer	The cooling power of air
Katharometer	Changes in composition of gases
Keratometer	Curvature of the cornea
Keraunograph	Records distant thunderstorms
Kinetoscope	Produces curves by combination of circular movements
Konimeter	Amount of dust in air
Koniscope	Dust in air
Kymograph	Records fluid pressure
Labidometer	Size of the head of a fetus
Lactometer	Tests relative density of milk
Lactoscope	Purity or richness of milk
Lanameter	Quality of wool
Laparoscope	Views interior of peritoneal cavity

Contd...

Instrument	Measurement/Operation
Laryngoscope	Examines interior of the larynx
Leptometer	Oil viscosity
Loxodograph	Device used to record ship's travels
Lucimeter	Light intensity
Luxmeter	Illumination
Lysimeter	Percolation of water through soil
Magnetograph	Records measurements of magnetic fields
Magnetometer	Intensity of magnetic fields
Manometer	Pressure of a liquid or gas
Marigraph	Records tide levels
Mecometer	Length
Megameter	Determines longitude by observing stars
Megascope	Projects an enlarged image
Mekometer	Range-finder
Meldometer	Melting points of substances
Meteorograph	Recording a variety of meteorological observations
Methanometer	Detecting presence of methane
Microbarograph	Record minute changes in atmospheric pressure
Microcalorimeter	Tiny quantities of heat
Micrograph	Writes on a very small scale
Micrometer	Very small distances
Micronometer	Short periods of time
Microscope	Magnifies small objects
Microseismograph	Records small or distant earthquakes
Microseismometer	Small or distant earthquakes
Mileometer	Records distance travelled in miles
Milliammeter	Records very small electrical currents
Myograph	Records muscular contractions
Myringoscope	Viewes the eardrum
Nephelometer	Cloudiness
Nephograph	Photograph clouds
Nephoscope	Observes direction and velocity of clouds
Nitrometer	Nitrogen and its compounds
Odograph	Records distance travelled
Odometer	Distance travelled

Contd...

Instrument	Measurement/Operation
Odontograph	Obtains curves for gear-teeth
Oenometer	Alcoholic strength of wine
Ohmmeter	Electrical resistance
Oleometer	Amount of oil in a substance
Olfactometer	Intensity of odour of a substance
Ombrometer	Rain-gauge
Oncometer	Change in size of internal organs
Oncosimeter	Variations in density of molten metal
Ondograph	Change in wave formations of electricity
Oometer	Eggs
Opacimeter	Opacity
Opeidoscope	Illustrates sound by means of light
Ophthalmometer	The eye
Ophthalmoscope	Views the interior of the eye
Opisometer	Curved lines
Optometer	Tests vision
Orchidometer	The size of the testicles
Oscillograph	Records alternating current wave forms
Oscillometer	Ship's rolling
Oscilloscope	Detect electrical fluctuations
Osmometer	Osmotic pressure
Otoscope	Examines the ear
Pachymeter	Small thicknesses
Pallograph	Ship's vibration
Pantochronometer	Combined sundial and compass
Pantograph	Copies drawing to a different scale
Passimeter	Issues automatic tickets
Pedometer	Distance travelled on foot
Peirameter	Resistance of road surfaces to wheel movement
Pelvimeter	The pelvis
Penetrometer	Firmness or consistency of substances
Permeameter	Permeability
Phacometer	Lenses
Phaometer	Old light intensity
Pharmacometer	Drugs

Contd...

Instrument	Measurement/Operation
Pharyngoscope	Inspects the pharynx
Phonautograph	Records sound vibrations
Phonendoscope	Device which amplifies small sounds
Phonometer	Sound levels
Phorometer	Instrument used to correct abnormalities in eye muscles
Photometer	Light intensity
Photopolarimeter	Intensity and polarization of reflected light
Phototachometer	The speed of light
Phototelegraph	Transmits drawings telegraphically
Phthongometer	Intensity of vowel sounds
Piezometer	Pressure or compressibility
Pitchometer	Angles of ship's propeller blades
Planigraph	Copies drawings at a different scale
Planimeter	Area of plane figures
Platometer	Area; planimeter
Plegometer	The strength of a blow
Plemyrameter	Variations in water level
Plethysmograph	Change in body part size due to blood flow
Pluviograph	Self-registering rain gauge
Pluviometer	Rain-meter
Pneometer	Measures respiration
Pneumatometer	Quantity of air breathed
Pneumograph	Records respiration
Polarimeter	Polarized light
Polariscope	Detects polarized light
Polygraph	Small changes in pulse and respiration
Porometer	Degree of porosity
Poroscope	Investigates porosity
Potentiometer	Electromotive forces
Potometer	Rate at which plants absorb water
Prisoptometer	Degree of astigmatism
Proctoscope	Examines the rectum
Psophometer	Audible interference of electrical current
Psychograph	Records spirit messages
Psychrometer	Air moisture or temperature

Contd...

Instrument	Measurement/Operation
Psychrometer	Dryness of the atmosphere
Pulsimeter	The pulse
Pycnometer	Specific gravity or density
Pyknometer	Specific gravities
Pyranometer	Solar radiation from the sky's whole hemisphere
Pyrgeometer	Radiation from earth
Pyrheliometer	Heating effect of sun
Pyrometer	Very high temperatures
Pyroscope	Intensity of radiant heat
Qualimeter	Penetrating power of X-ray beams
Quantimeter	Quantity of X-rays
Quantometer	Proportions of elements in metallic samples
Rachiometer	The spine
Radarscope	Detects radar signals
Radiogoniometer	Finds direction through radio signals
Radiometeorograph	Atmospheric conditions at high altitude
Radiometer	Radiation energy
Radioscope	Views objects using X-rays
Ratemeter	Counts rate of electronic counters
Recipiangle	Angles (Old instrument with two arms)
Reflectometer	Reflectance of radiant energy
Refractometer	Refraction of light
Respirometer	And studying respiration
Retinoscope	And viewing the retina
Rheometer	Current
Rhinoscope	Examines the nose
Rhythmometer	Speed of rhythms
Riometer	Absorbed cosmic radio waves
Rotameter	Length of curved lines
Rotameter	Liquid flow
Saccharimeter	Amount of sugar in a solution
Salinometer	Amount of salt in a solution
Scintillometer	Scintillation of star
Scintilloscope	Gamma rays emitted by a radioactive body
Sclerometer	Hardness

Contd...

Instrument	Measurement/Operation
Scoliometer	Curvature
Scotograph	Writes without seeing
Scotoscope	Detects objects in darkness
Seismograph	Records earthquakes
Seismometer	Earthquake intensity
Seismoscope	Detects earthquakes
Selenoscope	Views the moon
Sensitometer	Sensitivity of photographic material
Sepometer	Septic matter in the air
Serimeter	Tests quality of silk
Shuftiscope	Explores interior of dysentery case
Siccimeter	Liquid evaporation
Sideroscope	Uses magnets to detect presence of iron
Sigmoidoscope	Examines the interior of the rectum and sigmoid colon
Sillometer	Speed of ship
Skiascope	Eye's refraction from movement of shadows
Snooperscope	Views infrared radiation
Solarimeter	Solar radiation
Sonograph	Records and analyzes sound
Spectrofluorimeter	Measures/records fluorescence spectra
Spectrograph	Views a spectrum
Spectroheliograph	Takes pictures of the sun
Spectroheliokinematograph	Camera that takes pictures of the sun
Spectrohelioscope	Views solar disc in light of a single wavelength
Spectrometer	Wavelengths of light of a spectrum
Spectrophotometer	Speed of different parts of light spectrum
Spectroscope	Forms spectra by dispersing rays of light
Speedometer	Velocity
Spherometer	Curvature
Sphygmograph	Records pulse
Sphygmomanometer	Arterial blood pressure
Sphygmometer	Arterial blood pressure
Sphygmoscope	Makes arterial pulsations visible
Spinthariscope	Visually detects alpha particles
Spirograph	Records movements of breathing

Contd...

Instrument	Measurement/Operation
Spirometer	Lung capacity
Stactometer	Pipette with hollow bulb to counts drops
Stadiometer	The length of a curved line
Stagmometer	Number of drops in volume of liquid
Stalagmometer	Surface tension by drops
Statoscope	Small changes in atmospheric pressure
Stauroscope	Studies structure of crystals with polarized light
Stenometer	Distances
Stereometer	Specific gravity
Stereoscope	Views special three-dimensional photographs
Stethometer	Chest expansion during breathing
Stethoscope	Detects sounds produced by the body
Strabismometer	Degree of squinting
Strabometer	Strabismus in the eyes
Stroboscope	Studies motion using flashes of light
Stylometer	Columns
Swingometer	Swing in votes during an election
Sympiesometer	Pressure of a current
Synchroscope	Detects whether two moving parts are synchronized
Tacheometer	Rapidly measures survey points on a map
Tachistoscope	Rapidly shows images on a screen to test perception
Tachograph	Records speed of rotation
Tachometer	Speed of rotation
Taseometer	Stress in a structure
Tasimeter	Changes in pressure
Taximeter	Fee for hired vehicle
Telemeter	Strain or distance from observer
Telescope	Views objects at great distances
Telespectroscope	Analyzes radiation omitted by distant bodies
Telestereoscope	Views distant objects stereoscopically
Tellurometer	Uses microwaves to measure distance
Tenderometer	Tenderness of fruits and vegetables
Tensimeter	Vapour pressure
Tensiometer	Tension
Thalassometer	Tides

Contd...

Instrument	Measurement/Operation
Thermograph	Records changes in temperature
Thermometer	Temperature
Thermometrograph	Records changes in temperature
Thermoscope	Indicates change in temperature
Thoracoscope	Views the thorax and chest wall
Tiltmeter	Tilting of earth's surface
Tocodynamometer	Uterine contractions during childbirth
Tomograph	Views section of an object using X-rays
Tonometer	Pitch of musical tones
Topophone	Determines direction and distance of a fog-horn
Torsiograph	Records torsional vibrations on an object
Transmissometer	Transmission of light through a fluid
Trechometer	Determines distance travelled; odometer
Tremograph	Records involuntary muscular motion
Tribometer	Friction
Trigonometer	Solving triangles
Trocheameter	Counts wheel's revolutions
Tromometer	Slight earthquake shocks
Tropometer	Rotation
Turbidimeter	Turbidity of liquids
Turgometer	Turgidity
Typhlograph	Helps the blind write clearly
Udometer	Rainfall
Ultramicroscope	Views extremely small objects
Urethroscope	Views the interior of the urethra
Urinometer	Specific gravity of urine
Vaporimeter	Vapour pressure
Variometer	Magnetic declination
Velocimeter	Velocity
Velometer	Speed of air
Viameter	Revolutions of a wheel
Vibrograph	Records vibrations
Vibrometer	Vibrations
Viscometer	Viscosity
Visometer	Focal length of the eye

Contd...

Instrument	Measurement/Operation
Voltameter	Electrical current indirectly
Voltmeter	Electrical potential
Volumenometer	Volume of a solid
Volumeter	Volume of a liquid or gas
Wattmeter	Electrical power
Wavemeter	Wavelengths
Weatherometer	Weather-resisting properties of paint
Xanthometer	Colour of sea or lake water
Xylometer	Specific gravity of wood
Zymometer	Fermentation
Zymosimeter	Fermentation

Index

D

E

F

J

K

L

M

S

T

U